Advances in Intelligent Systems and Computing

Volume 1287

The series "Advances in Intelligent Systems and Computing" contains publications on theory, applications, and design methods of Intelligent Systems and Intelligent Computing. Virtually all disciplines such as engineering, natural sciences, computer and information science, ICT, economics, business, e-commerce, environment, healthcare, life science are covered. The list of topics spans all the areas of modern intelligent systems and computing such as: computational intelligence, soft computing including neural networks, fuzzy systems, evolutionary computing and the fusion of these paradigms, social intelligence, ambient intelligence, computational neuroscience, artificial life, virtual worlds and society, cognitive science and systems, Perception and Vision, DNA and immune based systems, self-organizing and adaptive systems, e-Learning and teaching, human-centered and human-centric computing, recommender systems, intelligent control, robotics and mechatronics including human-machine teaming, knowledge-based paradigms, learning paradigms, machine ethics, intelligent data analysis, knowledge management, intelligent agents, intelligent decision making and support, intelligent network security, trust management, interactive entertainment, Web intelligence and multimedia.

The publications within "Advances in Intelligent Systems and Computing" are primarily proceedings of important conferences, symposia and congresses. They cover significant recent developments in the field, both of a foundational and applicable character. An important characteristic feature of the series is the short publication time and world-wide distribution. This permits a rapid and broad dissemination of research results.

Indexed by SCOPUS, DBLP, EI Compendex, INSPEC, WTI Frankfurt eG, zbMATH, Japanese Science and Technology Agency (JST), SCImago.

All books published in the series are submitted for consideration in Web of Science.

More information about this series at http://www.springer.com/series/11156

Manoj Sahni · José M. Merigó ·
Brajesh Kumar Jha · Rajkumar Verma
Editors

Mathematical Modeling, Computational Intelligence Techniques and Renewable Energy

Proceedings of the First International Conference, MMCITRE 2020

Springer

Editors
Manoj Sahni
Department of Mathematics
School of Technology
Pandit Deendayal Petroleum University
Gandhinagar, India

Brajesh Kumar Jha
Department of Mathematics
School of Technology
Pandit Deendayal Petroleum University
Gandhinagar, India

José M. Merigó
University of Technology Sydney
Sydney, NSW, Australia

Rajkumar Verma
Department of Management Control
and Information Systems
University of Chile
Santiago, Chile

ISSN 2194-5357 ISSN 2194-5365 (electronic)
Advances in Intelligent Systems and Computing
ISBN 978-981-15-9952-1 ISBN 978-981-15-9953-8 (eBook)
https://doi.org/10.1007/978-981-15-9953-8

This Springer imprint is published by the registered company Springer Nature Singapore Pte Ltd.
The registered company address is: 152 Beach Road, #21-01/04 Gateway East, Singapore 189721, Singapore

Committees

Special thanks to the members of Honorary Committee, and International and National Advisory Committee for their support in organizing the 1st International Conference on "Mathematical Modeling, Computational Intelligence Techniques and Renewable Energy" (MMCITRE 2020).

Honorary Committee

Prof. S. Sundar Manoharan, Director General, Pandit Deendayal Petroleum University, Gandhinagar, Gujarat, India
Prof. Sunil Khanna, Director, School of Technology, Pandit Deendayal Petroleum University, Gandhinagar, Gujarat, India
Prof. T. P. Singh, Director Academic Affairs, Pandit Deendayal Petroleum University, Gandhinagar, Gujarat, India
Dr. Tarun Shah, Registrar, Pandit Deendayal Petroleum University, Gandhinagar, Gujarat, India
Shri Ankur Pandya, Chief Human Resource Officer, Pandit Deendayal Petroleum University, Gandhinagar, Gujarat, India
Shri Abhinav Kapadia, Chief Finance Officer, Pandit Deendayal Petroleum University, Gandhinagar, Gujarat, India

International Advisory Committee

Prof. Bhudev Sharma, Clark Atlanta University, USA
Prof. Junzo Watada, Universiti Teknologi PETRONAS, Malaysia
Prof. José M. Merigó, University of Chile, Chile and University of Technology Sydney, Australia
Prof. H. M. Srivastava, University of Victoria, Canada
Prof. Florentin Smarandache, University of New Mexico, USA

Prof. Tarald Kvalseth, University of Minnesota, USA.
Prof. P. M. Pardalos, University of Florida, USA
Prof. Enrique Z. Iriondo, Avda. Universidade, Spain
Prof. Enrique Herrera-Viedma, University of Granada, Spain
Prof. Marcin Paprzycki, Warsaw Management University, Poland
Prof. Dil B. Gurung, Kathmandu University, Nepal
Dr. Rajkumar Verma, University of Chile, Chile
Prof. Valentina Emilia Balas, University of Arad, Romania
Prof. De Baets Bernard, Ghent University, Belgium
Prof. Luis Martínez López, University of Jaén, Jaén, Spain
Prof. Vukman Bakic, University of Belgrade, Serbia
Prof. Huchang Liao, Sichuan University, Chengdu, China
Prof. Oscar Castillo, Tijuana Institute Technology, Tijuana, Mexico
Prof. Vincenzo Piuri, University of Milan, Italy
Prof. Zoran Radaković, University of Belgrade, Belgrade
Prof. Ravindra Pandey, Michigan Technological University, MI, USA
Prof. Syed Abdul Mohiuddine, King Abdulaziz University, Kingdom of Saudi Arabia
Prof. Weronika Radziszewska, Polish Academy of Sciences, Poland

National Advisory Committee

Prof. Bal Kishan Dass, University of Delhi, Delhi
Prof. S. C. Malik, M. D. University, Rohtak
Prof. H. C. Taneja, DTU, Delhi
Prof. P. V. Subrahmanyam, IIT Madras
Prof. S. Sundar, IIT Madras
Prof. Peeyush Chandra, IIT Kanpur
Prof. G. P. Raja Sekhar, IIT Kharagpur
Prof. Anuradha Sharma, IIIT Delhi
Prof. L. P. Singh, IIT BHU
Prof. Snehashish Chakraverty, NIT, Rourkela
Prof. S. K. Gupta, H. P. University, Shimla
Prof. R. P. Sharma, H. P. University, Shimla
Prof. K. R. Pardasani, MANIT, Bhopal
Prof. D.S. Hooda, GJUST, Hisar
Prof. S. P. Singh, DEI, Agra
Prof. Rashmi Bhardwaj, Guru Gobind Singh Indraprastha University, Delhi
Prof. A. K. Nayak, IIBM, Patna
Prof. Dinesh Singh, K. R. Mangalam University, Gurugram, Delhi
Prof. Bani Singh, IIT Roorkee
Prof. M. N. Mehta, SVNIT, Surat
Prof. Deepankar Sharma, K. N. Modi Institute
Prof. Sanjeev Sharma, JIIT, Noida

Prof. Neeru Adlakha, SVNIT Surat
Prof. Nita Shah, Gujarat University
Prof. Jagdish Chand Bansal, SAU, Delhi
Prof. Joydip Dhar, IIITM Gwalior
Prof. Kapil Sharma, SAU, Delhi
Prof. V. K. Gupta, Punjabi University, Patiala
Prof. Manish Gupta, DAIICT Gandhinagar
Prof. Manisha Ubale, Indus University, Ahmedabad
Dr. Navnit Jha, SAU, Delhi
Dr. Arvind Kumar Gupta, IIT Ropar
Dr. Asit Kumar Das, Reliance Industries Ltd., Jamnagar
Dr. Ritu Sahni, IAR, Gandhinagar
Er. Bharat S. Patel, IEI, Ahmedabad
Er. Niraj Shah, CSI, Ahmedabad

Thanks to all the members of the Organizing and Associated Committee for their kind support in organizing the 1st International Conference on "Mathematical Modeling, Computational Intelligence Techniques and Renewable Energy" (MMCITRE 2020).

Organizing Committee Members

Dr. Poonam Mishra
Dr. Md. S. Ansari
Dr. Bhasha Vachharajani
Dr. Shobhit Nigam
Dr. Jwngsar Brahma
Dr. Dishant Pandya
Dr. Ankush Raje
Dr. Chandra Shekhar Nishad
Dr. Abhijit Das
Dr. Neelam Singha
Dr. Samir Patel
Dr. Nishant Doshi

Associated Committee Members

Devanshi Dave
Mumukshu Trivedi
Parth Mehta
Hardik Joshi
Aanchit Nayak

Dhairya Shah
Prem Banker
Falak Patel
Vidhi Bhatt
Shaanya Singh
Ishan Mistry
Jahnavi Shah
Dev Shah
Devanshi M. Shah
Vinay Khilwani
Shristi Darbar
Priyanshi Patel
Krish Bhanushali
Sandeep Kumar Paul
Meghna Parikh
Kruti Upadhyay
Ashnil Mandaliya
Radhika Nanda
Harshil Patel
Devanshi Shah
Anushka Shah
Aditi Soni
Maithili Lohakare
Vaidehi Shah
Samip Sharma
Sanskar Bhuwania
Divam Kachoria
Anushka Sharma
Prachi Doshi
Meshva Patel

Preface

This book contains contributions of various mathematicians, physicists, engineers and other field researchers from six different countries and 15 different states of India, who have participated in the 1st International Conference on Mathematical Modelling, Computational Intelligence and Renewable energy (MMCITRE 2020) at Pandit Deendayal Petroleum University (PDPU), Gandhinagar, Gujarat, India, in association with Forum for Interdisciplinary Mathematics (FIM), Institution of Engineers (IEI)—Gujarat, and Computer Society of India (CSI)—Ahmedabad, and IFEHE National Creativity Aptitude Test (NCAT).

Pandit Deendayal Petroleum University (PDPU) 100 acre campus is located in Gandhinagar, Gujarat. PDPU offers multiple courses ranging from engineering, arts and management along with maximum exposure and opportunities to its students through various national and international exchange programmes with best universities worldwide. It has been established by GERMI as a private university through the State Act enacted on 4 April 2007. Since its establishment in 2007, the university has enlarged its scope by offering diversified courses in a very short period through various schools SOT, SPT, SPM and SLS to provide excellent academic programmes in technology, management, petroleum engineering, solar and nuclear energy and liberal education. It intends to broaden the opportunities for students and professionals to develop core subject knowledge which is duly complemented by leadership training interventions, thereby helping the students to make a mark in the global arena. This objective is being further addressed through a number of specialized and well-planned undergraduate, postgraduate and doctoral programmes as well as intensive research projects.

The Department of Mathematics has organized many national conferences, seminars and workshops in the previous years. This is the first time that department has organized international conference to provide opportunity for academicians and professionals from various countries and different parts of India to share their knowledge, research findings and educational practices among academic community. This international conference received a large number of research articles from all over the world. All of these papers are thoroughly refereed through double peer-review process by expertise of the respective disciplines, and then, only high-quality paper is selected for oral presentation during the conference. Finally, out of those presented papers, a total of 46 papers are selected on the basis of the quality of work of the experts from different disciplines, young researchers, academicians and students to put their work in the form of book chapters. It was taken into consideration that these chapters cover latest mathematical techniques useful not only for researchers, but also undergraduate and postgraduate students of physics, mathematics and engineering. In fact, this book is useful for the persons working in the energy and marketing sectors also. The article contains both new fundamental mathematical results and mathematical and computational techniques used in multidisciplinary applications.

The development of future scientist, researchers, coders, educationalist and industrialist is closely dependent on their ability of using mathematical tools in various real-life applications. The 1st International Conference on Mathematical Modelling, Computational Intelligence Techniques and Renewable Energy (MMCITRE-2020) has also been organized for keeping these things in mind that how new knowledge and recent development in all aspects of computational and mathematical techniques, and their applications should be spread among young researchers, educationalist and eminent scientist. Mathematics plays a vital significance to all the areas of people, as it is the backbone for the development of scientific and technical fields across all the spectrum of community. It is useful not only for future engineers but also for computer scientists, as this subject develops power of reasoning, creativity and critical thinking in almost all the areas of worldly problems.

The primary differentiation and distinguishing features of this book are latest mathematical results with all rigorous mathematical procedures, novel researches in the area of mathematical modelling of various real-life phenomena occurring in the education sector, business and marketing sectors, medical sector, especially in the energy sector for generation of renewable energy, their benefits and impact on the society and latest computational techniques to manage generation and smart storage of renewable energy. In conclusion, the book contains various challenging problems and their solutions from various expertises as they have strong background on mathematical modelling computational techniques and renewable energy. They have already explored various real-life challenging problems using advanced mathematical techniques. We wish that all the graduate students of mathematics, physics, engineering and all the other areas' scholars who are in search of new mathematical tools will be benefited from this book.

1st International Conference on Mathematical Modeling, Computational Intelligence Techniques and Renewable Energy (MMCITRE 2020) is supported by:

1. Forum for Interdisciplinary Mathematics (FIM)
2. National Creativity Aptitude Test (NCAT)
3. The Institution of Engineers (IEI, Ahmedabad)
4. The Ahmedabad Chapter of Computer Society of India (CSI)

FIM is an Indian-based international society of scholars working in mathematical sciences and its allied areas incepted in 1975 by a group of intellectuals, supported by mathematics teachers and also users of mathematics from different disciplines around the world led by Professor Bhu Dev Sharma, University of Delhi, and is a registered trust in India. First Gujarat Chapter of FIM was formed under the leadership of Dr. Manoj Sahni in the year 2016. The FIM Gujarat Chapter has organized workshops, seminars, short-term training programmes in PDPU and nearby universities in a short span of four years since its inception.

The Institution of Engineers (IEI) is one of the largest and vibrant professional societies of engineers, technologists and scientists worldwide. It encompasses all branches of engineering and this institution is making significant contribution to technical knowledge, skills and capacity building of the highest order through innovative approach for sustainable development.

The Ahmedabad Chapter of Computer Society of India started functioning during 1969, and today, it has expanded its wings for flying in the world of research and innovation by organizing seminars, conferences, public talks on different subjects, student awareness programmes and many more. Today, the Ahmedabad Chapter has 20 student branches with 854 student members and more than 8000 members, including India's most famous IT industry leaders, brilliant scientists and dedicated academicians.

IFEHE was started with a vision of contributing to the world of higher education by introducing a culture of continuous learning. IFEHE evaluates the best of the practices from all over the world, studies their scope of implementation in Indian scenario and helps in executing them by bringing the right kind of expertise. The Board of Advisors has professors from IITs and IIMs, who play a vital role in maintaining the quality.

Gandhinagar, Gujarat, India
February 2020

Manoj Sahni
Brajesh Kumar Jha
José M. Merigó
Rajkumar Verma

Contents

About the Editors

Manoj Sahni is working as Associate Professor at the Department of Mathematics, School of Technology, Pandit Deendayal Petroleum University, Gandhinagar, Gujarat, India. He has more than fifteen years of teaching and research experience. He holds a M.Sc. degree in Mathematics from Dayalbagh Educational Institute (Deemed University), Agra, M.Phil. from IIT Roorkee and a Ph.D. degree in Mathematics from Jaypee Institute of Information Technology (Deemed University), Noida, India. He has published 48 research papers in journals, conference proceedings and book chapters with reputed publishers like Springer and Elsevier. He also serves as a reviewer for many international journals of repute. He has contributed to the scientific committee of several conferences and associations. He has delivered many expert talks at national and international levels. He has organized many seminars, workshops and short-term training programs at PDPU and various other universities. He has also organized special Symposia in an International Conference (AMACS2018) on Fuzzy Set Theory: New Developments and Applications to Real Life Problems held at London in 2018. He is a member of many international professional societies including American Mathematical Society, Forum for Interdisciplinary Mathematics, Indian Mathematical Society, IAENG and many more.

José M. Merigó is Professor at the School of Information, Systems & Modelling at the Faculty of Engineering and Information Technology at the University of Technology Sydney (Australia) and Part-Time Full Professor at the Department of Management Control and Information Systems at the School of Economics and Business at the University of Chile. Previously, he was Senior Research Fellow at the Manchester Business School, University of Manchester (UK), and Assistant Professor at the Department of Business Administration at the University of Barcelona (Spain). He holds a master's and a Ph.D. degree in Business Administration from the University of Barcelona. He also holds B.Sc. and M.Sc. degrees from Lund University (Sweden). He has published more than 500 articles in journals, books and conference proceedings, including journals such as Information Sciences, IEEE Computational Intelligence Magazine, IEEE Transactions on Fuzzy Systems, European Journal of Operational Research, Expert Systems with Applications, International Journal of Intelligent Systems, Applied Soft Computing, Computers &

Industrial Engineering and Knowledge-Based Systems. He has also published several books with Springer and with World Scientific. He is on the editorial board of several journals including Computers & Industrial Engineering, Applied Soft Computing, Technological and Economic Development of Economy, Journal of Intelligent & Fuzzy Systems, International Journal of Fuzzy Systems, Kybernetes and Economic Computation and Economic Cybernetics Studies and Research. He has also been Guest Editor for several international journals, a member of the scientific committee of several conferences and a reviewer in a wide range of international journals. Recently, Thomson & Reuters (Clarivate Analytics) has distinguished him as a Highly Cited Researcher in Computer Science (2015–present). He is currently interested in decision making, aggregation operators, computational intelligence, bibliometrics and applications in business and economics.

Brajesh Kumar Jha is working as Assistant Professor at the Department of Mathematics, School of Technology, Pandit Deendayal Petroleum University, Gandhinagar, Gujarat, India. He has more than ten years of teaching and research experience. He holds a M.Sc. degree in Mathematics from Jiwaji University, Gwalior, and a Ph.D. degree in Mathematics from S.V. National Institute of Technology, Surat. He has published more than 30 research papers in journals and conference proceedings like Springer, Elsevier and World Scientific. He also serves as a reviewer for many international journals of repute. He has contributed to the scientific committee of several conferences and associations. He has delivered many expert talks at national and international levels. He has organized many seminars, workshops and short-term training programs at PDPU. He is a member of many international professional societies including Forum for Interdisciplinary Mathematics, GAMS, FATER and many more.

Rajkumar Verma is working as Postdoctoral Research Fellow at the Department of Management Control and Information Systems at the University of Chile. Before joining the University of Chile, he was Assistant Professor at the Department of Mathematics, Delhi Technical Campus, India. He holds a M.Sc. degree in Mathematics from Chaudhary Charan Singh University, Meerut, and a Ph.D. degree in Mathematics from the Jaypee Institute of Information Technology, Noida, India. He has published more than 40 research papers in journals and conference proceedings including the International Journal of Intelligent Systems, Journal of Intelligent & Fuzzy Systems, Kybernetika, Informatica, International Journal of Machine Learning and Cybernetics, International Journal of Uncertainty, Fuzziness, and Knowledge-Based Systems and Neural Computing and Applications. He has contributed to the scientific committee of several conferences and associations. He also serves as a reviewer for many international journals. He is a member of many international professional societies including American Mathematical Society, IEEE Computational Intelligence Society, Forum for Interdisciplinary Mathematics, Indian Mathematical Society, and Indian Society of Information Theory and Applications.

Mathematical Groundwork

Rubbling Number of $C_n \times C_n$ Under the Same Parity

Jitendra Binwal and Aakanksha Baber

Abstract Graph pebbling and graph rubbling are the shifting of two pebbles and attachment of one pebble on the specified vertex with respect to pebbling step and rubbling step. In this paper, we describe the rubbling number of $C_n \times C_n$ where $n \geq 2$ and under the assumption that cycle graph with n vertices is of same parity.

Keywords Pebbling number · Rubbling number · Cycle graphs · Graham's conjectures

1 Introduction

Consider an undirected graph or a graph $G = (V, E)$ encompass a set of vertices $V(G)$ and a set of edges $E(G)$ where incidence relationship is preserved [1, 2]. Graph pebbling was first introduced by Lagarias and Saks. Literature of graph pebbling was introduced by Chung. Graph pebbling comes from the foundations of graph theory, combinatorial number theory and group theory. Graph pebbling is a mathematical game played on a graph with pebbles on the vertices. Pebbles are represented by positive integers on the vertices of graph. Firstly, pebbles were introduced in "pebble game" on graph. Now, it is depicted as the new parameter which evaluates the graph on computer and widely used in the field of animations. For an example, consider the pebbles as liquid containers or tankers, then the loss of the pebble during a step or move is the cost of transportation and shipment of fuel takes place between the initial and final destination or vertices.

To calculate pebbling number of graph, we need an operation or step or move that is well known as pebbling step. Pebbling step is the shifting of two pebbles from an arbitrary vertex and attachment of one pebble on its adjacent vertex. Graph pebbling

J. Binwal · A. Baber (✉)
Department of Mathematics, School of Liberal Arts and Sciences, Mody University of Science and Technology, Lakshmangarh, Rajasthan, India
e-mail: baberaakanksha8@gmail.com

J. Binwal
e-mail: dr.jitendrabinwaldkm@gmail.com

M. Sahni et al. (eds.), *Mathematical Modeling, Computational Intelligence Techniques and Renewable Energy*, Advances in Intelligent Systems and Computing 1287,
https://doi.org/10.1007/978-981-15-9953-8_1

is the collection of pebbling steps on the n number of vertices on graph with different configurations is known as the pebbling number of graphs G that is $\pi(G)$ or $f(G)$.

The pebbling number of graph is equivalent to $(V(G))$ number of vertices of graph G is recognized as demonic. Consider $p(v)$ as the number of pebbles on vertex v [2–6].

Cartesian product of any two graphs $G = (V_G, E_G)$ and $H = (V_H, E_H)$ is defined by direct product of $G \times H$ where vertex set and edges set are Cartesian products such that [7–9] for $1 \leq i, j \leq n$,

$$V_{G\times H} = V_G \times V_H = \{(x_i, y_j) : x_i \in V_G, y_j \in V_H\},$$
$$E_{G\times H} = \left\{\left((x_i, y_j)\left(x_i', y_j'\right)\right) : x_i = x_i' \text{ and } \left(y_j, y_j'\right) \in E_H, \text{ or}(x_i, x_i') \in E_G \text{ and } y_j = y_j'\right\}$$

Cycle graph C_n on n vertices $\{x_1, x_2, \ldots, x_n\}$ is a closed path where every vertex x_i is adjacent only to x_{i+1} and x_{i-1} for $1 \leq i \leq n$ [3, 4, 8] .

Preposition 1 *The pebbling number of odd and even undirected cycles* [2, 4], *for* $k \geq 2$

$$f(C_{2k}) = 2^k \text{ and } f(C_{2k+1}) = 2\lfloor\frac{2^{k+1}}{3}\rfloor + 1.$$

Graham's Conjecture 1 *Graham's conjecture is stated from Cartesian products of two graphs. It is true for several graphs, G and S,* $f(G \times S) \leq f(G) \times f(S)$*; see* [2, 7].

Lemma 1 *The pebbling number of* $C_2 \times C_2$ *is* 4 *that is* $f(C_2 \times C_2) = 4$.

Since $C_2 \times C_2$ *is isomorphic to* C_4 *cycle graph. Hence,* $f(C_2 \times C_2) \cong f(C_4) = 4$*; see* [2].

Conjecture 1 *Statement: The pebbling number of* $C_3 \times C_3$ *is* 9*, i.e.,* $f(C_3 \times C_3) = 9$.

Proof Using a heuristic approach of all possible configurations of pebbles on the graph, consider the $C_3 \times C_3$ graph with 9 vertices. Since $C_3 \times C_3$ is symmetric, any vertex which contains 4 pebbles can send one pebble to the target vertex (x_2, x_2). Let us assume, the worst case scenario, $p(x_1, x_1) = 3$, $p(x_1, x_2) = 2$, $p(x_3, x_3) = 3$ and $p(x_3, x_1) = 1$.

(x_1, x_1) (3) $\rightarrow$ (x_3, x_1) $(1+1)$ $\rightarrow$ (x_3, x_2) $(1+1)$ $\leftarrow$ (x_3, x_3) (3)

$\downarrow$

(x_2, x_2) $(1+1)$

$\uparrow$

(x_1, x_2) (2)

Conjecture 2 *The pebbling number of graph* $C_5 \times C_5$ *is 25. It is demonic that is* $f(C_5 \times C_5) = 25$ [8].

Theorem 1 *Statement:* $f(C_n \times C_n) = n \times n$ *that is it is demonic.*

Proof We apply mathematical induction on n. Since, $C_n \times C_n$ has $(x_1, x_2, \ldots, x_n) \times (y_1, y_2, \ldots, y_n)$ that is $n \times n$ vertices.

Basis of induction: If $n = 2, 3$, then from Lemma 1 and Conjecture 1 the result is true.

Induction hypothesis: Let us assume that result is true for $n - 1$, i.e., $f(C_{n-1} \times C_{n-1}) = (n-1) \times (n-1)$.

Induction step: We need to prove that the result is true for n. Let the target vertex be $(x_{n/2}, y_{n/2})$. Using statement, $f(C_n \times C_n) = (n) \times (n)$

$$
\begin{aligned}
&= (n + 1 - 1) \times (n + 1 - 1) \\
&= ((n - 1) + 1) \times ((n - 1) + 1) \\
&= (n - 1) \times (n - 1) + (n - 1) + (n - 1) + 1 \\
&= f(C_{n-1} \times C_{n-1}) + 2(n - 1) + 1 \\
&= f(C_{n-1} \times C_{n-1}) + (2n - 1) \\
&= f(C_{n-1} \times C_{n-1}) + f(C_{2n-1})
\end{aligned}
$$

Since C_{2n-1} and $C_{n-1} \times C_{n-1}$ can send one pebble to target vertex $(x_{n/2}, y_{n/2})$ if target vertex is contained in C_{2n-1} and $C_{n-1} \times C_{n-1}$. Even if target vertex is not contained in both graphs, each graph C_{2n-1} and $C_{n-1} \times C_{n-1}$ can send one pebble to a common vertex (let us assume (x_i, y_j)) of graphs C_{2n-1} and $C_{n-1} \times C_{n-1}$. Note that the first paragraph of a section or subsection is not indented.

2 Rubbling in Graph

Graph rubbling or rubbling in graph is the supplement of graph pebbling. Often graph rubbling can be harder or weaker than pebbling. A rubbling has an additional transfer option in any object or thing which is widely used in case of small quantities. For an example, small tankers or containers are transferring fuel in long distances [2]. Now, rubbling number and pebbling number are also depicted as the new parameter which evaluates the graph on computer and widely used field of animations.

Similarly, rubbling on graph needs a move or step known as rubbling step. On graph, rubbling step is the shifting of one pebble from each neighboring vertex, i.e., s and v and attachment of one pebble on target vertex [2] (Fig. 1).

The rubbling number of graphs G or $\rho(G)$ is the minimum number of pebbles on graph G which are aimlessly scattered all over the graph on vertices, if target vertex is situated with one pebble from m number of pebbles, after some sequence of rubbling steps. For an example,

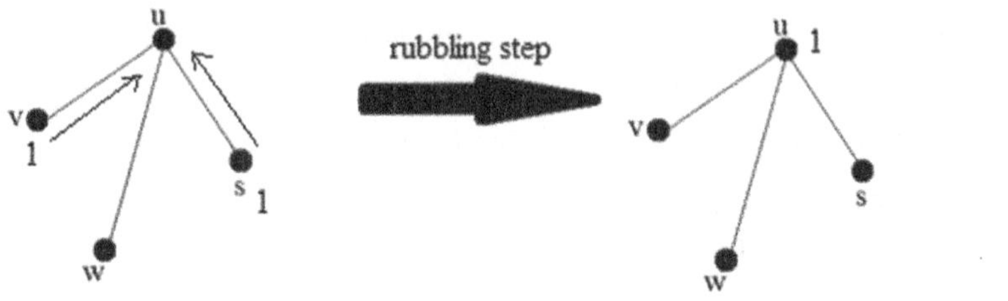

Fig. 1 Rubbling step of graph G

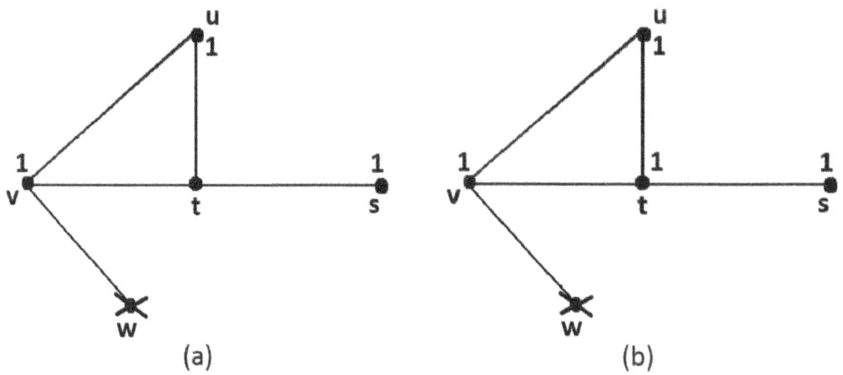

Fig. 2 Rubbling number $\rho(G)$ on graph G

Figure 2(a) has 3 pebbles and w vertex is not reachable. While Fig. 2(b) contains 4 pebbles shows an extra pebble results w vertex reachable.

Preposition 2 *The rubbling number of odd and even undirected cycles, for* $k \geq 2$ *see* [8]

$$\rho(C_{2k}) = 2^k \quad \text{and} \quad \rho(C_{2k+1}) = \lfloor \frac{7.2^{k-1} - 2}{3} \rfloor + 1.$$

Graham's Conjecture 2 *Graham's conjecture is not resulted true in case of graph rubbling* [2]. Preposition 3 *is an example.*

Preposition 3 *Since,* $C_2 \times C_2$ *is isomorphic to* C_4 *cycle graph; see* [2]. *Hence, then,* $\rho(C_2 \times C_2) = 4$.

The $\rho(C_3 \times C_3) = 5 > 4 = 2.2 = \rho(C_3).\rho(C_3)$.

Conjecture 3 *The* $\rho(K_2 \times C_4) = 3$, *where* K_2 *is a complete graph with 2 vertices; see* [8].

3 Main Results

Theorem 2 *Statement: The rubbling number of* $C_4 \times C_4$ *is* 6, *i.e.,* $\rho(C_4 \times C_4) = 6$.

Proof Applying a heuristic approach on the $C_4 \times C_4$ graph on all the possible configuration of pebbles on vertices of the graph to calculate its rubbling number. Since $C_4 \times C_4$ graph has 16 vertices, let us assume the target vertex be (x_1, x_1). The $C_4 \times \{x_1\}$ or $\{x_1\} \times C_4$ (such that, are symmetrical) is the subgraph of $C_4 \times C_4$ (Figs. 3 and 4).

There will be three possible cases:
Case (1): When $p(C_4 \times \{x_1\}) = 2$ or more.
Case (2): There is only 1 pebble on $C_4 \times \{x_1\}$.
Case (3): There is no pebble on $C_4 \times \{x_1\}$.
Case (1): When $p(C_4 \times \{x_1\}) = 2$ or more (Fig. 5).
Since, $C_4 \times \{x_1\}$ is a regular graph of 4 vertices. Therefore, $C_4 \times \{x_1\} \cong K_4$,

where $\rho(K_4) = 2$.

Case (2): There is only 1 pebble on $C_4 \times \{x_1\}$.

Let us assume $p(x_3, x_1) = 1$. Since, only one more pebble is required on $C_4 \times \{x_1\}$ to pebble the target vertex. Since, second and third column of $C_4 \times C_4$ is subgraph

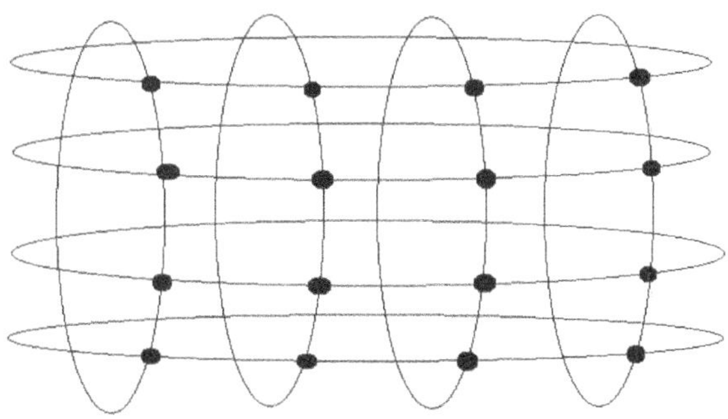

Fig. 3 Graph $C_4 \times C_4$

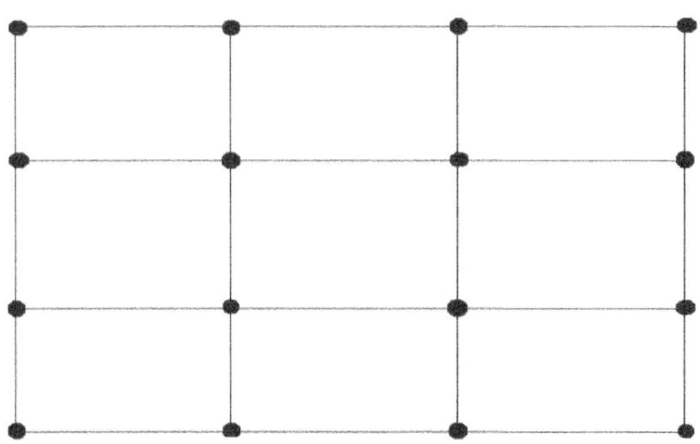

Fig. 4 Graph $C_4 \times C_4$ with outside edges omitted

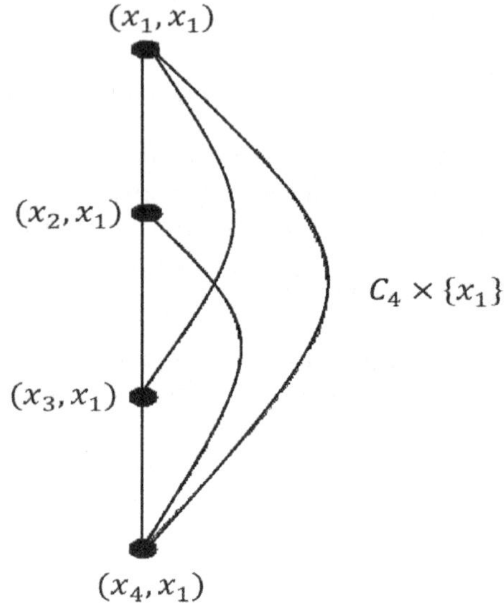

Fig. 5 Graph $C_4 \times \{x_1\}$

of $C_4 \times C_4$, which is isomorphic to $K_2 \times C_4$. Therefore, $\rho(K_2 \times C_4) = 3$ can send one pebble to $C_4 \times \{x_1\}$. Therefore, (x_1, x_1) can be pebbled.

Remaining fourth column of $C_4 \times C_4$ is a subgraph of $C_4 \times C_4$ and also isomorphic to K_4. Therefore, six pebbles are satisfied to pebble the target vertex (x_1, x_1).

Case (3): There is no pebble on $C_4 \times \{x_1\}$, i.e., $\rho(C_4 \times \{x_1\}) = 0$.

Remaining vertices of graph except $C_4 \times \{x_1\}$ and $\{x_1\} \times C_4$ vertices forms a subgraph $C_S \cong C_3 \times C_3$. Since, 5 pebbles are sufficient to pebble any one vertex of subgraph C_S. Now, from C_S, one pebble is sent to $C_4 \times \{x_1\}$. We require one more pebble.

If we add one more pebble to C_S subgraph, then 1 more pebble is moved to $C_4 \times \{x_1\}$.

$\therefore$ Target vertex (x_1, x_1) (1) $\leftarrow$ $C_4 \times \{x_1\}$ (1 + 1) $\uparrow$ C_S (6)

In worst case, if $p(x_4, x_4) = 3$, $p(x_3, x_3) = 1$, and $p(x_2, x_2) = 1$. Then target vertex (x_1, x_1) cannot be pebbled. Since, we require one more pebble on subgraph C_S to pebble the target vertex.

Theorem 3 *Statement: For $n \geq 2$ and n is of same parity then, $\rho(C_n \times C_n) = n+2$.*

Proof We apply mathematical induction on n.

Case (a): Basic of induction: For $n = 2, 3, 4$, result is true from Preposition 3 and Theorem 2.

Case (b): Induction hypothesis: Let us assume that result is true for n, i.e.,

$$\rho(C_n \times C_n) = n + 2.$$

Case (c): Induction step: To prove, result holds true for $n + 1$.

$$\begin{aligned}\rho(C_{n+1} \times C_{n+1}) &= (n+1) + 2\\ &= (n+2) + 1\\ &= \rho(C_n \times C_n) + 1\end{aligned}$$

Since $C_n \times C_n$ is subgraph of $C_{n+1} \times C_{n+1}$ that is vertices except $C_{n+1} \times \{x_1\}$ and $\{x_1\} \times C_{n+1}$. By same procedure in case (3) of Theorem 2, target vertex can be pebbled with $\rho(C_n \times C_n) + 1$ pebbles, i.e., $(n + 1) + 2$.

4 Conclusions

In this paper, we have calculated the rubbling number of $C_n \times C_n$ for $n \geq 2$ where n is of same parity. Pebbling and rubbling represent pebbles which are widely used in transportation of discrete items or objects, number theory and animations. Cycle graphs record all type of movements of path of an operator. The movements which are fast enough to trace by the human eye are traced by cycle graph techniques. It can be used for method improvement. Also, we have implemented the rubbling number of $C_4 \times C_4$ as well as the pebbling number of $C_n \times C_n$ for $n \geq 2$.

References

1. Chung, F.R.K.: Pebbling in hypercubes. SIAM J. Discrete Math. **2**, 467–472 (1989)
2. Sieben, N.: Introduction to graph pebbling and rubbling. Northern Arizona University (2015)
3. Gunda, G.: Pebbling on Directed Graphs. University of Dayton (2004)
4. Herscovici, D.S.: Graham's pebbling conjecture on products of cycles. J. Graph Theory **42**, 141–154 (2003)
5. Binwal, J., Tiwari, M.: Equality of pebbling numbers on directed graphs. J. Indian Acad. Math. **32**, (2010)
6. Binwal, J., Tiwari, M.: Pebbling on Isomorphic Graphs. Indian J. Math. and Math. Sci. **5**, 17–20 (2009)
7. Feng, R., Kim, J.Y.: Graham's pebbling conjecture on product of complete bipartite graphs. Sci. China Ser. A **44**, 817–822 (2001)
8. Herscovici, D.S., Higgins, A.W.: The pebbling number of $C_5 \times C_5$. Discrete Math. **187**, 123–135 (1998)
9. Xia, Z.J., Pan, Y.L.: Graham's pebbling conjectures on Cartesian product of the middle graphs of even cycles. Cornell University (2017)

Oscillating Central Force Field in Cylindrical-Polar Coordinates and Its Lagrange's Equation of Motion

Priyanshi Patel, Dhairya Shah, Jalaja Pandya, and Satyam Shinde

Abstract In this paper, differential orbit equation of the object is derived that undergoes rotational and translational motion with the effect of the drag oscillation, under the influence of central force field in cylindrical-polar coordinates. Further, it is shown that the magnitude of central force is the function of the radial distance and is not dependent on tangential oscillations. This determines that the drag oscillations do not have any effect on the central character of force. Due to this, one can examine that the system is torque free and conservative. Using the Euler–Lagrange equation of motion, the expression of radial energy is derived for the considered case. Also, one can see how the energy is distributed individually for the rotational motion and for perpendicular vibrations. The expression for the total energy of the system is in the form of summation of radial energy and oscillating energy possessed by the system.

Keywords Central force · Circular orbit · Cylindrical coordinate system · Drag oscillations · Lagrange's equation of motion

1 Introduction

Classical mechanics has always been a convenient resource to solve the dynamics of a particle/object and has been proven successful up to certain extent when we derive the equations of motions for it except when the speed of the object approaches the value of the universal constant of light. It serves as a handy tool with simple mathematics to solve real-life problems. Mechanics, a study of motion of material bodies, can itself divide into three sub-disciplines: kinematics, dynamics and statics. In dynamics, we introduce the concept of force. Here, an attempt has been made to

P. Patel · D. Shah
School of Liberal Studies, Pandit Deendayal Petroleum University, Gandhinagar, India

J. Pandya · S. Shinde (✉)
School of Technology, Pandit Deendayal Petroleum University, Gandhinagar, India
e-mail: satyamshinde@yahoo.com

M. Sahni et al. (eds.), *Mathematical Modeling, Computational Intelligence Techniques and Renewable Energy*, Advances in Intelligent Systems and Computing 1287,
https://doi.org/10.1007/978-981-15-9953-8_2

study the central problem dynamics where a physical system is considered to be in influence of a central force. Many applications of classical mechanics may be based directly on Newton's law of motion called Newtonian mechanics [1]. There are, however, number of other ways of formulating the principles of classical mechanics. The equation of Lagrange and Hamilton is such examples. These types of analytical mechanics are a new way to understand and interpret the mechanics of a physical system elegantly by using advanced mathematical skills and fundamentals of physics. A detailed information on above topics can be found in [2–5].

A force which is directed toward or away from the center of the system and whose magnitude is the function only of the distance from the center itself are called central forces [1]. The above force is classified in two categories: (1) attractive central forces, directed toward the origin by satisfying the condition $\overrightarrow{\mathcal{F}}(s) < 0$ and (2) repulsive central forces, directed away from the origin and falling into the condition of $\overrightarrow{\mathcal{F}}(s) > 0$ where s is the distance from center [1]. An example of an attractive central force is the universal law of gravitation acting between any two objects separated by a distance. On the atomic scale, the columbic attraction between two oppositely charged particles explains the attractive central force very well. Whereas the force between an alpha particle and proton is an example of repulsive central forces. Hence, central forces are able to explain many of phenomenon from atomic to cosmic scale.

In 2013, Edison et al. [6] derived Modified theory of Central Force Motion in which it was shown that how the magnitude of a central force changes due to drag oscillations considering six degrees of freedom in plane polar coordinates. Edison et al. [7] extended the work to derive the equation of motion of the object with the help of Lagrangian Mechanics and came up with the equation of energy of the system. The system was represented in plane polar coordinates with the help of four additional coordinates to describe the motion of the body.

In the current study, we use the cylindrical-polar coordinate system to show the motion of the body under the influence of drag oscillations, which is the simplified, compact and extended work of above theory where we can see that drag oscillation has no effect on the central character of force. This study can be used to solve the dynamics of any two-body problem.

2 Research Methodology

A set of generalized coordinates is any set of coordinates by which the position of the particles in a system may be specified [1]. Here, we have explained the motion of the body in cylindrical coordinate (considering it as a special case of generalized coordinate system) system considering the drag oscillations of the object. While trying to resolve this problem in cylindrical coordinate system, we require three coordinates defined as $(\rho, \emptyset, z)$,where ρ represents the radial distance to which the center of mass of system, and center of rotating body is connected. The origin O

of the system is considered to be center of mass of system which is stationary in order to achieve the circular trajectory of the motion [8]. The acceleration of the heavier body at the center is negligible [9]. The Ø coordinate represents the amount of rotation of the body around the fixed center O on a fixed axis. The angular position of the object with respect to $Ø = 0$ is given by Ø. Considering the areas swept out by s are the same for equal time intervals [8]. The law of equal areas holds for any central force and for both close and open orbits. Lastly, the oscillations are resolved in z direction. To make the system symmetric, we have considered the upward and downward displacements to be equal in magnitude. This also makes the system free from torques.

Hence, the motion of the object is easily explained by the solving it in three-dimensional system. This gives rise to the system which is conservative and symmetric. The radius is constant throughout the consideration due to circular motion of the object, and upward and downward oscillations are equal. Moreover, we have also derived the equation of orbit of the object under the influence of central force and expressions of radial and oscillating energy using Lagrange formulation.

We need following expressions in the subsequent work.

2.1 *Expression for Velocity and Acceleration [1]*

Velocity is the time derivative of position vector, and we know that the position vector in the cylindrical coordinate system is given by $\vec{s}$. Where $\vec{s}$ depends upon ρ and z.

Hence, $\vec{s}$ can be written as

$$\vec{s} = \rho\hat{\rho} + z\hat{z} \tag{1}$$

Taking the time derivative of Eq. (1), we get velocity $\vec{v}$ of the object.

$$\vec{v} = \frac{d\vec{s}}{dt} = \frac{d(\rho\hat{\rho} + z\hat{z})}{dt} = \frac{d(\rho\hat{\rho})}{dt} + \frac{d(z\hat{z})}{dt}$$

$$\vec{v} = \dot{\rho}\hat{\rho} + \rho\dot{\hat{\rho}} + \dot{z}\hat{z} + z\dot{\hat{z}}$$

$$\vec{v} = \dot{\rho}\hat{\rho} + \rho\dot{Ø}\hat{Ø} + \dot{z}\hat{z} \tag{2}$$

Similarly, for acceleration, we again take time derivative of velocity obtained in (2)

$$\vec{a} = \frac{dv}{dt} = \frac{d\left(\dot{\rho}\hat{\rho} + \rho\dot{Ø}\hat{Ø} + \dot{z}\hat{z}\right)}{dt}$$

$$\vec{a} = \frac{d(\dot{\rho}\hat{\rho})}{dt} + \frac{d\left(\rho\dot{Ø}\hat{Ø}\right)}{dt} + \frac{d(\dot{z}\hat{z})}{dt}$$

$$\vec{a} = \ddot{\rho}\hat{\rho} + \dot{\rho}\dot{\hat{\rho}} + \dot{\rho}\dot{\emptyset}\hat{\emptyset} + \rho\ddot{\emptyset}\hat{\emptyset} + \rho\dot{\emptyset}\dot{\hat{\emptyset}} + \ddot{z}\hat{z} + \dot{z}\dot{\hat{z}}$$
$$\vec{a} = \ddot{\rho}\hat{\rho} + \dot{\rho}\dot{\hat{\rho}} + \rho\ddot{\emptyset}\hat{\emptyset} + \dot{\rho}\dot{\emptyset}\hat{\emptyset} + \rho\dot{\emptyset}\dot{\hat{\emptyset}} + \dot{z}\dot{\hat{z}} + \ddot{z}\hat{z}$$
$$\vec{a} = \ddot{\rho}\hat{\rho} + \rho\ddot{\emptyset}\hat{\emptyset} + \dot{\rho}\dot{\emptyset}\hat{\emptyset} + \dot{\rho}\dot{\emptyset}\hat{\emptyset} - \rho(\dot{\emptyset})^2\hat{\rho} + \ddot{z}\hat{z}$$
$$\vec{a} = \left(\ddot{\rho} - \rho(\dot{\emptyset})^2\right)\hat{\rho} + (\rho\ddot{\emptyset} + 2\dot{\rho}\dot{\emptyset})\hat{\emptyset} + \ddot{z}\hat{z} \tag{3}$$

Equations (1), (2) and (3) are the equations of position vector, velocity and acceleration, respectively.

3 Work Done

In this section, we have analytically derived the equations of motion, which is the simple framework useful to solve some of the real life and celestial mechanics problems. For example, we can show the motion of two binary stars rotating around a common center with the help of reduced mass concept and reducing it to one body problem. Tracking the trajectories of the two stars can also be done. The same can also be applied to study the planetary motion.

3.1 Expression of Central Force

Since central force is radial force, the expression of force will depend on the position vector and will be equal to the mass of the body μ times acceleration.

$$\vec{\mathcal{F}}(s) = \mathcal{F}(|s|)\left(\hat{\rho}; \hat{\emptyset}; \hat{z}\right) = \mu\vec{a} \tag{4}$$

We have considered constant velocity in both angular and oscillatory direction that allows the magnitude of the force in respective directions to be null in the two cases.

$$\mathcal{F}(z) = \mu\ddot{z} = 0$$
$$\mathcal{F}(\emptyset) = \mu\left(\rho\ddot{\emptyset} + 2\dot{\rho}\dot{\emptyset}\right) = 0$$

Using (3),

$$\vec{\mathcal{F}}(s) = \vec{\mathcal{F}}(\rho) = \mathcal{F}(|\rho|)\hat{\rho} = \mu\left(\ddot{\rho} - \rho(\dot{\emptyset})^2\right)\hat{\rho}$$
$$\vec{\mathcal{F}}(s) = \mu\left(\ddot{\rho} - \rho(\dot{\emptyset})^2\right)\hat{\rho} \tag{5}$$

Hence, (5) is the equation of the central force.

3.2 Equation of Orbit [6]

To derive the equation of orbit, we take

$$\rho.b = 1 \tag{6}$$

Taking derivative of (6) with respect to b, we get

$$\frac{\partial \rho}{\partial b} = -\frac{1}{b^2}$$

Further taking time derivatives of ρ, we get

$$\dot{\rho} = \frac{d\rho}{dt} = \frac{\partial \rho}{\partial \emptyset}\left(\frac{d\emptyset}{dt}\right) = \dot{\emptyset}\frac{\partial \rho}{\partial \emptyset} = \frac{h}{\rho^2}\left(\frac{\partial \rho}{\partial \emptyset}\right)$$

Taking $\dot{\emptyset} = \frac{h}{\rho^2}$; where h is specific angular momentum.

$$\dot{\rho} = \frac{h}{\rho^2}\left(\frac{\partial \rho}{\partial b}\frac{\partial b}{\partial \emptyset}\right) = \frac{h}{\rho^2}\left(\frac{-1}{b^2}\right)\left(\frac{\partial b}{\partial \emptyset}\right) = \frac{-h}{b^2\rho^2}\left(\frac{\partial b}{\partial \emptyset}\right) = -h\frac{\partial b}{\partial \emptyset}$$

Similarly,

$$\begin{aligned}\ddot{\rho} &= -h\frac{\partial}{\partial \emptyset}\left(\frac{\partial b}{\partial \emptyset}\frac{d\emptyset}{dt}\right) = -h\dot{\emptyset}\left(\frac{\partial^2 b}{\partial \emptyset^2}\right)\\ &= -h\left(\frac{h}{\rho^2}\right)\left(\frac{\partial^2 b}{\partial \emptyset^2}\right) = -h^2b^2\left(\frac{\partial^2 b}{\partial \emptyset^2}\right)\end{aligned}$$

Using (5), the equation of central force becomes

$$\mathcal{F}(\rho) = \mu\left[-h^2b^2\left(\frac{\partial^2 b}{\partial \emptyset^2}\right) - \rho\frac{h^2}{\rho^4}\right]$$

Now, the equation of orbit of the body undergoing central force motion becomes

$$\begin{aligned}\frac{\mathcal{F}(\rho)}{h^2b^2} &= \frac{\mu}{h^2b^2}\left[-h^2b^2\left(\frac{\partial^2 b}{\partial \emptyset^2}\right) - \frac{h^2}{\rho^3}\right]\\ \frac{\mathcal{F}(\rho)}{\mu h^2b^2} &= -\left[\left(\frac{\partial^2 b}{\partial \emptyset^2}\right) + b\right]\end{aligned} \tag{7}$$

Equation (7) is the required partial differential equation of orbit of the body undergoing central force motion with the effect of drag oscillation.

3.3 Lagrange Equations of Motion for Calculation of Radial Energy

We can write kinetic energy K possessed by the object as,

$$K = \frac{1}{2}\mu v^2 = \frac{1}{2}\mu\left(\dot{\rho}^2 + \rho^2\dot{\emptyset}^2 + \dot{z}^2\right)$$

The Lagrangian of the system is given by

$$L = K - V(\rho) \tag{8}$$

where $V(\rho)$ is potential energy and is function of ρ.

The Euler–Lagrange equation of second kind can be written using (8) where n is generalized coordinate.

$$\frac{\mathrm{d}}{\mathrm{d}t}\frac{\partial L}{\partial \dot{n}} - \frac{\partial L}{\partial n} = 0$$
$$\frac{\mathrm{d}}{\mathrm{d}t}\frac{\partial(K - V(\rho))}{\partial \dot{n}} - \frac{\partial(K - V(\rho))}{\partial n} = 0$$
$$\frac{\partial(V(\rho))}{\partial \dot{n}} = 0;\ \frac{\partial(V(\rho))}{\partial n} = \frac{\mathrm{d}V}{\mathrm{d}n}$$
$$\frac{\mathrm{d}}{\mathrm{d}t}\frac{\partial(K)}{\partial \dot{n}} - \frac{\partial(K)}{\partial n} + \frac{\mathrm{d}(V(\rho))}{\mathrm{d}n} = 0$$

For, $n = (\rho, \emptyset, z)$; $\dot{n} = \left(\dot{\rho}, \dot{\emptyset}, \dot{z}\right)$.
Since central forces are radial forces, we take $n = \rho$.

$$\frac{d}{dt}\frac{\partial(K)}{\partial\dot{\rho}} - \frac{\partial(K)}{\partial\rho} + \frac{d(V(\rho))}{d\rho} = 0$$
$$\frac{\mathrm{d}}{\mathrm{d}t}\frac{1}{2}\mu(2\dot{\rho}) - \frac{1}{2}\mu\left(2\rho\dot{\emptyset}^2\right) + \frac{\mathrm{d}(V(\rho))}{\mathrm{d}\rho} = 0$$
$$\frac{\mathrm{d}}{\mathrm{d}t}\mu(\dot{\rho}) - \mu\left(\rho\dot{\emptyset}^2\right) + \frac{\mathrm{d}(V(\rho))}{\mathrm{d}\rho} = 0$$
$$\frac{\mathrm{d}}{\mathrm{d}t}\mu(\dot{\rho}) - \mu\rho\left(\frac{l^2}{\mu^2\rho^4}\right) + \frac{\mathrm{d}(V(\rho))}{\mathrm{d}\rho} = 0$$
$$\frac{\mathrm{d}}{\mathrm{d}t}\mu(\dot{\rho}) - \left(\frac{l^2}{\mu\rho^3}\right) + \frac{\mathrm{d}(V(\rho))}{\mathrm{d}\rho} = 0$$
$$\frac{\mathrm{d}}{\mathrm{d}t}\mu(\dot{\rho})\dot{\rho} - \left(\frac{l^2}{\mu\rho^3}\right)\dot{\rho} + \frac{\mathrm{d}(V(\rho)}{\mathrm{d}\rho}\dot{\rho} = 0$$

Force is negative gradient of potential energy[10] and can be written as: $\frac{\mathrm{d}(V(\rho)}{\mathrm{d}\rho} = -\mathcal{F}(\rho)$

$$\frac{\mathrm{d}}{\mathrm{d}t}\frac{1}{2}\left(\mu\dot{\rho}^2\right) - \left(\frac{l^2}{\mu\rho^3}\right)\frac{\mathrm{d}\rho}{\mathrm{d}t} - \mathcal{F}(\rho)\frac{\mathrm{d}\rho}{\mathrm{d}t} = 0$$
$$\frac{1}{2}\mathrm{d}\left(\mu\dot{\rho}^2\right) - \left(\frac{l^2}{\mu\rho^3}\right)\mathrm{d}\rho - \mathcal{F}(\rho)\mathrm{d}\rho = 0\,\mathrm{d}t$$
$$\int\frac{1}{2}\mathrm{d}\left(\mu\dot{\rho}^2\right) - \left(\frac{l^2}{\mu}\right)\int\frac{1}{\rho^3}\mathrm{d}\rho - \int\mathcal{F}(\rho)\mathrm{d}\rho = \int 0\,\mathrm{d}t$$
$$\frac{1}{2}\left(\mu\dot{\rho}^2\right) + \left(\frac{l^2}{2\mu\rho^2}\right) - \int\mathcal{F}(\rho)\mathrm{d}\rho = E_{\text{Radial}} \tag{9}$$

The expression (9) shows that how much energy is conveyed in rotational motion when the object is at radial distance away from the center considered. The equation of radial energy derived in (9) is certainly the function of radius and angular momentum l which is conserved for the particular case.

3.4 Expression of Oscillating Energy

Since the acceleration in z direction is zero, we can write the expression of force like this

$$\mu\ddot{z} = 0$$
$$\frac{\mu}{\rho}(\rho\ddot{z}) = 0$$

For $\frac{\mu}{\rho} \neq 0$; $\frac{\mathrm{d}(\rho\dot{z})}{\mathrm{d}t} = 0$

$$\int \mathrm{d}(\rho\dot{z}) = \int 0\,\mathrm{d}t$$
$$\rho\dot{z} = E_{\text{Oscillating}} \tag{10}$$

The Eq. (10) of oscillating energy shows the amount of energy conveyed for the vertical drag oscillations in $\hat{z}$ direction.

3.5 *Expression for Total Energy*

The total energy of the system is the summation of oscillating energy and radial energy.

Using (9) and (10),

$$E_{\text{Total}} = E_{\text{Oscillating}} + E_{\text{Radial}}$$

$$E_{\text{Total}} = \rho\dot{z} + \frac{1}{2}(\mu\dot{\rho}^2) + \left(\frac{l^2}{2\mu\rho^2}\right) - \int \mathcal{F}(\rho)\mathrm{d}\rho \tag{11}$$

The E_{Total} given in (11) is now an expression for combined energy of all the directions given as $\left(\hat{\rho};\hat{\varnothing};\hat{z}\right)$ specifically.

4 Conclusion

We have shown that how the central force problem is solved in three-dimensional cylindrical coordinates system, when a body goes under rotational and translational motion with vertical drag oscillations perpendicular to the plane about its own axis. The central force is found to be dependent only on the radial component of acceleration. This certainly shows that the drag oscillations have no effect on the magnitude of central force as the magnitude of acceleration is null in z direction. Moreover, the equation of both oscillating energy and radial energy is also a function of radius and is independent of the rest of the coordinates. The total energy comprises both radial and oscillating terms. The trajectory of the object is also found to be circular due to the constant velocity in the angular direction and conservation of angular momentum l.

References

1. Symon, K.: Mechanics, 2nd edn. Addison-Wesley, London, England (1960)
2. Cline, D.: Variational Principles in Classical Mechanics, 2nd edn. University of Rochester River Campus Libraries, New York (2018)
3. Taylor, J.: Classical Mechanics. University Science Books (2005)
4. Landau, L., Lifshitz, E.: Mechanics. 3rd edn. Butterworth-Heinemann (1976)
5. Guthrie, M., Wagner, G.: Demystifying the Lagrangian of Classical Mechanics. arXiv:1907.07069v2 (2020)
6. Enaibe, E., Enukpere, E., Idiodi, J.: The modified theory of central-force motion. J. Appl. Phys. **4**(2), 75–82 (2013)
7. Enaibe, E., Agbalagba, E., Francis, J., Maxwell, N.: Lagrange's equations of motion for oscillating central-force field. Theo. Math. Appl. **3**(2), 99–115 (2013)
8. Kleppner, D., Kolenkow, R..: An Introduction to Mechanics. 2nd edn. Cambridge University Press (2014)

9. Goldstein, H., Poole, C., Safko, J.: Classical Mechanics. 3rd edn. Addison-Wesley, (2000)
10. Morin, D.: Introduction to classical mechanics with problems and solutions. Cambridge University Press (2008)

Novel Results for the Factorization of Number Forms

Aziz Lokhandwala, Manoj Sahni, and Mumukshu Trivedi

Abstract Sophie Germain contributed a great work in the field of number theory, theory of elasticity and in the proof of Fermat's Last Theorem. While proving Fermat's Last theorem, she had proved many results on prime numbers. Beyond prime numbers, she had proved many results in the field of number theory. Today, her proved results are known as Sophie Germain theorems, and classification of special form of prime numbers derived by her is called Sophie Germain prime numbers. Many people are still working in this area. Sophie Germain identity has been used currently to find some of the number forms which are factorizable. But, one can only check decomposability of the equation of form x ^ 4 + 4y ^ 4 using this identity. In the present art of work, an attempt is made to search out the generalized number form, so that one can find out which number forms are factorizable and which are not. It has also been shown that from the derived result, one can easily obtain the Sophie Germain identity as a special case.

Keywords Number theory · Factorization · Combinatorics · Number forms · Sophie Germain identity

1 Introduction

French Mathematician, Philosopher and Physicist, Sophie Germain has contributed a lot of work in the field of number theory, theory of rings, elasticity and in the search of the proof of Fermat's Last Theorem [1–3] on which various mathematicians

A. Lokhandwala (✉)
School of Liberal Studies, Pandit Deendayal Petroleum University, Gandhinagar, India
e-mail: aziz.lbsc19@sls.pdpu.ac.in

M. Sahni · M. Trivedi
Department of Mathematics, Pandit Deendayal Petroleum University, Gandhinagar, India
e-mail: manoj.sahni@sot.pdpu.ac.in

M. Trivedi
e-mail: mumukshu.tphd15@sot.pdpu.ac.in

M. Sahni et al. (eds.), *Mathematical Modeling, Computational Intelligence Techniques and Renewable Energy*, Advances in Intelligent Systems and Computing 1287,
https://doi.org/10.1007/978-981-15-9953-8_3

were working in those times. Also, various other advancements have been done in field of study of polynomials [4] and congruences [5] by considering works of Sophie Germain. During the search of the proof of Fermat's Last Theorem [6, 7], she discovered a property regarding factorization of various number forms which is currently used in theory of rings. This property was communicated to Gauss with the help of letters [8], and now, it is known as Sophie Germain identity. The identity is stated in the form of functional equation and defined for any integers x, y as follows [8]:

$$x^4 + 4y^4 = (x^2 + 2xy + 2y^2)(x^2 - 2xy + 2y^2)$$

The above equation clearly justifies the importance of the identity with an insight to factorization of numbers which is quite useful in number theory. In the theory of rings, if the elements satisfy the condition $a^2 = a + a$, then those elements are called SS-element and a ring is called SS-ring if it has at least one non-trivial SS-element. In 2013, Chandra Sekhar Rao [9] observed that there is a relation between the set of SS-elements of SS-ring and the set of solutions of Sophie Germain Identity. To find out more such relations between theory of rings and factorization of number forms, we have tried to extend the Sophie German identity to the set of real numbers as their powers and also established various results related to other near rings to ring R.

2 Main Results

While trying to understand the behavior of Sophie Germain identity, we have discovered the following new relations which are the generalized forms of Sophie Germain identity.

Preposition 1 *For any integers x, y;*

$$x^{2^{4n-2}} + 2^{4n-2}y^{2^{4n-2}} = \left(x^{2^{4n-3}} + 2^{2n-1}y^{2^{4n-3}} + 2^n(xy)^{2^{4n-4}}\right) \left(x^{2^{4n-3}} + 2^{2n-1}y^{2^{4n-3}} - 2^n(xy)^{2^{4n-4}}\right) \quad (1)$$

The above proposition states that for $n \in R$, $x^{2^{4n-2}} + 2^{4n-2}y^{2^{4n-2}}$ is always factorable.

Proof Firstly, consider the right-hand side of Proposition 1 (1),

$$\text{RHS} = \left(x^{2^{4n-3}} + 2^{2n-1}y^{2^{4n-3}} + 2^n(xy)^{2^{4n-4}}\right)\left(x^{2^{4n-3}} + 2^{2n-1}y^{2^{4n-3}} - 2^n(xy)^{2^{4n-4}}\right)$$

Now, on evaluating the above equation,

$$= \left(x^{2^{4n-3}} + 2^{2n-1}y^{2^{4n-3}}\right)^2 - \left(2^n(xy)^{2^{4n-4}}\right)^2$$

$$= \left(x^{2^{4n-3}\times 2} + 2^{4n-2}y^{2^{4n-3}\times 2} + 2^{2n-1+1}(xy)^{2^{4n-3}}\right) - \left(2^{2n}(xy)^{2^{4n-4+1}}\right)$$

$$x^{2^{4n-2}} + 2^{4n-2}y^{2^{4n-2}} = \text{LHS}$$

Therefore, LHS = RHS, hence, Proposition 1 is true. Also the significance of terms on LHS and RHS can be given using the following observations:

(a) $x^{2^2} + 2^2y^{2^2}$ is factorable.
(b) $x^{2^6} + 2^6y^{2^6}$ is factorable.
(c) $x^{2^{10}} + 2^{10}y^{2^{10}}$ is factorable.
(d) $x^{2^{14}} + 2^{14}y^{2^{14}}$ is factorable.

Based on the above observations, we can find an arithmetic progression in powers of 2 on each term, given as follows,

$$2, 6, 10, 14, \ldots$$

From the above series in which the first term is 2 and common differences is 4, we can easily see that its general term can be written as $4n-2$.

Similarly, for each terms of the RHS of our above observations and using arithmetic progression in powers, we can find out the significance of all the terms in RHS in Eq. (1).

Corollary 1 *The Sophie German identity and some other identities are special case of above generalized Sophie German identity.*

Proof We prove it by taking different values of n.

Case 1: Let $n = 1$ in Preposition 1,

$$x^{2^{4(1)-2}} + 2^{4(1)-2}y^{2^{4(1)-2}} = \left(x^{2^{4(1)-3}} + 2^{2(1)-1}y^{2^{4(1)-3}} + 2^1(xy)^{2^{4(1)-4}}\right)$$
$$\left(x^{2^{4(1)-3}} + 2^{2(1)-1}y^{2^{4(1)-3}} - 2^1(xy)^{2^{4(1)-4}}\right)$$

Therefore,

$$x^4 + 4y^4 = (x^2 + 2xy + 2y^2)(x^2 - 2xy + 2y^2)$$

As we can see above, for $n = 1$ in our preposition 1, we get Sophie Germain identity.

Case 2: Let $n = 0.5$ in Preposition 1,

$$x^{2^{4(0.5)-2}} + 2^{4(0.5)-2}y^{2^{4(0.5)-2}} = \left(x^{2^{4(0.5)-3}} + 2^{2(0.5)-1}y^{2^{4(0.5)-3}} + 2^{0.5}(xy)^{2^{4(0.5)-4}}\right)$$
$$\left(x^{2^{4(0.5)-3}} + 2^{2(0.5)-1}y^{2^{4(0.5)-3}} - 2^{0.5}(xy)^{2^{4(0.5)-4}}\right)$$

Therefore,

$$x + y = \left(x^{\frac{1}{2}} + 2^{0.5}(xy)^{2^{-2}} + y^{\frac{1}{2}}\right)\left(x^{\frac{1}{2}} - 2^{0.5}(xy)^{2^{-2}} + y^{1/2}\right)$$

Case 3: Let $n = 2$ in Preposition 1,

$$= x^{2^{4(2)-2}} + 2^{4(2)-2}y^{2^{4(2)-2}} = \left(x^{2^{4(2)-3}} + 2^{2(2)-1}y^{2^{4(2)-3}} + 2^2(xy)^{2^{4(2)-4}}\right)$$
$$\left(x^{2^{4(2)-3}} + 2^{2(2)-1}y^{2^{4(2)-3}} - 2^2(xy)^{2^{4(2)-4}}\right)$$
$$x^{2^6} + 2^6y^{2^6} = \left(x^{2^5} + 2^3y^{2^5} + 2^2(xy)^{2^4}\right)$$
$$\left(x^{2^5} + 2^3y^{2^5} - 2^2(xy)^{2^4}\right)$$

Therefore,

$$x^{64} + 64y^{64} = (x^{32} + 8y^{32} + 4(xy)^{16})(x^{32} + 8y^{32} - 4(xy)^{16})$$

Case 4: Let $n = 3$ in Preposition 1,

$$= x^{2^{4(3)-2}} + 2^{4(3)-2}y^{2^{4(3)-2}} = \left(x^{2^{4(3)-3}} + 2^{2(3)-1}y^{2^{4(3)-3}} + 2^3(xy)^{2^{4(3)-4}}\right)$$
$$\left(x^{2^{4(3)-3}} + 2^{2(3)-1}y^{2^{4(3)-3}} - 2^3(xy)^{2^{4(3)-4}}\right)$$
$$= x^{2^{10}} + 2^{10}y^{2^{10}} = \left(x^{2^9} + 2^5y^{2^9} + 2^3(xy)^{2^8}\right)$$
$$\left(x^{2^9} + 2^5y^{2^9} - 2^3(xy)^{2^8}\right)$$

Therefore,

$$x^{1024} + 1024y^{1024} = (x^{512} + 32y^{512} + 8(xy)^{256})(x^{512} + 32y^{512} - 8(xy)^{256})$$

Preposition 2 *For any integers x, y;the following equation, which is a generalized form of Sophie Germain identity is always factorable,*

$$x^{p^p} + p^p y^{p^p},$$

where pbelongs to the set of all prime numbers.

Proof We prove it by taking different values of p.

Case 1: Let $p = 2$ which is the zeroth prime in Proposition 2;

$$x^{2^2} + 2^2y^{2^2}$$
$$x^4 + 4y^4 = (x^2 + 2xy + 2y^2)(x^2 - 2xy + 2y^2)$$

Therefore, for zeroth prime, Proposition 2 becomes the Sophie Germain identity which gives us an excellent account on factorizability of numbers.

Case 2: Let $p = 3$ which is the first prime in Preposition 2;

$$x^{3^3} + 3^3 y^{3^3}$$

We know that $a^3 + b^3 = (a + b)\left(a^2 - ab + b^2\right)$, therefore,

$$x^{27} + 27y^{27} = \left(x^9 + 3y^9\right)\left(x^{18} - 3x^9y^9 + 9y^{18}\right)$$

Case 3: Let $p = 5$ which is the second prime in Preposition 2;

$$x^{5^5} + 5^5 y^{5^5}$$

We know that $a^5 + b^5 = (a + b)\left(a^4 - a^3b - ab^3 + (ab)^2 + b^4\right)$, therefore $x^{3125} + 3125y^{3125}$ is also factorable.

Similarly, one can prove for other values of p.

3 Applications

Application 1

Chandra Sekhar Rao [9] proved that there is one-to-one correspondence between Sophie Germain identity and theory of rings. Here, we extend his ideas using Proposition 1 and 2. Let R be a commutative ring with identity named as MAM's ring which is a near ring to R and let the element of this be named as MAM's element. An element $a \in R$ such that $\frac{a^2}{4} = a + a$, then using Proposition 1, we can conclude the following results,

$$x^{64} + a^2y^{64} = \left(x^{32} + ay^{32} + \frac{a}{2}(xy)^{16}\right)\left(x^{32} + ay^{32} - \frac{a}{2}(xy)^{16}\right)$$

The above identity is a necessary condition for existence of such near ring to R. Therefore, here, we have formed a new ring which is near ring to R such that any element a from that set of ring must obey the condition $\frac{a^2}{4} = a + a$.

Many more near rings to R can be formed using Prepositions 1 and 2.

Application 2

To identify Smarandache Modularity property of near rings is of great interest for ring theorists. For that purpose, firstly, define Smarandache idempotent [10] (S-idempotent) of R as,

Definition Assume that N is a near ring. An element $0 \neq x \in N$ is a Smarandache idempotent [10] of R if,

(a) $x^2 = x$
(b) There exist $a \in N \backslash \{x, 1, 0\}\{if N has 1\}$. With $xa = a$ or $ax = a$, a is called as Smarandache co-idempotent (S-co-idempotent).

Assume a ring with no information about the existence of S-idempotent for that ring; in such a case, one can use Proposition 1 as follows:

$$x + a^2 y = \left(x^{\frac{1}{2}} + (a+a)^{0.5}(xy)^{2^{-2}} + ay^{\frac{1}{2}}\right)$$
$$\left(x^{\frac{1}{2}} - (a+a)^{0.5}(xy)^{2^{-2}} + ay^{1/2}\right)$$

If above equation holds true for some element a of that ring, then the ring has S-idempotent.

Application 3
Factorizing various number forms [11] is of great interest in number theory. It is known that the Sophie Germain identity is currently being used to identify that number forms like $n^4 + 4^n$ can never be prime for all $n > 1$. We can find many other relations like above using Proposition 1. Some of them are given below as follows,

(a) $p^{64} + 64$ can never be a prime for all p belonging to set of prime numbers.
(b) $p^{1024} + 1024$ can never be a prime for all p belonging to set of prime numbers.

Thus, we can say that $p^{2^{4n-2}} + 2^{4n-2}$ can never be prime for all $n \geq 1$.

Application 4
Preposition 1 and Preposition 2 are extremely useful to tackle factorization problems. For example, these two propositions are useful in the case when someone asks to find factors of the expressions given below:

(a) $5^4 + 4$ is Prime or not?
(b) $2^{27} + 27$ is Prime or not?
(c) $2^{3125} + 3125$ is Prime or not?

Application 5
Preposition 1 and Preposition 2 are adequate for solving integrals of the form $\frac{P(x)}{Q(x)}$, where $P(x)$, $Q(x)$ are some univariate polynomials such that $\deg(P(x)) < \deg(Q(x))$.

Open Problem(s)
Proposition 2 cannot be proved using induction method or using any other methods available in the current literature. One may attempt to prove it. Also, further properties of MAM's ring can be found, and many other near rings to R can be made using Proposition 1 and Proposition 2.

While observing Sophie Germaain identity and monic polynomials [12], in Proposition 1, if we take $x = 1$ and $y = 1$ and $p \equiv -1 \pmod 4$ for some prime p belonging to set of primes, we have,

$$1 + 2^{p-1} = \left(1 + 2^{\frac{p-1}{2}} + 2^{\frac{p+1}{4}}\right)\left(1 + 2^{\frac{p-1}{2}} - 2^{\frac{p+1}{4}}\right) \tag{2}$$

If $3|(p-1)$ and $10, 13\nmid(p-1)$, then above Eq. (2) always gives square-free values. It can be easily shown that factors in right-hand side of Eq. (2) have no factors in common under ordinary conditions.

4 Conclusions

Factorization of various number forms has been a topic of interest for number theorists around the globe. In the present scenario, we made an attempt to add a bit more to the cluster of storage by the medium of this article. Firstly, the "MAIN RESULTS" section states novel result(s) which are not observed in the past. To justify the propositions, sufficient numbers of examples are listed along with the prepositions. The noteworthy applications are presented in the section of "Applications". At last, a brief mentioning of the future scope and connections to other mathematical frameworks to this theory is in the "Open Problem(s)" section.

References

1. Dickson, L.E.: On the last theorem of Fermat. Messenger Math. **38**, 14–32 (1908)
2. Adleman, L.M., Heath-Brown, D.R.: The first case of Fermat's last theorem. Inventiones Mathematicae **79**(2), 409–416 (1985)
3. Vandiver, H.: Note on euler number criteria for the first case of Fermat's last theorem. Am. J. Math. **62**(1), 79–82 (1940)
4. Stevenhagen, P.: Class number parity for the *p*th cyclotomic field. Math. Comput. **63**(208), 773–784 (1994)
5. Vandiver, H.: Note on trinomial congruences and the first case of Fermat's last theorem. Annal. Math. Second Ser. **27**(1), 54–56 (1925)
6. Del Centina, A.: Letters of Sophie Germain preserved in Florence. Historia Mathematica **32**(1), 60–75 (2005)
7. Musielak, D.: Germain and Her Fearless Attempt to Prove Fermat's Last Theorem. arXiv:1904.03553 (2019)
8. Germain, S.: Cinq Lettres de Sophie Germain à Charles-Frédéric Gauss publiées par B. Boncompagni d'après les originaux possédés par la Société Royale de Sciences de Göttingen. Institut de photolithographie des frères Burchard (1880)
9. Chandra Sekhar Rao, K.S.: On Sophie Germain's Identity, Scientia Magna **9**(4), 13–15 (2013)
10. Vasantha Kandasamy, W.B.: Smarandache near rings, Infinite Study (2002)

11. Łukasik, R., Sikorska, J., Szostok, T.: On an equation of Sophie Germain. Res. Math. **73**, 60 (2018)
12. Bhargava, M., Shankar, A., Wang, X.: Square free values of polynomial discriminants I. arXiv: 1611.09806 (2016)

A Series Solution of Nonlinear Equation for Imbibition Phenomena in One-Dimensional Heterogeneous Porous Medium

Dhara T. Patel and Amit K. Parikh

Abstract In present article, we addressed the imbibition phenomena in a horizontal direction within a heterogeneous porous media that occurred during the procedure of secondary oil retrieval. Homotopy Perturbation Laplace Transform Method (HPLTM) is used to derive an infinite series form solution of the given differential equation. The solution reflects the saturation of injected water in a porous medium during imbibition. The numeric results were obtained via MATLAB.

Keywords Imbibition phenomenon · Laplace transformation · Homotopy perturbation method

1 Introduction

In the procedure of crude oil retrieval, there are three distinct phases such as first phase, intermediate phase, and final phase to retrieve crude oil. First phase retrieval is also called as primary oil retrieval. In this phase, crude oil is drawn via pump jacks and achieves about 10% of crude oil available in the basin. The procedure of crude oil retrieval phase continues until the pressure inside the well is no longer enough to draw crude oil. The rest of crude oil can be retrieved through driving water, gas, or some chemicals. So the given procedure is called the intermediate oil retrieval procedure which is also known as secondary oil retrieval. The secondary retrieval phase come to its end when a significant amount of the drawn fluid (water or gas) is generated from the wells of production and processing is no more economically viable. The sequential utilize of first phase retrieval and intermediate phase retrieval in a crude oil basin generates approximately 40% of the existing oil in place. Final phase retrieval is a method of drawing crude oil that has not already been drawn through the techniques of first or intermediate phase oil retrieval. In this process, the chemical/gas injection technique can be applied to reduce the viscosity of present

D. T. Patel (✉) · A. K. Parikh
Mehsana Urban Institute of Science, Ganpat University, Mehsana, Gujarat, India
e-mail: dharar89@yahoo.in

M. Sahni et al. (eds.), *Mathematical Modeling, Computational Intelligence Techniques and Renewable Energy*, Advances in Intelligent Systems and Computing 1287,
https://doi.org/10.1007/978-981-15-9953-8_4

oil and improve the flow. This phase is also known as enhanced oil retrieval. The enhanced oil retrieval method can draw approximately 60% of the available oil.

Imbibition procedure plays a principal character in the motion of fluids in ground & germination of seeds. This is one of the main significant phenomena for drawing crude oil in reservoirs during the process of the secondary oil retrieval.

"When a porous medium filled with some fluid, is brought in contact with another fluid which preferentially wets the medium then there is a natural flow of the wetting fluid into the medium and a counter flow of the native fluid from the medium. This phenomenon arising near to the common border due to the difference in wetting abilities of the fluids flowing in the medium is called imbibition phenomenon." Imbibition phenomenon has been explored by many authors such as E. R. Brownscombe and A. B. Dyes, J. W. Graham and J. G. Richardson, A. E. Scheidegger, C. C. Mattax and J. R. Kyte, P. M. Blair, R. W. Parsons and P. R. Cheney, A. P. Verma, M. N. Mehta, V. H. Pradhan.

Patel et al. [1] have discussed "A mathematical model of imbibition phenomenon in heterogeneous porous media during secondary oil recovery process" using a similarity transformation. Parikh et al. [2] have analyzed "Mathematical Model and Analysis of counter-current Imbibition in Vertical Downward Homogeneous Porous Media" using generalized separable method. Patel et al. [3] have discussed "A homotopy series solution to a nonlinear partial differential equation arising from a mathematical model of the counter-current imbibition phenomenon in a heterogeneous porous medium." Patel and Meher [4] have discussed "Approximate analytical study of counter-current imbibition phenomenon in a heterogeneous porous media." Patel and Desai [5] have discussed "Mathematical modeling and analysis of co-current imbibition phenomenon in inclined homogeneous porous media" using homotopy analysis method.

In present article, we have talked about the imbibitions phenomenon for two immiscible fluids motion in a heterogeneous porous medium. If water is inserted in oil-saturated heterogeneous porous medium, then imbibition takes place near the common border ($l = 0$) during secondary oil retrieval procedure. The aim is to estimate the water saturation at different lengths l and different time periods t in heterogeneous porous medium for imbibition. It is solved using Homotopy Perturbation Laplace Transform Method (HPLTM) with appropriate boundary and initial conditions.

2 Statement of the Problem

Here, we pick a cylindrical sample of heterogeneous porous media from the huge natural oil-saturated porous media that is given by Fig. 1. Then, we cut the vertical cross-section of this sample as a rectangle that is given by Fig. 2. There is three sides which are impermeable and one open end is permeable.

When water is brought into contact with oil-saturated heterogeneous porous matrix at the common border $l = 0$, then wetting fluid water flows naturally inside

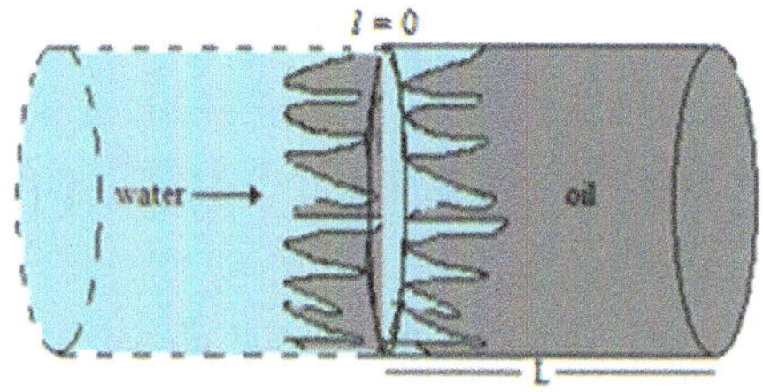

Fig. 1 Cylindrical sample

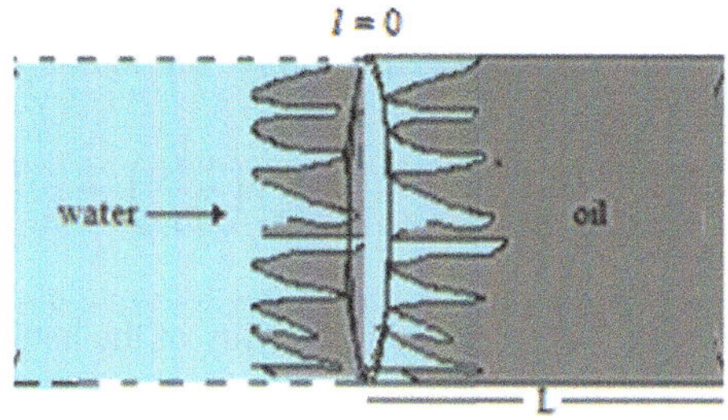

Fig. 2 Vertical cross-section

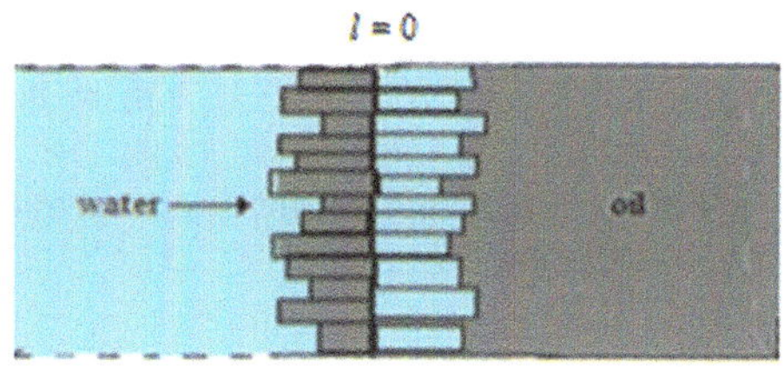

Fig. 3 Rectangle fingers

the medium and counter flow of resident fluid oil out of the medium. This is named as imbibition phenomenon. The shape and size of small fingers are distinct and irregular. For the study, small fingers are supposed to be a rectangle that is given by Fig. 3. This imbibition phenomenon occurs because of the variation in viscosity of both fluids at a common border $l = 0$. Inserted water is assumed to be more wetting than that of resident oil. Hence, during the imbibition process, the oil will be moved to a small length l. The resident fluid and immigrant fluid follow Darcy's law for laminar flow.

3 Mathematical Formulation of the Imbibition Phenomenon

As per the Darcy's law [6], the velocities of immigrant fluid water V_i and resident fluid oil V_r are given by

$$V_i = -k\left(\frac{k_i}{\eta_i}\frac{\partial p_i}{\partial l}\right) \tag{1}$$

$$V_r = -k\left(\frac{k_r}{\eta_r}\frac{\partial p_r}{\partial l}\right) \tag{2}$$

In the equations mentioned above, k_i and k_r are the relative permeability's of water and oil, respectively. The permeability k is variable, as porous medium is heterogeneous. p_i , p_r are the pressure of water and oil, respectively, whereas η_i , η_r are the viscosity of water and oil, respectively.

The continuity equation for immigrant fluid water is

$$\left(\frac{\partial V_i}{\partial l}\right) + \varphi\left(\frac{\partial S_i}{\partial t}\right) = 0 \tag{3}$$

where φ is the variable porosity of the heterogeneous porous media.

The fluid will run through interconnected capillaries because of capillary pressure. So it is represented by the pressure difference of the running fluids across the common border. Hence,

$$p_c = p_r - p_i \tag{4}$$

Mehta [7] has applied the direct correlation among capillary pressure (p_c) and water saturation (S_i) as

$$p_c = -\beta\, S_i \tag{5}$$

We suppose this authentic correlation because of Scheidegger and Johnson [8] among water saturation and relative permeability which is given by

$$k_i = S_i, k_r = 1 - \alpha\, S_i \quad (\alpha = 1.11) \tag{6}$$

The porosity and permeability of the heterogeneous porous material proposed by Patel and Mehta [1] are also used for certainty as

$$\varphi = \varphi(l, t) = (a_1 - a_2 l)^{-1} \tag{7}$$

$$k = k(l, t) = k_c(1 + a_3 l) \tag{8}$$

Due to the possibility of introducing impurity by water injection, porosity and permeability can change over time t. Where a_1 , a_2 and a_3 are some functions of time t. As porosity $\varphi(l, t)$ may not be greater than unity, we suppose $a_1 - a_2 \geq 1$ [1]

According to Scheidegger [9], the addition of the velocities of immigrant fluid water and resident fluid oil is nullity in the imbibition phenomenon.

$$V_i + V_r = 0 \tag{9}$$

Using Eqs. (1) and (2) into Eq. (9),

$$\left(\frac{k_i}{\eta_i}k\right)\left(\frac{\partial p_i}{\partial l}\right) = -\left(\frac{k_r}{\eta_r}k\right)\left(\frac{\partial p_r}{\partial l}\right) \tag{10}$$

From Eqs. (10) and (4)

$$\frac{k_i}{\eta_i}\frac{\partial p_i}{\partial l} + \frac{k_r}{\eta_r}\frac{\partial p_i}{\partial l} = -\frac{k_r}{\eta_r}\frac{\partial p_c}{\partial l} \tag{11}$$

Using Eq. (11), we have

$$\frac{\partial p_i}{\partial l} = -\frac{\frac{k_r}{\eta_r}\frac{\partial p_c}{\partial l}}{\left(\frac{k_i}{\eta_i} + \frac{k_r}{\eta_r}\right)} \tag{12}$$

Using Eq. (12) into Eq. (1), we obtain

$$V_i = \frac{k_i}{\eta_i}k\left[\frac{\frac{k_r}{\eta_r}\frac{\partial p_c}{\partial l}}{\left(\frac{k_i}{\eta_i} + \frac{k_r}{\eta_r}\right)}\right] \tag{13}$$

Replacing the V_i value with the continuity Eq. (3), we obtain

$$\varphi\left(\frac{\partial S_i}{\partial t}\right) + \frac{\partial}{\partial l}\left[k\frac{\frac{k_i}{\eta_i}\frac{k_r}{\eta_r}}{\frac{k_i}{\eta_i} + \frac{k_r}{\eta_r}}\frac{\partial p_c}{\partial l}\right] = 0 \tag{14}$$

For the flow mechanism of the investigation includes water and oil, so we have as per Scheidegger [9]

$$\frac{\frac{k_i}{\eta_i}\frac{k_r}{\eta_r}}{\frac{k_i}{\eta_i} + \frac{k_r}{\eta_r}} \approx \frac{k_r}{\eta_r} = \frac{(1 - \alpha S_i)}{\eta_r} \tag{15}$$

When we replace values from Eqs. (15) and (5) with Eq. (14), we obtain

$$\varphi\left(\frac{\partial S_i}{\partial t}\right) - \frac{\beta}{\eta_r}\frac{\partial}{\partial l}\left[k\,(1 - \alpha S_i)\frac{\partial S_i}{\partial l}\right] = 0 \tag{16}$$

To simplify Eq. (15) further, we take the standard relationship $k \propto \varphi$ indicated by Chen [10] $k = k_c\varphi$, where k_c is a proportionality constant.

If we replace this value in Eq. (15), we obtain

$$\varphi\left(\frac{\partial S_i}{\partial t}\right) - \frac{k_c\beta}{\eta_r}\frac{\partial}{\partial l}\left[\varphi(1-\alpha S_i)\frac{\partial S_i}{\partial l}\right] = 0 \tag{17}$$

Saturation of inserted water in heterogeneous porous medium for imbibition phenomenon is calculated using above partial differential equation (nonlinear).

Using dimensionless quantities

$$T = \frac{k_c\alpha\beta}{L^2\eta_r}t \ , \ Z = \frac{l}{L} \ , \ 0 \le Z \le 1$$

We have

$$\frac{\partial S_i}{\partial T} - \frac{1}{\varphi}\frac{\partial\varphi}{\partial Z}\left(\frac{1}{\alpha} - S_i\right)\frac{\partial S_i}{\partial Z} - \frac{\partial}{\partial Z}\left[\left(\frac{1}{\alpha} - S_i\right)\frac{\partial S_i}{\partial Z}\right] = 0 \tag{18}$$

To further simplify, we will take $\frac{1}{\alpha} = \delta$, so we obtain

$$\frac{\partial S_i}{\partial T} - \frac{1}{\varphi}\frac{\partial\varphi}{\partial Z}(\delta - S_i)\frac{\partial S_i}{\partial Z} - \frac{\partial}{\partial Z}\left[(\delta - S_i)\frac{\partial S_i}{\partial Z}\right] = 0 \tag{19}$$

Equation (19) is a partial differential equation (nonlinear) with several difficult terms. The term $\frac{1}{\varphi}\frac{\partial\varphi}{\partial Z}$ has been modified for the sake of convenience of the problem as

$$\begin{aligned}\frac{1}{\varphi}\frac{\partial\varphi}{\partial Z} = \frac{\partial}{\partial Z}(\log\varphi) &= \frac{\partial}{\partial Z}\log\left(\frac{1}{a_1 - a_2 ZL}\right)\\ &= \frac{\partial}{\partial Z}\left(\frac{a_2}{a_1}ZL + \frac{a_2^2}{a_1^2}Z^2L^2 + \cdots - \log a_1\right)\\ &\approx \frac{a_2}{a_1}L(\text{neglecting higher power of Z})\end{aligned}$$

Choosing a_1 and a_2 as given by Patel and Mehta [1]

$$\frac{a_2}{a_1} = (4T)^{\frac{-1}{2}}, T > 0 \tag{20}$$

Using the above value, Eq. (19) can be reduced to

$$\frac{\partial S_i}{\partial T} - \frac{L}{2\sqrt{T}}(\delta - S_i)\frac{\partial S_i}{\partial Z} - \frac{\partial}{\partial Z}\left[(\delta - S_i)\frac{\partial S_i}{\partial Z}\right] = 0 \tag{21}$$

To solve the nonlinear partial differential Eq. (21), suitable conditions should be selected. So we are considering [3]

$$S_i(Z, 0) = S_{i_0}e^Z \text{ at } T = 0 \quad \text{for} \quad Z > 0$$

$$S_i(0,T) = S_{i_0} \text{ at } X = 0 \quad \text{for} \quad T > 0 \tag{22}$$

4 Solution of Problem

We will derive the infinite series solution of (21) subject to initial condition and boundary condition which is given by Eq. (22), using Laplace Transform Homotopy Perturbation Method.

Applying the Homotopy Perturbation Method, He [11] constructed the function $\phi(Z,T;q)\colon \Omega \times [0,1] \to R[0,1] \to R$, which can be satisfied as follows:

$$(1-q)\left[\frac{\partial S_i}{\partial T} - \frac{\partial v_0}{\partial T}\right] + q\left[\frac{\partial S_i}{\partial T} - \frac{L}{2\sqrt{T}}(\delta - S_i)\frac{\partial S_i}{\partial Z} - \frac{\partial}{\partial Z}\left[(\delta - S_i)\frac{\partial S_i}{\partial Z}\right]\right] = 0$$

$$\frac{\partial S_i}{\partial T} = \frac{\partial v_0}{\partial T} + q\left[-\frac{\partial v_0}{\partial T} + \frac{L}{2\sqrt{T}}(\delta - S_i)\frac{\partial S_i}{\partial Z} + \frac{\partial}{\partial Z}\left((\delta - S_i)\frac{\partial S_i}{\partial Z}\right)\right] \tag{23}$$

where $v_0(Z,T) = S_{i_0}e^Z + \frac{ZT}{2}$ (which can be chosen freely) is initial approximation of Eq. (21), which satisfy the initial and boundary conditions. The transforming procedure of q, from 0 to 1, is like $\phi(Z,T;q)$ from $v_0(Z,T)$ to $S_i(Z,T)$, this is called deformation.

Taking Laplace transform on both sides, we obtain

$$sL\{S_i(Z,T)\} - S_i(Z,0) = L\left\{\frac{\partial v_0}{\partial T} + q\left[-\frac{\partial v_0}{\partial T} + \frac{L}{2\sqrt{T}}(\delta - S_i)\frac{\partial S_i}{\partial Z} + \frac{\partial}{\partial Z}\left((\delta - S_i)\frac{\partial S_i}{\partial Z}\right)\right]\right\} \tag{24}$$

$$L\{S_i(Z,T)\} - \frac{S_i(Z,0)}{s} = \frac{1}{s}L\left\{\frac{\partial v_0}{\partial T} + q\left[-\frac{\partial v_0}{\partial T} + \frac{L}{2\sqrt{T}}(\delta - S_i)\frac{\partial S_i}{\partial Z} + \frac{\partial}{\partial Z}\left((\delta - S_i)\frac{\partial S_i}{\partial Z}\right)\right]\right\} \tag{25}$$

Substituting all the given value and introducing inverse Laplace transformation on both ends, we obtain

$$S_i(X,T) = S_{i_0}e^Z + \frac{ZT}{2} + L^{-1}\left(\frac{1}{s}L\left\{q\left[-\frac{\partial v_0}{\partial T} + \frac{L}{2\sqrt{T}}(\delta - S_i)\frac{\partial S_i}{\partial Z} + \frac{\partial}{\partial Z}\left((\delta - S_i)\frac{\partial S_i}{\partial Z}\right)\right]\right\}\right) \tag{26}$$

Now according to homotopy perturbation method

$$S_i(Z,T) = S_0 + qS_1 + q^2S_2 + \cdots \tag{27}$$

Using Eq. (27), one can write Eq. (26) as

$$S_0 + qS_1 + \cdots = S_{i_0}e^Z + \frac{ZT}{2} + L^{-1}\left(\frac{1}{s}L\left\{q\left[\begin{array}{l} -\dfrac{\partial v_0}{\partial T} \\ +\dfrac{L}{2\sqrt{T}}(\delta - (S_0 + qS_1 + \cdots))\dfrac{\partial(S_0 + qS_1 + \cdots)}{\partial Z} \\ +\dfrac{\partial}{\partial Z}\left((\delta - (S_0 + qS_1 + \cdots))\dfrac{\partial(S_0 + qS_1 + \cdots)}{\partial Z}\right)\end{array}\right]\right\}\right)$$

On the expansion of the above equation and comparing the coefficient of various power of q, we have got

$$q^0\colon S_0(Z,T) = S_{i_0}e^Z + \frac{ZT}{2}$$

$$q^1\colon S_1(Z,T) = L^{-1}\left[\frac{1}{s}L\left\{\begin{array}{l} -\dfrac{\partial v_0}{\partial T} + \delta\dfrac{\partial^2 S_0}{\partial Z^2} - S_0\dfrac{\partial^2 S_0}{\partial Z^2} - \left(\dfrac{\partial S_0}{\partial Z}\right)^2 + \dfrac{L\delta}{2\sqrt{T}}\dfrac{\partial S_0}{\partial Z} \\ -\dfrac{L}{2\sqrt{T}}S_0\dfrac{\partial S_0}{\partial Z}\end{array}\right\}\right]$$

$$\begin{aligned} &= -\frac{T^3}{12} - \frac{ZT}{2} - \frac{LS_{i_0}e^Z ZT^{\frac{3}{2}}}{6} - \frac{LS_{i_0}e^Z T^{\frac{3}{2}}}{6} \\ &+ L\delta S_{i_0}e^Z\sqrt{T} - \frac{S_{i_0}e^Z ZT^2}{4} \\ &+ \delta S_{i_0}e^Z T - \frac{S_{i_0}e^Z T^2}{2} - LS_{i_0}^2 e^{2Z}\sqrt{T} \\ &- 2S_{i_0}^2 e^{2Z}T - \frac{LZT^{\frac{5}{2}}}{20} + \frac{L\delta T^{\frac{3}{2}}}{6}\end{aligned}$$

$$q^2\colon S_2(Z,T) = L^{-1}$$

$$\left[\frac{1}{s}L\left\{\begin{array}{l} \delta\dfrac{\partial^2 S_1}{\partial Z^2} - S_0\dfrac{\partial^2 S_1}{\partial Z^2} - S_1\dfrac{\partial^2 S_0}{\partial Z^2} - 2\dfrac{\partial S_0}{\partial Z}\dfrac{\partial S_1}{\partial Z} + \dfrac{L\delta}{2\sqrt{T}}\dfrac{\partial S_1}{\partial Z} \\ -\dfrac{L}{2\sqrt{T}}S_0\dfrac{\partial S_1}{\partial Z} - \dfrac{L}{2\sqrt{T}}S_1\dfrac{\partial S_0}{\partial Z}\end{array}\right\}\right]$$

$$\begin{aligned} &= -2L^2\delta\, S_{i_0}^2 e^{2Z} ZT - \frac{L^2\delta\, S_{i_0}e^Z ZT^2}{6} + \frac{5S_{i_0}e^Z T^4}{24} \\ &- \frac{5LS_{i_0}\delta e^Z ZT^{\frac{5}{2}}}{12} + \frac{107LS_{i_0}^2 e^{2Z}T^{\frac{5}{2}}}{60} + \frac{L^2 S_{i_0}e^Z ZT^3}{20} \\ &+ \frac{L^2\delta^2 S_{i_0}e^Z T}{2} + \frac{LS_{i_0}e^Z ZT^{\frac{3}{2}}}{6} + \frac{T^3}{6} - \frac{22L\delta\, S_{i_0}^2 e^{2Z}T^{\frac{3}{2}}}{3}\end{aligned}$$

$$
\begin{aligned}
&+\frac{S_{i_0}e^{Z}Z^{2}T^{4}}{32}+\frac{L^{2}S_{i_0}e^{Z}T^{3}}{45}+\frac{17LT^{\frac{9}{2}}}{1080}\\
&+\frac{6LS_{i_0}e^{Z}T^{\frac{7}{2}}}{35}+\frac{3L^{2}S_{i_0}^{3}e^{3Z}T}{2}+\frac{L^{2}S_{i_0}^{2}e^{2Z}T^{2}}{4}\\
&+\frac{5S_{i_0}^{2}e^{2Z}ZT^{3}}{3}-\frac{L^{2}\delta T^{3}}{45}+\frac{S_{i_0}e^{Z}T^{2}}{2}\\
&+\frac{L^{2}ZT^{4}}{160}+\frac{S_{i_0}e^{Z}ZT^{2}}{4}+\frac{LS_{i_0}e^{Z}T^{\frac{3}{2}}}{6}\\
&-\frac{11L\delta S_{i_0}e^{Z}T^{\frac{5}{2}}}{12}+\frac{7S_{i_0}^{2}e^{2Z}T^{3}}{3}+\frac{L^{2}S_{i_0}e^{Z}Z^{2}T^{3}}{72}\\
&+\frac{L^{2}S_{i_0}^{2}e^{2Z}ZT^{2}}{3}+\frac{LT^{\frac{5}{2}}Z}{10}+\frac{43LS_{i_0}e^{Z}T^{\frac{7}{2}}}{210}-\frac{L\delta T^{\frac{3}{2}}}{6}\\
&+\frac{\delta S_{i_0}e^{Z}T^{\frac{7}{2}}Z^{2}}{24}-\frac{S_{i_0}\delta e^{Z}ZT^{3}}{4}+9S_{i_0}^{3}e^{3Z}T^{2}\\
&-\frac{2S_{i_0}\delta e^{Z}T^{3}}{3}+\frac{3S_{i_0}e^{Z}ZT^{4}}{16}+\frac{\delta^{2}S_{i_0}e^{Z}T^{2}}{2}-6\delta S_{i_0}^{2}e^{2Z}T^{2}\\
&+\frac{47LS_{i_0}^{2}e^{2Z}ZT^{\frac{5}{2}}}{30}+8LS_{i_0}^{3}e^{3Z}T^{\frac{3}{2}}\\
&-\frac{L^{2}\delta S_{i_0}e^{Z}T^{2}}{4}+\frac{L\delta^{2}S_{i_0}e^{Z}T^{\frac{3}{2}}}{4}
\end{aligned}
$$

By placing all of the above values in Eq. (27), and putting $q = 1$, we obtain

$$
\begin{aligned}
S_i &= S_0 + S_1 + S_2 + \cdots\\
&= S_{i_0}e^{Z} - 2L^{2}\delta S_{i_0}^{2}e^{2Z}ZT - \frac{L^{2}\delta S_{i_0}Ze^{Z}T^{2}}{6}\\
&+\frac{5S_{i_0}e^{Z}T^{4}}{24}-\frac{5LS_{i_0}\delta Ze^{Z}T^{\frac{5}{2}}}{12}+\frac{107LS_{i_0}^{2}e^{2Z}T^{\frac{5}{2}}}{60}+\frac{L^{2}S_{i_0}e^{Z}ZT^{3}}{20}\\
&+\frac{L^{2}\delta^{2}S_{i_0}e^{Z}T}{2}+\frac{T^{3}}{12}-\frac{22L\delta S_{i_0}^{2}e^{2Z}T^{\frac{3}{2}}}{3}\\
&+\frac{S_{i_0}e^{Z}Z^{2}T^{4}}{32}+\frac{L^{2}S_{i_0}e^{Z}T^{3}}{45}+\frac{17LT^{\frac{9}{2}}}{1080}\\
&+\frac{6LS_{i_0}e^{Z}T^{\frac{7}{2}}}{35}+\frac{3L^{2}S_{i_0}^{3}e^{3Z}T}{2}\\
&+\frac{L^{2}S_{i_0}^{2}e^{2Z}T^{2}}{4}+\frac{5S_{i_0}^{2}e^{2Z}ZT^{3}}{3}-\frac{L^{2}\delta T^{3}}{45}+\frac{L^{2}ZT^{4}}{160}-\frac{11L\delta S_{i_0}e^{Z}T^{\frac{5}{2}}}{12}\\
&+\frac{7S_{i_0}^{2}e^{2Z}T^{3}}{3}+\frac{L^{2}S_{i_0}e^{Z}Z^{2}T^{3}}{72}+\frac{L^{2}S_{i_0}^{2}e^{2Z}ZT^{2}}{3}+\frac{LT^{\frac{5}{2}}Z}{20}+\frac{43LS_{i_0}e^{Z}T^{\frac{7}{2}}}{210}
\end{aligned}
$$

$$
+\frac{\delta S_{i_0}e^{Z}T^{\frac{7}{2}}Z^2}{24}-\frac{S_{i_0}\delta e^{Z}ZT^3}{4}+9S_{i_0}^3e^{3Z}T^2-\frac{2S_{i_0}\delta e^{Z}T^3}{3}+\frac{3S_{i_0}e^{Z}ZT^4}{16}
$$

$$
+\frac{\delta^2 S_{i_0}e^{Z}T^2}{2}-6\delta S_{i_0}^2e^{2Z}T^2+\frac{47LS_{i_0}^2e^{2Z}ZT^{\frac{5}{2}}}{30}
$$

$$
+8LS_{i_0}^3e^{3Z}T^{\frac{3}{2}}-\frac{L^2\delta S_{i_0}e^{Z}T^2}{4}
$$

$$
+\frac{L\delta^2 S_{i_0}e^{Z}T^{\frac{3}{2}}}{4}-2S_{i_0}^2e^{2Z}T+L\delta S_{i_0}e^{Z}\sqrt{T}+\delta S_{i_0}e^{Z}T-LS_{i_0}^2e^{2Z}\sqrt{T}+\cdots \tag{28}
$$

This is the appropriate solution of the imbibition phenomenon Eq. (21) that stands for the saturation of inserted water inhabited through the schematic fingers of average span at any length Z and for any fixed time level T.

5 Results and Discussion

The solution (28) represents the saturation of injected fluid in terms of the infinite series of Z and T in the secondary oil retrieval procedure during water injection. We have truncated the series hence it will give an approximate solution of imbibition phenomenon. We observed water saturation from the numerical table for some particular value of a parameter. The following constant values are taken from the standard literature for numerical calculation as follows.

$$
L = 2\,\text{m}\,,\ S_{i_0} = 0.1\,,\ \alpha = 1.11\,,\ \delta = 0.9
$$

We have considered a range of length Z from 0 to 1 according to dimensionless variable and time level T is considered as 0–0.5 as imbibition phenomenon occurs for a short period. For higher values of time T, the saturation of water S_i will become constant. Here, it is observed that Saturation of water S_i is growing as length Z grows. It is also growing for time T. Numeric and visual presentation of solution (28) have been derived by MATLAB.

Figure 4 exhibits the visual representation of S_i Vs. Z for fixed time $T = 0, 0.1, 0.2, 0.3, 0.4, 0.5$ (Table 1).

6 Conclusion

The governing formulation indicates the imbibition phenomenon for heterogeneous porous medium and it is determined by Homotopy Perturbation Laplace Transformation method. The solution is derived as an infinite series. It can be seen that the saturation of injected water grows in respect of time as well as grows in respect

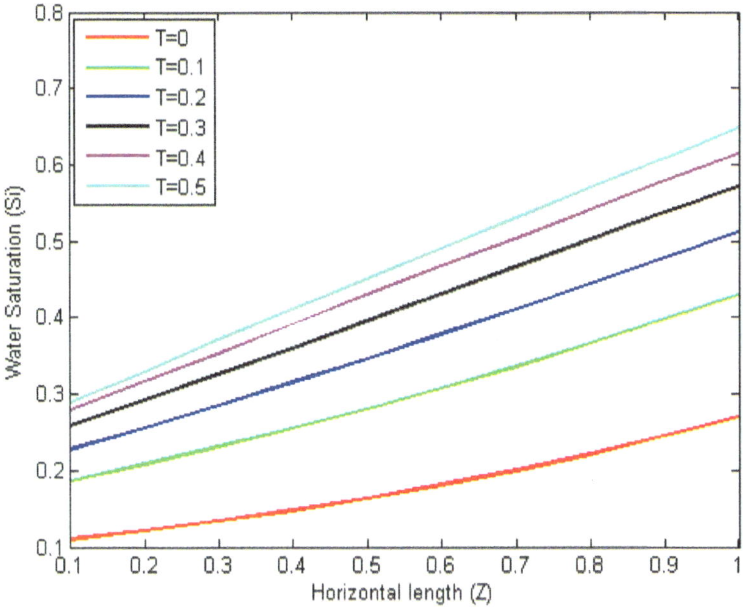

Fig. 4 Graph of water saturation (S_i) versus horizontal length Z for a fixed period T

Table 1 Water saturation (S_i) for various Z for a fixed period T

Z	$T = 0.0$	$T = 0.10$	$T = 0.20$	$T = 0.30$	$T = 0.40$	$T = 0.50$
$Z = 0.1$	0.110517	0.187699	0.228838	0.259046	0.279639	0.290731
$Z = 0.2$	0.12214	0.209399	0.256895	0.292199	0.316924	0.331368
$Z = 0.3$	0.134986	0.23237	0.285976	0.326046	0.354542	0.371998
$Z = 0.4$	0.149182	0.256659	0.316048	0.360482	0.392342	0.412455
$Z = 0.5$	0.164872	0.282304	0.347055	0.395382	0.430162	0.452587
$Z = 0.6$	0.182212	0.309332	0.378928	0.430609	0.46785	0.492289
$Z = 0.7$	0.201375	0.337761	0.411579	0.466026	0.505287	0.531551
$Z = 0.8$	0.222554	0.367598	0.444916	0.501517	0.54243	0.570528
$Z = 0.9$	0.24596	0.39884	0.478851	0.53702	0.579383	0.60965
$Z = 1$	0.271828	0.43148	0.513326	0.572584	0.616489	0.649778

of length for a heterogeneous porous medium which is consistent with the physical nature of the phenomena. As the saturation of water grows, oil pushes toward oil production well and this way, oil could be recovered through the secondary oil retrieval procedure.

References

1. Patel, K.R., Mehta, M.N., Patel, T.R.: A mathematical model of imbibitions phenomenon in heterogeneous porous media during secondary oil recovery process. Appl. Math. Model. **37**, 2933–2942 (2013)
2. Parikh, A.K., Mehta, M.N., Pradhan, V.H.: Mathematical model and analysis of countercurrent imbibition in vertical downward homogeneous porous media. Br. J. Math. Comp. Sci. **3**(4), 478–489 (2013)
3. Patel, K.K., Mehta, M.N., Singh, T.R.: A homotopy series solution to a nonlinear partial differential equation arising from a mathematical model of the counter-current imbibition phenomenon in a heterogeneous porous medium. Euro. J. Mech. B/Fluids **60**, 119–126 (2016)
4. Patel, H.S., Meher, R.: Approximate analytical study of counter-current imbibition phenomenon in a heterogeneous porous media. Appl. Math. Sci. **10**(14), 673–681 (2016)
5. Patel, M.A., Desai, N.B.: Mathematical modeling and analysis of co-current imbibition phenomenon in inclined homogeneous porous media, Kalpa Publications in Computing. Int. Conf. Res. Innov. Sci. Eng. Technol. Vallabh Vidyanagar **2**, 51–61 (2017)
6. Bear, J.: Dynamics of Fluids in Porous Media. American Elsevier Publishing Company, Inc. (1972)
7. Mehta, M.N.: Asymptotic Expansions of Fluid Flow Through Porous Media, Ph.D.Thesis, South Gujarat University, Surat (1977)
8. Scheidegger, A.E., Johnson, E.F.: The statistically behavior of instabilities in displacement process in porous media. Can. J. Phys. **39**(2), 326–334 (1961)
9. Scheidegger, A.E.: The Physics of Flow Through Porous Media. University of Toronto Press (1960)
10. Chen, Z.: Reservoir Simulation: Mathematical Techniques in Oil Recovery, pp. 1–25. SIAM, Philadelphia (2007)
11. He, J.H.: Homotopy perturbation technique. Comput. Meth. Appl. Mech. Eng. **178**, 257–262 (1999)

New Class of Probability Distributions Arising from Teissier Distribution

Sudhanshu V. Singh, Mohammed Elgarhy, Zubair Ahmad, Vikas Kumar Sharma, and Gholamhossein G. Hamedani

Abstract This article deals with the Teissier/Muth distributions. Some properties as well as four members of Teissier-G family are provided. Definite formulae for ordinary moments, mean deviation, quantile function and order statistics are given. Some characterizations owing to the truncated moments, hazard function and conditional expectations are derived. We consider the parametric estimation of the family along with one of its sub-model. The use of the Teissier–Weibull family is revealed depending on two real data sets.

Keywords Generalized distributions · Teissier/Muth distribution · Hazard function · Moments · Characterizations · Order statistics · Mean deviations · Maximum likelihood estimation

1 Introduction

Teissier distribution by [1] was proposed to model frequency of mortality due to aging. Laurent [2] studied its characterization based on the life expectancy and

S. V. Singh
Department of Mathematics, Institute of Infrastructure Technology Research and Management (IITRAM), Ahmedabad, Gujarat, India

M. Elgarhy
Vice Presidency for Graduate Studies and Scientific Research, University of Jeddah, Jeddah 23218, Saudi Arabia

Z. Ahmad
Department of Statistics, Quaid-I-Azam University, Islamabad 45320, Pakistan

V. K. Sharma (✉)
Department of Statistics, Institute of Science, Banaras Hindu University, Varanasi, India
e-mail: vikasstats@rediffmail.com

G. G. Hamedani
Department of Mathematics, Statistics and Computer Science, Marquette University, Milwaukee, WI 53201-1881, USA

M. Sahni et al. (eds.), *Mathematical Modeling, Computational Intelligence Techniques and Renewable Energy*, Advances in Intelligent Systems and Computing 1287,
https://doi.org/10.1007/978-981-15-9953-8_5

showed its applications to the demographic data sets. Muth [3] used this distribution for reliability analysis. [4] derived properties of the Teissier distribution. They referred it as the Muth distribution, while they might have missed articles [1, 2]. Hereafter, we call this distribution as the Teissier distribution. The cumulative distribution function (cdf) of the Teissier distribution is defined by

$$F(x) = 1 - e^{\alpha x - \frac{1}{\alpha}(e^{\alpha x} - 1)}, \quad 0 < \alpha \leq 1, x > 0. \tag{1}$$

The Teissier distribution may be defined by the hazard rate function (hrf)

$$h(x) = e^{\alpha x} - \alpha,$$

which is increasing in x. The Teissier distribution has not been paid much attention by the researchers. This distribution may be a good alternative over the one parameter exponential distribution for modeling data having the hrf.

During last two decades, researchers have paid attention to introduce new class of distributions or to generalize the distributions existing in the literature. Several approaches were proposed using the idea of combing two or more distributions to obtain a new family. Among others, the family (named as T-X) by [5] has been employed widely in the literature. The T-X family's cdf is defined as

$$F(x) = \int_{-\infty}^{W[G(x,\xi)]} r(u)du, \tag{2}$$

where $r(.)$ denotes the density function associated with the random variable (RV) T that is defined on $[0,\infty)$, and $G(x, \xi)$ stands for the cdf of RV X indexed by the parameter ξ.

This idea was first initiated by [6] in introducing beta-normal distribution, where they took $r(t) \sim$ beta, $G(x) \sim$ Normal and $W[G(x, \xi)] = G(x, \xi)$. Similarly, various other distributions are proposed using the T-X family with various choices of $W[G(x, \xi)]$ such as $-\log(1 - G(x, \xi))$, $\left\{\frac{G(x,\xi)}{1-G(x,\xi)}\right\}$ and $[G(x, \xi)]^{\alpha}$, $[G(x, \xi)]^{\alpha} > 0$. Recently, [7] introduced Muth-X family of distributions with several sub-models using uniform, Weibull and Lomax distributions using $W[G(x, \xi)] = -\log[G(x, \xi)]$.

The Kumaraswamy-generalized family by [8], gamma-G by [9], Weibull-G family by [10], Lomax-generated family by [11], Weibull-X family by [12], Kumaraswamy Weibull-G by [13], Garhy-generated family by [14], Weibull–Pareto by [15], Topp–Leone–X family by [16], Maxwell-X family by [17] and odd log-logistic family by [18] are few among others.

We are motivated toward the applications of the Teissier distribution discussed in the literature. One may note that combing two distributions may produce a flexible class of distribution having various varieties of the density and hazard shapes.

Therefore, this manuscript aims to develop a new class of the Teissier distributions combining it with the T-X family. We further study the properties in general and of some sub-models. A sub-model, Teissier–Weibull (TW) distribution, is discussed in detail with two real-life applications over the four generalizations of the Weibull distribution.

Next sections are organized as follows. We first define the Teissier-G (T-G) family in Sect. 2. General properties of T-G family are given in Sect. 3. The characterizations of the proposed family are shown in Sect. 4. Section 5 shows four different members of the T-G family. Estimation along with simulations for the TW distribution is discussed in Sect. 6. Performances of the maximum likelihood (ML) estimators of the TW parameters are also studied by the simulations. Section 7 shows two practical data sets for applications. Section 8 has final remarks.

2 Teissier Family of Distributions

This section defines the Teissier-generated family of distributions along with some of its basic properties. To present T-G family, we employ cdf (2) as well as Teissier distribution as a generator with $W[G(x,\xi)] = -\log(1 - G(x,\xi))$. The cdf of T-G family can be readily obtained by

$$F(x) = 1 - C\overline{G}(x;\xi)^{-\alpha}\exp\left\{-\frac{1}{\alpha}\overline{G}(x;\xi)^{-\alpha}\right\},\quad 0 < \alpha \le 1, x \in R \tag{3}$$

where $C = e^{1/\alpha}$ and $G(x;\xi)$ is the cdf indexed by the parameter ξ. The pdf corresponding to the cdf (3) is given by

$$f(x) = C[1 - [G(x,\zeta)]]^{-\alpha-1}$$
$$\exp\left\{-\frac{1}{\alpha}\overline{G}(x;\xi)^{-\alpha}\right\}\left([1 - G(x,\zeta)]^{-\alpha} - \alpha\right)g(x), x \in R. \tag{4}$$

The RV X following the pdf (4) shall be denoted by $X \sim$ T-G.

The density expansion derived here may be exploited in deriving the properties of the model. Using exponential expansion, $e^{ax} = \sum_{i=0}^{\infty}(i!)^{-1}a^i x^i$ in pdf (4), we get

$$f(x) = C\sum_{i=0}^{\infty}\frac{g(x)}{\alpha^i i!}$$
$$\left([1 - G(x,\xi)]^{-\alpha} - \alpha\right)(1 - G(x,\xi))^{-\alpha(i+1)-1}, x \in R. \tag{5}$$

Using the above expansion, $(1-w)^{-s} = \sum_{r=0}^{\infty} \binom{s+r-1}{r} w^r$, $|w| \langle 1, s \rangle 0$, in pdf (5), we have

$$f(x) = \sum_{k=0}^{\infty} \phi_k G(x; \xi)^k g(x; \xi), x \in R, \tag{6}$$

with $\phi_k = C \sum_{i=0}^{\infty} \left[\binom{\alpha(i+2)+k}{k} - \alpha \binom{\alpha(i+1)+k}{k} \right] \frac{\alpha^{-i}(-1)^j}{i!}$. The density function of the T-G family may have the form

$$f(x) = \sum_{k=0}^{\infty} J_k h_{k+1}(x), \tag{7}$$

where $h_a(x) = a\,G(x; \xi)^{a-1} g(x; \xi)$ and $J_k = \phi_k / k + 1$ which represents the exponentiated-G distribution with shape parameter a. Similarly, the T-G cdf may also be given by

$$[F(x)]^h = \sum_{z=0}^{\infty} s_z G(x; \zeta)^z, \tag{8}$$

where

$$s_z = \sum_{i=0}^{h} \sum_{j=0}^{\infty} \binom{h}{i} \binom{\alpha(i+j)+z-1}{z} \frac{C^i (-1)^{i+j}}{\alpha^j j!}.$$

The pdf (7) and cdf (8) will be employed to obtain several properties of the T-G family.

3 Properties

Here, we provide the basic properties of T-G family such as probability-weighted, incomplete and ordinary moments (PWM, IM, OM), distribution of order statistics and mean deviations (MDs). The PWM of the T-G family is computed by

$$\mathrm{P}_{r,h} = \int_{-\infty}^{\infty} x^r (F(x))^h f(x) \mathrm{d}x. \tag{9}$$

Hence, the PWM of T-G distribution is given by substituting (7) and (8) into (9), as follows

$$\mathrm{P}_{r,h} = \int_{-\infty}^{\infty} \sum_{k,z=0}^{\infty} x^r g(x;\xi) s_z \phi_k (G(x;\xi))^{k+z} \mathrm{d}x$$
$$= \sum_{k,z=0}^{\infty} s_z \phi_k P_{r,k+z}.$$

Based on (7), we get the rth OM given by

$$\mu_r' = \int_{-\infty}^{\infty} \sum_{k=0}^{\infty} \phi_k G(x;\xi)^k x^r g(x;\xi) \mathrm{d}x$$
$$= \sum_{k=0}^{\infty} \phi_k P_{r,k},$$

where $P_{r,k}$ denotes the PWM.

We know that the IM has vital role in applications for probability distribution. The qth IM is defined by

$$K_q(y) = \int_0^y \sum_{k=0}^{\infty} \phi_k G(x;\zeta)^k x^q g(x;\zeta) \mathrm{d}x.$$

Let μ (m) be the mean (median) of the T-G family of distributions. Then, the deviations about μ and m are $\delta_1 = 2\mu F(\mu) - 2P(\mu)$ and $\delta_2 = \mu - 2P(m), P(d) = \int_{-\infty}^{d} uf(u)\mathrm{d}u$, respectively.

Next we discuss the distribution of the order statistics (OS) that are useful in describing the distribution and estimation. Suppose that $X_{1:n}, X_{2:n}, \ldots, X_{n:n}$ are n OS associated with the random observations $X_1, X_2, \ldots, X_n$ from the proposed family of distributions. The density of the qth OS is

$$f_{q:n}(x) = \frac{1}{B(q, n-q+1)} \sum_{v=0}^{n-q} (-1)^v F(x)^{v+q-1} \binom{n-q}{v} f(x), \tag{10}$$

where $B(.,.)$ denotes beta function. Inserting (7) and (8) in (10), the pdf of the sth OS for T-G family is

$$f_{q:n}(x) = \frac{g(x;\xi)}{B(q, n-q+1)} \sum_{v=0}^{n-q} \sum_{k,z=0}^{\infty} \phi_k p_{z,v} G(x;\xi)^{k+z}, \tag{11}$$

with $p_{z,v} = s_z \binom{n-q}{v} (-1)^v$ and the rth moment as given by

$$E(X_{q:n}^r) = \int_{-\infty}^{\infty} x^r f_{q:n}(x) \mathrm{d}x$$
$$= \frac{1}{B(q, n-q+1)} \sum_{v=0}^{n-q} \sum_{k,z=0}^{\infty} \phi_k p_{z,v} P_{r,k+z}. \tag{12}$$

4 Characterizations of T-G Distribution

The present section gives characterizations (Ch) of the T-G distribution in distinct trends: (i) depending on the truncated moments (TM) ratio; (ii) using the hrf and (iii) depending on the conditional expectation of a function of the T-G-distributed RV. The Ch (i)–(iii) are presented. We first consider the Ch of T-G distribution using the TM ratio. We depend on Theorem 1 (see Glänzel [19, 20] for first Ch.)

Proposition 1 *Suppose that* $X: \Omega \to \mathbb{R}$ *is a RV with cdf given in* (3). *Define* $q_1(x) = \{\overline{G}(x; \zeta)^{-\alpha} - \alpha\}$ *and* $q_2(x) = q_1(x) \exp\{-\frac{1}{\alpha}\overline{G}(x; \zeta)^{-\alpha}\}$ *for* $x \in \mathbb{R}$. *The RV X is said to follow the cdf* (3) *iff the function* ξ *of RV X is*

$$\xi(x) = \frac{1}{2} \exp\left\{-\frac{1}{\alpha}\overline{G}(x; \zeta)^{-\alpha}\right\}, \quad x \in \mathbb{R}.$$

Proof For the proof, we refer the readers to Glänzel [19, 20].

Corollary 1 *The RV X has pdf* (4) *iff there exist functions* $q_1(x), q_2(x)$ *and* $\xi(x)$ *defined above satisfying the following condition*

$$g(x; \zeta)\overline{G}(x; \zeta)^{-\alpha-1} = \frac{q_1(x)\xi'(x)}{\xi(x)q_1(x) - q_2(x)}, \ x \in \mathbb{R},$$

with the general solution as given by

$$\xi(x) = \left[-\int \alpha\lambda G(x; \phi)^{\alpha\lambda-1} (q_1(x))^{-1} q_2(x) + K\right] \left(1 - G(x; \phi)^{\alpha\lambda}\right)^{-1},$$

where K *is a constant.*

In the next result, we come up with a Ch of T-G distribution using the hrf that satisfies the equation $\frac{f'(x)}{f(x)} = \frac{h'(x)}{h(x)} - h(x)$.

Proposition 2 *Suppose that* $X:\Omega \to \mathbb{R}$*is a RV with cdf given in* (3). *The RV X shall follow the cdf* (3)*iff the associated hrf* h_F*satisfies the differential equation as given by*

$$h^{/}(x) = \frac{g(x;\xi)^2\left\{(\alpha+1)\overline{G}(x;\xi)^{-\alpha} - \alpha\right\}}{\overline{G}(x;\xi)^2}, \quad x \in R.$$

Proof For the proof, we refer the readers to Glänzel [19, 20].

Next, we use a function $\psi(X)$ of the RV X and characterize the T-G family of distributions using the TM of $\psi(X)$, following the result from Hamedani [21].

Proposition 3 *Suppose that* $X:\Omega \to (e, f)$*is a RV with cdf F and* $\psi(x)$*be a differentiable function on* (e, f)*with* $\lim_{x\to e^+}\psi(x) = 1$. *For* $\delta = 1$, $E[\psi(X)/X \geq x] = \delta\psi(x), x \in (e, f)$*iff* $\psi(x) = (1 - F(x))^{\frac{1}{\delta}-1}$, $x \in (e, f)$.

Remark 1 For $(e, f) = R$, $\psi(x) = e^{1/\alpha^2}\overline{G}(x;\xi)^{-1}\exp\left\{-\frac{1}{\alpha^2}\overline{G}(x;\xi)^{-\alpha}\right\}$ and $\delta = \frac{\alpha}{\alpha+1}$, the results stated above define the Ch of the T-G distribution.

5 Illustrations of the T-G Family

This section provides four new models, namely Teissier uniform (TU), Teissier–Lomax (TL), Teissier linear failure rate (TLFR) and Teissier–Weibull (TW) among other members of the T-G family. Their flexibility is represented via their pdf and hrf shapes.

5.1 Teissier Uniform Distribution

The Teissier uniform (TU) distribution is obtained from the pdf (4) by taking the uniform distribution as the baseline distribution with pdf and cdf as given by

$$g(x;\theta) = \frac{1}{\theta}, 0 < x < \theta, G(x;\theta) = \frac{x}{\theta}, 0 < x < \theta.$$

Substituting pdf and cdfs above in (3), the TU is obtained and defined by the cdf

$$F_{\mathrm{TU}}(x) = 1 - \left\{\left(\frac{\theta - x}{\theta}\right)^{-\alpha}\exp\left\{-\left(\frac{1}{\alpha}\right)\left(\left(\frac{\theta - x}{\theta}\right)^{-\alpha} - 1\right)\right\}\right\},$$
$$0 < x < \theta, 0 < \alpha \leq 1. \quad (13)$$

The pdf and hrf corresponding to the cdf (13) can be derived.

5.2 Teissier–Lomax Distribution

If the cdf and pdf of a Lomax distribution are defined by $G(x;\xi) = 1 - (1+\lambda x)^{-\theta}$ and $g(x;\xi) = \theta\lambda(1+\lambda x)^{-(\theta+1)}$, where $\xi = (\theta, \lambda)$. Then, the cdf of the TL model becomes

$$F_{\mathrm{TL}}(x) = 1 - \left\{(1+\lambda x)^{\alpha\theta} \exp\left\{-\left(\frac{1}{\alpha}\right)\left((1+\lambda x)^{\alpha\theta} - 1\right)\right\}\right\},$$
$$x, \theta, \lambda > 0, 0 < \alpha \leq 1. \tag{14}$$

The density and hazard functions corresponding to the cdf (14) can be derived.

5.3 Teissier Linear Failure Rate Distribution

If the linear failure rate distribution is defined by the following cdf and pdf $G(x;\xi) = 1 - e^{-(\beta x^2+\theta x)}$ and $g(x;\xi) = (2\beta x + \theta)e^{-(\beta x^2+\theta x)}$, $x, \beta, \theta > 0$, where $\xi = (\theta, \beta)$. Then, the distribution function of the TLRF model is

$$F_{\mathrm{TLFR}}(x) = 1 - \exp\left\{\alpha\left(\beta x^2 + \theta x\right) - \left(\frac{1}{\alpha}\right)\left(e^{\alpha(\beta x^2+\theta x)} - 1\right)\right\},$$
$$\beta, \theta, \alpha, x > 0, 0 < \alpha \leq 1. \tag{15}$$

The density and hazard functions corresponding to the cdf (15) can be derived.

5.4 Teissier–Weibull Distribution

The Weibull density and cdf are given by $g(x;\xi) = \gamma\beta x^{\gamma-1}e^{-\beta x^{\gamma}}$ and $G(x;\xi) = 1 - e^{-\beta x^{\gamma}}$, $\beta, x, \gamma > 0$, $\xi = (\beta, \gamma)$, respectively. Then, the cdf of the TW is as follows

$$F_{\mathrm{TW}}(x) = 1 - \exp\left\{\alpha\beta x^{\gamma} - \left(\frac{1}{\alpha}\right)\left(e^{\alpha\beta x^{\gamma}} - 1\right)\right\}, x, \alpha, \beta, \gamma > 0. \tag{16}$$

The pdf and hrf of the TW model are

$$f_{\mathrm{TW}}(x) = \gamma\beta x^{\gamma-1} \exp\left\{\alpha\beta x^{\gamma} - \left(\frac{1}{\alpha}\right)\left(e^{\alpha\beta x^{\gamma}} - 1\right)\right\}\left(e^{\alpha\beta x^{\gamma}} - \alpha\right),$$
$$h_{\mathrm{TW}}(x) = \gamma\beta x^{\gamma-1}\left(e^{\alpha\beta x^{\gamma}} - \alpha\right).$$

The model defined in (16) will be used to analyze two real data. Also, from (16), it is quite clear that when $\beta = \gamma = 1$, the expression provided in (16) becomes identical to (1).

The rth moment and pth quantile of the TW distribution are, respectively, given by

$$E(X^r) = \frac{\left(e^{1/\alpha}\right)}{\left(\alpha^{\frac{r}{\gamma}+1}\beta^{\frac{r}{\gamma}}\right)}\left[C\left(\frac{r}{\gamma}+1\right)E_{-1}^{r/\gamma}(1/\alpha) - \alpha C\left(\frac{r}{\gamma}+1\right)E_{0}^{r/\gamma}(1/\alpha)\right],$$

$$Q_p = \left[-\left(1/\alpha^2\beta\right) + (1/\alpha\beta)\log(1-u) - (1/\alpha\beta)W_{-1}\left(\frac{(u-1)e^{-1/\alpha}}{\alpha}\right)\right]^{1/\gamma},$$

where C stands for the gamma function, E is generalized integro-exponential function and W_{-1} is the Lambert W function; see [4]. For the hazard function, we observe the following from the equation $\frac{\mathrm{d}}{\mathrm{d}x}h_{\mathrm{TW}}(x) = \left[(\gamma-1)(e^{\alpha\beta x^{\gamma}}-\alpha) + \alpha\beta\gamma x^{\gamma}e^{\alpha\beta x^{\gamma}}\right]\gamma\beta x^{\gamma-2} = 0$.

- If $\gamma = 1, h(x) = \beta\left(e^{\alpha\beta x} - \alpha\right)$ is increasing in x.
- If $\beta = \gamma = 1, h(x) = (e^{\alpha x} - \alpha)$ is increasing in x.
- If $\alpha \to 0, h(x) = \beta\gamma x^{\gamma-1} = \begin{cases} \text{Increasing} & \text{for } \gamma > 1 \\ \text{Constant} & \text{for } \gamma = 1 \\ \text{Decreasing} & \text{for } \gamma < 1 \end{cases}$
- If $\gamma \geq 1, h(x)$ is increasing in $x \in [t_0, \infty)$, where $t_0 = \left(\frac{\log\alpha}{\alpha\beta}\right)^{\gamma}$.
- If $\gamma < 1, h(x)$ is increasing in $x \in (0, t_0]$.

Figure 1 provides the shapes of the TW density and hrfs. Since the TW distribution is appealing to have good distributional properties, we discuss its estimation, simulation and applications in detail.

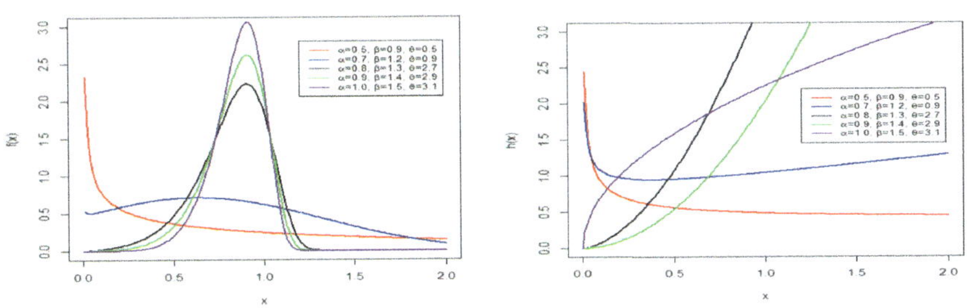

Fig. 1 Shapes of the TW distribution

6 Maximum Likelihood Method and Simulations

In this section, maximum likelihood (ML) estimators of the T-G family are developed using the complete samples of observations. Suppose that $x_1, x_2, \ldots, x_n$ is the observed sample from the T-G family with the parameter vector $\Phi = (\alpha, \xi)^T$. The log-likelihood function for Φ is given by

$$\ln L(\Phi) = \sum_{i=1}^{n} \ln\left[(1 - G(x_i;\xi))^{-\alpha} - \alpha\right] - (\alpha + 1)\sum_{i=1}^{n} \ln[1 - G(x_i;\xi)]$$
$$+ \sum_{i=1}^{n} \ln[g(x_i;\xi)] - \frac{1}{\alpha}\sum_{i=1}^{n}\left[(1 - G(x_i;\xi))^{-\alpha} - 1\right].$$

The ML estimator of $\Phi = (\alpha, \xi)^T$, say $\hat{\Phi} = (\hat{\alpha}, \hat{\xi})$, is obtained by maximizing the log-likelihood function given above. The log-likelihood of the TW distribution is given by

$$\ln L(\Phi) = n \ln \beta + n \ln \gamma + \alpha\beta \sum_{i=1}^{n} x_i^{\gamma} - \frac{1}{\alpha}\sum_{i=1}^{n}\left[e^{\alpha\beta(x_i)^{\gamma}} - 1\right]$$
$$+ (\gamma - 1)\sum_{i=1}^{n} \ln x_i + \sum_{i=1}^{n} \ln\left[e^{\alpha\beta(x_i)^{\gamma}} - \alpha\right].$$

The ML estimates of the TW parameters are the simulation of the following log-likelihood equations,

$$\frac{\sum_{i=1}^{n}\left(e^{\alpha\beta x_i^{\gamma}} - 1\right)}{\alpha^2} + \beta\sum_{i=1}^{n} x_i^{\gamma} - \frac{\sum_{i=1}^{n} \beta x_i^{\gamma} e^{\alpha\beta x_i^{\gamma}}}{\alpha} + \sum_{i=1}^{n} \frac{\beta x_i^{\gamma} e^{\alpha\beta x_i^{\gamma}} - 1}{e^{\alpha\beta x_i^{\gamma}} - \alpha} = 0,$$

$$\frac{n}{\beta} + \alpha\sum_{i=1}^{n} x_i^{\gamma} - \frac{\sum_{i=1}^{n} \alpha x_i^{\gamma} e^{\alpha\beta x_i^{\gamma}}}{\alpha} + \sum_{i=1}^{n} \frac{\alpha x_i^{\gamma} e^{\alpha\beta x_i^{\gamma}}}{e^{\alpha\beta x_i^{\gamma}} - \alpha} = 0,$$

$$\frac{n}{\gamma} + \sum_{i=1}^{n} \ln x_i + \alpha\beta\sum_{i=1}^{n} x_i^{\gamma} \ln x_i - \frac{\sum_{i=1}^{n} \alpha\beta x_i^{\gamma} e^{\alpha\beta x_i^{\gamma}} \ln x_i}{\alpha} + \sum_{i=1}^{n} \frac{\alpha\beta x_i^{\gamma} e^{\alpha\beta x_i^{\gamma}} \ln x_i}{e^{\alpha\beta x_i^{\gamma}} - \alpha} = 0.$$

In order to solve the above system of equations, we recommend to use Newton–Raphson method. This algorithm is available in many standard statistical software. Here, we use (Mathematica 9) package software.

Now, the simulations are provided to access the performances of ML estimates (MLEs) for TW distribution via Mathematica 9. We consider samples of sizes $n = 30$, 50, 70 and 80 for values of the TW parameters, $(\alpha,\ \beta,\ \gamma)$. We repeat the experiment 5000 times, and the measures like mean squared error (MSE) and bias for different sample sizes are computed. Numerical outcomes are shown in Table 1. We note from

Table 1 MSE, bias of the MLEs of the TW parameters,(α, β, γ)

n	Par.		MLE	BIAS	MSE		MLE	BIAS3	MSE4
30	α	0.2	0.212	0.012	0.137	0.2	0.213	0.013	0.134
	β	0.5	0.506	0.006	0.024	0.5	0.506	0.006	0.024
	γ	0.1	0.102	0.406	0.189	0.2	0.205	0.306	0.118
50	α	0.2	0.235	0.035	0.102	0.2	0.236	0.036	0.101
	β	0.5	0.514	0.014	0.017	0.5	0.514	0.014	0.017
	γ	0.1	0.099	0.414	0.187	0.2	0.199	0.314	0.116
70	α	0.2	0.230	0.030	0.078	0.2	0.231	0.031	0.076
	β	0.5	0.512	0.012	0.014	0.5	0.512	0.012	0.014
	γ	0.1	0.099	0.412	0.184	0.2	0.199	0.312	0.111
80	α	0.2	0.240	0.040	0.069	0.2	0.240	0.040	0.070
	β	0.5	0.515	0.015	0.012	0.5	0.515	0.015	0.012
	γ	0.1	0.098	0.415	0.183	0.2	0.197	0.315	0.110

the performances of all the estimators that the MSE is decreasing when sample size is increasing.

7 Applications

We analyze two real data aiming to explain the practicality of the considered family distributions. For this purpose, two data sets representing the failure times are considered. The data sets (see Murthy et al. [22]) represent the failure times of 84 windshield and 50 devices (see Arset [23]), respectively.

We now fit the TW distribution with four other well-known Weibull extensions, namely transmuted Weibull (TrW) [24], exponentiated Weibull (ExW) [25], alpha power transformed Weibull (APTW) [26] and Kumaraswamy Weibull (KuW) [27] distributions. We present the MLEs and the information measures that include log-likelihood (LL), Hannan–Quinn information criterion (HQIC), Bayesian information criterion (BIC), Akaike information criterion (AIC) and corrected AIC (CAIC). Also, Anderson–Darling (AD), Cramer von Mises (CM) and Kolmogorov–Smirnov (KS) tests are used. The analytical measures are given in Tables 2, 3 and 4.

From the results obtained, we can see that the TW distribution gives the smallest BIC, KS, AIC, HQIC and CIAC values than the other distributions. Since the smallest values of the above statistics correspond to the best model, the TW distribution is proved to be enough flexible and compatible to model such failure times data sets in comparison with proposed distributions. Similar concussions can be drawn from the graphical measures, Fig. 2. In these figures, one can see that the TW distribution cdf is very closer to the empirical estimate, which also endorses the usefulness of the TW distribution to failure data sets over other distributions.

Table 2 MLEs and standard errors (in the parenthesis) for all five distributions using data 1

Model	$\hat{\beta}$	$\hat{\gamma}$	$\hat{\alpha}$	$\hat{\lambda}$	$\hat{a}$	$\hat{b}$
TW	0.300 (0.0752)	1.235 (0.2168)	0.843 (0.1593)			
KuW	0.465 (0.076)	2.826 (0.2406)			0.451 (0.036)	0.095 (0.010)
ExW	0.019 (0.1035)	3.308 (3.8461)	0.632 (0.905)			
APTW	0.256 (0.1227)	1.790 (0.2865)	8.369 (8.114)			
TrW	0.175 (0.0768)	1.971 (0.2599)		0.659 (0.2539)		

Table 3 Analytical results of all five distributions for both data sets

	Data 2					Data 1				
	TW	KuW	ExW	APTW	TrW	TW	KuW	ExW	APTW	TrW
KS	0.17	0.21	0.23	0.18	0.19	0.09	0.09	0.09	0.1	0.1
AD	2.65	3.14	2.8	3.14	3.25	0.56	0.6	0.61	0.58	0.59
CM	0.42	0.55	0.44	0.51	0.53	0.08	0.08	0.1	0.1	0.1
LL	238.1	243.7	239.64	242.93	243.56	128.09	128.55	129.7	130	130.4
AIC	482.2	495.4	485.29	491.86	493.13	262.18	265.1	265.4	266	266.7
BIC	488	503.1	491.08	497.66	498.93	269.5	274.88	275.8	276.3	277
CIAC	482.7	496.3	485.8	492.37	493.64	262.47	265.6	265.7	266.3	267
HQIC	484.4	498.3	487.5	494.08	495.35	265.12	269.03	269.4	269.9	270

Table 4 MLEs and their respective standard errors in the parenthesis for data 2

Model	$\hat{\beta}$	$\hat{\gamma}$	$\hat{\alpha}$	$\hat{\lambda}$	$\hat{a}$	$\hat{b}$
TW	0.082 (0.0290)	0.682 (0.0850)	0.557 (0.1383)			
KuW	0.008 (0.1288)	0.921 (0.0934)			0.853 (0.0935)	3.002 (0.0987)
ExW	0.009 (0.0435)	1.437 (0.0961)	0.477 (0.0762)			
APTW	0.074 (0.0564)	0.780 (0.1535)	3.843 (3.4375)			
TrW	0.833 (0.0476)	0.0.50 (0.1164)		0.291 (0.2554)		

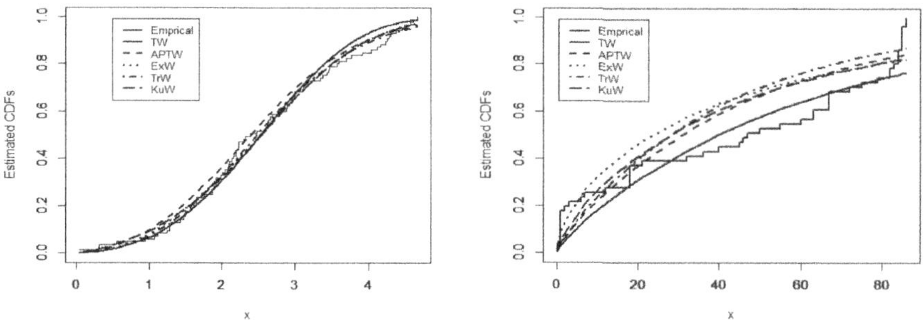

Fig. 2 Fitted cdfs of the TW and other competing distributions (left: data 1, right: data 2)

8 Conclusion

The Teissier-generated family of distributions is introduced in this article. The evolved family may be used for producing flexible probability distributions. Several structural properties of the T-G family are proposed. We presented four new sub-models such as Teissier uniform, Teissier–Lomax, Teissier linear failure rate and Teissier–Weibull distributions. Parameter estimators using the maximum likelihood procedure are obtained. Monte Carlo simulation study is made to study the behavior of the maximum likelihood estimates of the Teissier–Weibull parameters. We prove the potentiality of the stated family via two real data of lifetimes.

Acknowledgements "Dr. Sharma acknowledges the financial support from Science and Engineering Research Board, Department of Science & Technology, Government of India, under the scheme Early Career Research Award (file no.: ECR/2017/002416). Authors thank Prof. Amal S. Hassan for her help in improving the write-up of the paper."

References

1. Teissier, G.: Recherchessur le vieillissementetsur les lois de mortalite. Annales De Physiologieet De PhysicochimieBiologique **10**, 237–284 (1934)
2. Laurent, A.: Statistical distributions in scientific work, vol. 2. In: Model Building and Model Selection. Failure and Mortality from Wear and Aging. The Teissier Model, pp. 301–320. R. Reidel Publishing Company, Dordrecht Holland (1975)
3. Muth, J.E.: Reliability models with positive memory derived from the mean residual life function, vol. 2. In: C.P. Tsokos and I. Shimi (eds.) The Theory and Applications of Reliability, pp. 401–435. Academic Press, Inc., New York (1977)
4. Pedro, J., Jiménez-Gamero, M.D., Alba-Fernández, M.V.: On the muth distribution. Math. Modell. Anal. **20**, 291–310 (2015)
5. Alzaatreh, A., Lee, C., Famoye, F.: A new method for generating families of continuous distributions. Metron **71**, 63–79 (2013)
6. Eugene, N., Lee, C., Famoye, F.: The beta-normal distribution and its applications. Commun. Stat. Theor. Methods **31**(4), 497–512 (2002)

7. Almarashia, A.M., Elgarhy, M.: A new Muth generated family of distributions with applications. J. Nonlinear Sci. Appl. **11**, 1171–1184 (2018)
8. Cordeiro, G.M., Ortega, E.M., Nadarajah, S.: The Kumaraswamy Weibull distribution with application to failure data. J. Franklin Inst. **347**(8), 1399–1429 (2010)
9. Ristic, M.M., Balakrishnan, N.: The gamma-exponentiated exponential distribution. J. Stat. Comput. Simul. **82**(8), 1191–1206 (2012)
10. Bourguignon, M., Silva, R.B., Cordeiro, G.M.: The Weibull–G family of probability distributions. J. Data Sci. **12**, 53–68 (2014)
11. Cordeiro, G.M., Ortega, E.M.M., Popovic, B.V., Pescim, R.R.: The Lomax generator of distributions: Properties, minification process and regression model. Appl. Math. Comput. **247**, 465–486 (2014)
12. Alzaatreh, A., Ghosh, I.: On the Weibull-X family of distributions. J. Stat. Theor. Appl. **14**, 169–183 (2015)
13. Hassan, A.S., Elgarhy, M.: Kumaraswamy Weibull-generated family of distributions with applications. Adv. Appl. Stat. **48**, 205–239 (2016)
14. Elgarhy, M., Hassan, A.S., Rashed, M.: Garhy-generated family of distributions with application. Math. Theor. Model. **6**(2), 1–15 (2016)
15. Tahir, M.H., Gauss, M., Cordeiro, A.A., Mansoor, M., Zubair, M.: A new Weibull–Pareto distribution: properties and applications. Commun. Stat. Simul. Comput. **45**, 3548–3567
16. Sangsanit, Y., Bodhisuwan, W.: The Topp-Leone generator of distributions: properties and inferences. Songklanakarin J. Sci. Technol. **38**, 537–548 (2016)
17. Sharma, V.K., Bakouch, H.S., Suthar, K.: An extended maxwell distribution: properties and applications. Commun. Stat. Simul. Comput. **46**, 6982–7007 (2017)
18. Cordeiro, G.M., Alizadeh, M., Ozel, G., Hosseini, B., Ortega, E.M.M., Altun, E.: The generalized odd log-logistic family of distributions: properties, regression models and applications. J. Stat. Comput. Simul. **87**(5), 908–932 (2017)
19. Glänzel, W.: A Characterization Theorem Based on Truncated Moments and its Application to Some Distribution Families. Mathematical Statistics and Probability Theory, vol. B, Reidel, Dordrecht, pp. 75–84 (1987)
20. Glänzel, W.: Some consequences of a characterization theorem based on truncated moments. Stat. J. Theor. Appl. Stat. **21**(4), 613—618 (1990)
21. Hamedani, G.G.: On Certain Generalized Gamma Convolution Distributions II. Technical Report No. 484, MSCS, Marquette University (2013)
22. Murthy, D.P., Xie, M., Jiang, R.: Weibull Models. John Wiley & Sons, 505 (2004)
23. Aarset, M.V.: How to identify bathtub hazard rate. IEEE Trans. Reliab. **36**, 106–108 (1987)
24. Aryal, G.R., Tsokos, C.P.: Transmuted Weibull distribution: a generalization of the Weibull probability distribution. Euro. J. Pure Appl. Math. **4**(2), 89–102 (2011)
25. Mudholkar, G.S., Srivastava, D.K.: Exponentiated Weibull family for analyzing bathtub failure-rate data. IEEE Trans. Reliab. **42**(2), 299–302 (1993)
26. Dey, S., Sharma, V.K., Mesfioui, M.: A new extension of Weibull distribution with application to lifetime data. Ann. Data Sci. **4**(1), 31–61 (2017)
27. Cordeiro, G.M., Ortega, E.M., Silva, G.O.: The Kumaraswamy modified Weibull distribution: theory and applications. J. Stat. Comput. Simul. **84**(7), 1387–1411 (2014)

Development and Application of the DMS Iterative Method Having Third Order of Convergence

Riya Desai, Priyanshi Patel, Dhairya Shah, Dharil Shah, Manoj Sahni, and Ritu Sahni

Abstract In the present work, we suggest a fixed-point-based iterative method (the DMS iterative method) to attain numerical solutions of nonlinear equations of one variable arising in the real-world phenomena. We analytically obtain the order of convergence and the efficiency index of the developed (DMS) method which are 3 and 1.4422, respectively, both of which are higher than the famous Newton–Raphson (the N-R) method. The efficiency index of our method turns out to be the optimum value for one-point iterative methods for which number of function evaluations and order of convergence are same. Two examples from the physical sciences have been taken which reflect that the method used is solution-oriented and leads to the real root in the fewer number of iterations as compared to the Newton–Raphson method, which also complies with the analytical denouement. Graphs for both of the problems have been included which depict the same.

Keywords Cubic convergence · Efficiency index · Fixed-point method · Newton–Raphson method · Transcendental equation

1 Introduction

Numerical analysis is an elementary branch of mathematics that is widely used to solve nonlinear equations which are rigorously used to obtain the solutions of the real-life problems. In computer science, engineering and pure sciences, a frequently

R. Desai
Department of Mathematics, V.S. Patel College of Arts and Science, Bilimora, India

P. Patel · D. Shah (✉) · D. Shah
School of Liberal Studies, Pandit Deendayal Petroleum University, Gandhinagar, India
e-mail: dmaths8800@gmail.com

M. Sahni
School of Technology, Pandit Deendayal Petroleum University, Gandhinagar, India

R. Sahni
Department of Physical Sciences, Institute of Advanced Research, Gandhinagar, India

M. Sahni et al. (eds.), *Mathematical Modeling, Computational Intelligence Techniques and Renewable Energy*, Advances in Intelligent Systems and Computing 1287,
https://doi.org/10.1007/978-981-15-9953-8_6

occurring problem is to find the root of nonlinear equations of the form $f(x) = 0$. This involves conceptualization and development of new efficient numerical schemes for finding the approximate solutions (very close to exact solution) of such nonlinear equations. Many researchers have worked in this direction and developed several numerical methods for finding solutions to these equations. Some of the classical numerical methods are the bisection method, fixed-point iteration and Newton–Raphson (N-R) method. The main problem faced while solving the transcendental and algebraic equations is the high number of iterations. The researchers have made considerable efforts to develop new iterative schemes so that the numbers of iterations required to solve a problem can be reduced and helps to converge faster and hence tend toward the exact solution which facilitates us to understand the physical phenomenon in a meticulous way.

Iterative methods for solving nonlinear equations of type $f(x) = 0$ can be divided mainly in two parts: (1) Single-step iterative methods and (2) Multi-step iterative methods. Researchers have been developing both kinds of iterative methods since last few decades [1–7]. Many authors tried to modify the N-R method [7], whereas some took the fixed-point method as base to develop improved methods [3, 4]. Researchers use fixed-point method or Newton's method according to the situation and requirement. In 2019, Shah et al. [8] developed a modified accelerated iterative scheme for nonlinear equations using the amalgamation of these two classical iterative methods (NR and Fixed Point). Xiao et al. [9] in 2018 have developed a method for non-linear systems in which the rate of convergence is higher than some of available classical methods. In a recent work on fixed-point iteration method (FPIM), a technique to find optimum number of iterations for FPIM was developed by Shah et al. [10] in 2018. A work by Saqib et al. [3] in 2015 on nonlinear equations has invited a considerable attention and is based on the iterative method. A sixth-order scheme is developed by Chun et al. [5] in 2012 to reduce the number of iterations so that the required accuracy can be achieved. Some new iterative schemes are proposed by Noor [6] in 2007 for solving nonlinear equations. But majority of these methods were either (1) multi-step methods with higher number of function evaluation and therefore higher computational complexity or (2) they had lower efficiency index [11, 12].

In this paper, we attempt to develop an iterative method which is relatively less complex than multi-step methods and has higher efficiency index than the classical single-step methods. We then show that the developed (DMS) method has third-order convergence. We also show that the efficiency index of the DMS method is 1.4422 which is higher compared to the NR method having 1.4142.

2 Preliminaries

2.1 *Order of Convergence [13]*

Assume $\{x_n\}_{n=0}^{\infty}$ is a sequence that is converging toward α with $x_i \neq \alpha, \forall i$. If positive constants A and p exist such that

$$\lim_{n\to\infty} \frac{|x_{n+1} - \alpha|}{|x_n - \alpha|^p} = A,$$

then $\{x_n\}_{n=0}^{\infty}$ converges to α of order p with asymptotic error constant A. Here, p is said to be the order of convergence.

2.2 *Efficiency Index [12]*

If the order of convergence of the method is p and number of function evaluation is d, then the efficiency index of the method is given by

$$\text{EI} = p^{\frac{1}{d}}$$

3 Work Done

3.1 *Algorithm Development (Development of the DMS Method)*

Consider the following equation, (which is true for the fixed-point iteration method) [12]

$$\emptyset(\alpha) = \alpha \text{ With } \alpha = x_n + e_n$$
$$\therefore \emptyset(x_n + e_n) = x_n + e_n$$

By Taylor's series

$$\emptyset(x_n) + e_n\emptyset'(x_n) + \frac{e_n^2}{2!}\emptyset''(x_n) + 0(e_n^3) = x_n + e_n$$
$$\therefore \emptyset(x_n) + e_n\emptyset'(x_n) + \frac{e_n^2}{2!}\emptyset''(x_n) \approx x_n + e_n$$

On simplification of the above equation, we have,

$$\frac{e_n^2 \varnothing''(x_n)}{2} + \left(\varnothing'(x_n) - 1\right)e_n + (\varnothing(x_n) - x_n) \approx 0$$

Solving the equation for e_n, we have

$$e_n \approx \frac{(1 - \varnothing'(x_n)) \pm \sqrt{(1 - \varnothing'(x_n))^2 - 2\varnothing''(x_n)(\varnothing(x_n) - x_n)}}{\varnothing''(x_n)}$$

Again $\alpha = x_n + e_n$

$$\therefore \alpha \approx x_n + \frac{(1 - \varnothing'(x_n)) \pm \sqrt{(1 - \varnothing'(x_n))^2 - 2\varnothing''(x_n)(\varnothing(x_n) - x_n)}}{\varnothing''(x_n)}$$

As we know α is approximate value of the root and hence considering it as the new approximation x_{n+1}:

$$x_{n+1} = x_n + \frac{(1 - \varnothing'(x_n)) - \sqrt{(1 - \varnothing'(x_n))^2 - 2\varnothing''(x_n)(\varnothing(x_n) - x_n)}}{\varnothing''(x_n)}$$

3.2 *Order of Convergence*

To analytically get the order of convergence for the DMS iterative scheme, we start with,

$$\alpha + e_{n+1} = \alpha + e_n + \frac{(1 - \varnothing'(\alpha + e_n)) \pm \sqrt{(1 - \varnothing'(\alpha + e_n))^2 - 2\varnothing''(\alpha + e_n)(\varnothing(\alpha + e_n) - \alpha - e_n)}}{\varnothing''(\alpha + e_n)}$$

$$\therefore e_{n+1} = e_n + \frac{(1 - \varnothing'(\alpha + e_n)) \pm \sqrt{(1 - \varnothing'(\alpha + e_n))^2 - 2\varnothing''(\alpha + e_n)(\varnothing(\alpha + e_n) - \alpha - e_n)}}{\varnothing''(\alpha + e_n)}$$

$$\therefore e_{n+1} = e_n + \frac{(1 - \varnothing'(\alpha + e_n)) - \sqrt{(1 - \varnothing'(\alpha + e_n))^2 - 2\varnothing''(\alpha + e_n)(\varnothing(\alpha + e_n) - \alpha - e_n)}}{\varnothing''(\alpha + e_n)}$$

$$\therefore e_{n+1} = e_n + \frac{\left[\begin{array}{c} 1 - \varnothing'(\alpha) - e_n\varnothing''(\alpha) - \frac{e_n^2}{2}\varnothing'''(\alpha) - \frac{e_n^3}{6}\varnothing^{Iv}(\alpha) - 1 + \varnothing'(\alpha) - \frac{e_n^2}{2}\varnothing'''(\alpha) - \\ \left\{\frac{\varnothing^{Iv}(\alpha)}{3} - \frac{\varnothing''(\alpha)\varnothing'''(\alpha)}{6(1-\varnothing'(\alpha))}\right\}e_n^3 + 0\left(e_n^4\right) \end{array}\right]}{\varnothing''(\alpha) + e_n\varnothing'''(\alpha) + \frac{e_n^2}{2}\varnothing'^{v}(\alpha) + \frac{e_n^3}{6}\varnothing^{v}(\alpha) + \cdots}$$

On simplifying the expression reduces to,

$$e_{n+1} = \frac{\left[\frac{1}{2}\emptyset^{lv}(\alpha) - \frac{1}{6}\emptyset^{lv}(\alpha) - \frac{1}{3}\emptyset^{lv}(\alpha) + \frac{1}{6}\frac{\emptyset''(\alpha)\emptyset'''(\alpha)}{1-\emptyset'(\alpha)}\right]e_n^3 + (0)e_n^4}{\emptyset''(\alpha)\left[1 + e_n\frac{\emptyset'''(\alpha)}{\emptyset''(\alpha)} + \frac{e_n^2}{2}\frac{\emptyset^{lv}(\alpha)}{\emptyset''(\alpha)} + \frac{e_n^3}{6}\frac{\emptyset^{v}(\alpha)}{\emptyset''(\alpha)}\emptyset^{v}(\alpha) + \cdots\right]}$$

$$e_{n+1} = \frac{\left\{\left[\frac{1}{2}\emptyset^{lv}(\alpha) - \frac{1}{6}\emptyset^{lv}(\alpha) - \frac{1}{3}\emptyset^{lv}(\alpha) + \frac{1}{6}\frac{\emptyset''(\alpha)\emptyset'''(\alpha)}{1-\emptyset'(\alpha)}\right]e_n^3 + (0)e_n^4\right\} * \left[1 + e_n\frac{\emptyset'''(\alpha)}{\emptyset''(\alpha)} + \cdots\right]^{-1}}{\emptyset''(\alpha)}$$

Applying binomial expansion, the term upgrades as,

$$\begin{aligned} e_{n+1} &= \frac{1}{\emptyset''(\alpha)}\left\{\left[\frac{1}{6}\frac{\emptyset'''(\alpha)}{1-\emptyset'(\alpha)}\right]e_n^3 + (0)e_n^4\right\} \\ &\quad \left\{1 - \frac{\emptyset'''(\alpha)}{\emptyset''(\alpha)}e_n + \left(\frac{\emptyset'''(\alpha)}{\emptyset''(\alpha)}\right)^2 e_n^2 + \cdots\right\} \\ e_{n+1} &= \left[\frac{\emptyset'''(\alpha)}{6\emptyset''(\alpha)(1-\emptyset'(\alpha))}\right]e_n^3 \\ &\quad - \left[\frac{1}{6}\frac{(\emptyset'''(\alpha))^2}{(1-\emptyset'(\alpha))(\emptyset''(\alpha))^2}\right]e_n^4 + (0)e_n^5 \end{aligned}$$

From this, we have the following relation between e_{n+1} and e_n:

$$|e_{n+1}| \leq \left|\frac{\emptyset'''(\alpha)e_n^3}{6\emptyset''(\alpha)(1-\emptyset'(\alpha))}\right|$$

$$\therefore \left|\frac{e_{n+1}}{e_n^3}\right| \leq \left|\frac{\emptyset'''(\alpha)}{6\emptyset''(\alpha)(1-\emptyset'(\alpha))}\right|$$

Now consider,

$$\left|\frac{\emptyset'''(\alpha)}{6\emptyset''(\alpha)(1-\emptyset'(\alpha))}\right| = A,$$

where A is the asymptotic error constant.

$$\left|\frac{e_{n+1}}{e_n^3}\right| \leq A$$

Thus, from the definition of the order of convergence, the DMS has the cubic order of convergence.($p = 3$) (see Sect. 2.1).

3.3 Efficiency Index

Now, from Sect. 2.2, the efficiency index of the DMS method is

$$\mathrm{EI} = 3^{1/3} \approx 1.4422.$$

whereas the efficiency index of the N-R method is

$$\mathrm{EI} = 2^{1/2} \approx 1.4142.$$

For one-point iterative methods where $d = p$, efficiency index of $3^{1/3}$ is optimum [11]. As $2^{1/2} = 4^{1/4} \approx 1.4142$ and with $d = 4$, the EI gradually decreases. Hence, we have achieved the optimum efficiency index for the cases $d = p$.

4 Applications

4.1 Example 1. The Characteristic Equation of Transient Heat Conduction [14]

In lumped system analysis, to study the transient heat conduction in one dimension, the transcendental equation of the form $x\tan x = \mathrm{Bi}$ is used. Where Bi is called the Biot number.

The solution of transient heat transfer in one dimension can be given as:

$$x\tan x - 1 = 0,$$

on assuming the value of $\mathrm{Bi} = 1$.

Below is the graph that displays the number of iterations required for the DMS method to attain the accuracy ($|x_{n+1} - x_n|$) of 10^{-15} is 4 whereas the N-R method takes 9 iterations to achieve the same. The initial guess for both methods is $x_0 = 1.47$, and the solution is 0.860333589019 upto 12 digits of precision. Also, the fixed-point function used in the DMS method is $\emptyset(x) = \cot^{-1}(x)$.

For the DMS method and the absolute value of the function $f(x)$ after the final iteration, that is, after 4 iterations is $6.35 * 10^{-71}$, while with the NR method, the value of the function $f(x)$ after 9 iterations is $7.92 * 10^{-34}$. It shows that the DMS method takes less number of iterations with respect to the NR method to attain higher accuracy.

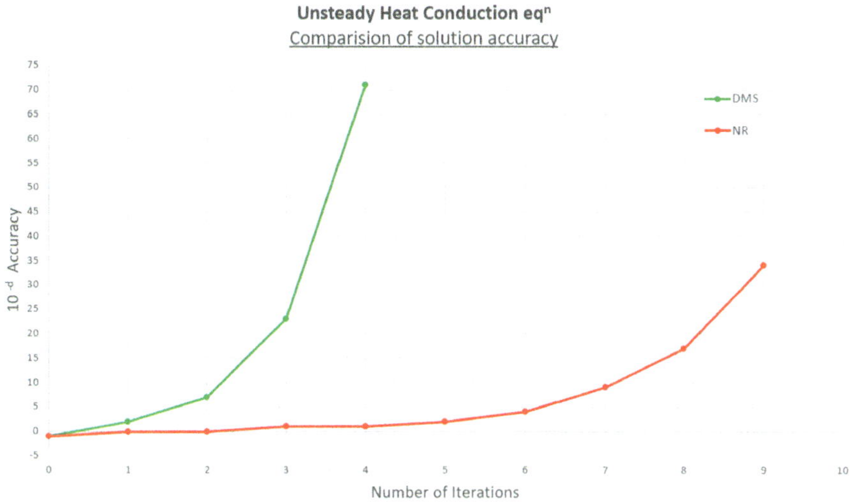

4.2 *Example 2. Kepler's Equation for Central Force [15]*

In orbital and celestial mechanics, Kepler's equation relates various geometric properties of the orbit of a body subjected to a central force.

$$M = E - e\sin(E),$$

where M is mean anomaly of an object in the orbit, E is eccentric anomaly of the object in orbit, and e is eccentricity of the orbit.

It is important to know the position of an object accurately in an elliptic Kepler orbit, so as to have an effective satellite communication, to avoid technical faults and collisions, and ultimately to benefit space research. To do so, we have taken the *orbital parameters* of NASA's Aqua satellite and plugged them in our method. From the figure, it is clear that our method achieves a greater accuracy in 'E' than NR method.

Here, $M = \frac{3{,}514{,}268}{10000}$ and $e = \frac{979}{10{,}000{,}000}$ are taken and hence the solution for E can be obtained from the following equation:

$$E - \left(\frac{979}{10{,}000{,}000}\right) * \sin(E) - \frac{3{,}514{,}268}{10{,}000} = 0.$$

The initial guess for both methods is $x_0 = 350$ and the solution is 351.426759044406 upto 12 digits of precision. Also, the fixed-point function used in the DMS method for the example 2 is $\emptyset(E) = \left(\frac{979}{10{,}000{,}000}\right) * \sin(E) + \frac{3{,}514{,}268}{10{,}000}$.

Like the example 1,the graph below displays that the number of iterations required for both the methods to attain the accuracy ($|x_{n+1} - x_n|$) of 10^{-15} is 3 but with the DMS method, function $f(E)$ attains absolute value $2.68 * 10^{-69}$, which is more accurate with respect to the NR method having the absolute value $3.57 * 10^{-39}$. This again displays that the DMS method is advantageous compared to the NR method.

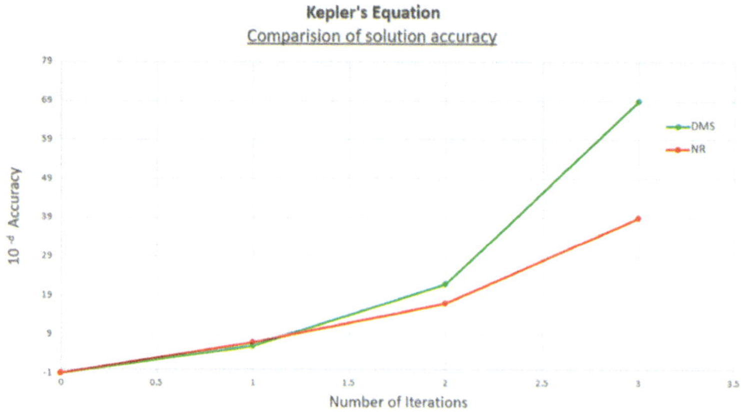

5 Conclusion

A work on the development of a new iterative scheme named as DMS method and based on known fixed-point iterative scheme is done in this paper, which is helpful in solving the nonlinear algebraic and transcendental equations of the form $f(x) = 0$. The DMS method has a third order of convergence with efficiency index of 1.4422. This method is more efficient than the methods like the fixed-point method and the Newton–Raphson method and has the optimum efficiency index among one-point iterative methods whose number of function evaluations and order of convergence are same. The DMS method turns out to be very accurate and provides the solution in less iteration and/or with higher accuracy than the NR method. The claim of faster convergence and higher accuracy has been verified using two distinct problems from the physical sciences.

References

1. Nazeer, W., Tanveer, M., Rehnam, K., Kang, S.M.: Modified new sixth-order fixed point Iterative Methods for solving nonlinear functional equations. Int. J. Pure Appl. Math. **109**(2), 223–232 (2016)
2. Asha, F.A., Noor, M.A.: Higher order iterative schemes for non-linear equations using decomposition technique. Appl. Math. Comput. **266**, 414–423

3. Saqib, M., Iqbal, M., Ahmed, S., Ali, S., Ismaeel, T.: New modification of fixed point iterative method for solving non-linear equations. Appl. Math. **6**, 1857–1863 (2015)
4. Kang, S.M., Rafiq, A., Kwun, Y.C.: A new second-order iteration method for solving non-linear equations. Abstr. Appl. Anal. **2013** (2013)
5. Chun, C., Neta, B.: A new sixth-order scheme for non-linear equations. Appl. Math. Lett. **25**, 185–189 (2012)
6. Noor, M.A.: New iterative schemes for non-linear equations. Appl. Math. Comput. **187**, 937–943 (2007)
7. Abbasbandy, S.: Improving Newton-Raphson method for non-linear equations by modified Adomian decomposition method. Appl. Math. Comput. **145**, 887–893 (2003)
8. Sahni, M., Shah, D., Sahni, R.: A new modified accelerated iterative scheme using amalgamation of fixed-point and Newton-Raphson method. J. Interdisc. Math. **22**(5), 679–688 (2019)
9. Xiao, X.Y., Yina, H.W.: Accelerating the convergence speed of iterative methods for solving non-linear systems. Appl. Math. Comput. **333**, 8–19 (2018)
10. Shah, D., Sahni, M.: DMS way of finding the optimum number of Iterations for fixed point Iteration method. In: Proceedings of the World Congress on Engineering, vol. I. WCE, London, U.K, 4–6 July (2018)
11. Ezquerro, J.A., Hernández, M.A., Romero, N.: Improving the efficiency index of one-point iterative processes. J. Comput. Appl. Math. **223**, 879–892 (2009)
12. Traub, J.F.: Iterative Methods for the Solution of Equations. American Mathematical Society, New Jersey (1982)
13. Burden, R.L., Faires, J.D., Burden, A.: M: Numerical Analysis. Cengage Learning, Boston (2016)
14. Cengel, Y., Ghajar, A.: Heat and Mass Transfer Fundamentals and Applications, 5th edn. McGraw-Hill Education, New York (2015)
15. Prussing, J., Conway, B.: Orbital Mechanics. Oxford University Press, New York (1993)

A Novel Hybrid Approach to the Sixth-Order Cahn-Hillard Time-Fractional Equation

Kunjan Shah and Himanshu Patel

Abstract This research focuses on the numerical analysis of the time-fractional sixth-order Cahn-Hillard equation by a novel hybrid approach using new integral and projected differential transform processes. Compared to other techniques, the new integrated projected differential transforms process, NIPDTM is much more effective and easier to handle. The results from the illustrative cases indicate the competence and consistency of the proposed procedure. The graphical result achieved through the presented method compared to the numerical integration method (NIM) and the q-homotopy analysis method (q-HAM) solution. The suggested method simplifies the calculation and makes it very simple to handle nonlinear terms with this method without using Adomian's & He's polynomial.

Keywords Integral transform method · Projected differential transform method · Sixth-order time-fractional Cahn-Hillard Equation

1 Introduction

The fractional calculus is a general type of calculus for integrating and differentiating non-integer-order operations[1, 2]. Almost at the same time as the classical ones developed, the scheme of fractional operators was proposed [3]. Fractional calculus theory developed rapidly after the nineteenth century, mainly as a basis for several functional disciplines, including fractional differential equations, fractional dynamics, etc. Nowadays, applications for the FC are very broad. It is fair to say that fractional methods and techniques remain largely unchanged in industrial engi-

K. Shah (✉)
Department of Mathematics, Center of Education,
Indian Institute of Teacher Education, Gandhinagar, Gujarat, India
e-mail: kunjankumars@iite.ac.in

H. Patel
Indian Institute of Teacher Education, Gandhinagar, Gujarat, India
e-mail: dr.prof.hcpatel@iite.ac.in

M. Sahni et al. (eds.), *Mathematical Modeling, Computational Intelligence Techniques and Renewable Energy*, Advances in Intelligent Systems and Computing 1287,
https://doi.org/10.1007/978-981-15-9953-8_7

neering and science at large. The practical explanation for introducing fractional calculus is that fractional models are much more robust than integer-order models, i.e., the fractional-order model seems to have more flexibility than the classical model. Recently, engineering and natural science researchers have been drawing attention to fractional calculus fractional derivatives and integration due to its broad application in chemical processes, mathematical biology, nanotechnology, etc. [4–12]

In the present paper, NIPDTM shows how the new integral transform can approximate nonlinear fractional equations by manipulating the projected differential transform process. The first suggestion for a new integral transform (NIT) is by Kashuri and Fundo's [13] for a new method of solving differential equations within the time domain. The modified edition of the Taylor series method proposed by Jang [14] in 2010 is named as the projected differential transform (PDT) method to evaluate IVP on linear and nonlinear values. The NIPDTM is a novel hybrid approach; in recent years, several researchers have been looking at solutions of DE's using different approaches mixed with the new integral transform, which are new integral homotopy perturbation transform method [15, 16], modified homotopy new integral transform method [17, 18].

Within this study, we take into account the sixth-order Cahn-Hillard time-fractional equation as:

$$D_t^{\alpha} u(x,t) = \mu u \, u_x + \left(u_{xx} - u^3 + u\right)_{xxxx}, \alpha \in (0,1], t > 0, \tag{1}$$

along with

$$u(x,0) = \mathfrak{f}(x), \tag{2}$$

$u(x,t)$ denotes concentration of one of the two components, α describing the fractional order of Cahn-Hillard equation, t represents time. Fractional derivatives are treated by means of Caputo [19]. When $\alpha = 1$, Eq. (1) will reduce to the traditional Cahn-Hillard equation. Equation (1) has stationary solution $u(x,t) = \pm\tanh\left[\frac{x}{\sqrt{2}}\right]$. Many authors such as [20–22] numerically investigated fractional Cahn-Hillard equation.

In the given article, we further apply the NIPDTM to address the Cahn-Hillard time-fractional equation. This approach's benefit is its ability to combine two main approaches for accurate and approximate solutions for fractional equations. A simple advantage of the proposed approach over the other method is that instead of using He's and Adomian's polynomials, the NIPDTM solves nonlinear problems.

2 Basic Definitions

Fundamental definitions of the fractional calculus that are often used in these papers are discussed in this section.

Definition 1 The Caputo fractional derivative of $u(t)$ given by [19, 23]:

$$\frac{d^\alpha u(t)}{dt^\alpha} = \begin{cases} u^{(n)}(t) & ,\alpha = n \in \mathbb{N}, \\ \frac{1}{\Gamma(n-\alpha)} \int\limits_0^t \frac{u^{(n)}(t)}{(t-\xi)^{\alpha-n+1}} dt & ,n-1 < \alpha \leq n, \end{cases} \tag{3}$$

where

$$\Gamma(\xi) = \int\limits_0^\infty t^{\xi-1} e^{-t} d\xi, Re(\xi) > 0. \tag{4}$$

Definition 2 The R-L $\alpha-$order operator of $u(t) \in C_\tau, \tau \geq -1$ is defined as[2, 19]:

$$\mathfrak{J}^\alpha u(t) = \frac{1}{\Gamma(\alpha)} \int\limits_0^t (t-\tau)^{\alpha-1} u(\tau) d\tau, \alpha,\ \tau > 0, \tag{5}$$

$$\mathfrak{J}^0 u(t) = u(t). \tag{6}$$

Definition 3 The NIT of a function $u(t), t \in (0, \infty)$ is defined by[13]:

$$\mathfrak{K}\left[u(t)\right](v) = \mathfrak{A}(v) = \frac{1}{v} \int\limits_0^\infty e^{-\frac{t}{v^2}} u(t) dt, v \in \mathbb{C}. \tag{7}$$

Definition 4 NIT of fractional derivative is given by[11, 13]:

$$\mathfrak{K}\left[D_t^\alpha u(x,t)\right] = \frac{\mathfrak{K}\left[u(x,t)\right]}{v^{2\alpha}} - \sum_{k=0}^{n-1} \frac{1}{v^{2(\alpha-k)-1}} \frac{\partial^k u(x,0)}{\partial t^k}. \tag{8}$$

3 Basic Concept for the NIPDTM

In order to explain the main methodology of this method, the general fractional nonlinear non-homogeneous PDE with the initial conditions considered as:

$$D_t^\alpha u + Ru + Nu = f(x,t), \tag{9}$$

$$u(x,0) = g(x),\ u_t(x,0) = h(x),$$

where $D_t^\alpha u$ represents the fractional derivative, R& N represents linear and nonlinear operator, respectively, and $f(x,t)$ is a continuous function. Introducing the NIT on Eq.(9), we get

$$\mathfrak{K}\left[D_t^\alpha u\right] + \mathfrak{K}\left[Ru\right] + \mathfrak{K}\left[Nu\right] = \mathfrak{K}\left[f(x,t)\right]. \tag{10}$$

Using the fractional differential property of NIT [11],

$$\frac{\mathfrak{K}[u]}{v^{2\alpha}} - \sum_{k=0}^{n-1} v^{2k-2\alpha+1} u^k(x,0) + \mathfrak{K}[Ru] + \mathfrak{K}[Nu] - \mathfrak{K}\left[f(x,t)\right] = 0. \tag{11}$$

From Eq.(11), we have

$$\mathfrak{K}[u] = \sum_{k=0}^{n-1} v^{2k+1} u^k(x,0) + v^{2\alpha}\mathfrak{K}\left[f(x,t)\right] - v^{2\alpha}\mathfrak{K}[R\,u] - v^{2\alpha}\mathfrak{K}[N\,u]. \tag{12}$$

Operating inverse NIT, we get

$$u(x,t) = F(x,t) - \mathfrak{K}^{-1}\left\{v^{2\alpha}\mathfrak{K}[Ru] + \mathfrak{K}[Nu]\right\}, \tag{13}$$

where $F(x,t)$ has prescribed a term that arises from the source term and the given initial conditions.
Now, employing the PDT,

$$u(x, m+1) = \mathfrak{K}^{-1}\left\{v^{2\alpha}\mathfrak{K}[A_m + B_m]\right\}, \tag{14}$$

where $u(x,0) = F(x,t)$ and $A_m, B_m, m \geq 0$ represents the PDT of Ru, Nu respectively. From Eq. (14), we get the following approximations:

$$\begin{aligned}
u(x,0) &= F(x,t) \\
u(x,1) &= \mathfrak{K}^{-1}\left\{v^{2\alpha}\mathfrak{K}[A_0 + B_0]\right\}, \\
u(x,2) &= \mathfrak{K}^{-1}\left\{v^{2\alpha}\mathfrak{K}[A_1 + B_1]\right\}, \\
u(x,3) &= \mathfrak{K}^{-1}\left\{v^{2\alpha}\mathfrak{K}[A_2 + B_2]\right\}, \\
&\dots
\end{aligned}$$

Finally, in the form of truncated series, the approximate analytical solution of $u(x,t)$ is given by

$$u(x,t) = \lim_{P\to\infty} \sum_{n=0}^{P} u(x,n), \tag{15}$$

which is usually converge often quickly.

4 Solution to Cahn-Hillard Time-Fractional Differential Equation

This section uses NIPDTM to achieve approximating solutions for the fractional Cahn-Hillard equation (1) under the appropriate conditions:

4.1 Case-1

Consider the following Cahn-Hillard fractional differential equation

$$\begin{aligned} D_t^{\alpha} u(x,t) &= \mu\, u\, u_x - 18u\, u_{xx}^2 - 36(u_x)^2\, u_{xx} \\ &\quad - 24u\, u_x\, u_{xxx} - 3\, u^2\, u_{xxxx} + u_{xxxx} + u_{xxxxxx}, \end{aligned} \tag{16}$$

along with

$$u(x,0) = \tanh\left[\frac{x}{\sqrt{2}}\right]. \tag{17}$$

Apply the NIT on Eq. (16), and using (17),

$$\begin{aligned} K\,[u(x,t)] &= v \tanh\left[\frac{x}{\sqrt{2}}\right] + v^{2\alpha} K[\mu\, u\, u_x - 18u\, u_{xx}^2 - 36\,(u_x)^2\, u_{xx} \\ &\quad - 24\, u\, u_x\, u_{xxx} - 3\, u^2\, u_{xxxx} + u_{xxxx} + u_{xxxxxx}] \end{aligned}$$

The inverse NIT yields

$$u(x, m+1) = \mathfrak{K}^{-1}\left\{v^{2\alpha}\mathfrak{K}\,[A_m + B_m]\right\},\ u(x,0) = \tanh\left[\frac{x}{\sqrt{2}}\right], \tag{18}$$

where

$$\begin{aligned} A_m &= \mu \sum_{i=0}^{m-1} u(x,i) u_x(x, m-i-1) \\ &\quad - 18 \sum_{i=0}^{m-1}\sum_{j=0}^{i} u(x,j) u_{xx}(x, i-j) u_{xx}(x, m-i-1) \\ &\quad - 36 \sum_{i=0}^{m-1}\sum_{j=0}^{i} u_x(x,j) u_x(x, i-j) u_{xx}(x, m-i-1) \\ &\quad - 24 \sum_{i=0}^{m-1}\sum_{j=0}^{i} u(x,j) u_x(x, i-j) u_{xxx}(x, m-i-1) \\ &\quad - 3 \sum_{i=0}^{m-1}\sum_{j=0}^{i} u(x,j) u(x, i-j) u_{xxxx}(x, m-i-1), \\ B_m &= u_{xxxx}(x, m-1) + u_{xxxxxx}(x, m-1) \end{aligned}$$

are projected differential transform of $\mu\, u\, u_x - 18\, u\, u_{xx}^2 - 36\,(u_x)^2\, u_{xx} - 24\, u\, u_x\, u_{xxx} - 3\, u^2\, u_{xxxx}$ and $u_{xxxx} + u_{xxxxxx}$ respectively.
From Eq. (18), we get

$$u(x,1) = \frac{t^{\alpha}\mu\,\mathrm{sech}\left[\frac{x}{\sqrt{2}}\right]^2\tanh\left[\frac{x}{\sqrt{2}}\right]}{\sqrt{2}\Gamma[1+\alpha]}. \tag{19}$$

$$\begin{aligned} u(x,2) = {} & \frac{t^{2\alpha\mu}}{32\Gamma[1+2\alpha]}\Big(-11952\sqrt{2}+14\mu+\left(7824\sqrt{2}+17\mu\right)\cosh\left[\sqrt{2}x\right] \\ & +\left(-384\sqrt{2}+2\mu\right)\cosh\left[2\sqrt{2}x\right] \\ & -\mu\cosh\left[3\sqrt{2}x\right]\Big)\,\mathrm{sech}\left[\frac{x}{\sqrt{2}}\right]^8\tanh\left[\frac{x}{\sqrt{2}}\right]. \end{aligned} \tag{20}$$

$$\begin{aligned} u(x,3) = {} & \frac{1}{2048\Gamma[1+\alpha]^2\Gamma[1+3\alpha]}t^{3\alpha}\mu\Big(\Big(131790057984\sqrt{2}-5547264\mu \\ & +798\sqrt{2}\mu^2+24\left(-6510256128\sqrt{2}-188448\mu+47\sqrt{2}\mu^2\right)\cosh\left[\sqrt{2}x\right] \\ & +3\left(11558757888\sqrt{2}+803328\mu+101\sqrt{2}\mu^2\right)\cosh\left[2\sqrt{2}x\right] \\ & -2696380416\sqrt{2}\cosh\left[3\sqrt{2}x\right]+1150464\mu\cosh\left[3\sqrt{2}x\right] \\ & -92\sqrt{2}\mu^2\cosh\left[3\sqrt{2}x\right]+56613888\sqrt{2}\cosh\left[4\sqrt{2}x\right] \\ & -229632\mu\cosh\left[4\sqrt{2}x\right]-78\sqrt{2}\mu^2\cosh\left[4\sqrt{2}x\right] \\ & -147456\sqrt{2}\cosh\left[5\sqrt{2}x\right]+5376\mu\cosh\left[5\sqrt{2}x\right]-12\sqrt{2}\mu^2\cosh\left[5\sqrt{2}x\right] \\ & +\sqrt{2}\mu^2\cosh\left[6\sqrt{2}x\right]\Big)\Gamma[1+\alpha]^2+32\mu\cosh\left[\frac{x}{\sqrt{2}}\right]^4\Big(57384+8\sqrt{2}\mu \\ & +\left(-55968+9\sqrt{2}\mu\right)\cosh\left[\sqrt{2}x\right]+7416\cosh\left[2\sqrt{2}x\right]-192\cosh\left[3\sqrt{2}x\right] \\ & -\sqrt{2}\mu\cosh\left[3\sqrt{2}x\right]\Big)\Gamma[1+2\alpha]\Big)\,\mathrm{sech}\left[\frac{x}{\sqrt{2}}\right]^{14}\tanh\left[\frac{x}{\sqrt{2}}\right] \end{aligned} \tag{21}$$

...

Using the initial condition (17) and Eqs.eqrefeq:1stapproxc1-eqrefeq:3rdapproxc1, the NIPDTM series solution is given as:

$$u(x,t)=\sum_{m=0}^{\infty}u(x,m), \tag{22}$$

4.2 Case-2

Consider the following Cahn-Hillard fractional differential equation

$$D_t^{\alpha}u=\mu\, u\, u_x-18uu_{xx}^2-36(u_x)^2u_{xx}-24uu_xu_{xxx}-3u^2u_{xxxx}+u_{xxxx}+u_{xxxxxx}, \tag{23}$$

along with

$$u(x,0)=e^{\lambda x}. \tag{24}$$

In the same manner as discussed in Case-1, the NIPDTM solution of Eq. (23) with condition (24) is given by:

$$\begin{aligned}
u(x,t)=&\lim_{P\to\infty}\sum_{n=0}^{P}u(x,n)\\
=&e^{x\lambda}\left(1+\frac{t^{\alpha}\lambda}{\Gamma[1+\alpha]}((1-81e^{2x\lambda})\lambda^3+\lambda^5+e^{x\lambda}\mu)\right.\\
&+\frac{t^{2\alpha}\lambda^2}{\Gamma[1+2\alpha]}((1-6804e^{2x\lambda}+151,875e^{4x\lambda})\lambda^6+(2-59,292e^{2x\lambda})\lambda^8+\lambda^{10}\\
&-6e^{x\lambda}(-3+182e^{2x\lambda})\lambda^3\mu+66e^{x\lambda}\lambda^5\mu+3e^{2x\lambda}\mu^2)\\
&+\frac{\lambda^6t^{3\alpha}}{\Gamma[1+3\alpha]}((1-551124e^{2x\lambda}+94921875e^{4x\lambda})\lambda^6\\
&+3(1-3254,256e^{2x\lambda}+791,015,625e^{4x\lambda})\lambda^8+(3-43,223,868e^{2x\lambda})\lambda^{10}+\lambda^{12}\\
&-96e^{x\lambda}(-3+2912e^{2x\lambda})\lambda^3\mu-96e^{x\lambda}(-23+46,592e^{2x\lambda})\lambda^5\mu+4224e^{x\lambda}\lambda^7\mu\\
&+243e^{2x\lambda}\mu^2+2187e^{2x\lambda}\lambda^2\mu^2)\\
&+\frac{e^{x\lambda}t^{3\alpha}\lambda^3}{\Gamma[1+\alpha]^2\Gamma[1+3\alpha]}\Big(\Big(-1093,955,625e^{5x\lambda}\lambda^9+5156,946e^{4x\lambda}\lambda^6\mu\\
&+2\lambda^6(1+\lambda^2)^2\mu+15e^{3x\lambda}\lambda^3(850,500\lambda^6+7411,500\lambda^8-739\mu^2)+12e^{2x\lambda}\mu(-3420\lambda^6\\
&-23,988\lambda^8+\mu^2)-9e^{x\lambda}\lambda^3(27\lambda^6+54\lambda^8+27\lambda^{10}-6\mu^2-22\lambda^2\mu^2)\Big)\Gamma[1+\alpha]^2\\
&-\Big(47,258,883e^{5x\lambda}\lambda^9-649,539e^{4x\lambda}\lambda^6\mu-\lambda^6(1+\lambda^2)^2\mu-30e^{3x\lambda}\lambda^3(10,125\lambda^6\\
&+10,125\lambda^8-76\mu^2)+3e^{x\lambda}\lambda^3(1+\lambda^2)(81\lambda^6+81\lambda^8-\mu^2)-2e^{2x\lambda}\mu(-930\lambda^6-930\lambda^8\\
&\left.+\mu^2)\Big)\Gamma[1+2\alpha]\Big)\right)
\end{aligned} \tag{25}$$

5 Graphical Results and Discussions

Within this section, we estimate the numeric results of u(x, t) for distinct fractional movements $\alpha = 0.25, 0.5, 0.75, 1$ as well as for various values of t and x. The graphical presentation obtained by using NIPDTM solution for the different values of t, x and α for (22) & (25) are shown in the Figs. 1, 3, 5, and 7(a)–(c), as well as those for different values of x and an at specific time $t = 0 : 25$ are demonstrated in Figs. 2, 4, 6, and 8.
It is seen from Fig. 2,4,6, and 8 the $u(x, t)$ rises with an increment in x & t for $\alpha = 1, 0.75, 0.5, 0.25$. It is seen from Fig 2, 4, 6, and 8 that, as α increase, the value of u(x, t) rises in both the cases. Finally, we notice that the estimated solution eqrefeq:finalsolcase1 and eqrefeq:finalsolcase2 is in complete fulfillment with the results derived by NIM [24] and q-HAM [24].

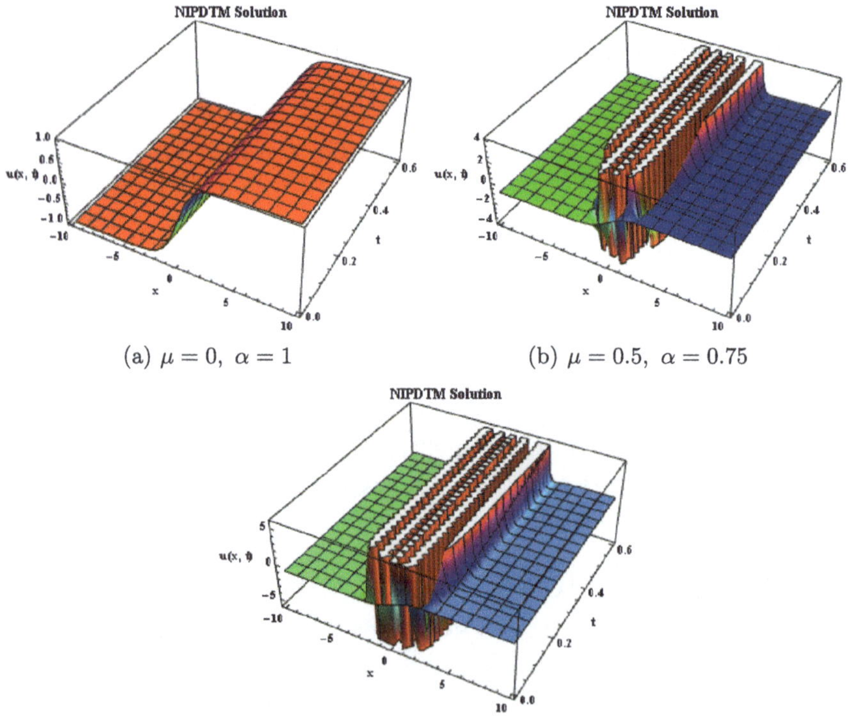

(a) $\mu = 0,\ \alpha = 1$ (b) $\mu = 0.5,\ \alpha = 0.75$

(c) $\mu = 1,\ \alpha = 0.5$

Fig. 1 Plot of $u(x, t)$ with r.to x and t when **a** $\mu = 0, \alpha = 1$ **b** $\mu = 0.5, \alpha = 0.75$ **c** $\mu = 1, \alpha = 0.5$ (Case-1)

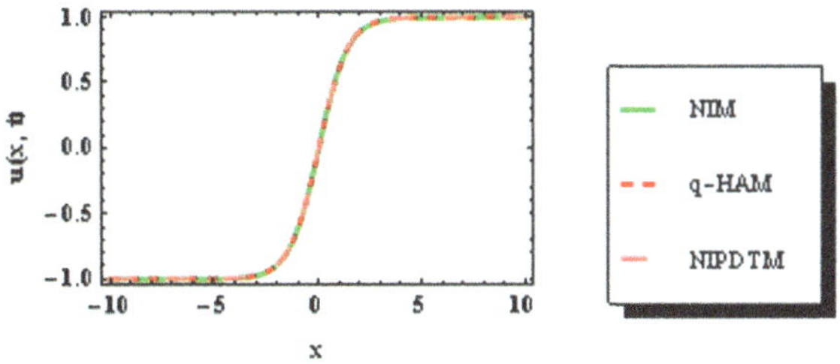

Fig. 2 On comparison between NIM [24], q-HAM [24], and NIPDTM solution of Cahn-Hillard fractional equation (Case-1)

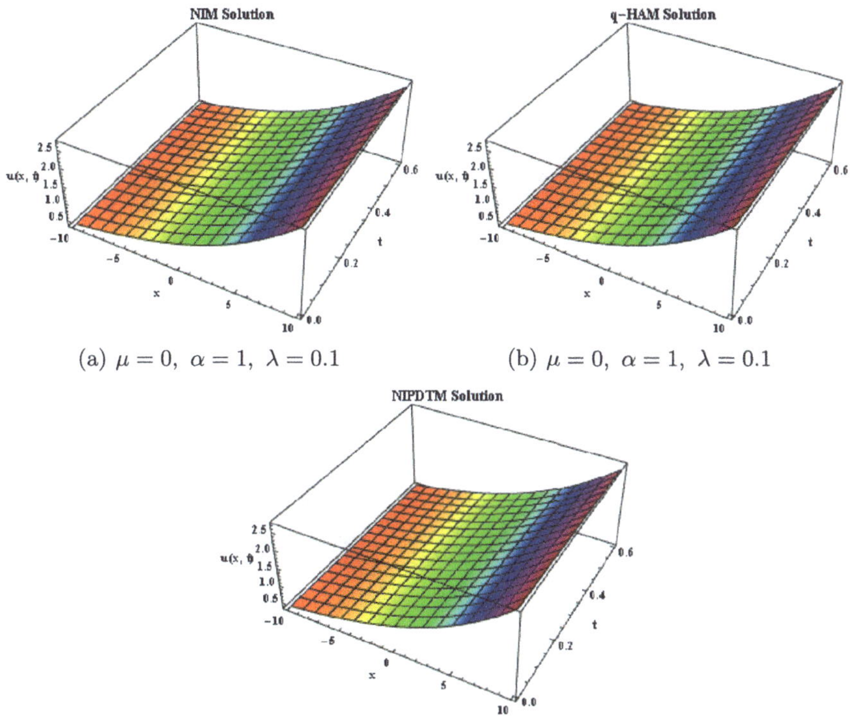

(a) $\mu = 0,\ \alpha = 1,\ \lambda = 0.1$ (b) $\mu = 0,\ \alpha = 1,\ \lambda = 0.1$

(c) $\mu = 0,\ \alpha = 1,\ \lambda = 0.1$

Fig. 3 Plot of $u(x, t)$ of **a** NIM, **b** q-HAM, and **c** NIPDTM with r.to x and t when $\mu = 0,\ \alpha = 1,\ \lambda = 0.1$ (Case-2)

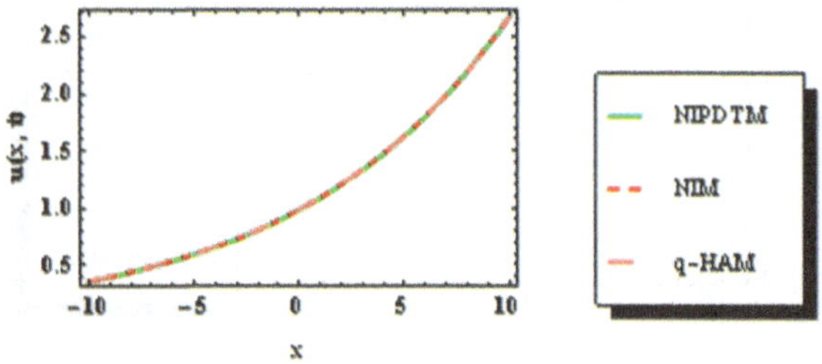

Fig. 4 On comparison between NIM [24], q-HAM [24], and NIPDTM solution of Cahn-Hillard fractional equation ($\mu = 0, \lambda = 0.1, \alpha = 1$) (Case-2)

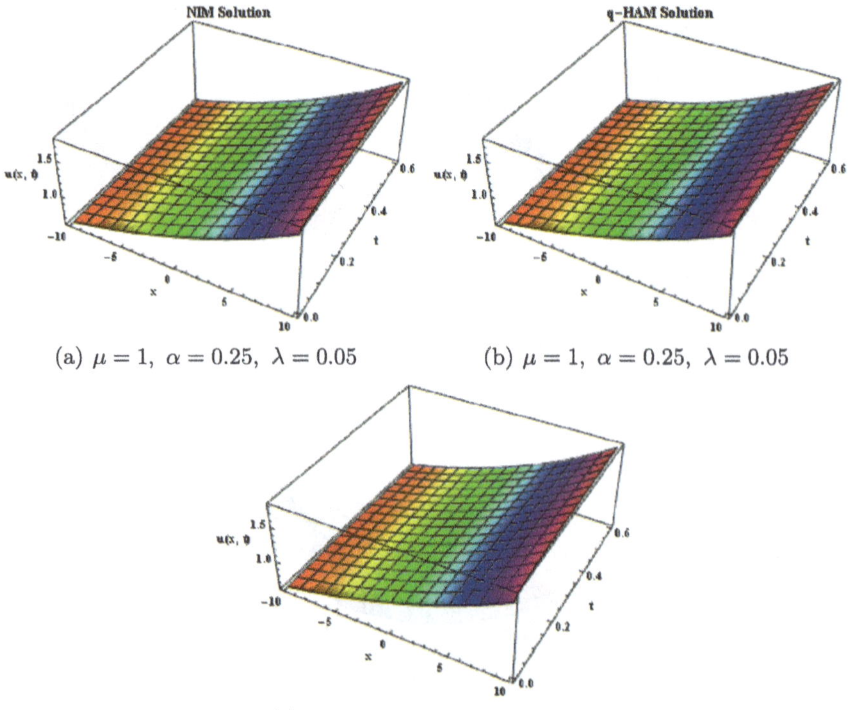

(a) $\mu = 1,\ \alpha = 0.25,\ \lambda = 0.05$

(b) $\mu = 1,\ \alpha = 0.25,\ \lambda = 0.05$

(c) $\mu = 1,\ \alpha = 0.25,\ \lambda = 0.05$

Fig. 5 Plot of $u(x, t)$ of **a** NIM, **b** q-HAM, and **c** NIPDTM with r.to x and t when $\mu = 1,\ \alpha = 0.25,\ \lambda = 0.05$ (Case-2)

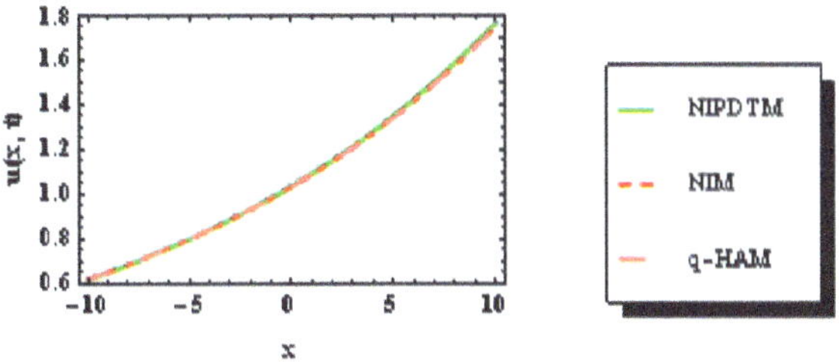

Fig. 6 On comparison between NIM [24], q-HAM [24], and NIPDTM solution of Cahn-Hillard fractional equation ($\mu = 1, \lambda = 0.05, \alpha = 0.25$) (Case-2)

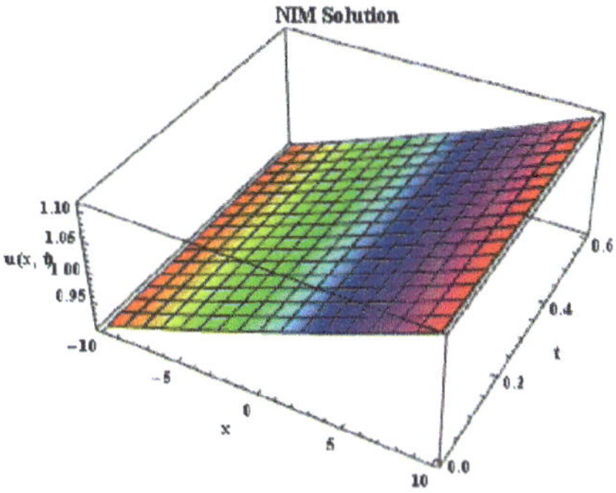

(a) $\mu = 0.5,\ \alpha = 0.75,\ \lambda = 0.01$

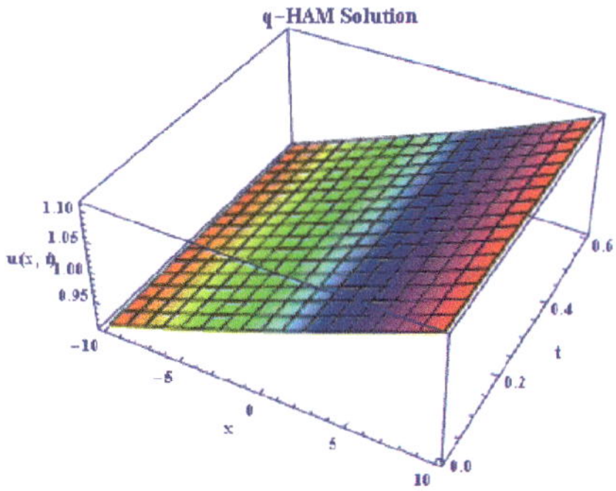

(b) $\mu = 0.5,\ \alpha = 0.75,\ \lambda = 0.01$

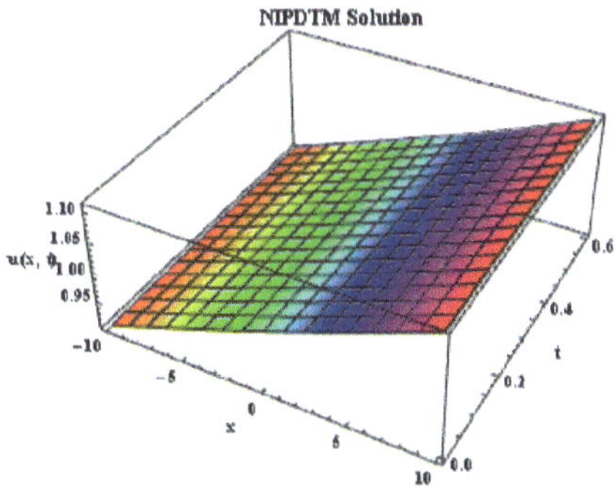

(c) $\mu = 0.5,\ \alpha = 0.75,\ \lambda = 0.01$

Fig. 7 Plot of $u(x, t)$ of **a** NIM, **b** q-HAM, and **c** NIPDTM with r.to x and t when $\mu = 0.5, \alpha = 0.75, \lambda = 0.01$ (Case-2)

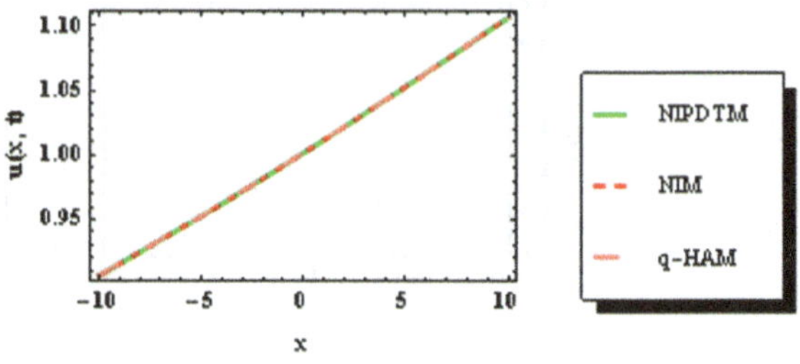

Fig. 8 On comparison between NIM [24], q-HAM [24], and NIPDTM solution of Cahn-Hillard fractional equation ($\mu = 0.5, \lambda = 0.01, \alpha = 0.75$) (Case-2)

6 Conclusion

The NIPDTM is successfully used in this analysis to solve fractional Cahn-Hillard equations. The results obtained through the use of the NIPDTM discussed here agree with NIM [24] and q-HAM [24] results. It offers solutions for converging series with components that are easily evaluated directly without the need to do linearization, disturbance, or limiting assumptions. We must note that the proposed novel approach is quite worthy of reducing the quantity of computational work relative to traditional methods while retaining the high precision of numerical performance. Finally, we conclude that NIPDTM is more effective and reliable in finding analytical solutions for broad classes of linear and nonlinear fractional equations.

References

1. Lazarević, M.P., Rapaić, M.R., Šekara, T.B., Mladenov, V., Mastorakis, N.: Introduction to fractional calculus with brief historical background. In: Advanced Topics on Applications of Fractional Calculus on Control Problems, System Stability and Modeling. WSAES Press, pp. 3–16 (2014)
2. Podlubny, I.: Fractional Differential Equations: An Introduction to Fractional Derivatives, Fractional Differential Equations, to Methods of their Solution and Some of their Applications. Elsevier (1998)
3. Alchikh, R., Khuri, S.A.: An iterative approach for the numerical solution of fractional bvps. Int. J. Appl. Comput. Math. **5**(6), 147 (2019)
4. Baleanu, D., Güvenç, Z.B., Tenreiro Machado, J.A. et al.: New Trends in Nanotechnology and Fractional Calculus Applications. Springer (2010)
5. Feng, M., Zhang, X., Ge, W.G.: New existence results for higher-order nonlinear fractional differential equation with integral boundary conditions. Bound. Value Probl. **2011**(1), 720702 (2011)
6. Güner, Ö., Bekir, A.: Exact solutions of some fractional differential equations arising in mathematical biology. International Journal of Biomathematics **8**(01), 1550003 (2015)
7. Han, Z., Li, Y., Sui, M.: Existence results for boundary value problems of fractional functional differential equations with delay. J. Appl. Math. Comput. **51**(1–2), 367–381 (2016)
8. Kumar, D., Seadawy, A.R., Joardar, A.K.: Modified Kudryashov method via new exact solutions for some conformable fractional differential equations arising in mathematical biology. Chin. J. Phys. **56**(1), 75–85 (2018)
9. Mohamed, M.S.: Analytical approximate solutions for the nonlinear fractional differential-difference equations arising in nanotechnology. Glob. J. Pure Appl. Math **13**, 7637–7652 (2017)

10. Odibat, Z., Momani, S.: The variational iteration method: an efficient scheme for handling fractional partial differential equations in fluid mechanics. Comput. Math. Appl. **58**(11–12), 2199–2208 (2009)
11. Shah, K., Singh, T., Kılıçman, A.: Combination of integral and projected differential transform methods for time-fractional gas dynamics equations. Ain Shams Eng. J. **9**(4), 1683–1688 (2018)
12. Xu, M., Sun, S.: Positivity for integral boundary value problems of fractional differential equations with two nonlinear terms. J. Appl. Math. Comput. **59**(1–2), 271–283 (2019)
13. Kashuri, A., Fundo, A., Liko, R.: New integral transform for solving some fractional differential equations. Int. J. Pure Appl. Math. **103**(4), 675–682 (2015)
14. Jang, B.: Solving linear and nonlinear initial value problems by the projected differential transform method. Comput. Phys. Commun. **181**(5), 848–854 (2010)
15. Shah, K., Singh, T.: The combined approach to obtain approximate analytical solution of instability phenomenon arising in secondary oil recovery process. Comput. Appl. Math. **37**(3), 3593–3607 (2018)
16. Singh, B.K.: Homotopy perturbation new integral transform method for numeric study of space-and time-fractional (n+ 1)-dimensional heat-and wave-like equations. Waves Wavelets Fractals **4**(1), 19–36 (2018)
17. Kumar, D., Singh, J., Baleanu, D.: A new analysis for fractional model of regularized long-wave equation arising in ion acoustic plasma waves. Math. Methods Appl. Sci. **40**(15), 5642–5653 (2017)
18. Kumar, D., Singh, J., Baleanu, D.: A new numerical algorithm for fractional fitzhugh-nagumo equation arising in transmission of nerve impulses. Nonlinear Dyn. **91**(1), 307–317 (2018)
19. Singh, J., Kumar, D., Kumar, S.: New treatment of fractional fornberg-whitham equation via laplace transform. Ain Shams Eng. J. **4**(3), 557–562 (2013)
20. Ainsworth, M., Mao, Z.: Analysis and approximation of a fractional Cahn-Hilliard equation. SIAM J. Numer. Anal. **55**(4), 1689–1718 (2017)
21. Hosseini, K., Bekir, A., Ansari, R.: New exact solutions of the conformable time-fractional Cahn-Allen and Cahn-Hilliard equations using the modified kudryashov method. Optik **132**, 203–209 (2017)
22. Liu, H., Cheng, A., Wang, H., Zhao, J.: Time-fractional allen-cahn and cahn-hilliard phase-field models and their numerical investigation. Comput. Math. Appl. **76**(8), 1876–1892 (2018)
23. Caputo, M.: Elasticita e Dissipazione. Zani-Chelli, Bologna (1969)
24. Akinyemi, L., Iyiola, O.S., Akpan, U.: Iterative methods for solving fourth-and sixth-order time-fractional Cahn-Hillard equation. Math. Methods Appl. Sci. **43**(7), 4050–4074 (2020)

Technological Paradigms and Optimization Methods

Cryptanalysis of IoT-Based Authentication Protocol Scheme

Nishant Doshi

Abstract Internet of things (IoT) is a technology in which resource constrained devices are placed in efficient manner for communication and do the assigned task. Indeed, due to nature of resource constrained devices which itself having low power, the entire system is more susceptible to not only computation overhead but also prone to attacker too. Before devices start sending data to server, the authentication is required through which either party can able to identify the other party. With this, the need of authentication model is required which also prone from attacker's viewpoint. Recently, in June 2020, Lee et al. mentioned the novel approach using three-factor authentication secure protocol in IoT paradigm. Conversely, this paper proves the Lee et al.'s method is susceptible to replay perfect forward secrecy as well as session-specific temporary information attacks.

Keywords Three factors · Authentication · IoT · Attacks

1 Introduction

The twenty-first century is the era for Industry 4.0. Indeed, Internet of things (IoT) is backbone component for all other components to work in collaborate within Industry 4.0. In IoT, one can use the resource starved devices and communication between them. There are several advantages of using these devices, i.e., low cost, easy replaceable, low maintenance, hostile in deployment and so on. Due to the pluses of these devices, in [1], the authors have estimated the use of more than 20 billion devices to be deployed around the globe as a part of IoT. On the hand, government regulatory authority International Telecommunication Union (ITU) recommends the need of around more than 1 million devices per km^2 as IoT paradigm for the full benefit of the 5G technology deployment.

However, with development and need of IoT devices, one has to concern the second side of it, i.e., security. As these devices are less powerful, the attacker can

N. Doshi (✉)
Pandit Deendayal Petroleum University, Koba Institutional Area, Gandhinagar, Gujarat, India
e-mail: Nishant.doshi@sot.pdpu.ac.in

M. Sahni et al. (eds.), *Mathematical Modeling, Computational Intelligence Techniques and Renewable Energy*, Advances in Intelligent Systems and Computing 1287,
https://doi.org/10.1007/978-981-15-9953-8_8

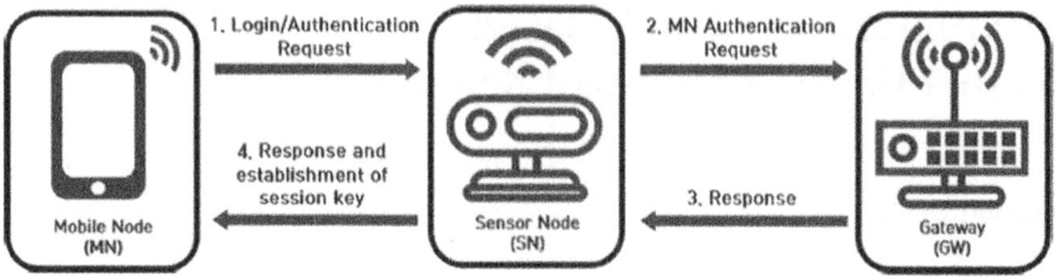

Fig. 1 IoT environment model [63]

take advantage and compromise the devices to gain certain data. Indeed, the way identified in research is to use the authentication protocols [2–7]. In [8], Lamport firstly addresses the issue of authentication and proposed the method that withstand against attacks. Afterward, in [9–62], the authors have proposed the various authentication methods and claimed to be secure as to predecessor method.

Recently, in [63], the Lee et al. mentioned the novel method for authentication in IoT environment model as depicted in Fig. 1.

1.1 Our Contributions

We have cryptanalysis method of Lee et al. and prove the vulnerabilities as follows

- Replay attack: Sending of same message cannot be identified by other entity and lead to computation overhead.
- Session-specific temporary information attack: Reveal of nonce term parameters lead to compromise of session key
- P(perfect) F(forward) S(secrecy): reveal of long-term secrets of involving entities lead to disclose of past and present session keys.

1.2 Paper Association

Preliminaries to be useful for Lee et al.'s method are given in Sect. 2. Section 3 describes in detail about Lee et al.'s method. Section 4 gives our cryptanalysis for Lee et al.'s method. Finally, Sect. 5 concludes with the future scope of this work. References are at the end.

Table 1 Notations

Symbol	Meaning
$\rightarrow$	Insecure channel
$\rightarrow$	Secure channel
MN_i	Mobile node
N_j	Sensor node
ID_i, PW_i	Identity and password for the MN_i
BIO_i	Biometric identity of MN_i
GW	Gateway node
NID_j	Identities for the N_j
T_1, T_2	Timestamp
n_x, r_x	Random nonce
SK	Shared session key between MN_i and N_j
$E_k()$, $D_k()$	Symmetric key algorithms using secret key k
$h()$	One way hash
$H()$	One way bio-hash
$\|\|$	Concatenation of two strings
$\oplus$	Binary XOR operation
K_G	Long-term secret key of GW
K_{GU}	Long-term secret key of MN_i
K_{GN}	Secret shared by GW and N_j

2 Preliminaries

In this section, we have given the notations which will be used in Lee et al. as well as in cryptanalysis of it (Table 1).

3 Method of Lee et al.

The method of Lee et al. is divided into following main stages.

3.1 Mobile Node Registration Stage

It will done by mobile node MN_i and gateway node GW in the system.

MN_i :	Select ID_i, PW_i, BIO_i Compute $PWB_i = h(PW_i \| \| H(BIO_i))$ Compute $MID_i = h(ID_i \| \| H(BIO_i))$
$MN_i \rightarrow GW$:	$\langle ID_i, PWB_i, MID_i \rangle$
GW :	Select random number r_{GU} and r_D Compute $RID_i = E_{KG}(ID_i)$ Compute $PID_i = E_{KG}(ID_i \| \| r_D)$ Compute $x_i = h(ID_i \| \| PWB_i)$ Compute $y_i = h(ID_i \\| PWB_i \\| r_{GU}) \oplus h(K_{GU} \| \| ID_i)$
$GW \rightarrow MN_i$:	$\langle PID_i, x_i, y_i, r_{GU} \rangle$
MN_i :	Store $\langle PID_i, x_i, y_i, r_{GU} \rangle$ into mobile device

3.2 Sensor Registration Stage

It will done between sensor node N_j and GW

N_j :	Choose r_j Compute $MP_j = h(K_{GN} \\| r_j \\| NID_j)$ Compute $MI_j = r_j \oplus h(NID_j \| \| K_{GN})$
$N_j \rightarrow GW$:	$\langle NID_j, MP_j, MI_j \rangle$
GW :	Compute $r_j^* = MI_j \oplus h(NID_j \| \| K_{GN})$ Compute $MP_j^* = h\left(K_{GN} \\| r_j^* \\| NID_j\right)$ See if $MP_j^* = MP_j$ holds Compute $x_j = h(NID \| \| K_{GN})$ Compute $y_j = x_j \oplus MP_j^*$
$GW \rightarrow N_j$:	$\left\langle y_j^* \right\rangle$
N_j :	Store $\left\langle y_j^* \right\rangle$ in memory

3.3 Login and Authentication Stage

In this phase user, MN_i will mutually confirm the sensor node N_j with session key SK_{ij}.

MN_i :	Insert $\mathrm{ID}_i, \mathrm{BIO}_i, \mathrm{PW}_i$ Compute $\mathrm{PWB}_i = h(\mathrm{PW}_i \| \| H(\mathrm{BIO}_i))$ $x_i^* = h(\mathrm{ID}_i \| \| \mathrm{PWB}_i)$ Check if $x_i^* = x_i$ holds Choose n_i Compute $A_i = y_i \oplus h(\mathrm{ID}_i \| \mathrm{PWB}_i \| r_{\mathrm{GU}})$ Compute $\mathrm{UN}_i = h(A_i \| \mathrm{PID}_i \| n_i)$ Compute $\mathrm{UZ}_i = n_i \oplus A_i$
$\mathrm{MN}_i \rightarrow N_j$:	$M_1 = \langle \mathrm{PID}_i, \mathrm{UN}_i \mathrm{UZ}_i, T_1 \rangle$
N_j :	Check if $T_{\mathrm{fresh}} - T_1 \leq \Delta T$ holds Choose n_j Compute $y_j \oplus h\left(K_{\mathrm{GN}} \| \| r_j \| \| \mathrm{NID}_j\right)$ Compute $A_j = h\left(x_j\right) \oplus n_j$ Compute $B_j = h(x_j \| \| n_j)$
$N_j \rightarrow \mathrm{GW}$:	$M_2 = \left\langle M_1, \mathrm{NID}_j, A_j, B_j \right\rangle$
GW :	Compute $x_j^* = h(\mathrm{NID}_j \| \| K_{\mathrm{GN}})$ Compute $n_j^* = h\left(x_j^*\right) \oplus A_j$ Compute $B_j^* = h(x_j^* \| \| n_j^*)$ Check if $B_j^* = B_j$ holds Compute $< \mathrm{ID}_i, r_D \geq D_{\mathrm{KG}}(\mathrm{PID}_i)$ Compute $A_i^* = h(\mathrm{ID}_i \| \| K_{\mathrm{GU}})$ Compute $n_i^* = \mathrm{UZ}_i \oplus A_i^*$ Compute $\mathrm{UN}_i^* = h(A_i^* \mathrm{PID}_i \| \| n_i^*)$ Check if $\mathrm{UN}_i^* = \mathrm{UN}_i$ holds Choose r_D^{new} Compute $F_j = h(\mathrm{ID}_i \| \| n_i^*)$ Compute $G_j = F_j \oplus x_j^*$ Compute $R_{ij} = n_j^* \oplus n_i^*$ Compute $H_j = h\left(x_j^* \middle\| n_j^* \middle\| n_i^* F_j\right)$ $\mathrm{PID}_i^{\mathrm{new}} = E_{\mathrm{KG}}\left(\mathrm{ID}_i, r_D^{\mathrm{new}}\right)$
$\mathrm{GW} \rightarrow N_j$:	$M_3 = \left\langle \mathrm{PID}_i^{\mathrm{new}}, G_j, R_{ij}, H_i \right\rangle$

(continued)

(continued)

MN_i :	Insert ID_i, BIO_i, PW_i Compute $PWB_i = h(PW_i\|\|H(BIO_i))$ $x_i^* = h(ID_i\|\|PWB_i)$ Check if $x_i^* = x_i$ holds Choose n_i Compute $A_i = y_i \oplus h(ID_i\|PWB_i\|r_{GU})$ Compute $UN_i = h(A_i\|PID_i\|n_i)$ Compute $UZ_i = n_i \oplus A_i$
N_j :	Compute $F_j^* = G_j \oplus x_j$ Compute $n_i^* = R_{ij} \oplus n_j$ Compute $H_j^* = h(x_j\|n_j\|n_i^*\|\|F_j^*)$ Check if $H_j^* = H_j$ holds Choose m_j Compute $L_j = h(NID_j\|\|n_i^*) \oplus m_j$ Compute $Sk_{ij} = h\left(F_j^*\|n_i^*\|m_j\right)$ Compute $SV_j = h(SK_{ij}\|T_1\|T_2)$
$N_j \rightarrow MN_i$:	$M_4 = \langle PID_i^{new}, L_j SV_j, T_2\rangle$
MN_i :	Check if $T_{fresh} - T_2 \leq \Delta T$ holds Compute $m_j^* = L_j \oplus h(NID_j\|\|n_i)$ Compute $SK_{ij} = h(h(ID_i\|n_i)\|n_i\|m_j^*)$ Compute $SV_i = h(SK_{ij}\|T_1\|T_2)$ If $SV_j = SV_i$ accept and store session key SK_{ij} else reject session

4 Cryptanalysis of Lee et al.'s Method

In this section, we will prove that the Lee et al.'s method is yet vulnerable to several attacks.

4.1 *Session-Specific Temporary Information Attack*

If compromise of short-term secrets (i.e., random nonce) cannot reveal the session key, then method is secure against this atatck. The attacker computes session key from nonce. In the Lee et al. method, the attack is performed as follows

- The attacker gets short-term secrets n_i, n_j and m_j
- Let us assume that attacker compromises (or part of) some sensor node N_A
- As N_A is having n_j, it computes $A_A = h(x_A) \oplus n_j$ and $B_A = h(x_A\|\|n_j)$

- Finally, from stored message M_1, N_A will send $M_2 = \langle M_1, \mathrm{NID}_A, A_A, B_A \rangle$ to GW
- GW will perform as mentioned in Sect. 3.2 and send M_3 to N_A.
- N_A will get $h(\mathrm{ID}_i||n_i)$ by computation of $G_A \oplus x_A$
- Compute $m_j = L_j \oplus h(\mathrm{NID}_j||n_i)$ from M_4
- Finally, N_A computers session key $\mathrm{SK}_{ij} = h(F_j\|n_i\|m_j)$

Thus, Lee et al.'s method is prone to the session-specific temporary information attack. ■

4.2 *Replay Attack*

If sending of same message is detected at early stage by the other entity for all messages, than method is withstand against this attack. As can seen from message exchange between N_j and GW, no any detection mechanism implemented for repetition of same message. In other words, GW fails to identify the repetition of M_2 and N_j fails to identify the repetition of M_3.

Thus, Lee et al.'s method is prone to the replay attack. ■

4.3 *Perfect Forward Secrecy*

If compromise of long-term secrets of involving entities does not compromise past or present session keys, then method is secure against this attack.

4.3.1 Conciliation of Credential of Sensor Node N_j

The attacker gets the credential of sensor node N_j, i.e., x_j. The attacker performs to get the session key SK_{ij}

- Compute $n_j = A_j \oplus h(x_j)$ from M_2
- Compute $F_j = G_j \oplus x_j$ from M_3
- Compute $n_i = R_{ij} \oplus n_j$ from M_3
- Compute $m_j = L_j \oplus h(\mathrm{NID}_j||n_i)$ from M_4
- Finally computers session key $\mathrm{SK}_{ij} = h(F_j\|n_i\|m_j)$

4.3.2 Conciliation of Mobile Node MN_i

The attacker has ID_i, A_i.

- Computes $n_i = \mathrm{UZ}_i \oplus A_i$ from M_1
- Computes $m_j = L_j \oplus h(\mathrm{NID}_j||n_i)$ from M_4

- Computes $F_j = h(ID_i||n_i)$
- Finally computers session key $\mathrm{SK}_{ij} = h\left(F_j||n_i||m_j\right)$

4.4 Conciliation of Credential of GW

It compromise the of GW, i.e., K_{GN}, K_G. It does as below to get session key SK_{ij}

- Compute $x_j^* = h(\mathrm{NID}_j||K_{\mathrm{GN}})$
- Compute $n_j^* = h\left(x_j^*\right) \oplus A_j$
- Compute $B_j^* = h(x_j^*||n_j^*)$
- Compute $< \mathrm{ID}_i, r_D \geq D_{\mathrm{KG}}(\mathrm{PID}_i)$
- Compute $A_i^* = h(\mathrm{ID}_i||K_{\mathrm{GU}})$
- Compute $n_i^* = \mathrm{UZ}_i \oplus A_i^*$
- Computes $m_j = L_j \oplus h(\mathrm{NID}_j||n_i^*)$ from M_4
- Computes $F_j = h(\mathrm{ID}_i||n_i^*)$
- Finally computers session key $\mathrm{SK}_{ij} = h\left(F_j\|n_i^*\|m_j\right)$

Thus, the method of Lee et al. is prone.

5 Conclusion

IoT is a platform for the fifth-generation users which comes with the great ease and flexibility. However, it also makes it necessary to provide security as well as authentication to sustain against the attacks. This paper analyzes the secure method by Lee et al. using three factors (i.e., identity, text password and biometric password). However, in this paper, we do the cryptanalysis of the Lee et al. method and found it to be vulnerable against replay attack, session-specific temporary information attack and perfect forward secrecy attack. In future, one can design more secure, robust and lightweight authentication method as compared to Lee et al. and apply it to real-time domain for ready to use in IoT paradigm.

References

1. Park, O., Hwang, H., Lee, C., Shin, J.: Trends of 5 g massive lot, electronics and telecommunications. Trends **31**(1), 68–77 (2016)
2. Series, M.: IMT vision–framework and overall objectives of the future development of IMT for 2020 and beyond. Recommendation ITU, pp. 2083–2090 (2015)
3. Ahmad, I., Shahabuddin, S., Kumar T., Okwuibe J., Gurtov, A., Ylianttila, M.: Security for 5 g and beyond. IEEE Commun. Surv. Tutorials (2019)
4. Perrig, A., Stankovic, J., Wagner, D.: Security in wireless sensor networks. Commun. ACM **47**(6), 53–57 (2004)

5. Mishra, D.: Efficient and secure two-factor dynamic ID-based password authentication scheme with provable security. Cryptologia **42**(2), 146–175 (2018)
6. Srinivas, J., Mukhopadhyay, S., Mishra, D.: A self-verifiable password based authentication scheme for multi-server architecture using smart card. Wirel Pers Commun **96**(4), 6273–6297 (2017)
7. Dhillon, P.K., Kalra, S.: Secure multi-factor remote user authentication scheme for internet of things environments. Int. J. Commun. Syst. **30**(16) (2017)
8. Lamport, L.: Password authentication with insecure communication. Commun. ACM **24**(11), 770–772 (1981)
9. Li, L.-H., Lin, L.-C., Hwang, M.S.: A remote password authentication scheme for multiserver architecture using neural networks. IEEE Trans. Neural Netw. **12**(6), 1498–1504 (2001)
10. Ramasamy, R., Muniyandi, A.P.: New remote mutual authentication scheme using smart cards. Trans Data Priv. **2**(2), 141–152 (2009)
11. Xu, J., Zhu, W.-T., Feng, D.G.: An improved smart card based password authentication scheme with provable security. Comput. Stan. Interfaces **31**(4), 723–728 (2009)
12. Banerjee, S, Mukhopadhyay, D.: Symmetric key based authenticated querying in wireless sensor networks. In: Proceedings of the First International Conference on Integrated Internet Ad Hoc and Sensor Networks, vol. 22. ACM (2006)
13. Du, W., Wang, R., Ning, P.: An efficient scheme for authenticating public keys in sensor networks. In: Proceedings of the 6th ACM International Symposium on Mobile Ad Hoc Networking and Computing, pp. 58–67. ACM (2005)
14. Chatterjee, S., Das, A.K.: An effective ECC-based user access control scheme with attribute-based encryption for wireless sensor networks. Sec. Commun. Netw. **8**(9), 1752–1771 (2015)
15. Chung, Y., Choi, S., Won, D.: Anonymous authentication scheme for intercommunication in the internet of things environments. Int. J. Distrib. Sens. Netw. **11**(11), 305785 (2015)
16. Park, Y., Park, Y.: Three-factor user authentication and key agreement using elliptic curve cryptosystem in wireless sensor networks. Sensors **16**(12), 21–23 (2016)
17. Wong, K.H, Zheng, Y., Cao, J., Wang, S.: A dynamic user authentication scheme for wireless sensor networks. In: Proceedings of the IEEE International Conference on Sensor Networks, Ubiquitous, and Trustworthy Computing (SUTC'06), vol. 1, pp. 244–251. IEEE Computer Society (2006)
18. Das, M.L.: Two-factor user authentication in wireless sensor networks. IEEE Trans. Wirel. Commun. **8**(3), 1086–1090 (2009)
19. Khan, M.K., Alghathbar, K.: Cryptanalysis and security improvements of 'two-factor user authentication in wireless sensor networks'. Sensors **10**(3), 2450–2459 (2010)
20. He, D., Gao, Y., Chan, S., Chen, C., Bu, J.: An enhanced two-factor user authentication scheme in wireless sensor networks. Ad hoc Sens. Wirel. Netw. **10**(4), 361–371 (2010)
21. Vaidya, B., Makrakis, D., Mouftah, H.T.: Improved two-factor user authentication in wireless sensor networks. In: 2010 IEEE 6th International Conference on Wireless and Mobile Computing, Networking and Communications, pp. 600–606. IEEE (2010)
22. Yeh, H.-L., Chen, T.-H., Liu, P.-C., Kim, T.-H., Wei, H.W.: A secured authentication protocol for wireless sensor networks using elliptic curves cryptography. Sensors **11**(5), 4767–4779 (2011)
23. Xue, K., Ma, C., Hong, P., Ding, R.: A temporal-credential-based mutual authentication and key agreement scheme for wireless sensor networks. J. Netw. Comput. Appl. **36**(1), 316–323 (2013)
24. Li, C.-T., Weng, C.-Y., Lee, C.C.: An advanced temporal credential-based security scheme with mutual authentication and key agreement for wireless sensor networks. Sensors **13**(8), 9589–9603 (2013)
25. Turkanović, M., Brumen, B., Hölbl, M.: A novel user authentication and key agreement scheme for heterogeneous ad hoc wireless sensor networks, based on the internet of things notion. Ad Hoc Netw. **20**, 96–112 (2014)
26. Farash, M.S, Turkanović, M., Kumari, S., Hölbl, M.: An efficient user authentication and key agreement scheme for heterogeneous wireless sensor network tailored for the internet of things environment. Ad Hoc Netw. **36**, 152–76 (2016)

27. Kumari, S., Das, A.K, Wazid, M., Li, X., Wu, F., Choo, K.-K.R., Khan, M.K.: On the design of a secure user authentication and key agreement scheme for wireless sensor networks, concurrency and computation. Prac. Exp. **29**(23), 3930 (2017)
28. Mishra, D., Das, A.K., Mukhopadhyay, S.: A secure user anonymity-preserving biometric-based multi-server authenticated key agreement scheme using smart cards. Exp. Syst. Appl. **41**(18), 8129–8143 (2014)
29. Jin, A.T.B., Ling, D.N.C., Goh, A.: Biohashing: two factor authentication featuring fingerprint data and tokenised random number. Pattern Recogn. **37**(11), 2245–55 (2004)
30. Chaudhry, S.A., Naqvi, H., Khan, M.K.: An enhanced lightweight anonymous biometric based authentication scheme for tmis. Multimed. Tools Appl. **77**(5), 5503–5524 (2018)
31. Khan, I., Chaudhry, S.A., Sher, M., Khan, J.I., Khan, M.K.: An anonymous and provably secure biometric-based authentication scheme using chaotic maps for accessing medical drop box data. J. Supercomput. **74**(8), 3685–3703 (2018)
32. Chaudhry, S.A.: A secure biometric based multi-server authentication scheme for social multimedia networks. Multimed. Tools Appl. **75**(20), 12705–12725 (2016)
33. Jiang, Q., Ma, J., Yang, C., Ma, X., Shen, J., Chaudhry, S.A.: Efficient end-to-end authentication protocol for wearable health monitoring systems. Comput. Electr. Eng. **63**, 182–195 (2017)
34. Kumari, S., Khan, M.K., Li, X.: A more secure digital rights management authentication scheme based on smart card. Multimed. Tools Appl. **75**(2), 1135–1158 (2016)
35. Wang, D., Gu, Q., Cheng, H., Wang, P.: The request for better measurement: a comparative evaluation of two-factor authentication schemes. In: Proceedings of the 11th ACM on Asia Conference on Computer and Communications Security, pp. 475–86. ACM (2016)
36. Jiang, Q., Zeadally, S., Ma, J., He, D.: Lightweight three-factor authentication and key agreement protocol for internet-integrated wireless sensor networks. IEEE Access **5**, 3376–3392 (2017)
37. Lee, H., Lee, D., Moon, J., Jung, J., Kang, D., Kim, H.: An improved anonymous authentication scheme for roaming in ubiquitous networks. PLoS one **13**(3), e0193366 (2018)
38. Sureshkumar, V., Amin, R., Anitha, R.: A robust mutual authentication scheme for session initiation protocol with key establishment. Peer-to-Peer Netw. Appl. **11**(5), 900–916 (2018)
39. Yang, L., Zheng, Z.: Cryptanalysis and improvement of a biometrics-based authentication and key agreement scheme for multi-server environments. PLoS One **13**(3), e0194093 (2018)
40. Banerjee, S., Odelu, V., Das, A.K, Srinivas, J., Kumar, N., Chattopadhyay, S.: A provably-secure and lightweight anonymous user authenticated session key exchange scheme for internet of things deployment. IEEE Internet Things J. (2019)
41. Li, X., Peng, J., Obaidat, M.S, Wu, F., Khan, M.K, Chen, C.: A secure three-factor user authentication protocol with forward secrecy for wireless medical sensor network systems. IEEE Syst. J. (2019)
42. Das, A.K., Kumari, S., Odelu, V., Li, X., Wu, F., Huang, X.: Provably secure user authentication and key agreement scheme for wireless sensor networks. Sec. Commun. Netw. **9**(16), 3670–3687 (2016)
43. Chang, C.-C., Le, H.D.: A provably secure, efficient, and flexible authentication scheme for ad hoc wireless sensor networks. IEEE Trans. Wirel. Commun. **15**(1), 357–366 (2015)
44. Yang, Z., Lai, J., Sun, Y., Zhou, J.: A novel authenticated key agreement protocol with dynamic credential for WSNS. ACM Trans. Sens. Netw. (TOSN) **15**(2), 22 (2019)
45. Banerjee, S., Chunka, C., Sen, S., Goswami, R.S.: An enhanced and secure biometric based user authentication scheme in wireless sensor networks using smart cards. Wirel. Pers. Commun. 1–28 (2019)
46. Wu, F., Li, X., Sangaiah, A.K., Xu, L., Kumari, S., Wu, L., Shen, J.: A lightweight and robust two-factor authentication scheme for personalized healthcare systems using wireless medical sensor networks. Future Gener. Comput. Syst. **82**, 727–737 (2018)
47. Das, A.K., Sutrala, A.K., Kumari, S., Odelu, V., Wazid, M., Li, X.: An efficient multi-gateway-based three-factor user authentication and key agreement scheme in hierarchical wireless sensor networks. Sec. Commun. Netw. **9**(13), 2070–2092 (2016)

48. Blanchet, B., Smyth, B., Cheval, V., Sylvestre, M.: Proverif 2.00: automatic cryptographic protocol verifier. In: User manual and tutorial, Version from, pp. 05–16 (2018)
49. Chaudhry, S.A., Khan, I., Irshad, A., Ashraf, M.U., Khan, M.K., Ahmad, H.F.: A provably secure anonymous authentication scheme for session initiation protocol. Sec. Commun. Netw. **9**(18), 5016–5027 (2016)
50. Karuppiah, M., Kumari, S., Li, X., Wu, F., Das, A.K., Khan, M.K., Saravanan, R., Basu, S.A.: Dynamic ID-based generic framework for anonymous authentication scheme for roaming service in global mobility networks. Wirel. Pers. Commun. **93**(2), 383–407 (2017)
51. Ryu, J., Lee, H., Kim, H., Won, D.: Secure and efficient three-factor protocol for wireless sensor networks. Sensors **18**(12), 4481 (2018)
52. Das, A.K.: A secure and effective user authentication and privacy preserving protocol with smart cards for wireless communications. Netw. Sci. **2**(1–2), 12–27 (2013)
53. Lu, Y., Li, L., Yang, X., Yang, Y.: Robust biometrics based authentication and key agreement scheme for multi-server environments using smart cards. PLoS One **10**(5), e0126323 (2015)
54. Wu, F., Xu, L., Kumari, S., Li, X., Khan, M.K., Das, A.K.: An enhanced mutual authentication and key agreement scheme for mobile user roaming service in global mobility networks. Ann. Telecommun. **72**(3–4), 131–144 (2017)
55. Burrows, M., Abadi, M., Needham, R.M.: A logic of authentication, proceedings of the royal society of london. Math. Phys. Sci. **426**(1871), 233–271 (1989)
56. Chatterjee, S., Roy, S., Das, A.K., Chattopadhyay, S., Kumar, N., Vasilakos, A.V.: Secure biometric-based authentication scheme using chebyshev chaotic map for multi-server environment. IEEE Trans. Dependable Sec. Comput. **15**(5), 824–839 (2016)
57. Jung, J., Kang, D., Lee, D., Won, D.: An improved and secure anonymous biometric-based user authentication with key agreement scheme for the integrated epr information system. PLoS One, **12**(1) (2017)
58. Odelu, V., Das, A.K., Goswami, A.: An effective and robust secure remote user authenticated key agreement scheme using smart cards in wireless communication systems. Wirel. Pers. Commun. **84**(4), 2571–2598 (2015)
59. Kang, D., Jung, J., Mun, J., Lee, D., Choi, Y., Won, D.: Efficient and robust user authentication scheme that achieve user anonymity with a Markov chain. Sec. Commun. Netw. **9**(11), 1462–1476 (2016)
60. Abbasinezhad-Mood, D., Nikooghadam, M.: Design and hardware implementation of a security-enhanced elliptic curve cryptography based lightweight authentication scheme for smart grid communications. Future Gener. Comput. Syst. **84**, 47–57 (2018)
61. Reddy, A.G, Das, A.K, Odelu, V, Yoo, K.Y.: An enhanced biometric based authentication with key-agreement protocol for multi-server architecture based on elliptic curve cryptography. PLoS One **11**(5), e0154308 (2016)
62. Kumari, S., Khan, M.K., Atiquzzaman, M.: User authentication schemes for wireless sensor networks: a review. Ad Hoc Netw. **27**, 159–194 (2015)
63. Hakjun, L., Dongwoo, K., Jihyeon, R., Dongho, W., Hyoungshick, K., Youngsook, L.: A three-factor anonymous user authentication scheme for Internet of Things environments. J. Inform. Sec. Appl. **52** (2020)

A Novel Approach of Polsar Image Classification Using Naïve Bayes Classifier

Nimrabanu Memon, Samir B. Patel, and Dhruvesh P. Patel

Abstract Polarimetric SAR (PolSAR) image classification is an increasing area of research in the field of remote sensing and computer vision. It is mainly used for land cover classification, which provides a basis for many applications like hydrology, natural hazards, urban planning, etc. In this paper, a novel approach is proposed to classify the polarimetric SAR (PolSAR) image using Naïve Bayes for land cover categorization. Naïve Bayes is a maximum a posteriori decision rule-based classifier which can perform better classification task than conventional supervised as well as unsupervised approaches for image classification. In the proposed method, an additional step of feature extraction, based on chi-square distribution is used for feature extraction and then the Naïve Bayes classifier is used to classify the image. The obtained results are then compared with traditional K-NN and K-means algorithms. The comparison is made in terms of accuracy assessment and quality analysis (visual analysis) of the obtained classified image. It showed that our proposed approach outperformed the other conventional classifiers in terms of qualitative analysis and has higher cross-validation accuracy than that of accuracy obtained through the K-NN algorithm.

Keywords PolSAR · SAR · K-means · K-NN · Naïve Bayes

N. Memon (✉) · D. P. Patel
Civil Engineering, School of Technology, Pandit Deendayal Petroleum University, Gandhinagar, Gujarat, India
e-mail: nimrabanu.hphd18@sot.pdpu.ac.in

D. P. Patel
e-mail: dhruvesh.patel@sot.pdpu.ac.in

S. B. Patel
Computer Science and Engineering, School of Technology, Pandit Deendayal Petroleum University, Gandhinagar, Gujarat, India
e-mail: samir.patel@sot.pdpu.ac.in

M. Sahni et al. (eds.), *Mathematical Modeling, Computational Intelligence Techniques and Renewable Energy*, Advances in Intelligent Systems and Computing 1287,
https://doi.org/10.1007/978-981-15-9953-8_9

1 Introduction

In many areas such as resource management, planning, and monitoring systems, the provision of reliable and up-to-date land use and land cover information is important. For example, for a number of agricultural, hydrological, and ecological models, the land cover is an essential input parameter. Recently, in the field of remote sensing, machine learning frameworks like artificial neural network (ANN) [1], convolutional neural network (CNN), particle swarm optimization (PSO), ant colony optimization (ACO), fuzzy sets and swarm intelligence (SI) are used in direction of image classification. Other techniques include nature-inspired algorithms such as biogeography-based optimization (BBO) [2, 3] and statistical/probablistic model. Probabilistic models are very common as a formalism for handling uncertainty, particularly those associated with Bayesian networks. Many authors used the Naïve Bayes to classify the image in remote sensing [4–7].

1.1 Why Naïve Bayes?

It is a part of a supervised classification family and it learns by looking at examples (here training sets) that are properly classified with respect to their labels. In machine learning, each example is a set of features (attributes) to describe that specific example. Because it is a learning-free approach and lacks tuned parameters, it achieves the result similar to the state-of-the-art image classification techniques [8].

1.2 Significance of Feature Extraction

There are drawbacks to Naïve Bayes process. The most important of these during testing is the high computational cost. The calculation time depends linearly on the number of functions extracted from the query image and linearly to logarithmically on the number of labeled features stored. A feature is a function of one or more measurements, the values of an object's quantifiable property, determined to quantify certain important properties of an object [9]. Some of the greatest benefits of feature extraction is that it significantly reduces the information (with respect to the original image) to understand the image's content. For good feature extraction, the feature extraction should be invariance so that feature extraction is faithful and applicable for any position, location, size, and scale variations. Chi-square statistic measures the lack of independence between the features and the land cover classes and can be compared to the chi-square distribution of certain degree of freedom to evaluate the extremeness [10]. In this paper, a modern approach based on Naïve Bayes algorithm and feature extraction was implemented to classify the polarimetric SAR image for land cover mapping. The obtained results were then compared with simple Naïve Bayes, k-nearest neighbor, and K-means algorithm in terms of accuracy, computation speed, and visual analysis. The study area and other details are provided in the following sections.

1.3 Geographic Extent of the Study Area

The research focuses on Mumbai region with its center located between 19.0760 °N and 72.8777 °E in Maharashtra, India, as shown in Fig. 1. The region has forests, fallow, mangrove forest, salt pans, urban and water bodies as major terrain features. In this research, we have covered all these classes in classification of the PolSAR image (Risat-1). The accuracy assessment is carried out of the classified areas. Table 1 contains the characteristics of the sensor and Fig. 1 shows the geographical extent of the research area.

Fig. 1 Geographic location of the study area

Table 1 Risat-1 sensor specifications

Parameters	Description
Date of acquisition	12th November, 2012
Incidence angle	35.986
Polarization	RH, RV (dual pol)
Pixel spacing	1.801
Line spacing	2.469
Sensor altitude	541.366

1.4 *Image Characteristics*

2 Compact Polarimetric Data Decomposition

Risat-1 data is the compact polarimetric (CP) data with two channels RH and RV. The main manifestation of CP data is the circular (R) transmit, and linear (H, V) received. The degree of polarization (m), circularity (χ), and orientation angle (ψ) Eqs. (1) to (6) are the three necessary and sufficient primary components to describe the partially polarized electromagnetic field. m-chi decomposition proposed by [11] is an excellent analysis tool used for hybrid polarized data; hence, it was used to generate image. It includes the calculation of Stokes parameters Eqs. (5) and (6), which were further utilized to generate R, G, and B channels Eqs. (7) to (9) of the image, respectively. Finally, the performance assessment of each adopted algorithm was done over various ground targets.

2.1 *Stokes Vector*

$$S_0 = \langle |R_H|^2 + |R_V|^2 \rangle \tag{1}$$

$$S_1 = \langle |R_H|^2 - |R_V|^2 \rangle \tag{2}$$

$$S_2 = 2\mathrm{Re}\langle R_H R_V^* \rangle \tag{3}$$

$$S_3 = -2\mathrm{Im}\langle R_H R_V^* \rangle \tag{4}$$

where $|R_H|$ and $|R_V|$ correspond to receive electric field intensities in both horizontal as well as vertical polarizations, respectively. '<>' represents the averages in 'time' or 'spatial' domain. S_0 indicates total intensity and S_3 is the signal intensity in circular polarization.

2.2 *Stokes Parameters*

- Degree of polarization

$$m = \frac{\sqrt{S_1^2 + S_2^2 + S_3^2}}{S_0} \tag{5}$$

- Degree of circularity

$$\sin 2\chi = \frac{-S_3}{m S_0} \tag{6}$$

2.3 M – Chi (χ) Decomposition

$$R = \sqrt{S_0 \times m \times \frac{1 - \sin 2\chi}{2}} \tag{7}$$

$$G = \sqrt{S_0 \times (1 - m)} \tag{8}$$

$$B = \sqrt{S_0 \times m \times \frac{1 + \sin 2\chi}{2}} \tag{9}$$

3 Machine Learning Models

Machine learning models, namely Naïve Bayes, K-NN, and K-means, are used to classify the image. Bayesian algorithms are famous in handling uncertainty in the data. Its structure is simple and is one of the most efficient classifiers used in remote sensing. And its predictive performance is likewise competitive with other classifiers established in recent years. K-nearest neighbors (K-NN) may be used both in classification and in regression. Because of its simple definition, comparatively less processing time and high predictive capacity, it is commonly used in classification problems. Similar to the Naïve Bayes algorithm, it is also part of the supervised machine learning family. K-means is the most common and popularly used classification algorithm in remote sensing. It is an unsupervised machine learning algorithm and is popular for classification due to its simplicity. It clusters the data into groups based on distance measure and thus is used for image segmentation and classification. In addition to these three algorithms, we have proposed a new approach to classify PolSAR image based on the Naïve Bayes algorithm. In this model, a preprocessing step of feature extraction using chi-square distribution was applied before the classification step. After this, the extracted features are given input to the Naive Bayes algorithm and classification task was performed.

3.1 Naive Bayes [4]

A Bayesian network with joint probability distribution $p(A_1, A_2, \ldots, A_n, C)$ with attributes $A = A_1, A_2, \ldots, A_n$ and $C = C_1, C_2, \ldots, C_s$ class variables (ground truth/labels to each land feature) can be constructed for classification problems. And the constructed model may be used to classify a given set of attributes $a = a_1, a_2, \ldots, a_n$ that corresponds to particular class $c \in C$ with maximum posterior probability 10

$$\max_{c \in C} p(c|a) \quad (10)$$

Naïve Bayes with the parent classification node C of all other nodes and joint probability distribution is defined as 11

$$p(a_1, a_2, \ldots, a_n, c) = p(c) \prod_{i=1}^{n} (a_i|c) \quad (11)$$

3.2 Chi-Square Function for Feature Extraction

The reason chi-square is used as a feature extractor is because SAR data follows Wishart distribution, and Wishart distribution leads to chi-square distribution with one degree of freedom [12].

We have used the scikit-learn library's chi-square function to select the best features between each class and non-negative feature based on chi-squared stats. This score is then used to select n features with the highest values for (χ^2) chi-squared test statistics, which must contain only non-negative characteristics such as booleans or group frequencies. The complexity of this algorithm is $O(n_{\text{classes}} \times n_{\text{features}})$.

4 PolSAR Image Classification

A PolSAR image consists of multidimensional vectors allocated to a particular pixel location (spatial region) representing the scattering response of each land feature with respect to a particular wavelength and polarization. Mathematically, these vectors can be represented as a matrix $F = f_{11}(x_1, y_1) \ldots f_{nm}(x_n, y_m)$ where, $f_{ij}(x_i, y_j) \in < \Re^i$, $i = 1, 2, 3, \ldots, m$; $j = 1, 2, 3, \ldots, n$ is the scattering information associated to each pixel location (x_i, y_j). For instance, the input image consists of three channels R, G, B corresponding to double bounce, volume, and odd/single bounce scattering, respectively. Double bounce is the effect of interaction of electromagnetic wave with two different surfaces like ground-trunk or road–building wall, etc. and thus indicates the land features like urban, etc. Volume is the result of multiple interactions by many scatterers like canopy, branches, stem, leaves, etc. and thus is mainly the indicator of forest or vegetation canopy. Surface scattering is the single bounce scattering on the ground/plain surface like water, fallow, etc. The vector components f_{ijk} (x_i, y_j), $k = 1, 2, \ldots, l$ reflect the response of the sensor in different polarization state and f_{ijk} (x_i, y_j) are the element in kth band (here $k = 3$).

4.1 Train-Test Split

The image comprises 46,392 pixels, out of which 75% of the pixels (34,794) were taken for training the algorithm, and 25% of the pixels (11,598) were kept for the testing phase. Both the samples were carefully selected to be non-overlapping to each other. Naïve Bayes algorithm is a supervised algorithm, and thus, ground truth data was required for each land cover class chosen. These ground truth samples are selected based on image interpretation knowledge and with reference to Google earth imagery as Google earth provides detailed information about the land features which is sufficient for land cover mapping.

5 Accuracy Assessment

Accuracy assessment always requires the comparison of the results with an external source or ground truth at the time of image acquisition. We collected ground truth from Google earth image and then based on train-test split accuracy assessment of each land cover class as well as the overall accuracy of each algorithm except K-means has been calculated and compared with each other. Since K-means is an unsupervised algorithm, we cannot have the quantitative analysis but we can compare the quality of results by visual inspection of the obtained results. Thus, we also performed the visual analysis of each obtained image in different areas having different land cover classes. Table 2 shows the cross-validation accuracy (CV = 10) of each algorithm. Tables 3, 4 and 5 contain quantitative information of each class in terms of the confusion matrix. The visual analysis of the resultant classified images is shown in Figs. 3 and 4.

Table 2 Accuracy obtained through the method of cross-validation

Overall accuracy (%)		
K-nearest neighbor	Naive Bayes	Proposed method
89.55	90.311	93.756

Table 3 Confusion matrix of K-NN

Land cover	Mangroves	Forest	Water	Urban	Fallow
Mangroves	1052	0	0	2	11
Forest	0	1919	49	11	396
Water	0	24	4370	0	0
Urban	17	10	0	2160	1
Fallow	15	275	0	0	1286

Table 4 Confusion matrix of Naive Bayes

Land cover	Mangroves	Forest	Water	Urban	Fallow
Mangroves	1026	28	0	2	9
Forest	6	1761	27	4	577
Water	0	64	4330	0	0
Urban	33	35	0	2119	1
Fallow	21	306	0	0	1249

Table 5 Confusion matrix of proposed method

Land cover	Mangroves	Forest	Water	Urban	Fallow
Mangroves	4076	96	0	5	27
Forest	24	7107	98	20	2384
Water	0	270	17,393	0	0
Urban	153	148	0	8288	1
Fallow	69	1163	0	0	5070

5.1 Cross-Validation

The cross-validation approach is used to determine how the effects of a model of statistical analysis can generalize to an input dataset. The aim of cross-validation is to identify a dataset to check the model in the training process to limit issues such as underfitting, overfitting, and to gain information on how the model generalizes on an independent dataset. It is primarily used in models where the goal is prediction, and one would like to guess how accurate a predictive model will be practically. It helps to assess the model's consistency, avoid overfitting and underfitting, and pick the model that will perform well on the unseen data. The confusion matrix is used to compare individual accuracy of each land cover class, while the overall accuracy is assessed using cross-validation score.

5.2 Visual Analysis

Urban and water classes are correctly classified in all the three classifier algorithms. We can see that individual accuracy of each class is higher in K-NN (Fig. 2). However, when visually analyzed as in Figs. 3 and 4, it clearly indicates that there exist many mix pixel classification due to their similar intensity levels and appearance in the image, e.g., fallow is mixed with forest class in many areas in both K-NN and Naïve Bayes but is much clearly identified in our proposed method. Similarly, in both K-NN and Naïve Bayes, pixels of mangrove forest class got mixed up into forest class. This is clearly visible in Fig. 4b, d. While in Fig. 4c, e, i.e., in K-means and proposed method images respectively, these pixels are very nicely aggregated. Both our proposed method and K-means gave a better classification of mangrove forest in all the spatial locations of mangroves in the reference image. Many researchers have shown potentiality of K-means for land cover classification in the field of remote sensing [13–15], and thus, it is applied on our dataset to create a reference classified image for comparison and hence not considered in the comparison. It is, of course, superior among all other algorithms used in the paper.

The main focus of the paper was to compare the results obtained through probabilistic as well as statistical classifiers, and thus, K-NN and Naïve Bayes were selected for comparison. We can see that K-NN results better than Naïve Bayes in both visual and quantitative analysis but in various areas as in mangroves and some areas in the forest due to height, the pixels are not correctly classified. But those areas are correctly classified in our proposed method as feature extraction of all important features was performed before classification. Both visual analysis and cross-validation results confirm the superiority of our proposed model.

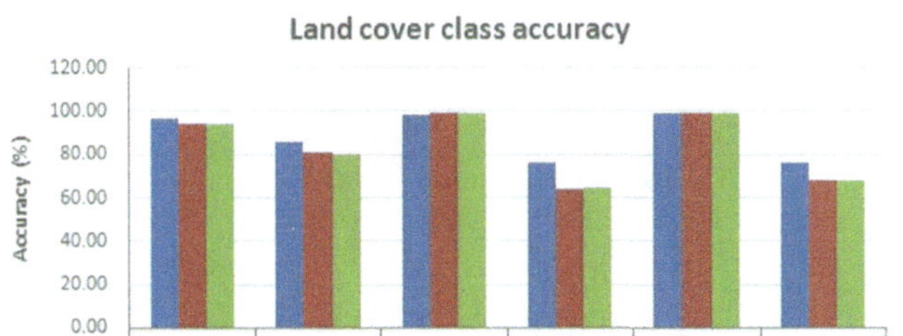

	Mangroves	Forest	Water	Saltpans	Urban	Fallow
K-NN	97.05	86.27	98.89	75.74	99.40	75.91
PNB	94.48	80.85	99.38	63.41	99.72	68.03
NB	94.31	80.55	99.44	64.81	99.70	67.76

Fig. 2 Individual accuracy of each land cover class

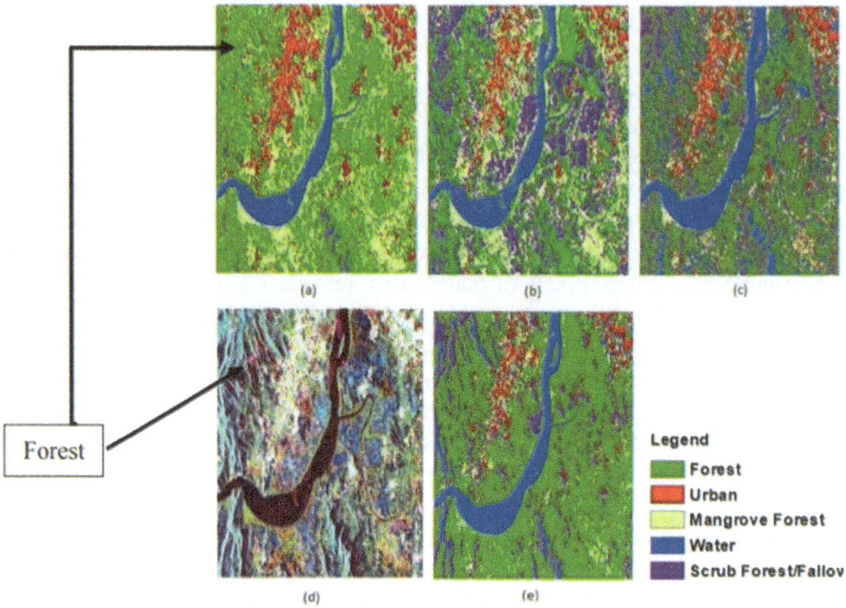

Fig. 3 **a** K-means, **b** proposed method, **c** K-NN, **d** reference image, and **e** Naive Bayes

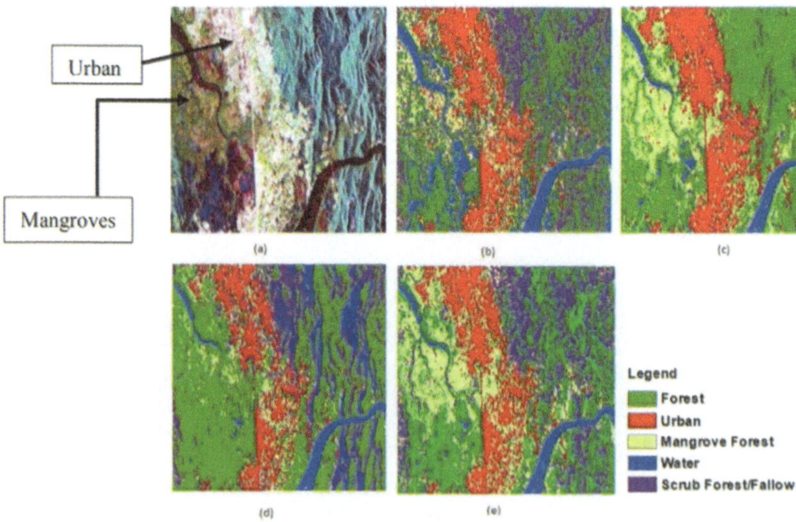

Fig. 4 **a** Reference image, **b** K-NN, **c** K-means, **d** Naive Bayes, and **e** proposed method

6 Conclusion

Quantitative analysis of individual land cover classes through confusion matrix suggests K-NN to be better algorithm for SAR dataset. However, the cross-validation technique suggests that both Naïve Bayes (90.311%) and proposed method (93.756%) performed better than K-NN (89.55%). Thus, to cross check we performed visual analysis of the classified result. This shows that many areas such as mangrove forest got merged with nearby forest class and various scrub/open forest areas were merged with fallow, and so on in K-NN classified image as well as in Naïve Bayes classified images. This was due to same intensities and their nearby association. Such areas have also been correctly identified in the result of K-means and, by comparing both the results, we can say that our proposed method has produced almost good results and is close to that of K-means than K-NN. Furthermore, cross-validation score suggests well generalization of our proposed method on unseen SAR dataset with better prediction than the other two classifiers, and also visual analysis supports this. Hence, we can conclude the supremacy of our proposed method over K-NN and Naïve Bayes result. Thus, Naïve Bayes classifier when used with chi-square may enhance the classification. Hence, if proper extraction technique is used with the traditional methods, this may lead to better results and may be used as a basis for various applications in earth science.

Acknowledgements The authors hold no conflict of interest in publishing this research work. It is part of sponsored project by Space Applications Center, Ahmedabad, at Pandit Deendayal Petroleum University (PDPU), Gandhinagar, under the project 'ORSP/R&D/ISRO/2016/SPXX.'

Authors are thankful to Ms. Hemani Parikh, Ph.D. scholar of Computer Science and Engineering Department, SOT, PDPU, for her guidance to understand machine learning frameworks.

References

1. Memon, N., Patel, S.B., Patel, D.P.: Comparative analysis of artificial neural network and xgboost algorithm for polsar image classification. In: International Conference on Pattern Recognition and Machine Intelligence, pp. 452–460. Springer (2019)
2. Gupta, S., Arora, A., Panchal, V.K., Goel, S.: Extended biogeography based optimization for natural terrain feature classification from satellite remote sensing images. In: International Conference on Contemporary Computing, pp. 262–269. Springer (2011)
3. Goel, L., Gupta, D., Panchal, V.: Hybrid bio-inspired techniques for land cover feature extraction: A remote sensing perspective. Appl. Soft Comput. **12**(2), 832–849 (2012)
4. Solares, C., Sanz, A.M.: Bayesian network classifiers. an application to remote sensing image classification. WSEAS Trans. Syst. **4**(4), 343–348 (2005)
5. Praveena, S.: Clustering and classification of images using abcfcm and naive bayes classifier. IJSR **5**(6), 199 (2016)
6. Farid, D.M., Zhang, L., Rahman, C.M., Hossain, M.A., Strachan, R.: Hybrid decision tree and naïve bayes classifiers for multi-class classification tasks. Exp. Syst. Appl. **41**(4), 1937–1946 (2014)
7. Park, D.C.: Image classification using naïve bayes classifier. Int. J. Comput. Sci. Electron. Eng. (IJCSEE)**4**(3), 135–139 (2016)

8. Timofte, R., Tuytelaars, T., Van Gool, L.: Naive bayes image classification: beyond nearest neighbors. In: Asian Conference on Computer Vision, pp. 689–703. Springer (2012)
9. Yadav, A., Yadav, P.: Digital Image Processing. Laxmi Publication Ltd., New Delhi (2011)
10. Moh'd A Mesleh, A.: Chi square feature extraction based svms arabic language text categorization system. J. Comput. Sci. **3**(6), 430–435 (2007)
11. Raney, R.K.: Hybrid-polarity sar architecture. IEEE Trans. Geosci. Remote Sens. **45**(11), 3397–3404 (2007)
12. Bilodeau, M., Brenner, D.: Wishart distributions. In: Theory of Multivariate Statistics, Chap. 7, pp. 85–97. Springer, New York, NY (1999)https://doi.org/10.1007/978-0-387-22616-37
13. Li, B., Zhao, H., Lv, Z.: Parallel isodata clustering of remote sensing images based on mapreduce. In: 2010 International Conference on Cyber-Enabled Distributed Computing and Knowledge Discovery, pp. 380–383. IEEE (2010)
14. Yang, M., Mei, H., Huang, D.: An effective detection of satellite image via K-means clustering on hadoop system. Int. J. Innov. Comput. Inf. Control **13**(3), 1037–1046 (2017)
15. Venkateswaran, K., Kasthuri, N., Balakrishnan, K., Prakash, K.: Performance analysis of K-means clustering for remotely sensed images. Int. J. Comput. Appl. **84**(12) (2013)

A Simple Optimization Algorithm for IoT Environment

Ishita Chakraborty and Prodipto Das

Abstract The optimization techniques in the network determine its performance in various environments. It plays a vital role in the current situation, where a huge amount of data is gathered from the Internet through various network-enabled devices called the Internet of Things (IoT). IoT is a network of devices that are able to share information between themselves and the Internet. The gathered data should be managed such that information loss should be minimal. Here, we have discussed the necessity of optimization and implemented the Genetic Algorithm (GA) to overcome the problems of routing, network traffic and energy management for better utilization of the available network resources. In this paper, we compare the results of the proposed algorithm with various network parameters.

Keywords Optimization · IoT · GA · Routing · WSN · Fitness function

1 Introduction

GA is based on Darwin's principle; 'the survival of the fittest' is used for various optimization problems. It can be applied to find the optimized path for routing of data packets [1]. Sensor nodes in wireless sensor networks (WSN) are used for making applications smart like intelligent defence network [2], smart environments to check the pollution, smart health care to sort out the medical issues in a remote area, etc. The batteries used in the sensors can supply energy for a short period of time and need to be recharged soon. Thus, optimum utilization of this resource is very much necessary [3]. Sachith Abeysundara et al. in [4] used GA to find the optimized path for road maps. A smaller WSN consists of a few numbers of sensors, therefore physically placing the sensor nodes in a small geographical area manually by examining their

I. Chakraborty (✉)
Department of Computer Science and Engineering, Royal Global University, Guwahati, India
e-mail: ishitachakraborty18@gmail.com

P. Das
Department of Computer Science, Assam University, Silchar, India

M. Sahni et al. (eds.), *Mathematical Modeling, Computational Intelligence Techniques and Renewable Energy*, Advances in Intelligent Systems and Computing 1287,
https://doi.org/10.1007/978-981-15-9953-8_10

Table 1 Performance of GA

Algorithm	Topology	T(Kbps)	D(ms)	E(J)	PDR(%)
GA	Mesh topology with fusion	5.1973	0.0328	91.8898	90.526
	Mesh topology without fusion	3.581	0.0407	98.4469	82.4442
GA	Hierarchical topology with fusion	2.6469	0.05891	90.6656	85.454
	Hierarchical topology without fusion	0.9701	0.0253	96.2232	73.5242

coverage area can be effective. But as IoT applications deal with a large number of sensors, the coverage problem of the sensor nodes should be sorted out for seamless connections [5]. To find a solution to a problem space, GA uses the population of chromosomes. With the signs of progress in the population due to the fitness function, chromosomes are allowed to reproduce among themselves and be nearer towards the solution.

A fitness function is used to regulate the fitness of each and every chromosomes. The highly fit chromosomes will be involved in the crossover process to get the offspring with the best results. The new family formed will be used to grow the population further if it produces improved solutions than the other weak population members. The crossover and mutation are used as reproduction techniques in GA. This practice is performed repeated continuously until and unless an optimal solution is achieved. GA is significant due to its parallel working capability with a different set of populations. Due to these characteristics, the entire problem space is searched in all directions. GA is suitable for situations, where the problem space is vast, and the time taken to explore is extensive [6, 7].

IoT has given rise to the advancement of information technology, with the new IoT enabling sensing devices to be connected to anything from anywhere, at anytime. [8, 9]. IoT-based networks will be successful and can be utilized in the real-world if the issues of security, connectivity, privacy, etc. are resolved.

Every single sensor comprises processing and communication fundamentals for monitoring the environment regarding different events for which it is specified. Information on the surroundings is gathered by the sensor node (SN) for delivering the data to the source node termed as the base station (BS). From the BS, the user can extract or collect the required information. In WSN, because of its low battery power, efficient routing happens to be a challenging research issue.

1.1 Node Positioning in WSN

From a networking perspective, a path can be defined as a connection that connects two or more nodes for communication to take place. The method of forwarding the data packets from a particular source to a specific destination is called routing. To reach the correct destination, data packets may traverse through several paths comprising of several intermediate nodes (IM). There is a need to choose the best possible path with the minimum number of SN to forward the data packets optimally. The selection of the path to reach the destination should comprise of those nodes that can accomplish the highest performance of the network [10].

There may be any number of paths that can exist between the source and the destination node. The optimized route concerning the connected cost is named as the minimum distance path (MDP). MDP uses a suitable algorithm to acquire the minimum distance path for a given network. MDP problems are used in various applications of communication, transportation and others. MDP problem can be solved with the Dijkstra's and Bellman-Ford algorithm, respectively. Due to some flaws like limitations in considering non-negative weighted edges, consideration of a single possible path and higher computational complexity has given rise to artificial neural networks and algorithms based on fuzzy logic. By using static sensors that have restricted movement, the coverage area sensors can be increased [11].

2 Methodology

The fitness function in GA is directed to acquire the most optimized path for routing of the packets in WSN. Therefore, the fitness function is formed by combining the range of the sensor node, minimum distance to reach its entire neighbour and the minimum angle formed by its neighbouring node with base station [12].

The fitness value is evaluated by finding the minimum distance between two vectors $\vec{i}$ and $\vec{j}$ using Eq. 1, and the minimum angle between two vectors $\vec{i}$ and $\vec{j}$ using Eq. 2.

$$d\left(\vec{i}, \vec{j}\right) = \left\|\vec{i} - \vec{j}\right\| = \sqrt{(i_1 - j_1)^2 + (i_2 - j_2)^2 + \ldots + (i_n - j_n)^2} \tag{1}$$

where n is the number of nodes in WSN.

$$\cos\theta = \frac{\left(\vec{i}.\vec{j}\right)}{\left(\left\|\vec{i}\right\|.\left\|\vec{j}\right\|\right)} \tag{2}$$

where

$$\left\|\vec{i}\right\| = \sqrt{i_1^2 + i_2^2}$$

$\left\|\vec{i}\right\|$ means the length of the vector $\vec{i}$

$\vec{i}.\vec{j}$ is the dot product of the two vectors

2.1 Path Optimization

To perform GA, we have considered two topologies. The first topology contains 50 nodes maintaining a mesh-like structure with scattered nodes around a centralized base station (BS). And the second topology consists of 21 nodes with a hierarchical structure with the BS located on the top. Both the topologies are shown in Figs. 1 and 2. To find the most optimized route for the above scenarios, the two nodes that are positioned closer to one other is considered as the parent node. The chromosome of the parent nodes crossover with each other to produce the population of the next generation. Among the families created, the best one is selected as the child node based on the fitness function. The fitness function is evaluated such that the amount of energy consumption is minimized, and the coverage area is maximized. Hence, for the next generation, the newly produced child node acts as one of the parents and selects another parent from its neighbourhood with minimum distance from it. This procedure is repeated continuously until the next destination is reached, and the entire space is searched. Figures 3 and 4 show the optimized path generated by genetic algorithm. Initially, the child node which has got the minimum distance from the BS is nominated.

Once the optimized path is being found out, data fusion is performed to evaluate the performance of the network. After optimization, it is observed that in the mesh topology out of 50 nodes in total, only 12 selective nodes need to be in the active state to transfer data to the sink. Only the active nodes will be consuming power. Similarly, in hierarchical topology out of 21 nodes, only 15 nodes are found active.

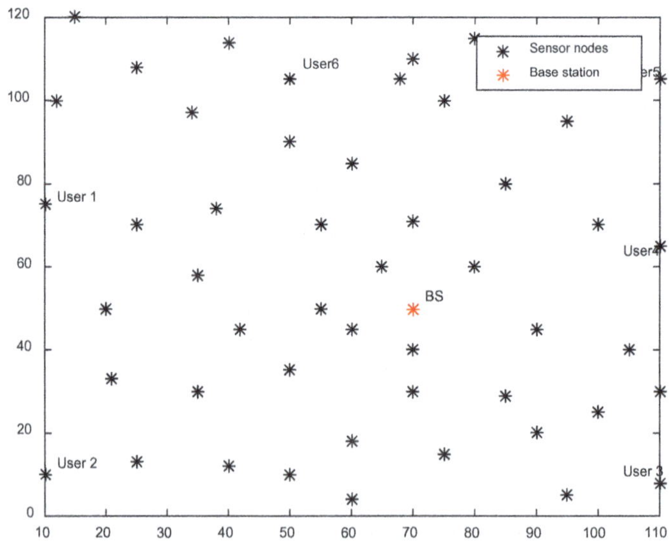

Fig. 1 Mesh topology

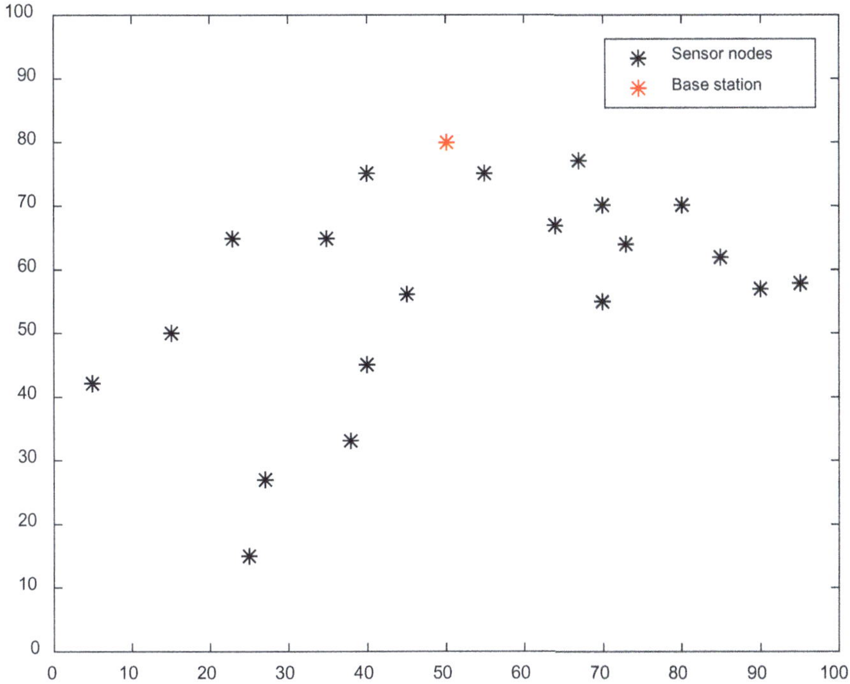

Fig. 2 Hierarchical topology

The simulation of the optimized path considering the active nodes for both the mesh and the tree topology is shown is Figs. 5 and 6 respectively.

3 Results and Discussion

This section particularizes the procedure for performing data fusion using network simulator version 2.35. For performance evaluation, we notice that in the proposed algorithm, the mesh and the hierarchical topology choose 12 and 15 nodes, respectively, for performing data fusion. The rest of the nodes will be in inactive mode. The active sensors are selected to increase the overall routine of the network with regards to energy, throughput and delay. The particulars of our network topologies are inherited from the author in [12]. MAC 802.15.4 is used as a wireless Ethernet in the data link layer. For the transport layer features, user datagram protocol (UDP) is used, and constant bit rate (CBR) traffic is used for the application layer. During data fusion, the maximum number of packets that can be sent by the transmitting node is given by Eq. 3. The number of packets received by the fusion node is calculated with Eq. 4. The throughput, energy, delay and packet delivery ratio (PDR) of the network

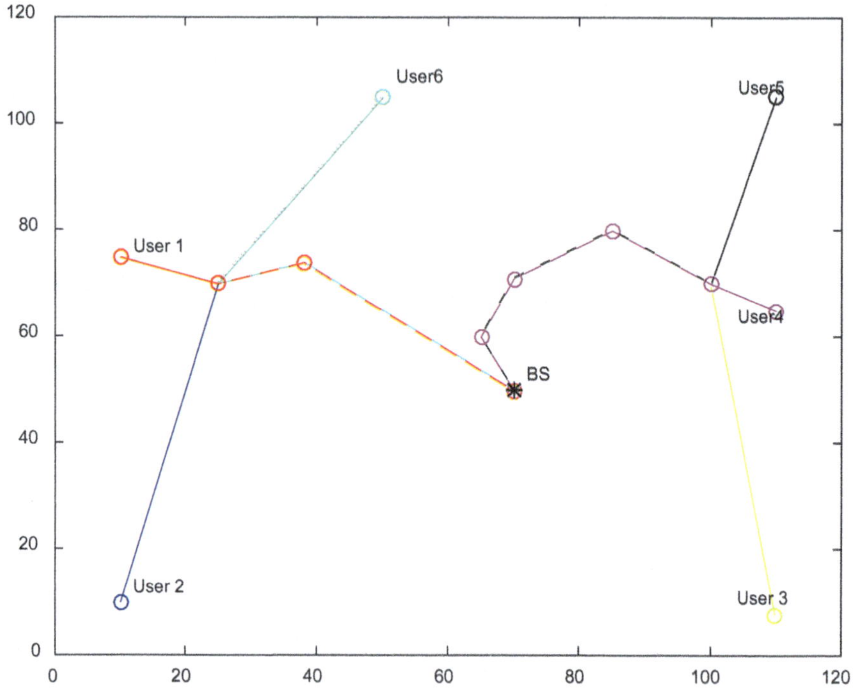

Fig. 3 Optimized path for mesh topology

is found using Eqs. 5, 6, 7 and 8, respectively.

$$\text{Max. } T_X\text{. Packets} = \text{Simulation duration} * \text{No. of Bytes/s} \tag{3}$$

$$\text{Max. Received packet by the fusion node} = \sum_{i=0}^{n-1} P_i \tag{4}$$

where P_i = Packets received from the preceding nodes

$$\text{Throughput}(T) = (\text{P}_\text{R}/1000 * 512)/1024 \tag{5}$$

where

P_R: Total number of packet received by a node

$$\text{Energy}(E) = (T_N * 10,000) - \text{EC}_T \tag{6}$$

where

T_N: Number of nodes in total

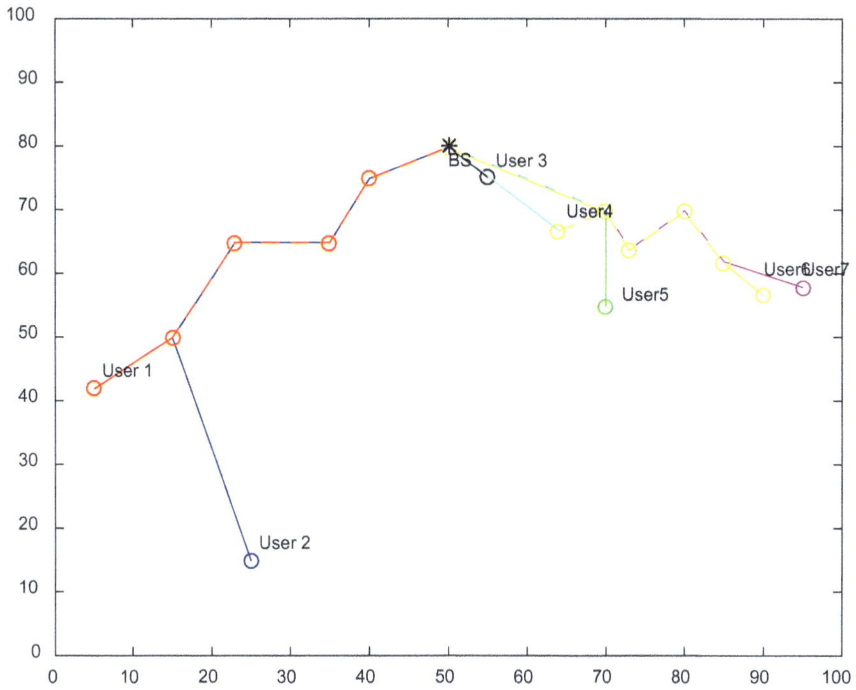

Fig. 4 Optimized path for hierarchical network

Fig. 5 Simulating the mesh topology

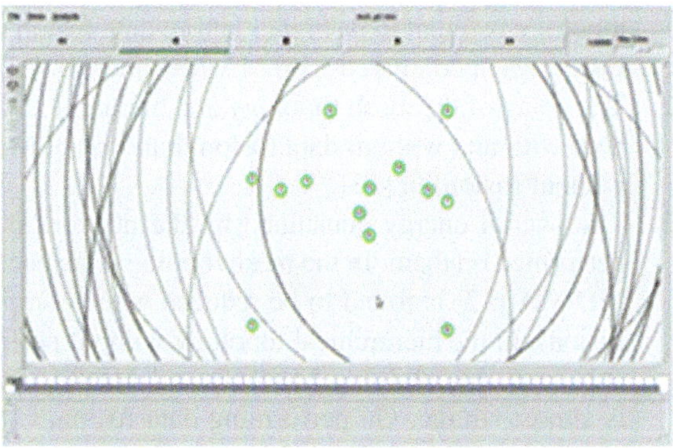

EC_T: Overall energy intake by the network

$$\text{Delay}(D) = T_{\text{PR}} - T_{\text{PS}} \tag{7}$$

$$\text{Packet Delivery Ratio(PDR)} = (\text{TR}_P/\text{TS}_P) * 100 \tag{8}$$

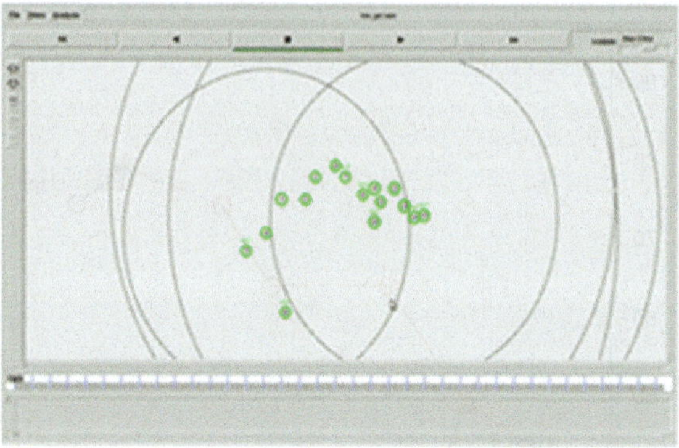

Fig. 6 Simulating hierarchical topology

where

T_{PR}: Amount of time taken by the receiver to receive the packets

T_{PS}: Amount of time taken by the sender to send the packets

where

TR_P: Total number of packets received

TS_P: Total number of packets sent.

By simulating the genetic algorithm for both the topologies, we observe the following points and compared the work done by the authors in [12].

1. The throughput obtained after running GA in both the mesh and hierarchical topology is found to be higher before data fusion and even after performing data fusion when compared with MA/MD and MD/MA as discussed in [12].
2. The delay of the mesh topology and hierarchical topology is seen higher in GA, both with and without data fusion than that of MA/MD and MD/MA (with and without fusion) in [12].
3. The overall energy consumed by the network is reduced when data fusion is performed on them. In the mesh topology, the energy consumed by MA/MD and MD/MA [12] is found to be reduced when compared to GA (with and without fusion). In the hierarchical topology without performing data fusion, the energy consumed by applying MA/MD and MD/MA algorithm in [12] and the GA is almost similar. On performing data fusion, GA consumes more energy than MD/MA and MA/MD explained in [12].
4. The packet delivery ratio (PDR) of the network using GA with fusion is higher than that of without fusion. In the first topology, when data fusion is performed on MD/MA and GA, MD/MA with fusion outperforms GA. Without data fusion PDR of MA/MD is higher compared to GA, but PDR of GA is higher than MD/MA [12]. In the second topology, PDR of GA is more than MD/MA with fusion, and PDR of GA is less without fusion when compared to MA/MD without fusion in [12].

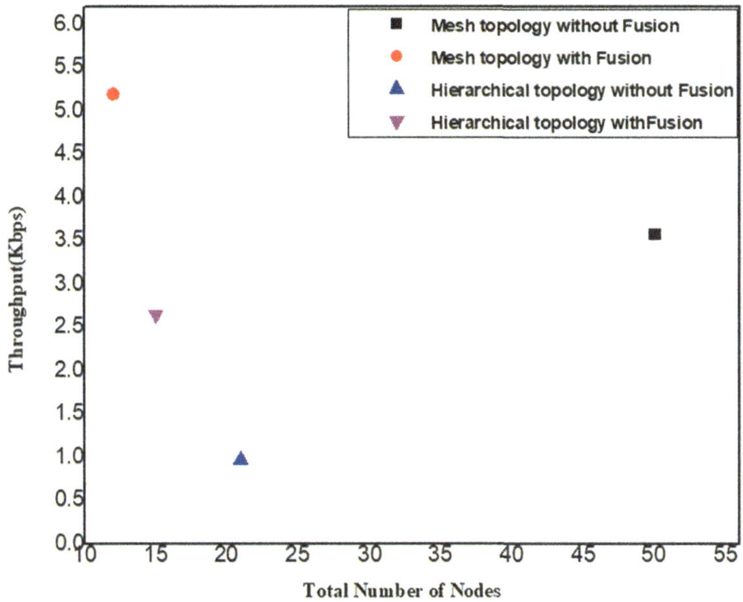

Fig. 7 Representation of throughput using GA (Kbps)

The resulting graphs (Figs. 7, 8, 9 and 10) and Table 1 indicates the performance of the system along with few of the network parameters for both the topologies.

4 Conclusion

In this paper, we have used the traditional genetic algorithm to find the optimized route for a multi-sink scenario, which can be used for the outgrowing Internet of Things technology. Since IoT expertise consists of a network of wireless sensors, the issues with its energy consumption, storage capabilities are a major concern. In order to overcome this alarming situation, a route discovery mechanism has been proposed. The path to the destination is found by using the fitness function as explained above. The proposed algorithm provides reliable data transmission. Both the mesh and the hierarchical network are analyzed with regards to some of the network parameters like throughput, energy consumption, packet delivery ratio and delay. The information collected by the sensors will be stored in an aws cloud database situated on cloud virtual machine to reduce the load on the system.

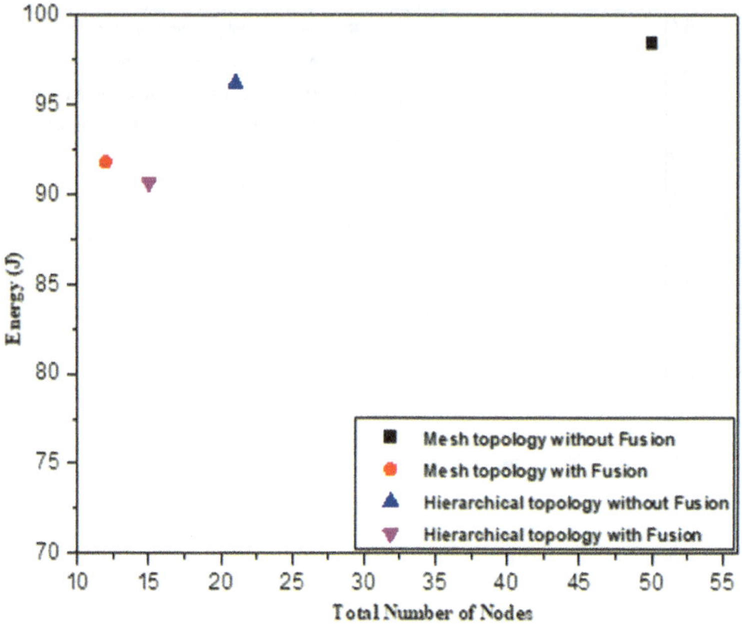

Fig. 8 Energy consumption (J) using GA

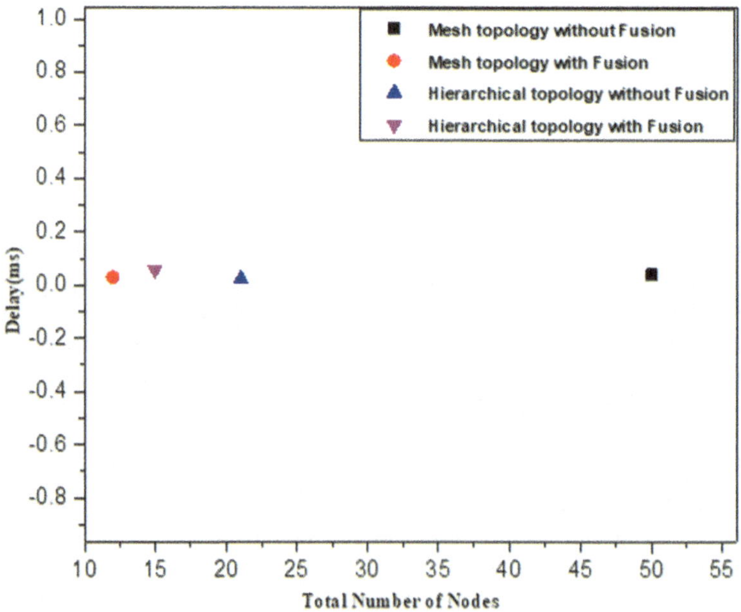

Fig. 9 Representation of delay (ms)

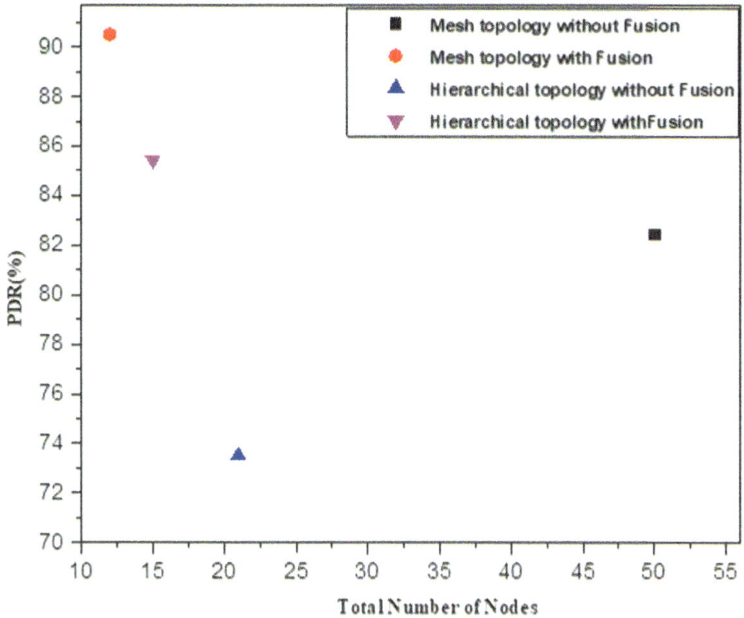

Fig. 10 Representation of PDR (%) using GA

References

1. Aggarwal.R, Mittal.A and Kaur.R.: Various Optimization techniques used in Wireless Sensor Networks.International Research Journal of Engineering and Technology 3(6), 2085–2090(2016)
2. Norouzi, A., Zaim, A.H.: Genetic Algorithm Application in Optimization of Wireless Sensor Networks. The Scientific World Journal **2014**, 1–15 (2014)
3. V. Oduguwa and R. Roy.:Bi-level optimisation using genetic algorithm.In IEEE International Conference on Artificial Intelligence Systems(ICAIS 2002), 322–327(2002)
4. Abeysundara, S., Giritharan, B., and Kodithuwakku, S.:A genetic algorithm approach to solve the shortest path problem for road maps.In International Conference on Information and Automation, 272–275(2005)
5. Y. Yoon and Y. Kim.: An Efficient Genetic Algorithm for Maximum Coverage Deployment in Wireless Sensor Networks.IEEE Transactions on Cybernetics 43(5), 1473–1483(2013)
6. E. Elbeltagi, T. Hegazy, D. Grierson.:Comparison among five evolutionary-based optimization algorithms.Advanced Engineering Informatics19(1), 43–53(2005)
7. Sooda.K and Nair, T.R.: A Comparative Analysis for Determining the Optimal Path using PSO and GA. International Journal of Computer Applications 32(4), 8–12 (2014)
8. Gubbi, J.; Buyya, R.; Marusic, S.: Palaniswami, M.:Internet of Things (IoT): A vision, architectural elements, and future directions.Future Generation Computer Systems 29(7), 1645–1660(2013)
9. Cruz-Piris, L.; Rivera, D.; Marsa-Maestre, I.; de la Hoz, E.; Velasco, R.J.: Access control mechanism for IoT environments based on modelling communication procedures as resources.Sensors 18, 917(2018)
10. F. Araujo, B. Ribeiro and L. Rodrigues.:A neural network for shortest path computation. IEEE Transactions on Neural Networks 12(5),1067–1073(2001)

11. S. S. Dhillon, K. Chakrabarty, and S. S. Iyengar.Sensor placement for grid coverage under imprecise detections.InInternational Conference on Information Fusion, 1581–1587(2002)
12. Chakraborty I., Chakraborty A., Das P. Sensor Selection and Data Fusion Approach for IoT Applications. In: Kalita J., Balas V., Borah S., Pradhan R. (eds) Recent Developments in Machine Learning and Data Analytics. Advances in Intelligent Systems and Computing, vol.740. Springer, Singapore (2019)

Robotic Grasp Synthesis Using Deep Learning Approaches: A Survey

Dhaval R. Vyas, Anilkumar Markana, and Nitin Padhiyar

Abstract Deep learning methods are derived and inspired from the structure and activities of a brain. In an intricate state, learning from the past experiences helps human to accomplish the task in efficient way. This paper addresses such deep learning approaches practiced in the robotic grasp detection for an unknown object. A grasp is defined as lifting of an object by the end effector of robotic manipulator without tumbling. To realize this task, a robot must be able to identify the grasp area of the target object autonomously by means of visual or touch feedback. During the last decade, there is an enormous improvements in hardware computational capabilities, and hence, deep learning methods play a vital role in making robots autonomous. Review of such deep learning methods, used in robotic grasping, is carried out in this work. Deep learning is a subset of machine learning with more neurons then artificial neural networks. Primarily four types of network architectures in deep learning are identified and discussed in this work. They are unsupervised pre-trained network, convolution neural network, recurrent neural network and recursive neural network architectures. Types and availability of general training datasets are discussed wherein deep learning models can be trained and benchmarked. Unavailability of extensive training datasets for the specific task application is a major hurdle in implementing deep learning methods. Performance of these deep learning methods is also debated in this work. The potential research directions are thus proposed in the paper.

Keywords Robotic grasping · Deep learning · Convolution neural network · Reinforcement learning

D. R. Vyas (✉) · A. Markana
School of Technology, Pandit Deendayal Petroleum University, Gandhinagar, India
e-mail: dhavalforever@gmail.com

A. Markana
e-mail: anil.markana@spt.pdpu.ac.in

N. Padhiyar
Indian Institute of Technology, Gandhinagar, India
e-mail: nitin@iitgn.ac.in

M. Sahni et al. (eds.), *Mathematical Modeling, Computational Intelligence Techniques and Renewable Energy*, Advances in Intelligent Systems and Computing 1287,
https://doi.org/10.1007/978-981-15-9953-8_11

1 Introduction

Human always apply engineering knowledge to extend their working abilities. Compared to human grasp abilities, Robotic grasp is still far behind, as human is trained by the years of experience and efficient knowledge transfer mechanism [1]. When robot grasp is derived through analytical approach, the grasp response is very slow, and model complexity is increased.

Recent advancements in machine learning (ML) approaches, a branch of artificial intelligence, have increased the interest of researchers, to implement in the field of robotics [2, 3]. This is mainly because of ML's capabilities to mimic how human learn from the experience or past dataset. Deep learning is a subset of ML. This approach consists of network of neurons having multiple layers [4]. This network is trained using existing dataset, and then, it is applied to some specific task. Deep learning methods are successfully applied in computer vision [5] and robotics. Many researchers have derived methods to address the robotic grasp problem using deep learning methods [6]. Deep learning utilizes several modified ML algorithms to make robot autonomous and perform task naturally as human. A large neural network performs parallel and simultaneous mathematical operations and derived heuristic relations between input and output. This heuristic relation is used in decision-making [7]. Owing to these capabilities, deep learning methods are more popular among researchers to address the robotic grasp problem based on classification and detection.

A robotic grasping is defined as successfully lifting an object without tumbling. To determine appropriate grasping pose, grasp detection system is used. Appropriate sensors ensure the proper inputs and feedback to deep learning architecture.

This paper reviews various deep learning methods used in detecting the robotic grasping pose. Section 2 describes details about representing the robotic grasping. Section 3 reviews the available dataset types for the training of deep network architecture. Section 4 reviews the various deep learning methods used for robotic grasping with pros and cons. Finally, the paper is concluded with further exploration of potential field for research.

2 Robotic Grasping

For robots, analytical approach through manual coding is carried out to perform specific tasks. It is based on the expert human knowledge, where robot performs some pre-defined task based on the programming instructions. This approach represents the kinematics relationship between robot parameters and coordinates of the interacting environment. To avoid failure in performing the task, feedback algorithm is suggested to perform this task. Such approach is not feasible when the environment is dynamic and there are fast variations. Due to this, continuous programming update is required in real time. Complex analytical approach is required to address this problem, as unpredictable environment for autonomous robots remains the open problem for researchers. Hence, self-learning is the probable solution to address this problem. Nonlinear regression, Gaussian process, Gaussian mixture models and support vector machines are other approaches to this context [1]. However, the training is limited to demonstrated profile in self learning approach.

Due to the advancement in hardware computing facilities, deep learning made significant impact in the applications like computer vision, robotics, environment understanding and natural language processing [8]. In computer vision, efficient results are obtained by deep learning methods. Recent research shows that performing robotic task in unstructured environment is still a challenging task and unexplored. Generalize robotic solutions where environment is highly dynamic, efficient algorithms are required. For human, it is relatively simple to mimic, but for the robot, it requires perception, planning and control [8].

Object grasping is the challenging task owing to novelty presents in the objects in terms of shape, size and pose. Currently, robot grasping is far behind the human grasping benchmarks. A robotic grasp synthesis is classified in the following three sub-systems [8]:

- **Grasp detection sub-system**: To detect grasp, pose of an object from their images, also known as perception.
- **Grasp planning sub-system**: Mapping the coordinates of image to real-world coordinates.
- **Control sub-system**: Determination of forward/inverse kinematics solution of grasp planning sub-system.

This paper reviews various deep learning approaches for grasp detection system as it is the key access point for grasp synthesis. Most literature uses convolution neural network (CNN), also referred as deep CNN in many literature. Training of such CNN requires huge number of dataset [9]. Data can be labelled or unlabelled depends upon whether supervised or unsupervised learning mechanism is used. Work by [6, 8, 10] uses transfer learning approach to simplify the network parameters and improve the results and minimize training time.

2.1 Robotic Grasp Representation

Previously grasp representation done with a point on 3D model and images of objects depending upon simulation environment utilized. Saxena et al. [11] use regression-based model, trained via supervised learning, using synthetic images as a training datasets to predict the points as function of successful grasping candidate. Figure 1 shows grasp point representation in Cartesian coordinates. Extending their work in [12], using discretization of 3D workspace, grasp point $g = (x, y, z)$ is determined. Zhang et al. [13] simplify the representation using 2D workspace to represent the point $g = (x, y)$. Major drawback of this method is unavailable information on the gripper orientation, which is critically required in successful grasping.

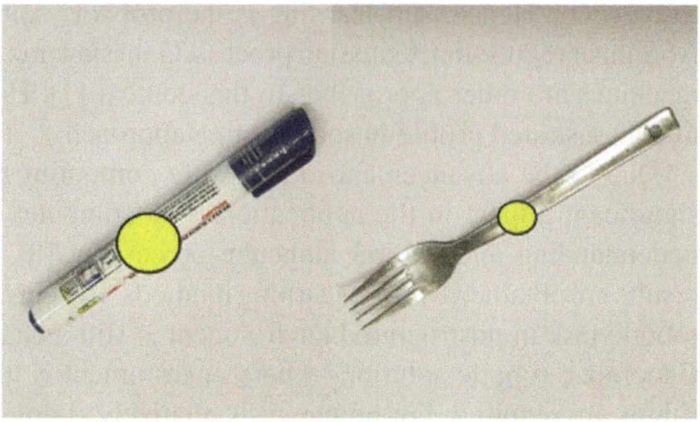

Fig. 1 Grasp point representation examples of a marker and a fork

In order to overcome these drawbacks, oriented rectangle representation has been proposed by [14]. This method has seven-dimensional representation including information of position, orientation and gripper opening width. Figure 2 of a stapler depicts such representation, where $\left(r_g, c_g\right)$ refers to upper left corner of rectangle, m_g and n_g are the dimension of rectangle and θ is angle between first age of rectangle and x-axis. Lenz et al. [6] proposed simplified five-dimensional representation. Later, Redmon et al. [10] have validated this representation with experimental results. Grasp parameter is described by (x, y, w, h, θ), where (x, y) is grasp centre, w and h are width and height of rectangle and θ is the angle with respect to x-axis. These methods are analogous to object detection methods in computer vision but with added gripper orientation parameter. Kumra et al. [8] used same representation proposed by Redmon et al. [10] for robotic grasping. This five-dimensional grasp representation is also used by Zhou et al. [2], Park et al. [15] and Trottier et al. [16]. An approach with four element grasp, represented as $g = (x, y, \theta, w)$, is proposed by Wang et al. [17], where gripper plate height h is excluded. Pinto et al. [18] excluded the gripper plate dimension (h, w) and utilized simple $g = (x, y, \theta)$ representation. Similar representation is also used in [19, 20]. z coordinate is included by Calandra et al. [21], and the grasp is represented by $g_z = (x, y, z, \theta)$. Such g_z representation is also referred by Murali et al. [3].

Literature survey suggests that selection of grasp representation method is purely based on the nature of application. In general, rectangle representation depicts all the information, i.e. grasping point, grasping orientation and gripper opening width for

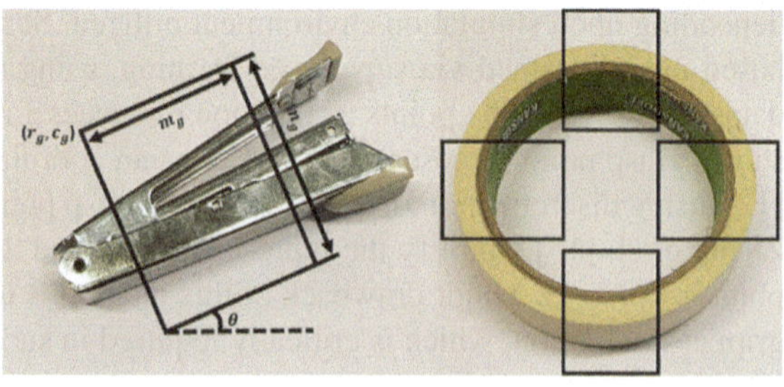

Fig. 2 Grasping rectangle representation examples of a stapler and a tape roll

Table 1 Grasp representation summary

Grasp type	Grasp parameters	Literature
Point representation	(x, y)	Zhang et al. [13]
	(x, y, z)	Saxena et al. [11]
Rectangle representation	$(r_g, c_g, m_g, n_g, \theta)$	Jiang et al. [14]
	(x, y, w, h, θ)	Lenz et al. [6], Redmon et al. [10], Kumra et al. [8], Zhou et al. [2], Park et al. [15], Trottier et al. [16]
	(x, y, w, θ)	Wang et al. [17]
	(x, y, θ)	Pinto et al. [18], Mahler et al. [19], Viereck et al. [20]
	(x, y, z, θ)	Calandra et al. [21], Murali et al. [3]

a successful accomplishment of a grasping task. Five-dimensional grasp originally proposed by Lenz et al. [6] is further required to be explored. Grasp representations discussed above are summarized in Table 1 for ready reference.

2.2 *Robotic Grasp Detection*

Conventional analytical methods required model parameters to model the grasp detection based on geometry of objects, force analysis and CAD model of objects [22]. Modelling of such parameters is a complex and time-consuming task. Methods discussed in [23] are developed using existing knowledge or by simulating grasping in virtual simulation environment. These methods rely on assumption that object parameters are known apriori. Therefore, it is difficult to derive the generalize solution using these methods. Caldera et al. [7] give the classification of grasp detection based on the detection with separate planner and direct image-to-action using visuomotor control policy shown in Fig. 3. Here, robotic grasp detection with separate planner gives either structured output for grasp candidate, i.e. (x, y, θ, h, w) or based on robustness function of grasp candidates to be chosen.

Lenz et al. [6] and Jiang et al. [14] proposed sliding window approach which detects structured output for the grasp candidate. Two step cascade method is proposed by [6] with two deep networks where first detection is re-evaluated by second network. First network with fewer features effectively filters unwanted grasp candidate faster. Second network with more features is slower but has to run filtered detection carried out by first network. Network is trained by Cornell grasping dataset shown in Fig. 4. This method yields 75% grasp accuracy with 13.5 s per frame.

Redmon et al. [10] proposed one shot detection method which is improvement over sliding window approach. Here, a deep network can simultaneously perform classification so that in single shot it detects the good graspable area for the object.

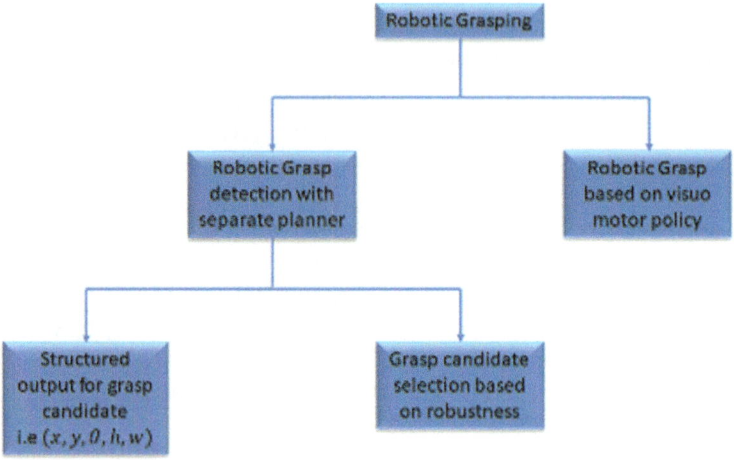

Fig. 3 Robotic grasping classification [7]

Fig. 4 Cornell grasping dataset sample [35]

This approach gives major improvement on detection speed of 76 ms compared to 13.5 s per frame with detection accuracy of 84.4%.

Similar approach is followed by Kumra et al. [8] with proposed multi-modal model. Where, one deep network extracts the features from the images, and second shallow deep network predicts the grasp configuration of the object. This method achieves 89.21% accuracy with the lower processing speed of 100 ms due to deeper network architecture. Although one shot detection approach uses regression-based model, Chu et al. [24] have defined the learning problem as classification-based. For implementation of the deep network, ResNet [25] is used. Authors reported 94.4% accuracy with 0.25 s time from image to plan in real world. This work is proposed on the concept given by Guo et al. [26] and Zhou et al. [2]. But classification-based approach has limitation that the output is bounded for pre-defined training dataset [18].

Grasp robustness function is proposed and evaluated in the literature for grasp detection. Higher score to this function represents the successful grasp. Binary classification for the successful and unsuccessful grasp point is also developed using binary variables 1 or 0, respectively. Pas et al. [27] utilize binary classification when the input is partially occluded RGBD image or point cloud. With this method, 93% success rate is achieved in dense clutter environment. Park et al. [15] proposed classification-based grasp detection approach that uses spatial transformer network (STN). This method also allows partial observation of intermediate results such as grasp location and orientation of grasp candidates. This method achieved 89.6% accuracy with 23 ms per frame. Lu et al. [28] propose the training of deep network as a function of visual information and grasp configuration. Proposed method is data efficient and fast enough to deploy in real-world application. Gradient ascent optimization is implemented inside deep network with back-propagation algorithm. These methods are useful when partial information for inputs is available [15].

Johns et al. [29] proposed robust grasp function and report 75.2% accuracy with minor gripper uncertainty and 64.8% accuracy with major gripper pose uncertainty. Mahler et al. [30] proposed grasp robustness function using 6.7 million cloud points and analytic grasp metrics Dex-Net 2.0 dataset. Grasp quality convolution neural network (GQ-CNN) is trained with this dataset to predict the successful grasp quickly. Proposed method achieves 98.1% grasp accuracy on ten novel objects which may be reduced when experimented with larger novel objects.

Learning visuomotor control policy for grasping application opens new avenue for research in this domain as this method does not require separate grasp planner. Mahler et al. [19] have proposed discrete time partially observable Markov decision process on Dex-Net 2.1. This method is used to grasp the object from the dense clutter. Using transfer learning approach, author achieved accuracy of 92.4%. But with 20 novel objects, each has five trials method achieved 70% accuracy in grasping. Zeng et al. [31] proposed multiple motion primitives for suction and grasping: (1) suction down (2) suction side (3) grasp down (4) flush grasp. Author reported maximum accuracy of 88.6% when dealing with the mixed objects, i.e. known, novel, objects in clutter both known and novel. Zhang et al. [13] proposed to use reinforcement learning [32] to train deep Q network by synthetic images. This method achieved 51% success rate as there is large variation in synthetic images and real-world objects. To summarize, most of the work is concentrated on the structured grasp representation. Suggested research direction is on the use of combination of visual and tactile data as inputs.

3 Training Dataset

Literature review reveals that use of deep learning methods requires large volume of labelled as well as unlabelled data for training [33]. Recently, researcher uses either third party dataset or methods to generate automated dataset [18, 30]. Also, unavailability of task-specific data is a huge challenge in training [34]. Although few methods have been suggested by [34], use of simulated data and CAD model availability is impractical. However, network pre-training on limited simulated data can be carried out to reduce the training time [8].

The Cornell grasping dataset (CGD) from [35] is very much popular among researcher [8, 10, 36]. CGD was created by 1035 different images with 280 objects with multiple images taken in different orientation and poses. A sample set is shown in Fig. 4. They also release the code for training which requires robot operating system (ROS) [37] and open computer vision (OpenCV) [38] to be pre-installed. This dataset was specifically designed for parallel plate gripper arm. CGD appeared in number of article which proves its diversity for creating generalized application. It is extensively used in [6, 8, 10, 15, 24, 26, 36, 39]. Wang et al. [17] used Washington RGB-D dataset [40] due to its comprehensive collection of RGB-D images.

Murali et al. [3] have performed random grasp and collected data for invalid grasp pose. Experiment has been performed on 52 different objects. Pinto et al. [18] proposed reinforcement learning-based approach to collect the data. Their approach is time consuming as they perform 50,000 grasp trials with 700 robotic hours. Levine et al. [41] further improve the result by 900,000 grasping trials using eight robots.

Tobin et al. [34] proposed simulation methods to generate domain-specific dataset or synthetic data using 3D mesh models. Dataset relies on domain adaption of the object from simulated world to real world. Bousmalis et al. [42] generated 9.4 million data instances by combining simulated data and realistic data. Grasp success rate about 78% is achieved. Similarly, Viereck et al. [20] use visuomotor policy learning methods to generate image-to-action pairs. Mahler et al. [30] generate dataset called Dex-Net 2.0 which contains 6.7 million point cloud and analytic grasp metrics.

Saxena et al. [11] have experimented with depth images of five different objects in training. Overall success rate achieved is 90% with mean absolute error of 1.8 cm. Jiang et al. [14] stated that availability of multi-modal data is useful in detecting edges of the objects which can clearly differentiate the viable grasping and non-viable grasping region. Lenz et al. [6] also supported this work and used multi-modal RGB-D data for training. Murali et al. [3] use tactile sensing to complement the visual sensor. They also use multi-modal data to train the deep network. Kumra et al. [8] train the deep network using uni-modal data and achieved accuracy of 88.84%.

4 Convolution Neural Network

A deep CNN is fully connected multiple layers of neurons to extract the information from the applied input. Goodfellow et al. [9] expressed that deep network replicates the human brain thinking process. Multiple architectures are proposed for deep CNN during the last half decade. Most of them use ImageNet [43] tests for benchmarking. All CNN structure follows the general structure as shown in Fig. 5. Literature shows that task-specific features are extracted using fully connected partial network or shallow network and deep CNN can extract low level to high level features via trade-off in processing speed. Recent advancement is carried out in parallel layer approach owing to increase in parallel processing capabilities of the hardware.

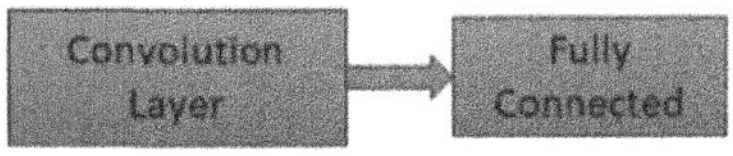

Fig. 5 General convolution network

Redmon et al. [10] implemented AlexNet initially to detect the robotic grasp as shown in Fig. 6. The direct regression model achieved accuracy of 84.4% which was trained on RG-D images and multigrasp model achieved accuracy of 88%. Later, this work is followed by Watson et al. [36], with one fold cross validation to achieve the accuracy of 78%. Recent dense CNN promises to extract more advance features from dataset. Szegedy et al. [44] expressed that there are several drawbacks when increasing depth of CNN. To train deep CNN, labelling for each images of dataset must be unique. In most cases, number of trainable parameters automatically increases when depth of CNN increases. He et al. [25] proposed ResNet architecture with skip connection. Example of sample residual block is shown in Fig. 7. Szegedy et al. [45] proposed Inception-ResNet architecture by combining [25] with [46]. Kumra et al. [8] used dense 50-layer network version of ResNet [34] and input as RGB-D images. They proposed multi-modal and uni-modal network architecture to achieve grasp accuracy of 89.21%, as shown in Fig. 8. Zhou et al. [2] used ResNet-50 and ResNet-101 network as feature extractor. Chu et al. [24] used same ResNet-50 network to propose multiple grasps at once using image wise splitting.

Sliding window approach proposed by Lenz et al. [6] has two cascade CNN for detecting higher level of features and to detect valid grasp. To initialize weights of hidden layer, sparse auto encoder [33] is used. Lu et al. [28] propose custom architecture in multi-fingered grasp and reported 75.6% grasp accuracy.

Xia et al. [47] and Polydoros et al. [48] use deep CNN to compute inverse and forward kinematics of redundant robotic manipulator. Szegedy et al. [44] have concluded that quality of feature extraction using deep methods relies on depth of the network and number of training dataset. Redmon et al. [10] replaced blue channel of RGB data with depth data and normalize the depth data to default RGB colour range [0, 255]. Watson et al. [36] normalized RGB values to [0, 1]. Pinto et al. [18] adopted

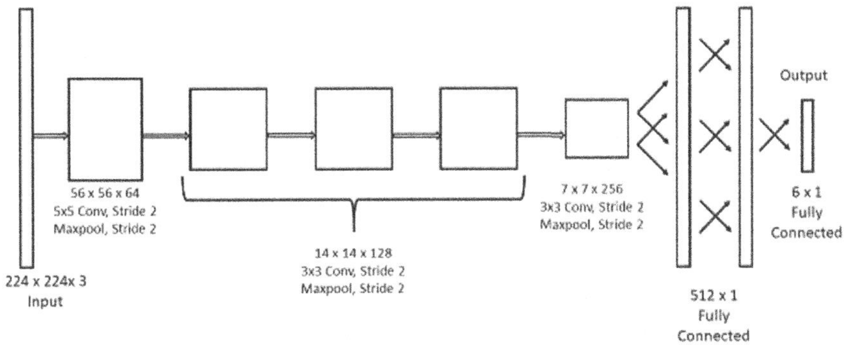

Fig. 6 Direct regression grasp model [10]

Fig. 7 Residual block example [25]

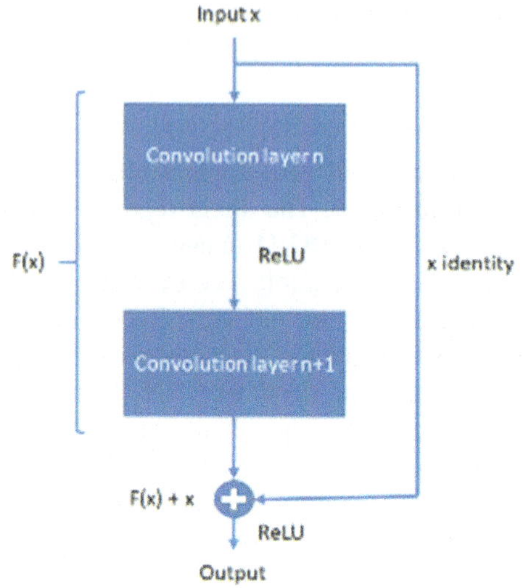

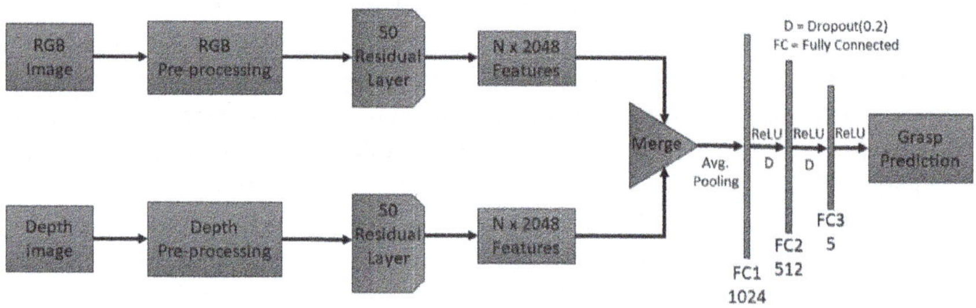

Fig. 8 Multi-modal grasp predictor [8]

same approach and resized input images to 227 × 227 using similar architecture of AlexNet [49, 50] as shown in Fig. 9.

To overcome the problem of overfitting, many researchers trained their network on larger dataset which is known as pre-training in simulated environment. Then, it is again trained on limited task-specific or domain-specific dataset. This method is referred as transfer learning approach. Most successful grasp detection work used this approach and achieved accuracy near to 90% [18]. Redmon et al. [10] used AlexNet [50] CNN and reported 84.4% grasp accuracy. Similar approach is adopted by [36]. A similar transfer learning approach is adopted by Kumra et al. [8] with deeper ResNet-50 architecture. With uni-modal grasp predictor as shown in Fig. 10, grasp accuracy of 88.84% is achieved.

Research required to find optimal number of units for fully connected layer of CNN. Redmon et al. [10] used two fully connected layers that have 512 units each. Kumra et al. [8] used one layer of fully connected which has 512 units for uni-modal architecture and two fully connected with 512 layers each for multi-modal

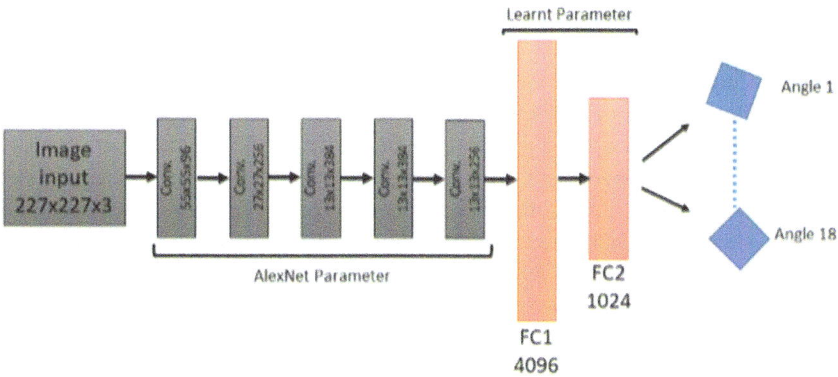

Fig. 9 Eighteen way binary classifier [18]

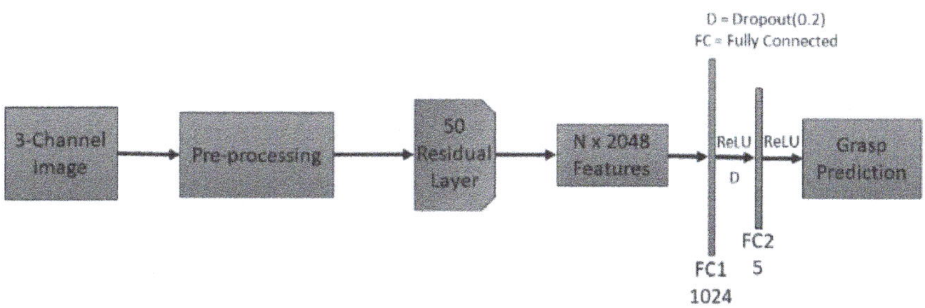

Fig. 10 Uni-modal grasp predictor [8]

architecture. Most of the researchers used stochastic gradient descent (SGD), but it was not optimal optimizer and more advance optimizer required for grasping detection. Comprehensive review of optimizer is given by Ruder [51].

5 Conclusions and Future Research Directions

Applications of deep learning methods in the field of robotics are still an emerging field of research as it requires more computational power and large dataset for training. Various approaches to represent the grasp, i.e. point grasp representation, square representation are discussed in detail. The sliding window approach, where input images are scanned, is not suitable for real world application owing to its slow processing time. One shot detection method predicts structured grasp directly from an image and produces graceful results. Most of the literature used Cornell grasping dataset for training which contains RGB-D images of household objects. Deep convolution neural network adopted for grasp detection is also discussed with various architectures. Classification-based and regression-based approaches are adopted for the robotic grasp detection. Uni-modal grasp predictor with one fully connected layer and multi-modal grasp predictor with two fully connected layers are discussed.

Larger dataset is required for training to extract more features from inputs. Hence, robust simulated approach required to generate task-specific dataset. There is also a need to use and explore reinforcement learning for training as it does not require pre-labelled data. Though it consumes more time in learning, but recent advancement in computational power makes feasible to implement in robotic grasp synthesis. Use of optimal number of units in CNN layer is also a thrust area of research. Deep learning methods to learn from the visuomotor policies are not extensively explored yet and are still a potential green area of research.

References

1. Billard, A., Calinon, S., Dillmann, R.: Learning from humans. In: Springer Handbook of Robotics, 2nd edn. (2016)
2. Zhou, X., Lan, X., Zhang, H., Tian, Z., Zhang, Y., Zheng, N.: Fully Convolutional Grasp Detection Network with Oriented Anchor Box. IEEE Conf. Intell. Robot. Syst. (2018)
3. Murali, A. et al.: Learning to Grasp Without Seeing. arXiv:abs/1805.0 (2018)
4. LeCun, Y., Bengio, Y., Hinton, G.: Deep learning. Nature **521**, 436–444 (2015)
5. Ruiz-del-Solar, J., Loncomilla, P., Soto, N.: A Survey on Deep Learning Methods for Robot Vision, pp. 1–43 (2018). arXiv:abs/1803.1
6. Lenz, I., Lee, H., Saxena, A.: Deep learning for detecting robotic grasps. In: 1st International Conference on Learning Representations, ICLR 2013—Work. Track Proc. (2013)
7. Caldera, S., Rassau, A., Chai, D.: Review of deep learning methods in robotic grasp detection. Multimodal Technol. Interact. **2**, 57 (2018)
8. Kumra, S., Kanan, C.: Robotic grasp detection using deep convolutional neural networks. IEEE Int. Conf. Intell. Robot. Syst. 769–776 (2017)
9. Goodfellow, I., Bengio, Y., Courville, A.: Deep Learning. MIT press (2016)
10. Redmon, J., Angelova, A.: Real-time grasp detection using convolutional neural networks. Proceedings of IEEE International Conference on Robotics and Automation, June 2015, pp. 1316–1322 (2015)
11. Saxena, A., Driemeyer, J., Kearns, J., Ng, A.Y.: Robotic grasping of novel objects. Adv. Neural Inf. Process. Syst. 1209–1216 (2007)
12. Saxena, A., Driemeyer, J., Ng, A.Y.: Robotic grasping of novel objects using vision. Int. J. Rob. Res. **27**, 157–173 (2008). https://doi.org/10.1177/0278364907087172
13. Zhang, F., Leitner, J., Milford, M., Upcroft, B., Corke, P.: Towards vision-based deep reinforcement learning for robotic motion control. In: ACRA (2015)
14. Jiang, Y., Moseson, S., Saxena, A.: Efficient grasping from RGBD images: Learning using a new rectangle representation. In: Proceedings of IEEE International Conference on Robotics and Automation (2011)
15. Park, D., Chun, S.Y.: Classification based Grasp detection using spatial transformer network. arXiv:abs/1803.0 (2018)
16. Trottier, L., Giguere, P., Chaib-Draa, B.: Sparse dictionary learning for identifying grasp locations. In: Proceedings of 2017 IEEE Winter Conference on Applications of Computer Vision (2017)
17. Wang, Z., Li, Z., Wang, B., Liu, H.: Robot grasp detection using multimodal deep convolutional neural networks. Adv. Mech. Eng. **8**, 1–12 (2016)

18. Pinto, L., Gupta, A.: Supersizing self-supervision: Learning to grasp from 50 K tries and 700 robot hours. In: Proceedings of IEEE International Conference on Robotics and Automation, June 2016, pp. 3406–3413 (2016)
19. Mahler, J., Goldberg, K.: Learning deep policies for robot bin picking by simulating robust grasping sequences. In: Proceedings of 1st Annual Conference on Robot Learning, vol 78, pp. 515–524 (2017)
20. Viereck, U., Pas, A. ten, Saenko, K., Platt, R.: Learning a visuomotor controller for real world robotic grasping using simulated depth images, pp. 1–10 (2017). arXiv:abs/1706.0
21. Calandra, R., et al.: More than a feeling: learning to grasp and regrasp using vision and touch. IEEE Robot. Autom. Lett. **3**, 3300–3307 (2018)
22. Bicchi, A., Kumar, V.: Robotic grasping and contact: a review. In: Proceedings 2000 ICRA. Millennium Conference. IEEE International Conference on Robotics and Automation. Symposia Proceedings vol 1, pp 348–353 (2000)
23. Bohg, J., Morales, A., Asfour, T., Kragic, D.: Data-driven grasp synthesis—a survey. IEEE Trans. Robot. **30**, 289–309 (2014). https://doi.org/10.1109/TRO.2013.2289018
24. Chu, F.J., Xu, R., Vela, P.A.: Real-world multiobject, multigrasp detection. IEEE Robot. Autom. Lett. **3**, 3355–3362 (2018). https://doi.org/10.1109/LRA.2018.2852777
25. He, K., Zhang, X., Ren, S., Sun, J.: Deep residual learning for image recognition. In: Proceedings of the IEEE Computer Society Conference on Computer Vision and Pattern Recognition, December 2016
26. Guo, D., Sun, F., Liu, H., Kong, T., Fang, B., Xi, N.: A hybrid deep architecture for robotic grasp detection. In: Proceedings of IEEE International Conference on Robotics and Automation, pp. 1609–1614 (2017)
27. Ten Pas, A., Gualtieri, M., Saenko, K., Platt, R.: Grasp pose detection in point clouds. Int. J. Rob. Res. **36**, 1455–1473 (2017). https://doi.org/10.1177/0278364917735594
28. Lu, Q., Chenna, K., Sundaralingam, B., Hermans, T.: Planning multi-fingered grasps as probabilistic inference in a learned deep network, pp. 1–16 (2018). arXiv:abs/1804.0
29. Johns, E., Leutenegger, S., Davison, A.J.: Deep learning a grasp function for grasping under gripper pose uncertainty. IEEE Int. Conf. Intell. Robot. Syst. (2016)
30. Mahler, J. et al.: Dex-Net 2.0: Deep learning to plan robust grasps with synthetic point clouds and analytic grasp metrics. Robot. Sci. Syst. **13** (2017)
31. Zeng, A. et al.: Robotic pick-and-place of novel objects in clutter with multi-affordance grasping and cross-domain image matching. Int. J. Rob. Res. (2019)
32. Mnih, V., et al.: Human-level control through deep reinforcement learning. Nature **518**, 529–533 (2015)
33. Goodfellow, I. et al.: Measuring invariances in deep networks. In: Advances in neural information processing systems, vol. 22. pp. 646–654. Curran Associates, Inc. (2009)
34. Tobin, J. et al.: Domain randomization for transferring deep neural networks from simulation to the real world. In: 2017 IEEE/RSJ International Conference on Intelligent Robots and Systems, pp. 23–30 (2017)
35. CornellUniversity: Robot learning lab: Learning to grasp, http://pr.cs.cornell.edu/grasping/rect_data/data.php. Accessed on 17/09/2019
36. Joe Watson, J.H., Iida, F.: Real-world, real-time robotic grasping with convolution neural network. In: Conference Towards Autonomous Robotic Systems, pp. 627–636 (2017)
37. Quigley, M., Conley, K., Gerkey, B.P., Faust, J., Foote, T., Leibs, J., Wheeler, R.C., Ng, A.Y.: ROS: an open-source robot operating system. In: ICRA 2009 (2009)
38. Bradski, G.R., Kaehler, A.: Learning OpenCV—computer vision with the OpenCV library: software that sees. O'Reilly Media, Inc. (2008)
39. Park, D., Seo, Y., Chun, S.Y.: Real-time, highly accurate robotic grasp detection using fully convolutional neural network with rotation ensemble module (2019)
40. Washington-Dataset, web:rgbd-dataset.cs.washington.edu. Accessed on 19/09/2019
41. Levine, S., et al.: Learning hand-eye coordination for robotic grasping with deep learning and large-scale data collection. Int. J. Rob. Res. **37**, 421–436 (2018)

42. Bousmalis, K. et al.: Using simulation and domain adaptation to improve efficiency of deep robotic grasping. In: Proceedings of IEEE International Conference on Robotics and Automation, pp. 4243–4250 (2018)
43. Russakovsky, O. et al.: ImageNet large scale visual recognition challenge. CoRR. abs/1409.0 (2014)
44. Szegedy, C. et al.: Going deeper with convolutions. CoRR. (2014)
45. Szegedy, C. et al.: Inception-v4, inception-resnet and the impact of residual connections on learning. In: Thirty-First AAAI Conference on Artificial Intelligence (2017)
46. Szegedy, C. et al.: Rethinking the inception architecture for computer vision (2016)
47. Xia, Y., Wang, J.: A dual neural network for kinematic control of redundant robot manipulators. IEEE Trans. Syst. Man, Cybern. Part B. **31**, 147–154 (2001)
48. Polydoros, A.S. et al.: Real-time deep learning of robotic manipulator inverse dynamics. In: 2015 IEEE/RSJ International Conference on Intelligent Robots and Systems (2015)
49. Gonzalez, T.F.: Handbook of approximation algorithms and metaheuristics. In: Handbook of Approximation Algorithms and Metaheuristics, pp. 1–1432 (2007)
50. Krizhevsky, A., Sutskever, I., Hinton, G.: Imagenet classification with deep convolutional neural. In: Neural Information Processing Systems. pp. 1–9 (2014)
51. Ruder, S.: An overview of gradient descent optimization algorithms (2016)

A Masking-Based Image Encryption Scheme Using Chaotic Map and Elliptic Curve Cryptography

Mansi Palav, Ruchi Telang Gode, and S. V. S. S. N. V. G. Krishna Murthy

Abstract The paper introduces an efficient asymmetric algorithm for encrypting images utilizing chaotic maps and elliptic curve cryptography. In the proposed image encryption method, there is a combination of elliptic curve ElGamal encryption method with masking of image through chaotic map for enhancing security and efficiency. The proposed algorithm can encrypt images of any size. Security analysis is carried through to validate efficiency, sturdiness and feasibility of the algorithm. Compare to recent related encryption schemes, this scheme has better ability to resist crypt analytical and statistical attacks.

Keywords Image encryption · Mask · Chaotic map · Elliptic curve cryptography · ElGamal encryption

1 Introduction

Nowadays, image has become an important medium of communication because of the fact that images can be used in less word and depicts more information in less time. Due to the growing importance of image security in the last few years, several encryption schemes have been designed to secure images from cryptographic attacks. There are some important characteristics of images like high capacity, reluctance and heavy correlation coefficient between adjacent pixels, which makes enciphering of image distinct from the text. The two categories of modern encryption schemes are symmetric also called as private key cryptography and asymmetric or public key

M. Palav (✉) · S. V. S. S. N. V. G. Krishna Murthy
Department of Applied Mathematics, DIAT, Pune, Maharashtra, India
e-mail: mansipalav28@gmail.com

S. V. S. S. N. V. G. Krishna Murthy
e-mail: sgkmurthy@gmail.com

R. T. Gode
Department of Mathematics, NDA, Pune, Maharashtra, India
e-mail: telang.ruchi82@gmail.com

M. Sahni et al. (eds.), *Mathematical Modeling, Computational Intelligence Techniques and Renewable Energy*, Advances in Intelligent Systems and Computing 1287,
https://doi.org/10.1007/978-981-15-9953-8_12

cryptography. Key features of symmetric key-based cryptosystems are fast speed and low computational complexity. Confidentiality is very important in symmetric key cryptography as the key which is utilized for both encryption as well and decryption process is same. Some well-known symmetric key cryptosystems [12] are RC4, Salsa20, AES, DES, etc. Unlike symmetric key cryptography, in asymmetric key cryptography, distinct keys are chosen for encryption and decryption. The keys are depended on mathematical functions which are one way and whose inverse is difficult to calculate. Asymmetric key cryptography protects insecure communication between various users by solving the problem of insecure key distribution. Some standard schemes are RSA, ElGamal, ECC, and Diffie–Hellman [12]. Elliptic curve cryptography (ECC) [5] is among the significant asymmetric encryption technologies. Based on elliptic curves and public key cryptosystem, elliptic curve ElGamal cryptosystem was introduced [5] which is widely used in image encryption. For providing high-security ECC pair up with many other techniques, as in [15], ECC is paired with Arnold transformation technique; also in [1], RSA method is combined with Markov random field (MRF) and ECC to provide high security. In [4], ECC technique is applied to encrypt color figures. Some recent image ciphering methods also used chaos techniques like logistic chaotic map [3], tent map [9], sine map [10], hyper-chaotic system [14], and Arnold cat map [15]. Chaotic maps are often unpredictable and sensitive to starting conditions, because of these properties, chaotic maps are used in different fields. In [6], authors compare performance of non-chaotic, chaotic, and hyper-chaotic techniques for encryption. In [8], the technique used is discrete fractal map for image encryption; here, cipherimage is obtained using secret key and ECC-based Menezes Vanstone method. In [11], a hyper chaos-map designed an image encryption method which consist of three main parts. In the first part, using row–column algorithm, columns and rows are encrypted, and then in the second part, an image is divided into four parts and masking is applied on each quarter of image, and then in the third part, diffusion is used. Further [16] designed a chaos map image encryption scheme combining modulation, permutation, and diffusion model paired up with information entropy. Based on the above-mentioned review in our paper, an efficient image encrypting technique combining ECC ElGamal cryptosystem with masking-based chaos system is proposed.

In our image encryption technique, both transformation and substitution are combined to make the encryption process more effective and secured. Substitution is brought about by ElGamal encryption in which each point in the transformed matrix is mapped to a point of the defined elliptic curve. Transformation is brought by a method of masking based on asymmetric tent map, and this step enhances the randomness among pixel values by a permutation of the obtained elliptic curve points. Our encryption can be handle image of any size; however, in this paper, concentration is on gray image and the plaintext image is of size of 512×512. The algorithm targets at providing a better security for the encryption of images by reducing the correlation and increasing the entropy value among the pixels of the enciphered picture. Statistical analysis such as histogram and correlation and cryptanalytic attacks such as key sensitivity analysis, differential attack analysis, avalanche effect analysis as well as time complexity analysis are executed on the original and the enciphered image. All the experimental results validate the security of the encryption scheme. Further, the better influence of our scheme is depicted by comparing it with related work [8, 10, 11].

2 Preliminary Section

2.1 Introduction to Elliptic Curve Cryptography ECC

The elliptic curve named as E defined upon a real field F is the subset of $F \times F$ satisfying the Weierstrass's equation $y^2 + axy + by = x^3 + cx^2 + dx + h$ given as $E = \{(x, y) : y^2 + axy + by = x^3 + cx^2 + dx + h\}$ where $a, b, c, d, h \in F$ together with a special point at infinity (often denoted by ∞). The use of elliptic curves is due to their unique structure because a simple addition operation defined on E converts E into a finite abelian group.

Elliptic Curve over Prime Fields From cryptographic point of view, it is adequate to consider non-singular elliptic curve given by the equation of the form $y^2 = x^3 + ax + b$ over a finite field Z_p; here $p > 3$ is prime and $a, b \in Z_p$. Then, condition $(4a^2 + 27b^3) \neq 0$ signifies that the curve has no "singular points"; the set of all points on elliptic curve is denoted by E_p and defined as

$$E_p(a, b) = \{(x, y) : y^2 = (x^3 + ax + b) \bmod p\} \tag{1}$$

including the point at infinity (generally denoted by ∞). All the entities in the elliptic curve cryptosystem agree upon a, b, p, G, n which are called domain parameters of ECC. Here, the point G is the base and n is the order of G. Base point corresponds to a point in $E_p(a, b)$ whose order is largest in the group $E_p(a, b)$.

Elliptic Curve Arithmetic Let $P = (x_1, y_1)$ and $Q = (x_2, y_2)$ be points on elliptic curve $E_p(a, b)$ defined by Eq. (1); then $R = (x_3, y_3) = P + Q$ is defined by $x_3 = (\lambda^3 - x_1 - x_2) \bmod p$ and $y_3 = (\lambda(x_1 - x_3) - y_1) \bmod p$ where

$$\lambda = \begin{cases} \frac{y_2 - y_1}{x_2 - x_1} \bmod p \text{ if } P \neq Q \\ \frac{3x_1^2 + a}{2y_1} \bmod p \text{ if } P = Q \end{cases}$$

1. $P + \infty = P$ for all values of $P = (x, y) \in E_p(a, b)$. Namely, $E_p(a, b)$ has ∞ as its identity element.
2. $P + Q = \infty$ for all values of $P = (x, y) \in E_p(a, b)$ and $Q = (x, -y) \in E_p(a, b)$; in other words, inverse of (x, y) is simply $(x, -y)$.

$E_p(a, b)$ forms an abelian group with binary operation as addition of points.

Multiplication of Points Multiplication of points over elliptic curve can be considered as repeated addition. So, the above rules and formulas can also be applied to multiplication by adding repeatedly.

2.2 *Elliptic Curve Discrete Logarithm Problem (ECDLP)*

The reliability of ECC is dependent on the complexity of solving EDCLP. Let R and S be two points on the curve for which $cR = S$, where c is a scalar. There is computational infeasibility to obtain c for given R and S. For large value, this scalar c is known as the discrete logarithm of S to base R and acts as a private key, and the result obtained after multiplying the scalar c with the base point R is the obtained public key.

ElGamal Elliptic Curve Encryption/Decryption ElGamal encryption and decryption system are asymmetric key encryption algorithms which can be applied over elliptic curve as follows.

G is a selected base point of $E_p(a, b)$ known both to the sender and the receiver. Suppose P_m is the plaintext which is to be encrypted, taking the private keys of the sending party and receiving party as a and b, respectively; P_A and P_B are their respective public keys where $P_A = aG$ and $P_B = bG$. Let C_m be the ciphertext obtained after ElGamal encryption. Let us name the sender as A and the receiver as B.

ElGamal Elliptic Curve Encryption

- Step 1: A chooses a random positive integer c between 1 and $p - 1$ where p is the prime number corresponding to prime field Z_p.
- Step 2: A multiplies c with G. This forms the first element of ciphertext C_m
- Step 3: A multiplies c with P_B and adds it to the plaintext. This forms the second element of C_m. So $C_m = (C_1, C_2) = (cG, P_m + cP_B)$.

ElGamal Elliptic Curve Decryption

- Step 1: B multiplies the first element of C_m with its private key. So, $kGb = kP_B$.
- Step 2: B subtracts the above obtained result from the second element of received C_m to get P_m.

2.3 *Chaotic Map*

Chaotic maps [7, 8] are used to generate random sequence of number to scramble or diffuse the original image pixels. Unpredictability of the orbital evolution, sensitivity toward starting conditions as well as control parameters are some characteristics of a chaotic map. Chaotic map provides high encryption rate. In this paper, one-dimensional asymmetric tent map is used to generate mask which is defined as follows

$$x_{i+1} = \begin{cases} \frac{x_i}{\mu} & \text{if } 0 < x_i \le \mu \\ \frac{1-x_i}{1-\mu} & \text{otherwise} \end{cases} \quad \text{where } \mu \in (0, 1)$$

Here, μ is known as control parameter. (Here, range of above mention map is (0, 1) i.e., $x_i \in (0, 1)$).

2.4 Masking

Some operations on an image pixel operations depend upon the values of neighboring pixels, and such sub-image having the same dimension as of neighborhood is called mask or filter of image. The values of mask are known as coefficient instead of pixels. Filtering or masking process consists of moving mask in image from one point to other point at each point (x, y), and the response of mask at that point is evaluated by applying predefined relation. The relation used in the proposed algorithm can be expressed in the following form:

$$Q = \sum_{k=-1}^{1} \sum_{l=-1}^{1} w(k, l) f(x + k, y + l)$$

here, k and l are integers.

$$b = Q \bmod (256) \text{ and } g(x, y) = (f(x, y) + b_1) \bmod 256$$

where b_1 is obtained from b, by shifting each bit of b to right. Since $x = 0, 1, \ldots, m - 1$ and $y = 0, 1, 2, \ldots, n - 1$ and $f(x, y)$ and $g(x, y)$ be pixel value at position (x, y) of original and new image (image is of size $m \times n$), w is the mask (Fig. 1).

Reverse making means reverse of masking. Since for masking, we start from upper left most corner; on the other hand, for reverse masking, we start from lower right most corner. For reverse masking, the only change in function is $f(x, y) = g(x, y) - b_1$. For masking (and reverse masking), each time we use updated pixel value to encrypt (or decrypt) the next pixel value.

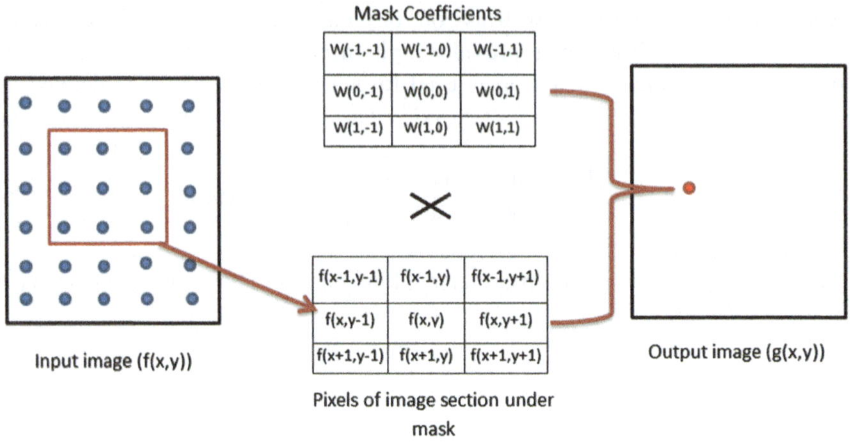

Fig. 1 Masking

3 Proposed Algorithm

3.1 Key Generation of Chaotic Map

The given algorithm has two keys; one is secret key and other is key of ECC ElGamal algorithm. Let k_1 be secrete key used to generate key of chaotic map.

Calculate:

$$(x, y) = \begin{cases} cP_B = c(Gb) \text{ for encryption} \\ bC_1 = b(cG) \text{ for decryption} \end{cases}$$

Then $X = \lfloor \frac{255x}{p} \rfloor$ and $Y = \lfloor \frac{255y}{p} \rfloor$, $K = \text{XOR}(X, Y)$; then

$$h_1 = (Kk_1) \bmod 1 \text{ and } h_2 = \left(\frac{c}{k_1}\right) \bmod 1 \tag{2}$$

Using h_1 and h_2 which are obtained in Eq. (2) are used as initial values for chaotic map and generate two masks, one for masking and other for reverse masking.

3.2 Generation Mask

Using chaotic map and initial value h_1 obtained from Eq. (2), generate a sequence of nine numbers $\{l_1, l_2, \ldots, l_9\}$ and then transform that number to 0 to 255 using the following transformation map:

$$D_i = \lfloor (l_i + d) \times 255 \rfloor \quad \text{for } i = 1, 2, \ldots, 9$$

$$d = \begin{cases} 0 & \text{for } i = 1 \\ (l_{i-1} \times 255) \bmod 1 & \text{for } i = 2, 3, \ldots, 9 \end{cases}$$

Then, arrange the sequence in the following way such that it will return 3 × 3 matrix

$$\begin{bmatrix} D_1 & D_2 & D_3 \\ D_4 & D_5 & D_6 \\ D_7 & D_8 & D_9 \end{bmatrix}$$

Then, make center 0. (i.e., assign $D_5 = 0$). (Using the same procedure, we generate the second mask which is obtained by using h_2 as initial value.)

Here, we explain the algorithm for encrypting and decrypting image of size 512 × 512; however, the method can be extended to encrypt and decrypt images of all sizes.

Encryption

Suppose M is the matrix of the original image which is of size 512 × 512 and k_1 is secret key of B which belongs to (0, 1).

1. Sender chooses an elliptic curve $E_p(a, b)$, and the points on the curve are generated. Image M is mapped to row vector S of length 262,144 using mapping method. This process corresponds to substitution.
2. ElGamal encryption is applied on points obtained in the above step. This produces the encrypted image of size 512 × 512.
3. Using a secret key k_1 and public key P_B of B, generate keys of chaotic map and create two 3 × 3 mask using chaotic map where h_1 is initial point for generating first mask and h_2 is initial point for generating the second mask.
4. Apply that masking technique using the first mask and then apply reverse masking technique using the second mask. (For masking and reverse masking, use zero padding) (Fig. 2).

Decryption

1. Using a secret key k_1 and private key b of B, generate keys of chaotic map and create two 3 × 3 mask using a chaotic map where h_1 is initial point for generating the first mask and h_2 is initial point for generating the second mask.
2. Apply that masking technique using the first mask and then apply reverse masking technique using the second mask. (For masking and reverse masking, use zero padding).
3. ElGamal decryption is applied on image obtained from the above step.
4. The collected points on the curve are mapped again to the pixels by applying inverse of mapping algorithm (Fig. 3).

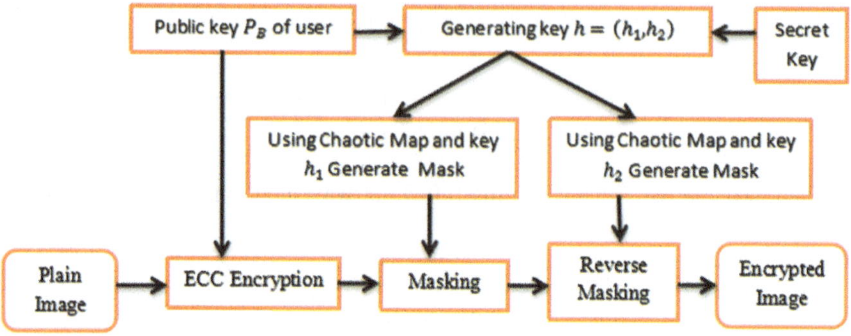

Fig. 2 Block diagram of encryption algorithm

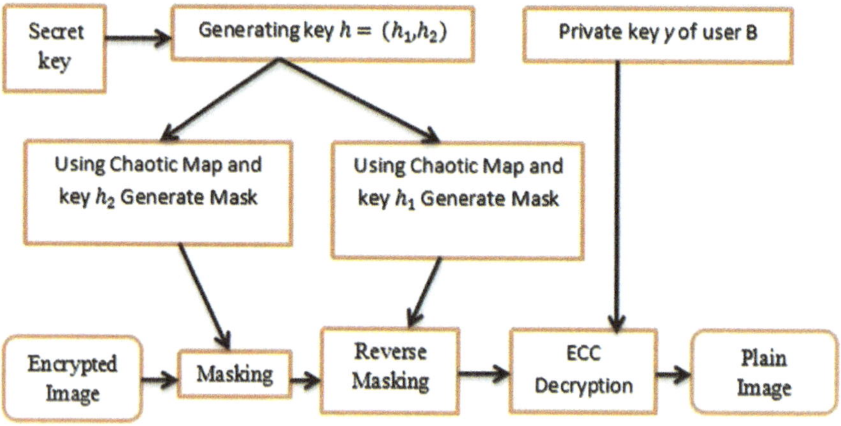

Fig. 3 Block diagram of decryption algorithm

4 Security Analysis and Simulations

4.1 Implementation

This process is applied on 512×512 grayscale images and implemented on 64-bit, Intel® core^{TM} i5-8250U 1.60 GHz. The programming is done in MATLAB. The elliptic curve used for generation of curve points is $y^2 = x^3 + x + 1$ over a prime field Z_{263}. The prime field is chosen as Z_{263} as we need at least 256 different points for one-to-one mapping of the image pixels to the elliptic curve points. Here, the chosen generator is $G = (199, 261)$, $P_B = (137, 107)$, private key of user B is $b = 79$, secret key is $k_1 = 0.39685$, and control parameter of chaotic map is $\mu = 0.225$.

4.2 Statistical Analysis

Histogram Analysis A graphical description of frequency distribution among pixel intensity of an image is represented by histogram. Ideally, histogram of the cipher-image should be distributed uniformly and should not have any similarity with the histogram of plain image. Our scheme gives good results with respect to histogram analysis (Table 1).

Information Entropy Information entropy of encrypted image should be high. If entropy of encrypted image is less, then it is likelihood to predict actual image. The standard entropy value of the enciphered image should be nearer to 8. See our results in Tables 2 and 6, respectively (Fig. 4).

Variances Uniformness of histogram is assessed by variance. Variances show deviation of histogram from its mean value. The variance of an image is given as $\mathrm{Var}(W) = \frac{1}{2 \times 256^2} \sum_{x=0}^{255} \sum_{y=0}^{255} \left(w_x - w_y\right)^2$ where $W = (w_0, w_1, \ldots, w_{255})$ is the histogram values vector and w_x and w_y are number of pixel having values x and y, respectively. The variance of our method is less which depicts effectiveness to resist statistical attacks. Results given in Table 3.

Correlation Coefficients In image, any pixel is strongly correlated with other pixel in its neighborhood; therefore, encrypted image should be such that it does not have

Table 1 Actual and enciphered image with their histograms

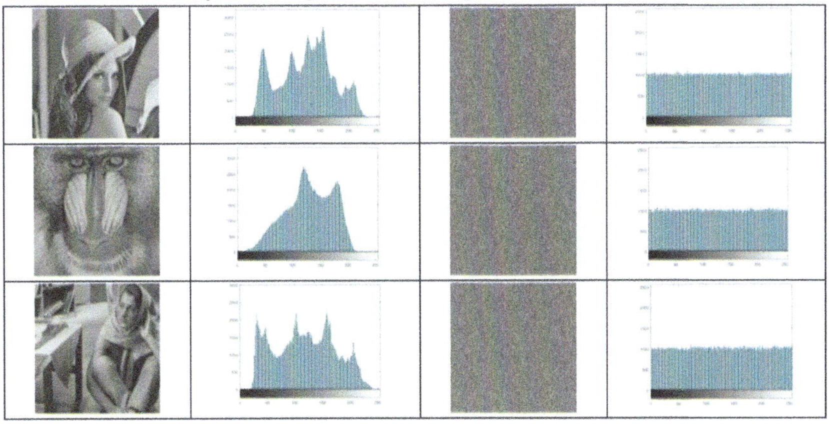

Table 2 Information entropy results

Image	"Lena"		"Baboon"		"Barbara"	
	Actual	Enciphered	Actual	Enciphered	Actual	Enciphered
Entropy Value	7.3871	7.9994	7.3579	7.9994	7.4664	7.9994

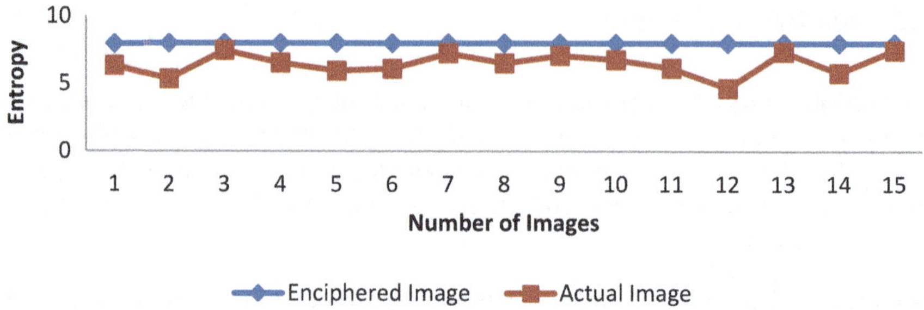

Fig. 4 Entropy analysis of the proposed algorithm

Table 3 Results of variance

Image	"Lena"		"Baboon"		"Barbara"	
	Actual	Enciphered	Actual	Enciphered	Actual	Enciphered
Variance	630,730	904.7891	627,520	885.6641	43,3040	933.0234

any correlation with neighborhood pixel. Zero value of correlation coefficient represents no relation between adjacent pixel values. Therefore, correlation coefficient should be closed to 0. Our results are depicted in Table 4. See Tables 5 and 6 for numerical values.

Table 4 Correlation distribution of plain "Lena" image and encrypted "Lena" image

Image	Correlation		
	Vertical	Horizontal	Diagonal
Actual	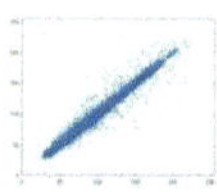	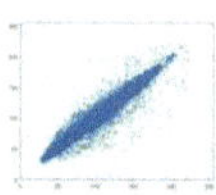	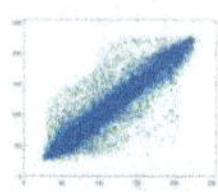
Enciphered	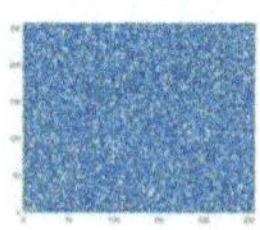	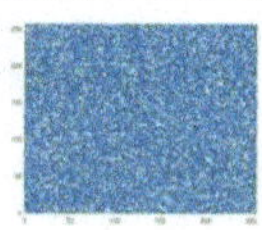	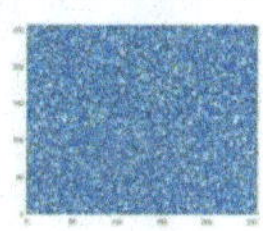

Table 5 Correlation coefficients actual image and enciphered image

Image	Correlation coefficient					
	Vertical		Horizontal		Diagonal	
	Actual	Enciphered	Actual	Enciphered	Actual	Enciphered
"Lena"	0.9853	−0.0014	0.9722	0.0013	0.9591	0.0007
"Baboon"	0.7587	−0.0034	0.8665	−0.00063	0.7262	0.00078
"Barbara"	0.9607	−0.00064	0.8979	−0.0032	0.8858	−0.00096

Table 6 Entropy and correlation coefficients: Black-and-white image

Image		Entropy Value	Correlation Coefficient		
			Vertical	Horizontal	Diagonal
Black	Actual	0	–	–	–
	Enciphered	7.9994	0.0021	0.0051	0.00009
White	Actual	0	–	–	–
	Enciphered	7.9992	0.0015	0.0025	−0.000006

4.3 Analysis of Differential Attack

Differential attack is a form of chosen-plain text attack. The capacity of withstanding differential attack is assessed by comparison of two enciphered images, which indicates that changing a bit of the original images, the enciphered images should be totally different. This attack is evaluated by two parameters UACI and NPCR.

NPCR and UACI Let C_1 and C_2 are enciphered images corresponding to the actual images with only one pixel difference. Here, images are of size $m \times n$.

- NPCR is number of pixels changes in encrypted image.

$$NPCR = \frac{\sum_{x=0}^{m-1} \sum_{y=0}^{n-1} D_k(x, y)}{m \times n} \times 100$$

where $D_k = \begin{cases} 0 \text{ if } C_1(x, y) = C_2(x, y) \\ 1 \text{ if } C_1(x, y) \neq C_2(x, y) \end{cases}.$

- UACI is unified average changing intensity, and its formula is given by

$$\text{UACI} = \frac{1}{m \times n} \times \sum_{x=0}^{m-1} \sum_{y=0}^{n-1} \frac{|C_1(x, y) - C_2(x, y)|}{F}$$

where F denotes the largest given pixel value. For our case, $F = 255$. Encryption algorithm is considered as good for average values of UACI and NPCR greater than 33 and 99.6, respectively. Here, UACI and NPCR are calculated by choosing 100 pixel position randomly and then taking average of that. See Table 7 (Fig. 5).

Table 7 NPCR, UACI of different images

Image	"Lena"	"Baboon"	"Barbara"
NPCR value	99.6156	99.6155	99.6105
UACI value	33.4786	33.4639	33.4464

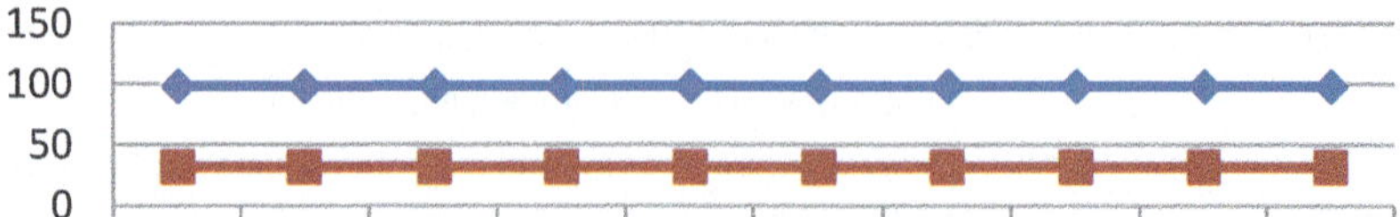

	1	2	3	4	5	6	7	8	9	10
NPCR	99.616	99.616	99.611	99.613	99.61	99.611	99.615	99.615	99.615	99.616
UACI	33.479	33.464	33.446	33.479	33.445	33.463	33.465	33.468	33.481	33.493

Fig. 5 NPCR and UACI analysis of the proposed algorithm

4.4 Key Space Analysis

Key space means the collection of all possible encryption keys. The two related terms to analyze key space are the key sensitivity and the number of keys.

Number of Keys Key space size can protect from brute force attack. Encryption process is safe from brute force attack if key space is greater that 2^{100} [7, 10]. In the proposed algorithm, we have two keys, one secret key k_1 and other one is key of ECC ElGamal algorithm. Using these two keys, the initial value for chaotic map is calculated. Also, chaotic map value is depending on control parameter μ. For the computational precision to be around 2^{52}, the key space size becomes 2^{248} which is much greater than 2^{100}.

Key Sensitivity Analysis In robust cryptosystem, if one changes key slightly, then cipherimage should change completely, and this is called as key sensitivity. Also, the original image should not be recovered even if decryption key changes slightly. Larger values of UACI and NPCR show better key sensitivity of cryptosystem. Table 7 shows our scheme which provides good results with respect to key sensitivity analysis.

4.5 Avalanche Effect Analysis

Any small alteration in keys or original image may bring prominent changes in encrypted image; such property is called as avalanche effect. This is given as

$$\text{Avalanche} = \frac{\text{Number of bits which are changed}}{\text{Total number of all bits}} \times 100\%$$

According to standards of avalanche effects if there is change of one bit in actual image, then rate of change in cipherimage should be minimum than 50% [8]. Another term which measures avalanche effect is mean square error (MSE). Its formula is "$\text{MSE} = \frac{1}{m \times n} \sum_{i=0}^{m-1} \sum_{j=0}^{n-1} |E_1(i, j) - E_2(i, j)|^2$" where $m \times n$ is size of image and E_1, E_2 are enciphered images with actual image having difference of one bit only. If

Table 8 MSE and avalanche values of various images

Image	Actual image is changed by one bit		Key is changed by one bit	
	Avalanche (%)	MSE (dB)	Avalanche (%)	MSE (dB)
"Lena"	67.0318	42.6986	66.9668	42.6445
"Baboon"	67.0250	42.6663	67.1984	42.8636
"Barbara"	66.9107	42.6241	66.8501	42.5980

Table 9 Comparison: performance of some recent encryption schemes on Lena (512 × 512) image

Performance factor		Liu et al. [8]	Luo et al. [10]	Norouzi, et al. [11]	Proposed algorithm
Image size		Non-square and square	Square	Non-square and square	Non-square and square
Sensitive to secret keys		Yes	Yes	Yes	Yes
Entropy		7.9993	7.9993	7.9980	7.9994
Variance		Not given	980.8	Not given	904.7891
Correlation coefficient	Horizontal	−0.0140	0.0019	0.0007	0.0013
	Vertical	−0.0086	−0.0024	0.02045	−0.0014
	Diagonal	−0.0034	0.0011	−0.00025	0.0007
Differential attack	NPCR	99.60	99.6113	99.6124	99.6156
	UACI	33.48	33.4682	33.4591	33.4786
Avalanche effect	Avalanche	Not given	50.0334	Not given	67.0318
	MSE	Not given	40.3882	Not given	42.6986

value of MSE is greater than 30 dB, then we can conclude that encryption algorithm is good. Table 8 shows avalanche effect, MSE results for one bit difference in actual image and the key. Our algorithm has greater avalanche effect and MSE values than the standard ones, which verify that scheme has avalanche effect up to expectations (Table 9).

4.6 Time Complexity Analysis

Efficient encryption algorithm must provide high security with fast speed. The time required for encrypting 512 × 512 "Lena" image is given in Table 10 (Table 11).

Table 10 Processing time of the proposed algorithm

Process	EC-ElGamal encryption	Masking	Reverse masking	Other	Total
Time[second]	1.892	0.631	0.642	1.516	4.681
Percentage	40.4187	13.48	13.715	32.3862	100

Table 11 Comparison: running time consumption [unit: second]

Image size	128 × 128	256 × 256	512 × 512
Proposed	0.371	1.148	4.681
Chen et al. [2]	3.254312	5.556790	8.974393
Luo et al. [10]	0.351779	1.170844	4.73389

5 Conclusion

In the proposed algorithm, image is encrypted using ECC and masking method based on asymmetric tent chaotic map. The proposed algorithm can work on image of any size and has higher security because of large size key space, and the scheme also gives satisfactory values of NPCR and UACI. Encrypted image obtained by this algorithm has low correlation coefficient, high entropy and has strong capability to resist histogram-based attacks. The comprehensive experimental analysis shows that our method shows improvement in performance with respect to time consumption, security as well as efficiency compared to recent image encryption methods.

References

1. Chen, C.C.: RSA Scheme with MRF and ECC for DATA Encryption. IEEE (2004)
2. Chen, J., Zhang, Y., Qi, L., Fu, C., Xu, L: Exploiting chaos-based compressed sensing and cryptographic algorithm for image encryption and compression. Opt. Laser Technol. 238–248 (2017)
3. Huang, S.Y.: Colour image encryption based on logistic map ping and double random-phase encoding. IET Image Process. **11**, 211–216 (2017)
4. K.S., M.S.: Image encryption using efficient elliptic cruve cryptography. IJIRCC. (2007)
5. Koblitz, A. M., Vanstone.: The state of elliptic curve cryptography. Desig. Codes Cryptogr. 173–193 (2000)
6. Kumar, B.M., Karthikka, P., Dhivya, N., Gopalakrishnan, T.: A performance comparison of encryption algorithms for digital images. IJERT **3**
7. Kumari, M., Gupta, S., Sardana, P.: A Survey of Image Encryption Algorithm. Springer. Cross Mark (2017)
8. Liu, Z., Xia, T., Wang, J.: image encryption technology based on fractional two-dimensional triangle function combination discrete chaotic map coupled with Menezes-Vanstone elliptic curve cryptosystem. Hindawi Discrete Dynamics in Nature and Society (2018)
9. Luo, Y., Cao, L., Qiu, S., Lin, H., Harkin, J., Liu, J.: A chaotic map control-based and the plain image-related cryptosystem. Nonlinear Dyn. 2293–2310 (2016)
10. Luo, X.O., Liu, L.C.: An image encryption method based on elliptic curve ElGamal encryption and chaotic systems. IEEE Access (2019)

11. Norouzi, S.S., Mirzakuchaki, M.M.: A novel image encryption based on row-column, masking and main diffusion processes with hyper chaos. Multimed. Tools Appl. (2013)
12. Omar, M.S.M.R., Zeebaree, F.Y.H.A.: A survey and analysis of the image encryption methods. Int. J. Appl. Eng. Res. 13265–13280 (2017)
13. Singh, A.T., Singh J.: Probabilistic data encryption using elliptic curve cryptography and Arnold transformation. In: International on conference on I-SMAC (I-SMAC 2017)
14. Wang, H.L.Z.: A novel image encryption algorithm based on genetic recombination and hyper-chaotic systems. Nonlinear Dyn. **83**, 333–346 (2016)
15. Ye, G., Wong, K.-W.: An efficient chaotic image encryption algorithm based on a generalized Arnold map. Nonlinear Dyn. **69**, 2079–2087 (2012)
16. Ye, G., Pan, C., Huang X., Zhao, Z., He, J.: A chaotic image encryption algorithm based on information entropy. Int. J. Bifurcation Chaos. **28** (2018)

Automatic Speech Recognition of Continuous Speech Signal of Gujarati Language Using Machine Learning

Purnima Pandit, Priyank Makwana, and Shardav Bhatt

Abstract In this work we perform automatic recognition of continuous speech signal spoken in Gujarati language using machine learning (ML) technique. For this purpose, from continuous speech signal of sentence we first extract words using short term auto-correlation (STAC) method. Since the signals for each word are large in size, the dimension reduction is done using feature extraction algorithm: mel-frequency discrete wavelet coefficient (MFDWC). Then these features are trained using ML algorithm for recognition of speech.

Keywords Automatic speech recognition (ASR) · Gujarati language · Short term auto-correlation (STAC) · Mel-frequency discrete wavelet coefficient (MFDWC) · Machine learning (ML)

1 Introduction

ASR is a technique of building an intelligent machine which recognises a naturally spoken human speech [1]. This is done by training a machine using vital features hidden inside the digital signal of recorded speech. Some well-known ASR software are Alexa, Cortana, Google assistant and Siri. ASR is very useful for giving instructions to machines like mobile phones, computers and other gadgets by speech

P. Pandit
Department of Applied Mathematics, Faculty of Technology and Engineering, The Maharaja Sayajirao University of Baroda, Vadodara 390001, Gujarat, India
e-mail: pkpandit@yahoo.com

P. Makwana (✉)
Applied Science and Humanities Department, Parul Institute of Engineering and Technology (Diploma Studies), Parul University, Limda–Waghodia, Vadodara 391760, Gujarat, India
e-mail: priyankmak@gmail.com

S. Bhatt
School of Engineering and Technology, Navrachana University, Vasna-Bhayli Road, Vadodara 391410, Gujarat, India
e-mail: shardavb@gmail.com

M. Sahni et al. (eds.), *Mathematical Modeling, Computational Intelligence Techniques and Renewable Energy*, Advances in Intelligent Systems and Computing 1287,
https://doi.org/10.1007/978-981-15-9953-8_13

commands. This facilitates a user to overcome the use of hardware like keyboards, mouse and touch screens etc. This is very beneficial to the blind people and people having other disabilities who finds difficult to use keyboards, mouse and touch screen devices etc. They can use mobile phones, computers and other gadgets using their speech.

India is a country of more than 120 scheduled and unscheduled languages [2]. Most of the Indian languages has a greater number of phonemes, fricatives, retroflex consonants as compared to the western languages, which makes it more challenging to recognise a speech in an Indian language. But it is very beneficial to the people who knows only the mother tongue.

According latest available census data, there are 55 million Gujarati language speakers [2]. Moreover, Gujarat state is a business hub of India, which makes use of communication devices more frequent within Gujarat. Unfortunately due to 79.31% literacy rate [2] in Gujarat, not every people can use these gadgets due to lack of their English knowledge, particularly in rural areas. Hence ASR in Gujarati language can be useful to such people, too.

ASR design consists of two main algorithms [3] namely: feature extraction and recognition. Feature extraction is widely done using mel-frequency cepstral coefficients (MFCC) [4]. For the recognition part, various researchers have applied many techniques. Recently, ML techniques are widely used by several authors across the world for audio signals like speech and music. In [5], authors used ML technique for continuous speech spoken in Indian national language Hindi. MFCC are used for feature extraction and further hidden Markov model, ML technique and recurrent neural networks are used for the recognition. In [6], authors used ML techniques for several Indian and foreign languages which excludes the Gujarati language. Here also feature extraction is done using MFCC. Advanced ML technique like Deep neural network is applied in this work. In [7], authors used support vector machine for audio signal. Here audio signal is not a speech signal, it is in terms of music.

Our work involves speech recognition of continuous words spoken in Gujarati language. Moreover, unlike most of the other works, we are using wavelet based feature extraction technique MFDWC as it is based on wavelets and wavelets are better for non-stationary signals like a speech signal which has abrupt changes [8]. Moreover, ML technique, ANN is used in our work.

In our previous works, we have done experiments on speech recognition of isolated digits spoken in Gujarati language [9–12]. Since continuous sentences have more applicability as compared to isolated digits, in this work we have used continuous sentences, converted them into isolated words and then processed further. In Sect. 2, we discuss about mathematical and signal processing techniques as well as ML techniques used in our work. In Sect. 3, our experiments are explained in detail along with the discussion on results obtained. Section 4 concludes the paper with references at the end.

2 Algorithm for Speech Recognition of Continuous Sentence

2.1 Feature Extraction

Short term auto-correlation method to divide continuous sentences in words.

Auto-correlation method is used to find similarity between a signal and a shifted version of itself. If two signals are different then this method is known as cross correlation. The general formula for determining auto-correlation between two shifted parts of sequence $x(m)$ is

$$r_{xx}(k) = \sum_{m=-\infty}^{\infty} x(m)x(m+k) \tag{1}$$

Here, $r_{xx}(k)$ is the kth coefficient of auto-correlation between original signal $x(m)$ and same signal shifted by k samples $x(m+k)$. The value $k = 0$ gives energy of the signal and the value $k = 1$ gives summation of products of original signal with shifted version of itself by one sample.

Most of the signal processing formulae are derived for stationary signals. But speech signals are non-stationary. So first, it is required to divide the signal into several number of short frames. This makes signal stationary within particular short frame. Then the auto-correlation of samples of each short frame with different values of k is determined using (1). Hence, it is known as short term auto-correlation (STAC).

To remove noise and silence part of speech signal, the 2nd coefficient of auto correlation is used i.e. coefficient of correlation with one lag ($k = 1$). On computing STAC, it is observed that some values of auto-correlation are high and some are close to zero. From this we can identify voiced frame, unvoiced frame and noisy frame. The above explained process can be formally described using following steps.

Step 1: Pre-emphasising.

Let, $x(n)$ be a recorded speech in digital form. Then the pre-emphasised signal is obtained using

$$s(n) = x(n) - \alpha x(n-1) \tag{2}$$

where $\alpha = 0.95$. This amplifies higher frequencies so signal becomes spectrally flat and less susceptible to noise [1].

Step 2: Normalising and framing.

Next, the signal is normalised using maximum norm. After that, it is divided into non-overlapping frames of 25 millisecond rectangular window. This step is required to make speech signal stationary within each frame.

Step 3: Extracting words from sentence using STAC.

Now auto-correlation, given by (1), is calculated for each short frame. It helps us to decide whether given frame is voiced, unvoiced or noisy. Using this, the continuous signal of speech is divided into words. Once the words are extracted from the sentence, we proceed to find feature vector for each word since, the samples for the words are large in dimension. This also allows to keep only the vital information for each word.

Mel-frequency discrete wavelet coefficients.

For feature extraction, MFDWC [8] is used here. This algorithm reduces the dimension of the speech signal by removing unwanted or excess information from it and keeping the important ones. It is done using following steps [13]:

Step 1: Framing the word into overlapping frames.

Speech signals of word consists of abrupt variations. Hence for further analysis it is divided into several small parts known as frames (256 samples per frame), with each frame overlapping on its adjacent frame (100 samples overlapping). If $s(n)$ is a speech signal representing word, then after framing we get $s_i(n)$ where i is frame index and n represents sample number.

Step 2: Applying Hamming window over each frame.

Windowing is applied to gradually drop the amplitudes towards the ends of the frames to avoid noise. This minimises the signal discontinuities. Hamming window is used here which is given by

$$w(n) = (1 - \alpha) - \alpha \cos\left(\frac{2\pi n}{N - 1}\right), 0 \leq n \leq N - 1 \tag{3}$$

Value of α lies between 0 and 1. We have used $\alpha = 0.46$.

Step 3: Determining Fourier transform and power spectrum.

Next, each frame of word is shifted to frequency domain to get information of frequency content of frames. For this, Fourier transform is used:

$$s_i(k) = \sum_{n=1}^{N} s_i(n) w(n) e^{\frac{-2j\pi kn}{N}} \tag{4}$$

Here, $1 \leq k \leq N/2$. Next, power spectrum of each frame is determined using

$$p_i(k) = \frac{1}{N} |s_i(k)|^2 \tag{5}$$

Step 4: Constructing a mel-scaled triangular filter-bank.

Human ear does not perceive frequency on linear scale. So, frequencies are converted to mel-scale using

$$M(f) = 1125 \ln\left(1 + \frac{f}{700}\right) \tag{6}$$

To convert frequency information, to mimic what human ear perceives, a filter-bank consisting of 20 triangular filters is used which is given by

$$z_m(k) = \begin{cases} \frac{k-f(m-1)}{f(m)-f(m-1)} & f(m-1) \leq k \leq f(m) \\ \frac{f(m+1)-k}{f(m+1)-f(m)} & f(m) \leq k \leq f(m+1) \\ 0 & \text{otherwise} \end{cases} \tag{7}$$

Here, $f(m)$ represents mel-scaled frequencies. There are less filters in high frequency region because human ears are less sensitive to that region.

Step 5: Calculating mel filter-bank energies.

The mel filter-bank defined in (7) is multiplied with power spectrum defined in (5) and then the energy of mel filter-bank is determined using

$$\tilde{p}_m = \sum_{k=0}^{K/2} p_i(k) z_m(k) \tag{8}$$

This is also known as periodogram estimated of power spectrum.

Step 6: Calculating wavelettransform.

In the last step, the discrete wavelet transform (DWT) of log of energies of mel filter-bank is determined. This gives mel-frequency discrete wavelet coefficients for each frame of a word [13].

$$\tilde{c}_n = DWT(\log(\tilde{p}_m)) \tag{9}$$

For each word, these feature vectors are obtained which are used for recognition using ML technique.

2.2 Recognition Using Machine learning

Machine learning methods consists of mathematical algorithms to train machines to do tasks automatically. It consists of many algorithms like ANN, RNN, and radial basis function networks etc., which are useful for various tasks like function approximation, classification, pattern recognition etc. In our work, we have used ANN [14] with back-propagation algorithm [14] for training. The following section gives brief introduction of ANN.

Artificial Neural Networks.

ANN are inspired from mathematical modelling of biological neurons present in human brain. Human brain learns phenomena by observing the surrounded things. Similarly, ANN learns using a training procedure by observing provided input—output patterns. Here training procedure is classified as supervised learning if the inputs and corresponding desired outputs are provided, or unsupervised learning if only inputs are provided. The inputs are embedded in a layer called input layer. Outputs are also embedded separately in a layer called output layer. There is one more layer between them called hidden layer, which is introduced to increase complexity of network in order to train non-linear relationship between inputs and outputs. Nodes in different layers are called neurons. They consist of activation function applied on a weighted sum.

Let $\vec{x}$ be input neurons, $\vec{y}$ be output neurons, $\vec{z}$ be hidden neurons and $\vec{d}$ be desired outputs. Let w_{ji} be weights linking input neuron i to hidden neuron j and w_{lj} be weights linking hidden neuron j to output neuron l. Assuming activation f_h in hidden layer and f_o in output layer, the neurons in hidden layer and output layer are determined using $f_h\left(\sum_i w_{ji}x_i\right)$ and $f_o\left(\sum_j w_{lj}z_j\right)$ respecvely. One such ANN is shown in Fig. 1.

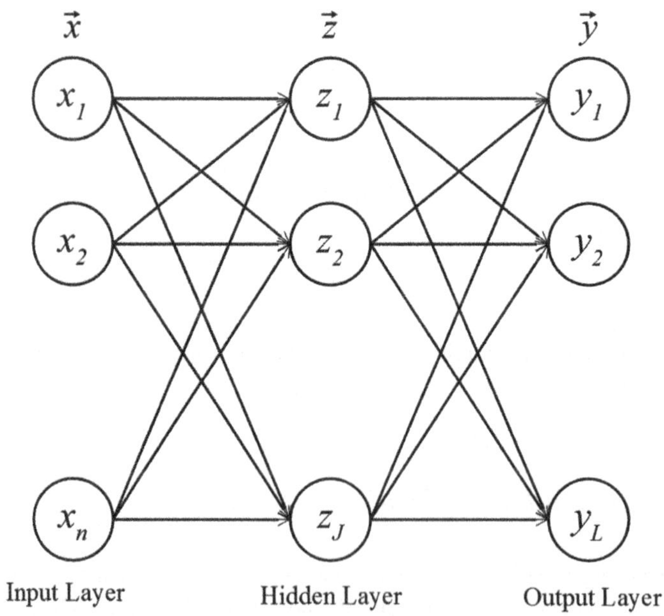

Fig. 1 ArtificialNeuralNetwork with one hidden layer

Neural network establishes input–output relationship by adjusting weights using back-propagation algorithm [14]. Change in weights of hidden layer is given by

$$\Delta w_{ji} = \eta_h \left[\sum_l (d_l - y_l) f_o^{'} \left(\sum_j w_{lj} z_j \right) z_j \right] f_h^{'} \left(\sum_i w_{ji} x_i \right) x_i \tag{10}$$

Change in weights of output layer is given by

$$\Delta w_{lj} = \eta_o (d_l - y_l) f_o^{'} \left(\sum_j w_{lj} z_j \right) z_j \tag{11}$$

Here, η_h is learning rate used in hidden layer and η_o is learning rate used in output layer. The weight updating using more recent Adam algorithm [15] is given by,

$$\Delta w = w - \frac{\eta}{\sqrt{\frac{\beta_2 v_{t-1} + (1-\beta_1)(\nabla w_t)^2}{1-\beta_2^t}} + \in} \frac{\beta_1 m_{t-1} + (1 - \beta_1) \nabla w_t}{1 - \beta_1^t} \tag{12}$$

3 Experimental Works

3.1 Data Description

The experiments consist of recording of 6 speakers. The recording is done for 5 sentences spoken in Gujarati language shown in Fig. 2. The recording was done using microphone of usual headphone in normal room noise using Audacity software with sampling rate of 16,000 samples per second and mono channel. Hence there are 6 wave files. These wave files are imported using python programming language. The graph of one such wave file is shown in Fig. 3. In our initial experiments, we considered single recording of each speaker. Further it was required to increase

નદીવહેછે.
સૂરજપૂર્વમાંઊગેછે.
બતકપાણીમાંતરેછે.
કમળખીલેછે.
નિરજજાગેછે.

Fig. 2 Five sentences in Gujarati formed from 17 words

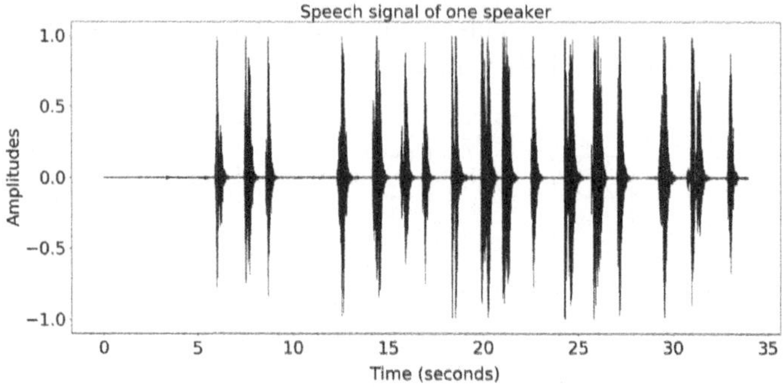

Fig. 3 Graph of speech signal by one speaker

dataset for better generalisation of ANN. Therefore, instead of single recording, two recordings of each speaker were considered for training.

3.2 Dividing Continuous Speech in Words Using STAC

Our first task is to divide these continuous sentences into isolated words. This is done using STAC method as explained in Sect. 2.1, Eq. (1). We obtained STAC as shown in Fig. 4. The sentences are divided into 17 words. One such word "નદી" is shown in Fig. 5. Hence now we have 17 vectors corresponding to 17 words for each of the 6 speakers, giving us 102 words.

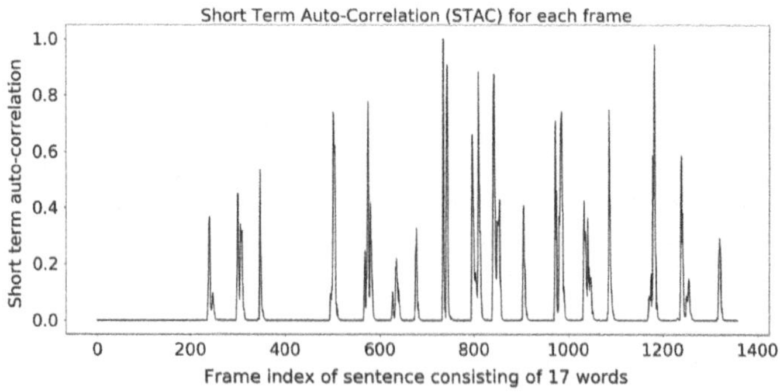

Fig. 4 Short term auto-correlation for each frame to detect voiced part of signal

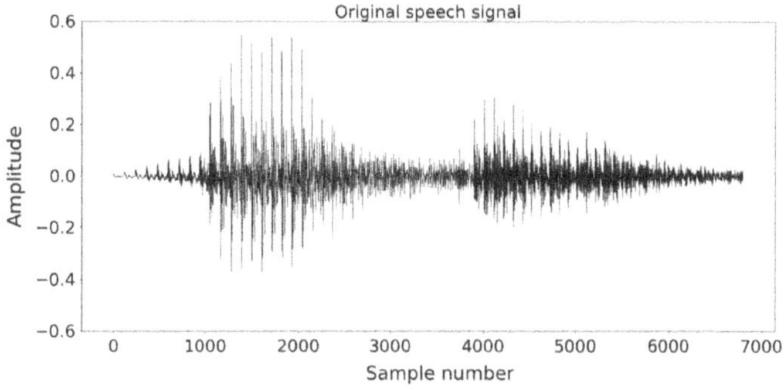

Fig. 5 Speech signal of word "નદી" spoken in Gujarati language

3.3 *Feature Extraction Using MFDWC*

In next step, vital features are extracted from each of the 17 words of 6 speakers using feature extraction technique MFDWC. This method consists of six steps as explained in Sect. 2.1 using Eqs. (2)–(8). Applying these steps on one of the frames of sample word, we have following figures. Figure 6 shows thirteenth frame of word "નદી". Applying Hamming window on it, the new amplitudes are as shown in Fig. 7. Here it is be observed that the amplitudes towards the two ends of the frames are dropped due to Hamming window.

The Fourier transform and power spectrum for thirteenth frame is as shown in Fig. 8.

Finally, applying discrete wavelet transform on log of periodogram estimates of power spectrum, we obtain MFDWC feature vector for thirteenth frame as shown in Fig. 9.

Such features are obtained for all words. These features are used for recognition part, which is done using machine learning technique.

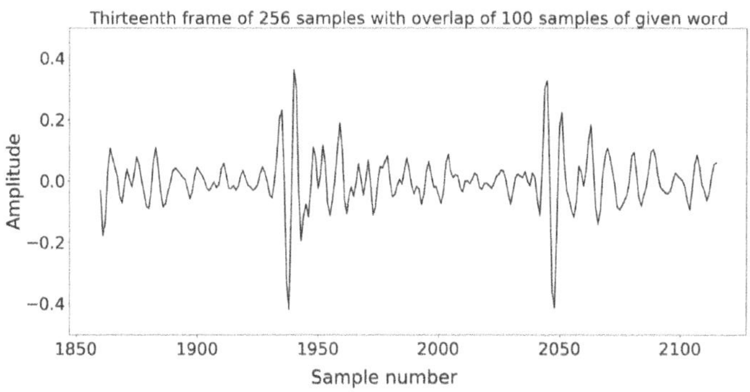

Fig. 6 Thirteenth frame of speech signal of word "નદી"

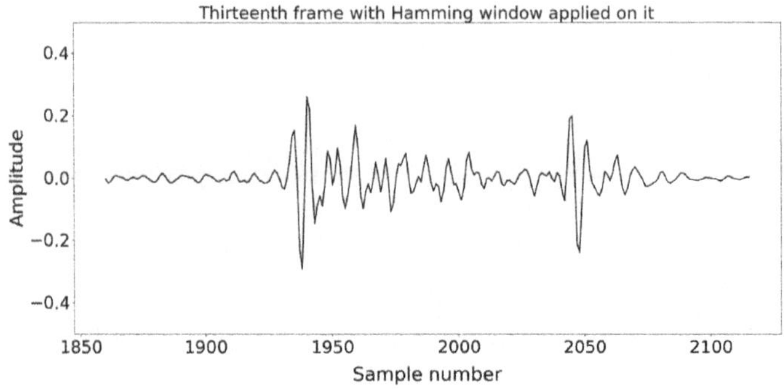

Fig. 7 Hamming window of thirteenth frame of the speech signal

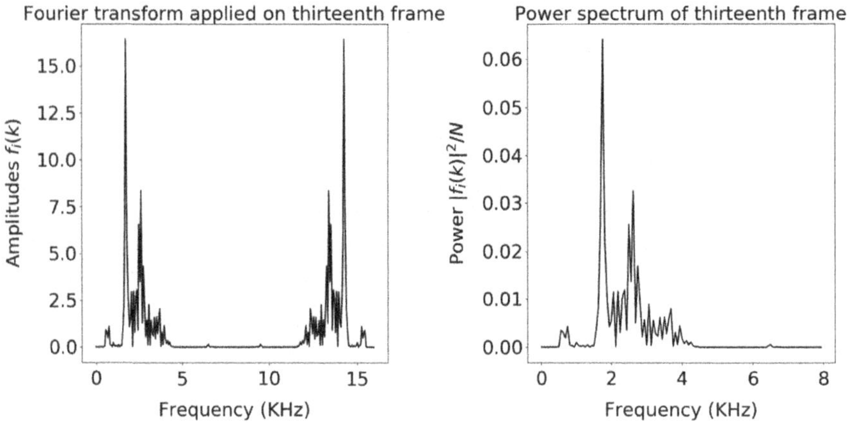

Fig. 8 Fourier transform and power spectrum of thirteenth frame of speech signal

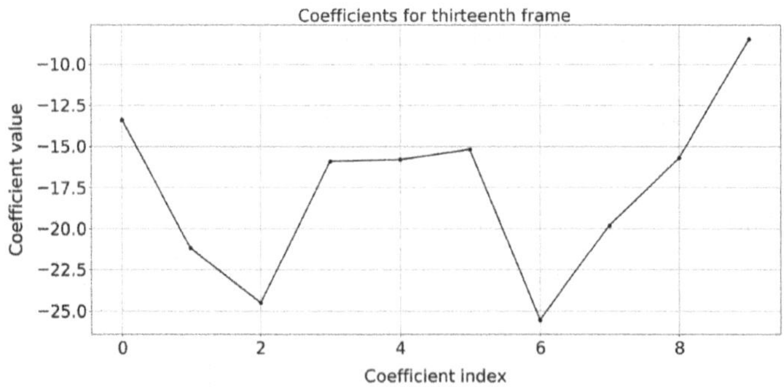

Fig. 9 MFDWC coefficients of thirteenth frame

Table 1 Summary of experiments

Algorithm	Activation function	Number of iterations	Training time (s)	Training loss	Testing accuracy (%)
BP	Sigmoid	956	134	0.00742	84.62
Adam	ReLU	1098	29	0.00042	76.92

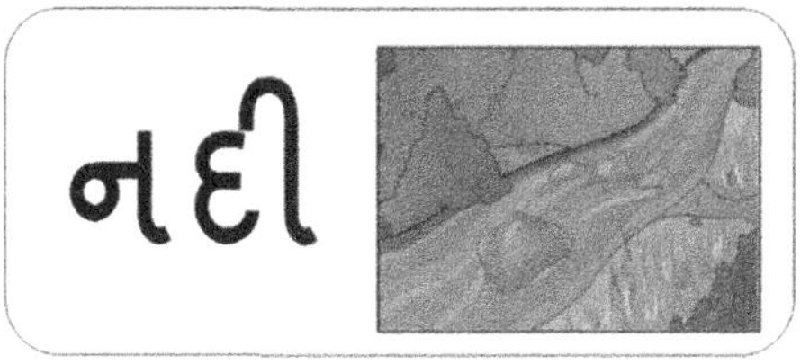

Fig. 10 Recognised word in Gujarati and its pictorial form

3.4 Recognition Using Machine Learning

We have used ANN to train the machine. Initially the network was trained with single recording of each speaker, which gave accuracy less than 70%. So, dataset was increased by considering multiple recording of each speakers. Thus, new dataset had 102 patterns for training a feed-forward architecture of ANN. This network comprises of input layer having 676 neurons representing feature vectors, hidden layer having 1000 neurons and an output layer having one neuron for recognition of words. Hence network specification is $N^2_{676,1000,1}$. Various wavelets were used in feature extraction step. Wavelet decomposition at level 2 for wavelet Daubechies–6 gave us best results. The networks were trained using back-propagation algorithm [14] with sigmoid activation function and Adam algorithm [15] with rectified linear unit (ReLU) activation function [16]. The results obtained for both networks is summarised in Table 1.

The trained networks were tested with unknown speech consisting of sentences different from the sentences trained but having same words. After being successfully recognized, our system gives recognised sentence output in Gujarati language text. Moreover, for correctly recognised isolated word, pictorial representation of it is also displayed in outputas shown in Fig. 10 to facilitate people who knows only Gujarati language and people with disabilities.

4 Conclusion

In this paper, we have done experiments on ASR of continuous sentence spoken in Gujarati language using feature extraction technique MFDWC and recognition using ML technique: ANN. The continuous sentence is first divided into words using STAC method. Then from each word, vital features are extracted. Finally, these features are trained using ML technique for recognition. Initially, recognition accuracy obtained was below 70%. We observed that the reason of lower accuracy was that the neural network needs more patterns for better generalisation of input signals. Hence, the database was increased by considering two recordings of each speaker. After increasing database, the recognition accuracy obtained was 84.62% using Back propagation algorithm, and 76.92% using Adam algorithm. However, Adam algorithm trained much faster.

Future work includes expanding database by considering a greater number of speakers, multiple recording of speakers and speakers from different parts of Gujarat to include more variations in training.

References

1. Rabiner, L.R., Juang, B.-H., Yegnanarayana, B.: Fundamentals of speech recognition. Pearson Education (2010)
2. Census of India 2011: http://censusindia.gov.in/
3. Juang, B.H., Rabiner, L.R.: Automatic speech recognition—a brief history of the technology. 1–24 (2005)
4. Davis, S.B., Mermelstein, P.: Comparison of parametric representations for monosyllabic word recognition in continuously spoken sentences. IEEE Trans. Acoust. **28**, 357–366 (1980). https://doi.org/10.1109/TASSP.1980.1163420
5. Dua, M., Aggarwal, R.K., Biswas, M.: Discriminatively trained continuous Hindi speech recognition system using interpolated recurrent neural network language modeling. Neural Comput. Appl. **31**, 6747–6755 (2019). https://doi.org/10.1007/s00521-018-3499-9
6. China Bhanja, C., Laskar, M.A., Laskar, R.H., Bandyopadhyay, S.: Deep neural network based two-stage Indian language identification system using glottal closure instants as anchor points. J. King Saud Univ. Comput. Inf. Sci. (2019). https://doi.org/10.1016/j.jksuci.2019.07.001
7. Goel, S., Pangasa, R., Dawn, S., Arora, A.: Audio acoustic features based tagging and comparative analysis of its classifications. In: 2018 11th International Conference Contemporary Computing IC3 2018, pp. 1–5 (2018). https://doi.org/10.1109/IC3.2018.8530512
8. Tufekci, Z., Gowdy, J.N.: Feature extraction using discrete wavelet transform for speech recognition. In: Conference Proceedings—IEEE SOUTHEASTCON. pp. 116–123 (2000)
9. Pandit, P., Bhatt, S.: Automatic speech recognition of Gujarati digits using dynamic time warping. Int. J. Eng. Innov. Technol. **3**, 69–73 (2014)
10. Pandit, P., Bhatt, S., Makwana, P.: Automatic speech recognition of Gujarati digits using artificial neural network. In: Proceedings of 19th Annual Cum 4th International Conference of GAMS On Advances in Mathematical Modelling to Real World Problems. pp. 141–146. Excellent Publishers (2014)
11. Pandit, P., Bhatt, S.: Automatic speech recognition of Gujarati digits using radial basis function network. In: International Conference on Futuristic Trends in Engineering, Science, Pharmacy and Management. pp. 216–226. A D Publication (2016)
12. Pandit, P., Bhatt, S.: Automatic speech recognition of Gujarati digits using wavelet coefficients. J. Maharaja Sayajirao Univ. Baroda. **52**, 101–110 (2017)
13. Tufekci, Z., Gowdy, J.N., Gurbuz, S., Patterson, E.: Applied mel-frequency discrete wavelet coefficients and parallel model compensation for noise-robust speech recognition. Speech Commun. **48**, 1294–1307 (2006). https://doi.org/10.1016/j.specom.2006.06.006
14. Rumelhart, D.E., Hinton, G.E., Williams, R.J.: Learning representations by back-propagating errors. Nature **323**, 533–536 (1986). https://doi.org/10.1038/323533a0

15. Kingma, D.P., Ba, J.L.: Adam: A method for stochastic optimization. In: International Conference on Learning Representations. pp. 1–15 (2015)
16. Maas, A., Hannun, A., Ng, A.: Rectifier nonlinearities improve neural network acoustic models. In: ICML Workshop on Deep Learning for Audio, Speech and Language Processing (2013)

Effectiveness of RSM Based Box Behnken DOE over Conventional Method for Process Optimization of Biodiesel Production

Kartikkumar Thakkar, Ammar Vhora, Pravin Kodgire, and Surendra Singh Kachhwaha

Abstract The conventional way of performing experiments has many drawbacks in terms of time and resources investments and lack of accuracy in analysis. Use of scientific tools for design of experiments (DoE) is a better approach for experimentation such that, minimum number of experiments can successfully give requisite aspects of analysis. Currently, many researchers still use the conventional way for performing experiments which results in higher utilization of resource and time investment, while the same objectives can be achieved by combined application of DoE with response surface methodology (RSM) which utilizes minimal efforts, time and resources. The experimental data from a published research paper have been considered as a case study in which conventional approach was used to decide to conduct experimental runs for transesterification process. The authors reported the optimized reaction conditions viz. methanol to oil molar ratio (6:1), catalyst amount (1 wt. %), reaction temperature (60 °C), and reaction time (15 min) for biodiesel production (98.1% yield) using hydrodynamic cavitation technique. They used a method of varying one factor at a time while the other three factors were constant. Twelve sets of experiments with total 176 number of sample observations were performed with varying factor conditions. Compared to this fifteen-run based on

K. Thakkar · A. Vhora · S. S. Kachhwaha
Department of Mechanical Engineering, Pandit Deendayal Petroleum University, Gandhinagar, India
e-mail: kartik.tphd16@sot.pdpu.ac.in

A. Vhora
e-mail: ammar.vmtmt18@sot.pdpu.ac.in

S. S. Kachhwaha
e-mail: surendra.singh@sot.pdpu.ac.in

P. Kodgire (✉)
Department of Chemical Engineering, Pandit Deendayal Petroleum University, Gandhinagar, India
e-mail: pravin.kodgire@sot.pdpu.ac.in

K. Thakkar · A. Vhora · P. Kodgire · S. S. Kachhwaha
Centre of Biofuel and Bioenergy Studies (CBBS), Pandit Deendayal Petroleum University, Gandhinagar, India

M. Sahni et al. (eds.), *Mathematical Modeling, Computational Intelligence Techniques and Renewable Energy*, Advances in Intelligent Systems and Computing 1287,
https://doi.org/10.1007/978-981-15-9953-8_14

Box-Behnken technique with RSM has been used in the present study to predict the optimized reaction condition. The accuracy of the predicted yield and operating parameters were 99.99% and 99.8% respectively. Thus Box-Behnken DoE method with RSM is clearly a better approach as compared to the conventional one in terms of better resource utilization and time saving.

Keywords Box-Behnken · Design of experiments · Response surface methodology (RSM) · Hydrodynamic cavitation · Biodiesel

1 Introduction

Frequent evolution of different products around us is the result of experience from the past, feedback from customers, change of requirement as well as vision and perception of humans for the betterment of the future. It is also known that change is the only constant as per the nature of the universe [1]. However, the random evolution of any product or development of new products without systematic investigation may lead to wastage of time and resources. Therefore, for the systematic evaluation of any product in any scientific field, thorough research is required which consumes resources and time. Products and processes are normally derived from research in the engineering and scientific disciplines. Investment of resources and time for the research may be viable up to a certain extent as per the return in terms of the betterment of the product or outcome of the system is concerned. Therefore, the main aim of the research should be obtaining the best possible output from the system or product which is only possible after analyzing factors affecting the response or output.

Conventionally, optimization in research is carried out by changing one variable at a time to analyze the effect of it while other variables are kept constant. However, this approach cannot be helpful to study the interaction effects of the variables on the response. Therefore, the conventional approach cannot depict the complete effect of the variables on response [2]. The conventional approach in research can also lead to the consumption of more resources and time due to a large number of sample preparation and its analysis. Further, the uncertainty of the optimized conditions is higher as it cannot predict the global parameter setting for optimum output as it is dependent on the starting condition [3].

Multivariable optimization can be carried out using experimental designs that suggest a set of experiments to be carried out to analyze the complete effects of different factors on response. It also reduces the degree of biases in experiments [4]. The design of experiments (DoE) uses statistical methods to identify important factors and can reduce uncontrollable variables. There are mainly two approaches in DoE viz. full factorial design and response surface methodology (RSM). DoE in conjunction with RSM can reduce the experimental load by suggesting only a set of critical experiments that can analyze all the factors and their effects [5]. DoE can fulfill different research objectives such as identification of important factors, estimation of the impact of those factors on the specified response(s), identification of interaction effects, and development of a mathematical model to obtain optimum factor conditions and achieve the highest possible response [6]. The first step for DoE is the identification of important factors followed by a selection of the range of the factors. The range of the factors can be selected based on preliminary experiments, experience, and literature study. Different values such as −1, 0, and 1 are assigned to the factor levels, indicating low, medium, and high which may be quantitative or qualitative in actuality. Response (Y) of any system can be mathematically defined as the function of different factors $X_1, X_2, X_3 \ldots X_n$ (refer Eq. (1)).

$$Y = f(X_1, X_2, X_3, \ldots, X_n) \tag{1}$$

Considering the quadratic model represented by an empirical approximate generalized equation can be written as given in Eq. (2) [7].

$$Y = b_0 + \sum_{i=1}^{n} b_i X_i + \sum_{i=1}^{n} b_{ii} X_i^2 + \sum_{i=1}^{n} \sum_{j=i}^{n} b_{ij} X_i X_j + \varepsilon \tag{2}$$

where Y is the output/response variable, X_i and X_j are the coded factors, b_0 is a constant coefficient to fit the intercept, b_i is the linear coefficient, b_{ii} is the quadratic coefficient, b_{ij} is interaction coefficient and ε is an error in prediction of output/response variable.

Many researchers till now have used RSM coupled with different techniques, i.e. artificial neural network (ANN) [8] and Placket-Burman [9], but the most commonly used techniques for different research applications are Box-Behnken [10] and central composite design (CCD) [11, 12]. The mathematical model developed through these techniques is generally analyzed using analysis of variance (ANOVA). ANOVA is a statistical technique that can be used to analyze the effect of different factors/terms (main effect (x_i)/quadratic (x_i^2)/interaction effect ($x_i x_j$) terms) present in the developed mathematical model. The significance of each term can be tested using an F-test in hypothesis testing. Generally, 95% of confidence interval corresponding to a 5% significance level is considered, however it may vary depending upon the accuracy required in the prediction using the mathematical model. Many researchers have used RSM technique and ANOVA has been performed for its analysis, however, there aren't enough studies available for the critical analysis, resource reduction, and time-saving achieved through the application of RSM technique.

The main aim of the present study is to compare analysis obtained through the conventional technique with that of obtaining through Box-Behnken with RSM. Moreover, it is also important to analyze time-saving and resource reduction achieved through effective utilization of the RSM technique which can improve the quality and speed of the research.

Table 1 Factors and levels for DoE

Factors	Denotation	Levels		
		−1	0	+1
Oil: MeOH molar ratio	$X_1 = A$	1:5	1:6	1:7
Catalyst (% w/w)	$X_2 = B$	0.75	1	1.25
Time (min)	$X_3 = C$	10	15	20

2 Methodology

2.1 Data Collection

Published data of [13] which follow the conventional way of experimentation, i.e. varying one factor at a time for the optimization of reaction parameters of biodiesel production through hydrodynamic cavitation technique were considered for the current study. They used waste cooking palm olein oil for transesterification reaction with methanol in the presence of a homogeneous catalyst (KOH). Effect of operating parameters on methyl ester conversion was analyzed and a comparison of hydrodynamic cavitation technique with the mechanical stirring technique was given by using the conventional method of experimentation.

The quantitative data of the experiments and results were extracted using web plot data generator software from the graphs given in [13]. The factors identified for DoE were kept the same as those mentioned by [13]. The levels of the factors were decided from the range of experiments performed by the authors as given in Table 1. Run numbers 1, 6, 9, and 15 which could not be extracted from the paper were interpolated using the critical trend analysis from the graphs in the paper. List of experiments considered for RSM (Box-Behnken DoE) along with methyl ester conversion (Y) as extracted from [13] is given in Table 2. Predicted conversions given in Table 2 were obtained through the regression equation developed through the present study, which is explained in sub-sequent Sect. 3.1.

2.2 Statistical Parameters

Different statistical parameters viz. sum of squares, mean sum of square of errors (MSE), F-value, degrees of freedom (dof), p-value, correlation coefficient (R^2) adjusted correlation coefficient (R^2_{adj}) and standard error (SE) were calculated which gives critical statistical inference for the analysis of results obtained through DOE and RSM. For the effective analysis of the results, all these parameters are needed to be understood thoroughly for proper analysis.

Degrees of freedom of regression equation are the number of terms used to represent the regression equation. For example, if in the regression equation obtained

Table 2 List of observations considered for DoE

Run order	A	B	C	Predicted conversion (%)	Experimental conversion (%)
1	5	1.25	15	80.12	80.89
2	6	1.25	10	74.14	74.27
3	7	1	20	89.68	90.57
4	6	1	15	97.58	98.25
5	6	1.25	20	88.43	90.06
6	5	0.75	15	63.02	65.54
7	5	1	10	56.31	55.43
8	6	0.75	20	80.52	80.41
9	7	0.75	15	89.25	88.49
10	6	1	15	97.58	97.00
11	5	1	20	66.10	63.71
12	7	1	10	68.85	71.25
13	6	0.75	10	64.19	62.57
14	6	1	15	97.58	97.50
15	7	1.25	15	90.01	87.50

in a 3^3 full factorial design, there are 10 total terms (main effects and interaction) and one constant term, then the degrees of freedom of the regression equation is 10 as one degree of freedom is imparted by each term. In DoE, a total of 27 observations are required to be taken which imparts 16 extra degrees of freedom (subtracting one dof for the constant term) which forms degrees of freedom for error. Thus the error term has 16° of freedom.

Sum of squares gives the total variability of observed/predicted data around its mean when we talk about the total/regression sum of squares and can be expressed as given in Eqs. (3 and 4). While the error sum of squares can be expressed as a summation of squares of residuals (refer Eq. (5)).

$$\text{Total sum of squares}(\text{SS}_\text{T}) = \sum_{i=1}^{n} (Y_i - \bar{Y})^2 \tag{3}$$

$$\text{Regression sum of squares}(\text{SS}_\text{R}) = \sum_{i=1}^{n} \left(Y_i' - \bar{Y}'\right)^2 \tag{4}$$

$$\text{Error sum of squares}(\text{SS}_\text{E}) = \sum_{i=1}^{n} \left(Y_i' - Y_i\right)^2 \tag{5}$$

where Y_i= observed data, Y_j' = predicted data, $\bar{Y}$ = mean of actual data, $\bar{Y}'$ = mean of predicted data, n = number of observation

Adjusted sum of squares is the type of error sum of squares that is calculated considering all the terms in the regression model other than the given term whose adjusted sum of square value is needed to be determined. The adjusted sum of squares is also referred to as the partial sum of squares [14].

Mean sum of squares of any term is the ratio of the adjusted sum of squares of that term to the total dof of the regression equation. It signifies the difference in data due to the presence of that term in the regression equation.

$$MS_R = \frac{SS_R}{\text{dof of regression}} \tag{6}$$

$$MS_E = \frac{SS_E}{\text{dof of error}} \tag{7}$$

F value is the ratio of the adjusted mean sum of square to the error mean sum of squares for the given term. It can be used to check the significance of the term by comparing it to the F value in its distribution and thus finding the P-value (probability for $F > F_{cr}$). For comparing calculated F value with critical F value from F-Table, degrees of freedom for the numerator and denominator is required. For example, if we want to compare the F value of interaction term X_2, then degree of freedom in the numerator and denominator will be 1 and 16 respectively. Where 1 is the degree of freedom for the given term and 16 is the degree of freedom of error term. It is checked against the required confidence interval.

$$F \text{ value} = \frac{MS_R}{MS_E} \tag{8}$$

P-value can be defined as the probability that the calculated F value will be less than tabulated critical F value (F_{cr}) for the given degrees of freedom and confidence interval. F value needs to be higher for the factor to be significant to respond. Thus P value should be minimum than the significance level ,which is given as one minus confidence interval. If the confidence interval for the statistics is required to be 95%, then a significance level of p-value is, $1 - \text{C.I.} = 1 - 0.95 = 0.05$. Thus P-value calculated from ANOVA needs to be smaller than 0.05 for the term to be significant.

Coefficient of determination (R^2) gives the measure of the variability of the predicted response obtained through the regression model. The total variability of the response values can be indicated by the term "total sum of squares" where as regression sum of squares indicates the amount of variability explained by the regression model. Coefficient of determination is a relative measure indicating the ability of the regression model to explain the variation in the response and can be given by the ratio of regression sum of squares and total sum of squares.

$$\text{Coefficient of determination, } R^2 = \frac{SS_R}{SS_T} \tag{9}$$

The coefficient of determination always increases by adding more terms in the model. An increase in R^2 value doesn't signify the superiority of a new model and can lead to overfitting and wrong interpretation of data as there may be the presence of insignificant terms that should be avoided. For better statistics, adjusted $R^2(R^2_{adj})$ can be defined as in Eq. (10).

$$R^2_{adj} = 1 - \frac{MS_E}{MS_T} \tag{10}$$

Unlike R^2 (Coefficient of determination), adjusted R^2 doesn't always increase with the addition of extra terms in the model. If insignificant terms are added in the model, it decreases the value of R^2_{adj}. Thus R^2_{adj} is an important term for the analysis of the regression model.

Standard error is the standard deviation of the sample and was calculated using Eq. (11). Standard deviation is the square root of variance.

$$S = \sqrt{\frac{\sum(Y' - Y)^2}{n - 1}} \tag{11}$$

Using the statistical parameters explained above significant terms in the regression equation affecting the yield of biodiesel can be identified.

3 Results and Discussion

3.1 Regression Equation and Analysis of Variance

The regression equation in terms of coded factors developed through the curve fitting using the least square method is given in Eq. (12).

$$\begin{aligned} Y(\%) = {} & 97.58 + 9.03\mathrm{A} + 4.46\mathrm{B} + 7.65\mathrm{C} - 11.78\mathrm{A}^2 - 5.20\mathrm{B}^2 \\ & - 15.56\mathrm{C}^2 - 4.08\mathrm{AB} + 2.76\mathrm{AC} - 0.5125\mathrm{BC} \end{aligned} \tag{12}$$

Predicted conversion of biodiesel (%) obtained through Eq. (12) is given in Table 2. It can be observed from Fig. 1a that the predicted data fit well with experimental data and error is within ±5%. Figure 1b shows the acceptance of the model with the value of studentized residual within a permissible limit of 95% confidence interval. It can be observed from Fig. 1c that the conversion data follow the reference line and is normally distributed which indicates that the test of the model is significant [15]. Fig. 1d shows the Pareto chart of the standardized effect and it is noticeable that all the linear and quadratic terms in the regression equations are found to be significant. The most significant term affecting the conversion of biodiesel is a C^2, while

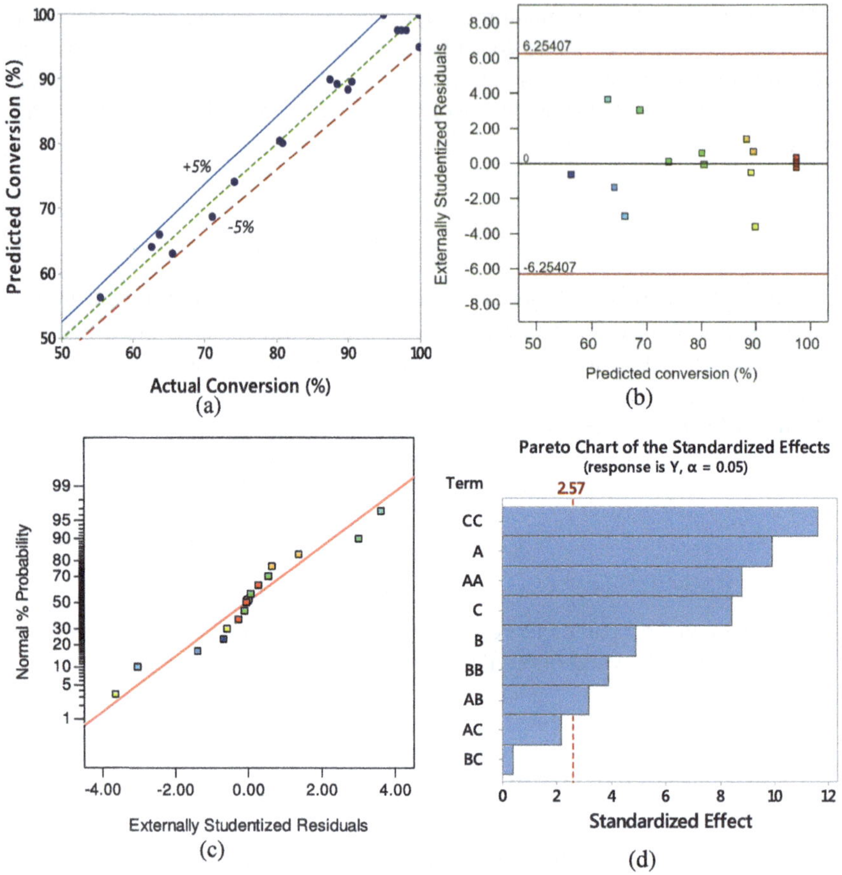

Fig. 1 **a** Comparison of actual and predicted conversion (%) **b** externally studentized residuals **c** normal probability plot of residuals **d** Pareto chart of the standardized effects

interaction terms AC and BC are insignificant as can be observed from Fig. 1d. The developed regression model is significant with R^2 and R^2_{adj} value of 0.99 and 0.97 respectively, moreover the p-value for the regression model ($p < 0.05$) also justifies the significance of the regression equation (refer Table 3). There are total 9 terms in the regression equation (refer Eq. (12)) out of which three terms (A, B, C) represents main effects, three terms (AB, AC, BC) represents interaction effects and remaining three (A^2, B^2, C^2) represents quadratic effects of the factors. It is noteworthy that, the terms AC and BC are not significant in the regression model which shows the absence of interaction effect between MeOH: oil molar ratio (A) and reaction time (C) as well as catalyst loading (B) and reaction time (C). However, the interaction effect between Methanol to oil molar ratio (A) and catalyst loading (B) is significant as a combination of these two forms methoxide solution which is responsible for

Table 3 ANOVA parameters

Source	Sum of squares	dof	Mean square	F-value	p-value	Remark
Model	2728.12	9	303.12	45.98	0.0003	Significant
A	652.33	1	652.33	98.95	0.0002	Significant
B	159.40	1	159.40	24.18	0.0044	Significant
C	468.64	1	468.64	71.09	0.0004	Significant
AB	66.75	1	66.75	10.13	0.0245	Significant
AC	30.47	1	30.47	4.62	0.0843	Insignificant
BC	1.05	1	1.05	0.1594	0.7062	Insignificant
A^2	512.63	1	512.63	77.76	0.0003	Significant
B^2	99.66	1	99.66	15.12	0.0115	Significant
C^2	894.01	1	894.01	135.62	< 0.0001	Significant
Residual	32.96	5	6.59			
Std. error	0.7917	2	0.3958			
Total	2761.08	14				
$R^2 = 0.99$	$R^2_{\text{adj}} = 0.97$					

transesterification reaction. Therefore, the relative proportion of methanol and catalyst loading is important to study to optimize the reaction condition. However, the conventional way of experimentation lacks an analysis of such interaction effects. DoE also helps to find the contribution of each term for the response (Y) which is valuable information for the research in any field.

3.2 *Main Effects and Interaction Effects*

The perturbation plots show the main effects of all three factors on biodiesel conversion as can be seen from Fig. 2. From the trend of MeOH: oil molar ratio (A), it can be followed that there is a quick increase in biodiesel conversion with an increase in A up to midlevel 6:1, however further increase in A above 6.5:1 decreases biodiesel conversion. It can also be followed that variation in catalyst loading (B) has the least effect on the conversion percentage (Y) compared to the other two factors (A and C). The effect of reaction time (C) on conversion (%) is also observed to be significant with its steeper slope up to mid level. The conversion increases significantly with an increase in reaction time (C) up to mid level, however, there is a reduction in conversion when reaction time is increased beyond 16 min. Maximum biodiesel conversion of 98.25% was obtained at optimized reaction condition of 6.1:1 MeOH: oil, 1% (w/w) of catalyst loading and 16 min of reaction time from the partial differentiation of developed regression equation which is similar to those obtained in the referred paper, however, a higher number of samples using conventional experimentation were analyzed by authors to conclude similar results.

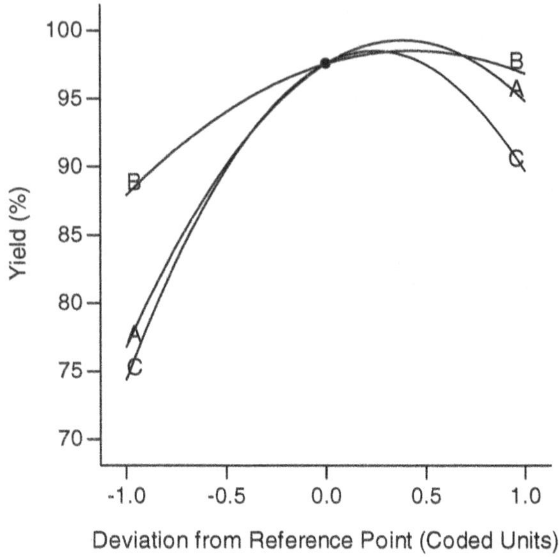

Fig. 2 Perturbation plots showing the main effects of the factors

The authors could not comment on the interaction effect of the factors due to the limitation imparted by conventional technique while RSM based Box-Behnken technique could successfully analyze interaction effect which can be interpreted from the surface plots given in Fig. 3. It is easy to follow from Fig. 3b, c that the trend of biodiesel conversion for changes in reaction time and methanol: oil molar ratio are similar which indicates the absence of interaction effect of these two factors. The same can also be observed in the ANOVA table (refer to Table 3). The same can also be said for reaction time and catalyst loading. However, the interaction effect of methanol to oil molar ratio and catalyst loading is significant which can be observed from Fig. 3a. Any change in the value of methanol: oil molar ratio reflects a change in the optimal value of catalyst loading to obtain maximum yield due to interaction effect of the factors which is due to the fact that the proportion of methanol and catalyst loading decides the formation of methoxide solution which is responsible for transesterification reaction. Thus, interaction effects are also important to understand the principle behind the reaction mechanisms and helps to obtain optimum factor level setting to achieve maximum conversion which is not possible using the conventional method of experimentation.

3.3 Comparison of RSM with the Conventional Technique

A generalized procedure for experimentation and sample analysis was considered for both the techniques of experimentation. For a single sample time allocated for pretreatment, reaction, separation, methanol recovery, washing, and GC analysis

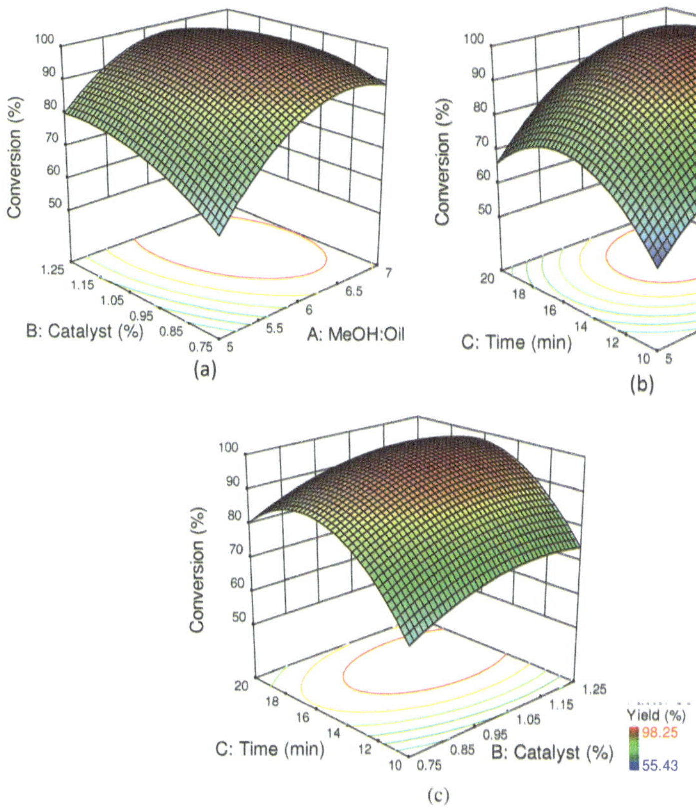

Fig. 3 Surface plots representing interaction effects of **a** catalyst and methanol: oil **b** reaction time and methanol: oil **c** reaction time and catalyst loading

was considered. The average time consumed to analyze a single sample was found to be about 1 h. The authors in the referred study analyzed 176 samples which may consume about 22 days considering 8 h/day as working hours. It is easy to follow from Table 2 that with the application of DoE coupled with the Box-Behnken technique, only 15 such samples are required to be prepared and analyzed. However, the time required to prepare and analyze such samples is higher compared to samples prepared from conventional technique since each experimental set has a unique combination of factor levels which require separate experiment compared to the conventional technique which requires only single experiment for the complete range of reaction time. The average time required to prepare samples of DoE coupled with the Box-Behnken technique was found to be 3 h/sample. Thus, the Box-Behnken technique requires 6 days of total experimentation time, which is 3.67 folds less compared to the conventional technique. Similarly, the cost of analysis of the sample can also be decreased by about 12 folds with a decrease in the number of samples (by 12 folds) to be analyzed.

4 Conclusions

DoE coupled with RSM techniques is reliable and accurate to predict optimum process parameters. It also leads to better interpretation and analysis of the results along with saving in research time (3.67 folds) and resource allocation (12 folds). Therefore, researchers of today's generation should be more focused to employ different DoE tools for more organized research with minimum resource and time allocation.

References

1. Stenling, A., Ivarsson, A., Lindwall, M.: The only constant is change: analysing and understanding change in sport and exercise psychology research. Int. Rev. Sport. Exerc. Psychol. **10**, 230–251 (2017)
2. Bezerra, M.A., Santelli, R.E., Oliveira, E.P.: Response surface methodology (RSM) as a tool for optimization in analytical chemistry. Talanta **76**, 965–977 (2008)
3. Dejaegher, B., Vander, Heyden Y.: Experimental designs and their recent advances in set-up, data interpretation, and analytical applications. J. Pharm. Biomed. Anal. **56**, 141–158 (2011)
4. Yu, P., Low, M.Y., Zhou, W.: Design of experiments and regression modelling in food flavour and sensory analysis: a review. Trends Food Sci. Technol. **71**, 202–215 (2018)
5. Bravo-Linares, C., Ovando-Fuentealba, L., Mudge, S.M., Loyola-Sepulveda, R.: Application of response surface methodology to oil spill remediation. Fuel **103**, 876–883 (2013)
6. Mousavi, L., Tamiji, Z., Khoshayand, M.R.: Applications and opportunities of experimental design for the dispersive liquid–liquid microextraction method–a review. Talanta **190**, 335–356 (2018)
7. Omorogie, M.O., Naidoo, E.B., Ofomaja, A.E.: Response surface methodology, central composite design, process methodology and characterization of pyrolyzed KOH pretreated environmental biomass: mathematical modelling and optimization approach. Model Earth Syst. Environ. **3**, 1171–1186 (2017)
8. Elfghi, F.M.: A hybrid statistical approach for modeling and optimization of RON: a comparative study and combined application of response surface methodology (RSM) and artificial neural network (ANN) based on design of experiment (DOE). Chem. Eng. Res. Des. **113**, 264–272 (2016)
9. Ahuja, S.K., Ferreira, G.M., Moreira, A.R.: Application of Plackett Burman design and response surface methodology to achieve exponential growth for aggregated shipworm bacterium. Biotechnol. Bioeng. **85**, 666–675 (2004)
10. Dharma, S., Masjuki, H.H., Ong, H.C.: Optimization of biodiesel production process for mixed Jatropha curcas–Ceiba pentandra biodiesel using response surface methodology. Energy Convers. Manag. **115**, 178–190 (2016)
11. Behera, S.K., Meena, H., Chakraborty, S., Meikap, B.C.: Application of response surface methodology (RSM) for optimization of leaching parameters for ash reduction from low-grade coal. Int. J. Min. Sci. Technol. **28**, 621–629 (2018)
12. Thakkar, K., Shah, K., Kodgire, P., Kachhwaha, S.S.: In-situ reactive extraction of castor seeds for biodiesel production using the coordinated ultrasound–microwave irradiation: process optimization and kinetic modeling. Ultrason. Sonochem. **50**, 6–14 (2019)
13. Chuah, L.F., Yusup, S., Aziz, A.R.A.: Intensification of biodiesel synthesis from waste cooking oil (Palm Olein) in a hydrodynamic cavitation reactor: effect of operating parameters on methyl ester conversion. Chem. Eng. Process Process Intensif **95**, 235–240 (2015)

14. Montgomery, D.C.: Design and analysis of experiements, 8th ed. Wiley (2015)
15. Sharma, A., Kodgire, P., Kachhwaha, S.S.: Biodiesel production from waste cotton-seed cooking oil using microwave-assisted transesterification: optimization and kinetic modeling. Renew. Sustain. Energy Rev. **116**, 109394 (2019)

Dealing with COVID-19 Pandemic Using Machine Learning Technique: A City Model Without Internal Lockdown

Sushil Chandra Dimri, Umesh Kumar Tiwari, and Mangey Ram

Abstract COVID-19 pandemic is a serious threat to human being all over the world. No vaccine for COVID-19 has been discovered yet. This virus attack on human civilization is not the first, and it would not be the last. Now, the time has come that we must have to learn how to live under such unforeseen type of virus attacks. A little change in health infrastructure and use of ICT can make us capable to deal with such type of viral pandemic. A reliable test kit can help us to isolate the infected people from the local population and then treat them specially. The zone free from infected persons is the green zone which is free to go for their daily life routines as it is. This will support a nation, an individual economically, mentally, and we will feel normalcy without any panic in the city. This paper presents a model which helps to isolate infected person from the population, create different zones and deal each zone separately. The paper suggests that no need to lockdown the city internally, even controlled communication with other cites is also possible.

Keywords Kernel · Quarantine · Critical · Pandemic · Paraboloid · COVID-19

1 Introduction

As of today, corona virus 2019, popularly called as COVID-19 and announced as pandemic [1], has spread almost every part of the world and infected almost every

S. C. Dimri (✉) · U. K. Tiwari · M. Ram
Department of Computer Science and Engineering, Graphic Era Deemed to be University, Dehradun 248002, India
e-mail: dimri.sushil2@gmail.com

U. K. Tiwari
e-mail: umeshtiwari22@gmail.com

M. Ram
e-mail: drmrswami@yahoo.com

U. K. Tiwari · M. Ram
Department of Mathematics, Graphic Era Deemed to be University, Dehradun 248002, India

M. Sahni et al. (eds.), *Mathematical Modeling, Computational Intelligence Techniques and Renewable Energy*, Advances in Intelligent Systems and Computing 1287,
https://doi.org/10.1007/978-981-15-9953-8_15

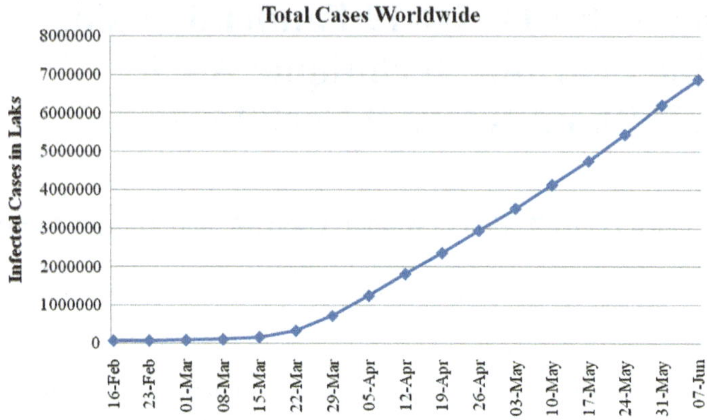

Fig. 1 Corona infected cases worldwide [2]

age group of human being. At the time of this work, there are more than 7 million cases worldwide. Major countries infected by this pandemic disease include US more than 2.3 million cases, Italy 1.8 million cases, Brazil 0.5 million cases and India 0.2 million cases. More than 0.35 million causalities are reported till date. Following Fig. 1 shows total cases worldwide infected by COVID-19 disease [2].

There is no vaccine for the COVID-19 till date. To prevent people from the COVID-19's effect, governments of various countries are following 'preventive measures', including application of complete lockdown state wide as well as country wide. Locking down a city, state or country affects the economical, social and mental wellness of people. Lockdown not only puts infected people insecure but also the whole society comes under threat.

In this paper, we suggest a model as a solution to deal with the problem of lockdown due to the attack of COVID-19 virus. This model helps to isolate infected, potential person from the rest population and create different zones and deals each zone separately. Introduction of the work is given in Sect. 1. Section 2 discusses some related work in the field of dealing with COVID-19. Section 3 discusses the proposed model, and Sect. 4 implements a case study of the proposed model. Section 5 concludes the work.

2 Related Work

World's largest nationwide lockdown is implemented by government of India and is extended up to 30th June 2020. The implementation of the lockdown in such a populated country requires huge preparedness as it is going to affect billions of people. 'There has been a mass exodus of migrant workers and concerns are rising about starvation among people who work in the informal economy' [3]. In their work, authors urged the governments of different states to apply some 'universal basic income' for deprived and vulnerable citizens of the state who are affected by lockdowns [4].

In their work, Rothan and Siddappa [5] suggested to take extensive measures to minimize personal contacts to decrease the effects of COVID-19. Their suggestion is to control the human-to-human interactions so that the transmissions can be controlled at various levels of society [6–8]. In the work [9], authors highlight the scope and beliefs of science and technology to design and implement the social distancing. Authors discussed regarding the role of information and trust in media and institutions in determining physical distancing responses. Gopalkrishna et al. [10] given a sentiment analysis of national lockdown due to the pandemic. To implement their studies, authors have used social media platforms like Twitter. This study discusses the sentiments of people during lockdown.

Sulaman et al. [11] discuss the environmental effects of COVID-19. They discusses that the lockdown throughout the world due to the pandemic is helpful to decrease the pollution, though this temporary. Bhuiyan et al. [12] describe the severe side effects of lockdown in the form of committing suicide by people who are not able to seek food, medical facilities or have the challenge of survival. In addition, authors' discuss some serious interruptions of necessary services because of the lockdown like adequate supply of food, manufacturing and accessibility. Zahir et al. [13] presented a study that discusses the harmful psychological consequences and related disaffects of COVID-19. According to the work, 'infected by COVID-19 or similar is not prerequisite to develop psychological problems and disorders, i.e. anxiety, depression, alcohol use disorder rather circumstantial effects, locked down in own home for infinite time, infection of family and friends, death of closed one all these could worsen the overall mental health well-being' [13].

Authors [14] conducted a survey with ten thousand participants to assess the effects of households' spend and other economic expectations. In that survey, authors find that about fifty per cent of respondents face loss of income and other monetary losses due to the corona virus. In their work [15], authors illustrated a survey-based analysis in their work. This survey was done with 1260 ophthalmologists as participants while the 21 days lockdown in India. Results of the survey show that 'majority of ophthalmologists in India were not seeing patients during the COVID-19 lockdown, with near-total cessation of elective surgeries' [15].

3 Proposed City Model

In this section, we are giving a model to counter such pandemic situations. Let a city A has population n, and we assume that each person has mobile phone with Internet connectivity. The city data centre collects information from each and every resident of the city on following *six* aspects and creates information set for each individual.

The aspects are as follows:

X_1: Travel history

X_2: Any medical problem
X_3: Age > 65
X_4: Medical staff/Sanitization staff
X_5: Contact with infected person
X_6: COVID-19 related symptoms

where all x_i are Boolean values.
i.e. $x_i \in \{0, 1\}$ for $1 \leq i \leq 6$
x_i is either 0 or 1, 0 for false and 1 for true.
After getting this information from each individual, we have '*n*' set of values like: $[x_1, x_2, x_3, x_4, x_5, x_6]$, which is like a six digit binary number.
For example, [0 1 0 0 0 1]
Now, city data centre has this data, and conclusion has to be designed about the residents of the city that what should be the appropriate action so that the city handles this pandemic situation of COVID-19. For each dataset k-

$$W_{ik} = \frac{\left(2^2 X_3 + 2^1 X_2 + 2^0 X_1\right)}{10}$$

$$W_{jk} = \frac{\left(2^2 X_6 + 2^1 X_5 + 2^0 X_4\right)}{10}$$

Divide each dataset into two parts each of *three* bits and then convert them into equivalent decimal numbers, now each dataset k $(1 \leq k \leq n)$.
We have two decimal numbers $\left(w_{ik}, w_{jk}\right)$
where $w_{ik}, w_{jk} \geq 0$
For each data set k, we have set of decimal values $\left(w_{ik}, w_{jk}\right)$ that are shown in Fig. 2.
Now, using the kernel trick to convert this 2D observation in 3D, we take a point $\left(w_{ik}, w_{jk}, z_k\right)$ in 3D for each k, $1 \leq k \leq n$ where $z_k = w_{ik}^2 + w_{jk}^2$

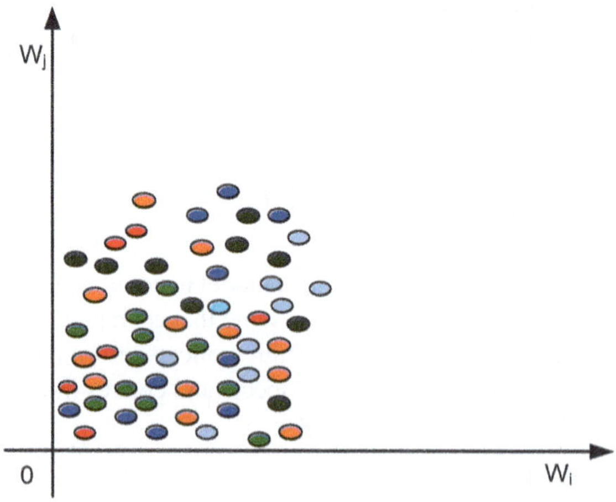

Fig. 2 Set of decimal values (w_{ik}, w_{jk})

Therefore, point $\left(w_{ik}, w_{jk}\right)$ is mapped to $\left(w_{ik}, w_{jk}, w_{ik}^2 + w_{jk}^2\right)$ for each k that is two dimension to three dimension.

This trick of kernel will separate the data point on 2D linearly by converting them into 3D.

$$z_k = w_{ik}^2 + w_{jk}^2$$
$$0 \le w_{ik} \le 0.7$$
$$0 \le w_{jk} \le 5.6$$

For data points in 3D

$\left(w_{ik}, w_{jk}, w_{ik}^2 + w_{jk}^2\right) = (0, 0, 0)$, all these data points in green zone.

For those data points for which $z = 0$, they all are in green zone.

For those data points for which $0 < z < 1$, they all are in home quarantine.

For those data points for which $1 < z < 6.25$, they all are in isolation.

For those data points for which $6.25 < z < 32$, they all are in hospitalization category.

Green zone: Nothing required

Home quarantine: Kit [which includes- mask, sanitizer, basic drugs]

Isolation: Medical assistance + Kit

Hospitalization: All medical facility + Kit

Critical: All medical facility including ICU and ventilator + Kit

For those data points which are COVID-19 confirm will be in critical ICU zone.

For those data points for which ($z > 6.25$) do the continuous laboratory testing $\psi([x])$ on weekly basis that is

$\psi([x_1, x_2, x_3, x_4, x_5, x_6]) = \{0(\text{negative})\}$, further *four* more test with weekly time interval.

If $\psi([x_1, x_2, x_3, x_4, x_5, x_6]) = \{1(\text{positive})\}$, then transfer to the critical zone.

If 5X (times) $\psi([x_1, x_2, x_3, x_4, x_5, x_6]) = \{0(\text{negative})\}$, then transfer data point x to green zone.

For example, taking a data point [1 0 0 1 1 0]

Mapping all the data points from 2D to 3D and we get a circular paraboloid as shown in Fig. 3. The top portion of data point is in critical zone, middle needs isolation, lower middle requires quarantine or home quarantine and bottom data points are in safe green zone. We can set numerical values to classify these zones separately and separate entry class of data points (persons) linearly.

Bottom data points which are in safe zone do not require any internal lockdown. Every week updation is required in upper half, and the updation will transfer above data points to lower bottom part.

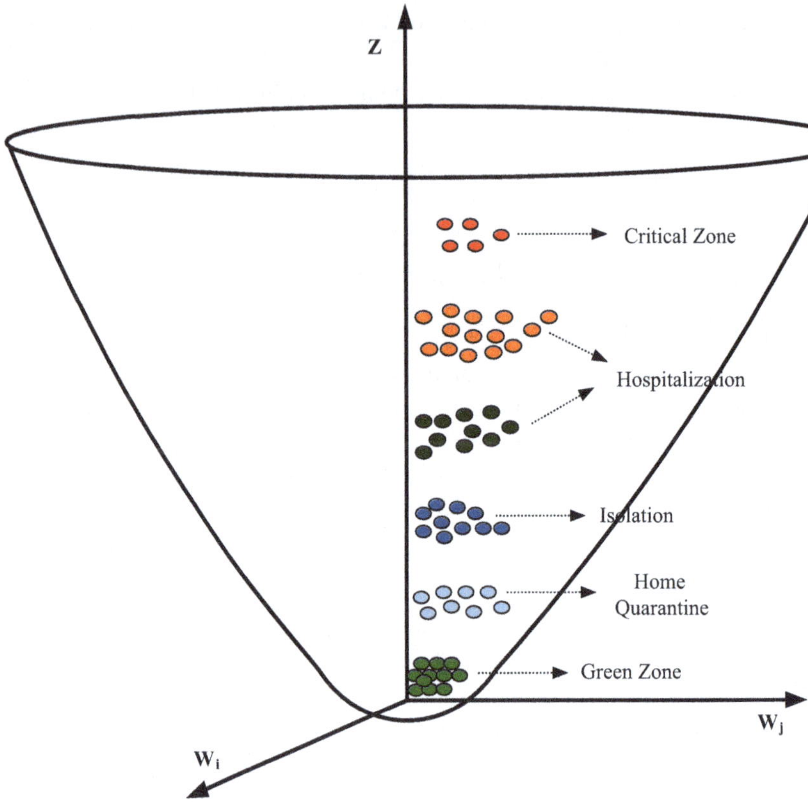

Fig. 3 Set of decimal values (w_{ik}, w_{jk})

4 Case Study

Numerical testing for $n = 1{,}000{,}000$ (Ten lakh city population)

In India, population is 130 crore, and number of infected people till date is 60,000; that is, $1{,}300{,}000/6 = 0.000461\%$. So expected infected people on above population is approximately 5 (i.e. 4.61).

It is not difficult to provide good health care to five people over a population of 1,000,000.

If we assume that total suspects of COVID-19 virus in the city are around 1%, then this number would be around 10,000. This quantity can be distributed among quarantine, home quarantine, isolation and hospitalization. Each facility regarding their health would take the load of 2500 people which is manageable. Transfer of data points is in both the directions that is from bottom to up and from top to bottom. Since the recovery rate is much higher than the mortality rate, so the transfer of data points from bottom to up is lesser than top to down. This ultimately results in increase of size of green zone. And after a period of time, this cycle will function in a constant rate that is manageable with established systems. Remember it is pandemic which requires more devotion and alertness.

5 Conclusion

This paper proposes a model for a city which divides the city in different zones. In this work, we describe a machine learning technique that helps to create various zones and methods to deal with each zone. Green zone is the zone, where no restrictions on public life are required. Once the home quarantine, isolation, hospitalization and critical zones are separated from green zone, then there is no need to put lockdown in in-affected area. A machine learning technique is used to isolate the infected person from the population. 2D data points are mapped to 3D space and then separate them linearly.

References

1. World Health Organization Corona virus 2020. https://www.who.int/healthtopics/coronavirus#tab=tab_1. Last accessed 2020/05/20
2. Worlmeter home page. https://www.worldometers.info/coronavirus/. Last accessed 2020/05/20
3. The Lancet: India under COVID-19 lockdown. Elsevier Public Health Emergency Collection, 395(10233), 1315 (2020). https://doi.org/10.1016/s0140-6736(20)30938-7
4. Pulla, P.: Covid-19: India imposes lockdown for 21 days and cases rise, BMJ 368 (2020). https://doi.org/10.1136/bmj.m1251
5. Hussin, A. R., Siddappa N. B.: The epidemiology and pathogenesis of coronavirus disease (COVID-19) outbreak. J Auto. **109**(102433) (2020)
6. Anderson, R., Hans, H., Don, K., Déirdre, H.: How will country-based mitigation measures influence the course of the covid-19 epidemic? Lancet **395**(10228), 931–934 (2020)
7. Bai, Y., Lingsheng, Y., Tao, W., Fei, T., Dong-Yan, J., Lijuan, C., Meiyun, W.: Presumed asymptomatic carrier transmission of COVID-19. JAMA Res Lett (2020)
8. Viner, R.M., Simon, J.R., Helen, C., Jessica, P., Joseph, W., Claire, S., Oliver, M., Chris, B., Robert, B.: School closure and management practices during coronavirus outbreaks including covid-19: a rapid systematic review. The Lancet Child & Adolescent Health
9. Adam, B., Valentin, K., David, V., Austin, L.: Belief in science influences physical distancing in response to COVID-19 lockdown policies, working Paper-No. 2020-56, Becker Friedman Institute, Apr 2020. Electronic copy available at: https://ssrn.com/abstract=3587990
10. Gopalkrishna, B., Vibha, Giridhar, B.: Sentiment analysis of nationwide lockdown due to COVID 19 outbreak: evidence from India. Asian J Psychiatr. **12**(102089) (2020)
11. Sulaman, M., Xingle, L.: COVID-19 pandemic and environmental pollution: a blessing in disguise?. Sci. Total Environ **728**(138820) (2020)
12. Israfil, B., Najmuj, S., Amir, H. P., Mark, D., Mohammed, A.: COVID-19-related suicides in bangladesh dueto lockdown and economic factors: case study evidence from media reports. Int J Mental Health Addict. https://doi.org/10.1007/s11469-020-00307-y
13. Zahir, A., Oli, A., Zhou, A., Sang, H., Liu, S., Akbaruddin, A.: Epidemic of COVID-19 in China and associated psychological problems. Asian J. Psychiatry **51**(102092) (2020)

14. Olivier, C., Yuriy, G., Michael, W.: The cost of the Covid-19 crisis: lockdowns, macroeconomic expectations, and consumer spending. working paper 27141. http://www.nber.org/papers/w27141. Nber Working Paper Series
15. Akshay, G.N., Rashmin, A.G., Sundaram, N.: Effect of COVID-19 related lockdown on ophthalmic practice and patient care in India: results of a survey. Indian J. Ophthalmol. **68**(5), 725–730 (2020)

Dynamic SentiPhraseNet to Support Sentiment Analysis in Telugu

Santosh Kumar Bharti, Reddy Naidu, and Korra Sathya Babu

Abstract In the scarcity of Telugu, annotated dataset makes sentiment analysis task challenging for researchers in the recent times. A rapid growth was seen in development of annotated datasets in Telugu for sentiment analysis. SentiWordNet (SWNet) is one of them where they mapped a sentiment score to every word. However, we found that there is a limitation in unigram words of SWNet. As several unigram words are ambiguous and it is unable to contribute in sentiment analysis task without the context of word in the given sentence. To resolve such limitations of SWNet, this article proposed SentiPhraseNet (SPNet). SPNet is a collection of sentiment phrases (such as bigram and trigram to get the context of word in the given sentence). Additionally, the proposed approach is compared with existing other approaches as well. The SPNet is also extended for dynamic support which learns the unknown phrases automatically while testing the sentiment of Telugu sentences with SPNet. The proposed approach is outperformed the other existing approaches and attains an accuracy of 90.9% after testing five sets of Telugu sentences in five trials.

Keywords Natural language processing · Sentiphrasenet · Sentiment analysis · Telugu · Dynamic support · News data and NLTK Indian data

1 Introduction

In recent past, Indian languages such as Telugu, Gujarati, Panjabi were the good resource for for sentiment analysis. The several researchers have been inclined toward this domain for detection of sentiment at sentence and document level. One of the main reasons of this is, the Indians are very much influenced toward their native

---

S. K. Bharti (✉) · K. S. Babu
Pandit Deendayal Petroleum University, Gandhinagar, Gujarat 382007, India
e-mail: sbharti1984@gmail.com

R. Naidu · K. S. Babu
National Institute of Technology Rourkela, Rourkela, Odisha768009, India

M. Sahni et al. (eds.), *Mathematical Modeling, Computational Intelligence Techniques and Renewable Energy*, Advances in Intelligent Systems and Computing 1287,
https://doi.org/10.1007/978-981-15-9953-8_16

language in social media chat. It produces a large volumes of regional data in every passing day. These data consist of tweets, reviews, blogs, news, etc.

In India, Telugu is one of the popular languages and it mostly spoken by south India community approx. 75 million native speakers [1]. The everyday data produced by this community users stand second largest data just after Hindi compare to other Indian languages. Annotated dataset availability for Telugu is negligible in the Web due to its morphological complexity which makes sentiment analysis task challenging for the researchers.

In recent times, Das et al. [2] introduced a Telugu SWNet (A bags-of-unigram-words that are tagged with their sentiment polarity in term of positive, neutral and negative.) to detect sentiment polarity from Telugu sentences or documents. The SWNet has a fixed count of unigram words and it leads the several limitations while analyzing sentiment using SWNet:

1. It is not capable to handle unknown words in the given sentences for testing due to fixed count of unigram words.
2. Unigram words are often failed in detection of appropriate sentiment value in several occasion as shown in Fig. 1.
3. The unigram words are often proven as ambigous in the absence of the context of words in the given sentence which leads to misclassification of sentiment value. Few such examples are shown in Fig. 2.

In this article, we proposed the concept of SentiPhraseNet (SPNet) to resolve the ambiguity problem of SWNet's unigram words by replacing bigram and trigram phrases to correctly identify the sentiment of a given sentence as shown in Fig. 1. Similarly, to resolve the contextual ambiguity problem of SWNet, we use the context of surrounding words. Such instances are shown in Fig. 2.

#	Telugu Sentences	SentiWordNet	SentiPhraseNet
1	సైబర్ నేరాల పై ***అవగాహన తక్కువే***, డిజిపి సాంబశివరావు వెల్లడి	అవగాహన (Awareness) (Positive)	అవగాహన తక్కువే (Negative)
2	నగదురహిత చెల్లింపుల పై ***అవగాహన కార్యక్రమాలు***, కళాశాలలకు, విశ్వవిద్యాలయాలకు యూజీసీ సూచన	అవగాహన (Awareness) (Positive)	అవగాహన కార్యక్రమాలు (Positive)
3	అది ప్రక్రియలను, ఫలితాలను ***నిర్లక్ష్యం చేసింది***.	నిర్లక్ష్యం (Neglected) (Negative)	నిర్లక్ష్యం చేసింది (Negative)
4	నీరు ప్రగతిని ***నిర్లక్ష్యం చేయొద్దు***, సిఎం చంద్రబాబు వెల్లడి	నిర్లక్ష్యం (Neglected) (Negative)	నిర్లక్ష్యం చేయొద్దు (Positive)
5	***దార్శనిక నాయకురాలిని కోల్పోయాం***, జయలలిత మృతికి పారిశ్రామిక సంఘాల సంతాపం	దార్శనిక (Philosophical) (Positive)	దార్శనిక నాయకురాలిని కోల్పోయాం (Negative)
6	***గిరిజనుల త్యాగాలకు న్యాయమెక్కడ***? చంద్రబాబు సర్కారుపై ధ్వజమెత్తిన జగన్	త్యాగాలకు (Sacrifices) (Positive)	గిరిజనుల త్యాగాలకు న్యాయమెక్కడ (Negative)
7	ఆదాయం 10 లక్షలు దాటితే ***గ్యాస్ రాయితీ ఉండదు***, పన్ను చెల్లింపుదారుల వివరాలు పెట్రోశాఖకు ఇవ్వనున్న ఐటి విభాగం	రాయితీ (Discount) (Positive)	గ్యాస్ రాయితీ ఉండదు (Negative)
8	***పెరిగిన చలి తీవ్రత***, వాతావరణ శాఖ వెల్లడి	పెరిగిన (Increased) (Positive)	పెరిగిన చలి తీవ్రత (Negative)

Fig. 1 Example of ambiguous situations that fits for SPNet's bigram and trigram phrases over SWNet's unigram words to detect correct sentiment

Telugu Sentences	SentiWordNet	SentiPhraseNet
ఓటుకి నోటు కేసులో ***ఆరోపణ ఎదుర్కొన్న చంద్రబాబు***	ఆరోపణ (Allegation) (Ambiguous)	ఆరోపణ ఎదుర్కొన్న చంద్రబాబు (Allegations faced by Chandrababu) (Negative)
చంద్రబాబు మీదున్న ***ఆరోపణ రుజువు చేస్తే*** దేనికైనా సిద్ధం అన్న తెలుగుదేశం పార్టీ.	ఆరోపణ (Allegation) (Ambiguous)	ఆరోపణ రుజువు చేస్తే (If proves the allegations) (Positive)
పచ్చ కళ్లు. ***పిచ్చి రాతలు***. కృష్ణపట్నంపై ఈనాడు కట్టుకథలు వైఎస్సార్సీపీ మండిపాటు	పిచ్చి (Madness) (Ambiguous)	పిచ్చి రాతలు (Mad Writings) (Negative)
చంద్రబాబు అంటే నాకు ***పిచ్చి ప్రేమ***, తెలిపిన మంత్రి ఉమా మహేశ్వరరావు	పిచ్చి (Madness) (Ambiguous)	పిచ్చి ప్రేమ (Madness of love) (Positive)

Fig. 2 Example of ambiguous situations due to SWNet's unigram words that replaced by bigram and trigram phrases of SPNet to resolve the ambiguity in sentiment analysis

In this approach, a small amount of manually annotated phrases are required to build SPNet. Later, while testing the sentiment, it learns the unknown phrases automatically from Telugu sentences. This dynamic nature of Telugu SPNet is capable of resolving above-mentioned issues of existing SWNet. Also, an algorithm is proposed for sentiment classification using SPNet which classify a sentence into either negative, positive or neutral.

The rest part of this article is arranged as follows: Sect. 2 focused on literature survey. The proposed work of this article is described in Sect. 3. The experimental results are drawn in Sect. 4. Finally, Sect. 5 concludes the article.

2 Literature Survey

In this section, we deals with the related work on identification of sentiment for Indian Languages like Telugu, Gujarati, Punjabi, Tamil, etc.

In order to detect sentiment in social media data scripted in Indian languages, Patra et al. [3] organized an event shared task on sentiment analysis for Indian languages (SAIL) to develop dataset for Indian languages. Researchers have given overwhelming response and participated in massive number with their task of sentiment analysis on text scripted in Indian languages. Prasad et al. [4] exploited decision tree approach to identify sentiment in Indian languages while Kumar et al. [5] introduced the concept of regularized least square (RLS) with randomized feature learning (RFL) to detect sentiment of various Indian languages tweets. Sarkar et al. [6] exploited unigrams, bigrams, trigrams, etc. as feature set to train multinomial Naive Bayes classifier to recognize the sentiment polarity from tweets scripted in Indian languages.

An ontology-based novel approach for finding Hindi named entity recognition (NER) was introduced by Jain et al. [7]. They have followed association rules on the corpus of popular Hindi newspapers. Their main approach is to mine the association rules with respect to dictionary (TYPE 1), bigram (TYPE 2) and feature (TYPE

3) rules. They have performed the experiment with the combination of TYPEs and achieved higher performance. In his subsequent work [8], they used tweets extracted from Twitter to detect NER in health care domain.

The formation of SWNet was done using a game called "Dr. Sentiment" to detect sentiment for Indian languages. Das et al. [9] adopted this gaming with the involvement of internet users for various categories such as age, gender, culture.

After development of dataset for sentiment analysis in Indian languages, as of my best knowledge, a very few work reported so far on identification of sentiment on dataset scripted in Telugu language [10–12]. Mukku et al. [10] exploited the raw corpus developed by Indian Languages Corpora Initiative (ILCI). They used several classical machine learning algorithm such as support vector machine, logistic regression, decision tree, Naive Bayes. to build automatic sentiment analyzer for Telugu language. Naidu et al. [12] utilized Telugu SentiWordNet in Telugu news dataset for sentiment analysis.

3 Proposed Work

In this section, a pipelined process of constructing Telugu SPNet was discussed as shown in Fig. 3. It begins with dataset collection followed by part-of-speech tagging. Further, a method is devised to extract the phrases and add it to the SPNet. Finally, the sentiment annotation (such as positive, negative and neutral) is performed on extracted phrase manually to detect sentiment polarity in testing sentences or documents using proposed SPNet.

3.1 Data Collection

This article used to collect Telugu datasets from several Internet sources, namely Telugu e-Newspapers (namely Eenadu, Sakshi, Andhrajyothy and Vaartha), Twitter

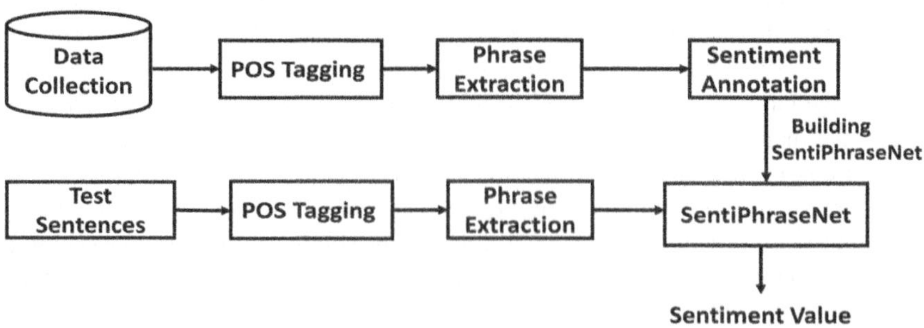

Fig. 3 Framework for sentiment identification in Telugu data

and NLTK Indian Telugu Data for building SPNet. Initially, the SPNet is constituents with 5500 Telugu sentences in the various categories, namely NEWS headlines, tweets and historical articles. Similarly, for testing purpose, approximately twenty-five thousand news sentences were collected.

3.2 Parts-of-Speech Tagging

To identify the correct parts-of-speech information in the collected Telugu dataset, a hidden Markov model (HMM)-based Telugu POS tagger [13] is used.

3.3 Phrase Generation Algorithm

The Algorithm 1 is devised to build the proposed SPNet, where all the 5500 Telugu sentences were given as the input to the Algorithm 1.

Algorithm 1: $Extraction\_of\_Phrases\_for\_SPNet(EPS)$

Input: Telugu dataset ($\mathbb{C}$)
Output: Phrases of SPNet
Notation: JJ: Adjective, VB: Verb, RB: Adverb, N: Noun, S: sentence, $\mathbb{C}$: corpus, TF: tag file, PT: POS tag, SSP: Sentence-wise set of phrase, EPF: Extracted Phrase file
Initialization : $EPF = \{\emptyset\}$

```
while S in C do
    TF = Find_POS_Information (S)
    while T in TF do
        if (Pattern of (RB + N) || (JJ + VB ) || (N + JJ) || (JJ + N) || (N + RB ) || (RB + VB) || (RB + JJ + N) ||
        (VB + JJ + N) || (VB + RB + JJ) || (VB + N + N) || (N + VB + N) || (N + N + VB) is Matched) then
            (PatSet) = Extract_matched_pattern_phrase (TF)
            (SSP) = Phrase[PatSet]
        end
        else
            Input Sentence is Neutral
        end
    end
    EPF ← EPF ∪ (SSP)
end
```

The Algorithm 1 scans sentences one by one from input dataset and identifies its POS information for every sentence in the dataset and store it in a separate file. Further, it extracts following bigram and trigram patterns of POS tags such as, (adverb followed by noun) or (adjective followed by verb) or (noun followed by adjective) or (adjective followed by noun) or (noun followed by adverb) or (adverb followed by verb) or (adverb followed by adjective followed by noun) or (verb followed by adverb followed by adjective) or (verb followed by noun followed by noun) or (noun followed by verb followed by noun) or (noun followed by noun followed by verb). If any of such pattern found, then the original words of that tag pattern are added to the SPNet and this process will continue till all the last sentence of the input dataset. If any of the above-mentioned pattern does not matched in any of the input sentence, then we declare the sentence having neutral sentiment value.

3.4 Sentiment Classification Using SPNet

Algorithm 2 is devised to detect sentiment polarity using proposed SPNet. It takes testing set and SPNet as the input and produced the number of negative phrase count, positive phrase count and neutral phrase count of sentence wise. Finally, it based on the count value sentiment polarity is decided as negative count is more then negative sentiment and positive count is more then positive sentiment and so on.

If negative and positive phrase count are equal. Then, we need to calculate sentiment polarity score based on predefined weight of tags like adjective:3, adverb:2, noun:1 and verb:1. In this case, highest sentiment score phrase count is decided as sentiment value.

4 Experimental Results

The experimental results of proposed SentiPhraseNet are discussed here. To know the capabilities of proposed SPNet, a set of 9000 sentences (scripted in Telugu language) are taken as testing purpose. After the experiment, proposed SPNet attains an accuracy of 90.9%. The attained accuracy can be further improved by applying dynamic support to the SPNet where all the unknown testing words will be added to the SPNet testing. Finally, the performance (accuracy) of proposed SPNet is compared with SWNet and various existing ML Techniques and shown in Table 1. It is observed that SPNet outperformed other existing techniques for sentiment analysis in Telugu datasets.

While updating SPNet dynamically, we have observed that few positive and negative phrases are misclassified into neutral class. The error analysis of misclassification in neutral class is shown in Table 2. After the fifth trial of testing sentiment, a cumulative total of around 31,000 phrases were identified unknown. While updating SPNet dynamically, we observe that 3480 phrases were misclassified into the neutral class due to wrong translation from Telugu to English. Out of 3480, 1212 phrases were found positive, 1019 phrase were negative and remaining 1249 were neutral.

In this article, TextBlob is used to translate the Telugu phrases into English equivalent phrases as well as sentiment analysis in English phrases. It is a Python-based open-source module, and it is easy to implement.

Algorithm 2: *Sentiment_Polarity_Detection_using_SPNet*

Input: Testing dataset (C1), *SPNet*
Output: *classification := sentimentpolarity*
Notation: *JJ*: Adjective, *VB* : Verb, *RB* : Adverb, *N*: Noun, *S*: sentence, *TF*: Tag file, *PatSet*: Pattern set, *SSP*: Sentence-wise set of phrases, (C1): Testing set, *UKPL* : Unknown phrase list, *SPS*: Sentiment polarity score
Initialization : *UKPL* = {∅}

```
while S in (C1) do
    TF = Find_POS_Information (S)
    countP = 0
    while Tag in TF do
        if (Pattern of (RB + N) || (JJ + VB ) || (N + JJ) || (JJ + N) || (N + RB ) || (RB + VB) || (RB + JJ + N) ||
        (VB + JJ + N) || (VB + RB + JJ) || (VB + N + N) || (N + VB + N) || (N + N + VB) is Matched) then
            ⟨PatSet⟩ = Extract_matched_pattern_phrase (TF)
            ⟨SSP⟩ = Phrase [PatSet]
        end
    end
    Neg_c = 0, Pos_c = 0, Neu_c = 0
    while (phrs in SSP) do
        Match with SPNet.
        if (phrs matched with positive_sentiment phrase list ) then
            Pos_c ← Pos_c + 1
        end
        else if (phrs matched with negative_sentiment phrase list ) then
            Neg_c ← Neg_c + 1
        end
        else if (phrase matched with neutral_sentiment list ) then
            Neu_c ← Neu_c + 1
        end
    end
    if ( CountP == Neg_c) then
        Testing sentence is negative.
    end
    else if ( CountP == Pos_c ) then
        Testing sentence is positive.
    end
    else if ( CountP == Neu_c) then
        Testing sentence is neutral
    end
    else if ((Pos_c > 0) && (Neu_c > 0) && (Neg_c == 0)) then
        Testing sentence is positive.
    end
    else if ((Neg_c > 0) && (Neu_c > 0) && (Pos_c == 0)) then
        Testing sentence is negative.
    end
    else if ((Pos_c > 0) && (Neg_c > 0) &&(Neu_c == 0)) then
        SPS = Find_Sentiment_Polarity_Score (SSP, PatSet)
        if (SPS > 0) then
            Testing sentence is positive
        end
        else if (SPS < 0) then
            Testing sentence is negative
        end
        else
            Testing sentence is ambiguous.
        end
    end
    else
        UKPL ← UKPL ∪ SPS
    end
end
```

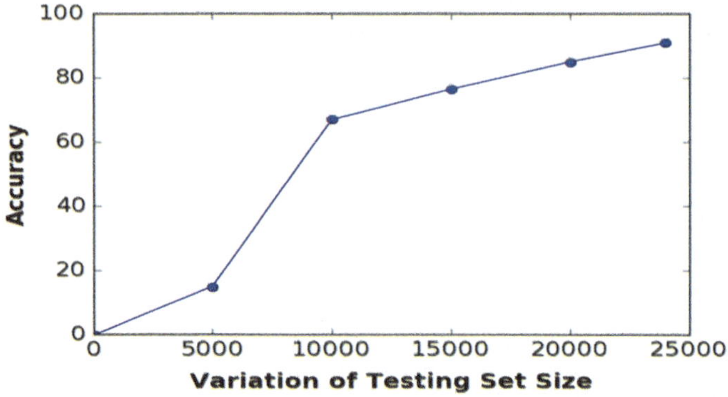

Fig. 4 Accuracy variation with increasing set of testing datasets

Table 1 Accuracy comparasion of SPNet, SWNet and various existing ML techniques

Classification algorithm	Binary	Ternary
	Accuracy (%)	Accuracy (%)
Support vector machine	53	42
Logistic regression	89	78
Naive Bayes	65	55
Random forest	87	75
Multi-layer perceptron neural network	53	66
Decision tree	85	77
Using SWNet	–	79
Using SPNet (proposed)	–	**90.9**

Table 2 Error analysis of misclassified neutral class in SPNet

Using TextBlob module	Misclassified into neutral
Number of phrases	3480
Positive	1212
Negative	1019
Actual neutral	1249

5 Conclusion and Future Work

Dataset availability for Telugu sentiment analysis is negligible in the Web. To resolve such problem, this article builds a SentiPhaseNet (SPNet) in Telugu language for sentiment analysis. While sentiment analysis in Telugu, if any word is marked as unknown in SPNet, then the proposed SPNet is extended to update unknown phrases to SPNet dynamically. Using the proposed SPNet, the system attains an accuracy of 90.9% after five trials of testing approx. 25,000 Telugu sentences. It is shown that after every trial, the accuracy has improved due to the dynamic nature of SPNet. The proposed approach is compared with existing techniques and it found that SPNet is performing better than SWNet for sentiment analysis in Telugu language.

In future, this research can be improved by enhancing the quality of machine translation and automatic sentiment analysis. It also covers for sentiment analysis in other Indian languages.

References

1. Census: States of India by Telugu Speakers (2001). https://en.wikipedia.org/wiki/StatesofIndiabyTeluguspeakers
2. Das, A., Bandyopadhyay, S.: Sentiwordnet for Indian languages. In: Asian Federation for Natural Language Processing, pp. 56–63, China (2010)
3. Patra, B.G., Das, D., Das, A., Prasath, R.: Shared task on sentiment analysis in Indian languages (sail) tweets-an overview. In: International Conference on Mining Intelligence and Knowledge Exploration, pp. 650–655, Springer (2015)
4. Prasad, S.S., Kumar, J., Prabhakar, D.K., Pal, S.: Sentiment classification: An ap- proach for indian language tweets using decision tree. In: International Conference on Mining Intelligence and Knowledge Exploration, pp. 656–663. Springer (2015)
5. Kumar, S.S., Premjith, B., Kumar, M.A., Soman, K.: Amrita cen-nlp@ sail2015: Sentiment analysis in indian language using regularized least square approach with randomized feature learning. In: International Conference on Mining Intelligence and Knowledge Exploration, pp. 671–683. Springer (2015)
6. Sarkar, K., Chakraborty, S.: A sentiment analysis system for Indian language tweets. In: International Conference on Mining Intelligence and Knowledge Exploration, pp. 694–702. Springer (2015)
7. Jain, A., Tayal, D., Arora, A.: Ontohindi neran ontology based novel approach for Hindi named entity recognition. Int. J. Artif. Intell. (IJAI) **16**(2), 106–135 (2018)
8. Jain, A., Arora, A.: Named entity system for tweets in hindi language. Int. J. Intell. Inf. Technol. (IJIIT) **14**(4), 55–76 (2018)
9. Das, A., Bandyopadhay, S.: Dr sentiment creates sentiwordnet (s) for Indian languages involving internet population. In: Proceedings of Indo-Wordnet Workshop (2010)
10. Mukku, S.S., LTRC, I., Choudhary, N., Mamidi, R.: Enhanced sentiment classification of telugu text using ml techniques. In: 25th International Joint Conference on Artificial Intelligence, pp. 29–34 (2016)
11. Mukku, S.S., Mamidi, R.: Actsa: Annotated corpus for telugu sentiment analysis. In: Proceedings of the First Workshop on Building Linguistically Generalizable NLP Systems, pp. 54–58 (2017)
12. Naidu, R., Bharti, S.K., Babu, K.S., Mohapatra, R.K.: Sentiment analysis using telugu sentiwordnet. In: 2017 International Conference on Wireless Communications, Signal Processing and Networking (WiSPNET), p. 666–670. IEEE (2017)
13. Reddy, S., Sharoff, S.: Cross language pos taggers (and other tools) for indian languages: An experiment with kannada using telugu resources. In: Proceedings of IJCNLP workshop on Cross Lingual Information Access: Computational Linguistics and the Information Need of Multilingual Societies. Chiang Mai, Thailand (2011)

Evolutionary Computation and Simulation Techniques

Numerical Solution of Counter-Current Imbibition Phenomenon in Homogeneous Porous Media Using Polynomial Base Differential Quadrature Method with Chebyshev-Gauss-Lobatto Grid Points

Amit K. Parikh and Jishan K. Shaikh

Abstract In this paper, we proposed analytical based Polynomial differential quadrature process to obtain numerical solutions of imbibition phenomenon with help of basic condition. We are applying Chebyshev-Gauss-Lobatto grid points for numerical solution. Polynomial based differential quadrature method reduced the non-linear Partial differential equation into a set of first order linear differential equations. We are applying RK4 process for obtaining solution of set of equations. We have concluded numerical result comparison between non-uniform and uniform grid points. As per outcome numerical result, Chebyshev grid points give enhanced definiteness and stationary numerical solutions as comparison to the uniform grid points.

Keywords Imbibition phenomenon · Polynomial based differential quadrature method · Chebyshev-Gauss-Lobatto grid

1 Introduction

When we are using simple natural compression method for oil recovery from oilfield, oil is produced without using any external forces at wells. But in this process oil recovered is less than 13% from the oil reservoir. Therefore, to recover remaining part of oil from the oilfield we are using water flooding process. In this process we usually inject liquid in oilfield. Mainly imbibition phenomenon appears at the time of water flooding process. This phenomenon arises, when porous media is occupied by some liquid and drifts into the connection with another liquid. There is natural

A. K. Parikh · J. K. Shaikh (✉)
Mehsana Urban Institute of Sciences, Ganpat University, Mehsana-Kherva, Mehsana 384012, Gujarat, India
e-mail: jishanshaikh10@gmail.com

A. K. Parikh
e-mail: amit.parikh.maths@gmail.com

M. Sahni et al. (eds.), *Mathematical Modeling, Computational Intelligence Techniques and Renewable Energy*, Advances in Intelligent Systems and Computing 1287,
https://doi.org/10.1007/978-981-15-9953-8_17

flow of saturating liquid into the medium. This result in counter-current imbibition process which is involves flowing saturating and non-saturating fluids in opposite directions. It is one of the effective oil regaining processes. It occurs when liquid is inserted to move oil from the oilfield to the manufacture well at the time of water flooding technique. In spontaneous phenomenon, vessel power makes saturating liquid to drain into rock and non-saturating liquid gets vanished. Ratio and Range of imbibition depend on the thickness of the saturating and non-saturating liquids. The researchers have expected that flow rates of liquid and oil are equivalent and not in a same direction at the time of this phenomenon. The rock properties like porosity and permeability remain uniform almost everywhere in homogeneous porous medium.

Imbibition is a procedure through which a saturating liquid shifts a non-saturating liquid which begins to saturate a medium, through vessel enforcement. Assume a medium is fully saturated with a non-saturating liquid, and some saturating liquid is introduced on its surface. There will be an impulsive flow of saturating liquid into the medium, causing displacement of the non-saturating liquid.

This is another essential recovery processes at the time of oil extraction from the hydrocarbon reservoir. A lot of [1–4] researchers have been conducted on the phenomenon of homogeneous, heterogeneous and cracked porous medium. An admirable analysis of the latest evolution of spontaneous imbibition has been provided by Morrow and Mason [5]. Yadav and Mehta [6] developed mathematical model along with similarity result of counter-current in banded porous matrix. The vast application along with great importance of the imbibition phenomenon has been evident in petroleum applied science. It is further relevant in ground mechanism, water distillation, ceramics applied science in addition to oil extraction process.

Differential quadrature process is an analytical method for obtaining solution of differential equations. In this approach, we are using weighting coefficient of first and second order partial derivatives. Here, we assume M grid $X_1 < X_2 < X_3 < \cdots < X_M$ on the real axis. For numerical solution of this problem we have to assume non-uniform grid points [7].

Discretization of the partial derivatives of first and second order at a point X_i is given by Eqs. (1) and (2) [8].

$$\left(\frac{\partial S}{\partial X}\right)_{(X=X_i)} = a_{i1}S(X_1) + a_{i2}S(X_2) + a_{i3}S(X_3) + \cdots + a_{iN}\,S(X_M), \text{ for } i = 1, 2, 3 \ldots, M \tag{1}$$

$$\left(\frac{\partial^2 S}{\partial X^2}\right)_{(X=X_i)} = b_{i1}S(X_1) + b_{i2}S(X_2) + b_{i3}S(X_3) + \cdots + b_{iN}S(X_M), \text{ for } i = 1, 2, 3 \ldots, M \tag{2}$$

C_{ij} and D_{ij} are represent weighting coefficients of first and second order partial derivatives respectively. We are applying following base functions to obtain the value of C_{ij} and D_{ij}.

$$h_k(X) = \frac{(X - X_1)(X - X_2)(X - X_3)\dots(X - X_M)}{(X - X_k)(X_k - X_1)(X_k - X_2)(X_k - X_3)\dots(X_k - X_M)},$$
$$k = 1, 2, 3, \dots, M \tag{3}$$

$$T^{(1)}(M_i) = \prod_{k=1,\,k\neq i}^{M} (X_i - X_k), i = 1, 2, 3, \dots, M \tag{4}$$

Applying Eqs. (3) and (4), formula of C_{ij} is given by

$$C_{ij} = \frac{T^{(1)}(X_i)}{(X_i - X_j)T^{(1)}(X_j)}, i \neq j, i, j = 1, 2, 3 \dots, M \tag{5}$$

And

$$C_{ii} = -\sum_{j=1}^{M} C_{ij}, i = 1, 2, 3 \dots, M \tag{6}$$

Similarly formula of D_{ij} is given by

$$D_{ij} = 2C_{ij}\left[C_{ii} - \frac{1}{X_i - X_j}\right], \text{for } i \neq j, i, j = 1, 2, 3, \dots, M \tag{7}$$

And

$$D_{ii} = -\sum_{j=1, j\neq i}^{M} D_{ij}, i = 1, 2, 3, \dots, M \tag{8}$$

2 Types of Grid Points

There are different kinds of grid points (1) Uniform grid (2) Non-uniform (Chebyshev-Gauss-Lobatto grid) and (3) Roots of Chebyshev polynomial can be chose as grid points.In the approximation analysis, we used following grid points.

Uniform grid. In uniform grid points, we are using same step sizes.

$$X_i = X_1 + ih, i = 1, 2, 3, \dots, M, \text{ where } X_1 = a \text{ and } h = \frac{b - a}{M} \tag{9}$$

Non-uniform grid. Non-uniform grid points are defined by

$$X_i = c + (0.5) * \left(1 - \cos\frac{((i-1)*\pi)}{M-1}\right)(d-c),\ i = 1, 2, 3, \ldots, M \quad (10)$$

We are going to use PDQM for obtaining numeric solution of non-linear PDE using non-uniform grid points.

$$\frac{\partial S}{\partial T} = \frac{\partial}{\partial X}\left(S\frac{\partial S}{\partial X}\right) \quad (11)$$

Above Eq. (11) is required governing non-linear PDE [9].

We had already solved above governing equation by using PDQ method with the help of uniform grid points. In this paper, we are going to obtain result by using non-uniform grid points. Here we want to compare result which is obtained by PDQ method with uniform and non-uniform grid points.

Discretization of DQM.

Equation (11) can be written as [10, 11]

$$\frac{\partial S}{\partial T} = \left(\frac{\partial S}{\partial X}\right)^2 + S\left(\frac{\partial^2 S}{\partial X^2}\right) \quad (12)$$

Discretize Eq. (12) by applying the Eqs. (13)–(15) [12]

$$\frac{\partial S}{\partial T} = \frac{dS(X_i)}{dT}, i = 1, 2, \ldots, M \quad (13)$$

$$\frac{\partial S}{\partial X} = \sum_{j=1}^{M} C_{ij} S(X_j), i = 1, 2, \ldots, M \quad (14)$$

$$\frac{\partial^2 S}{\partial X^2} = \sum_{j=1}^{M} D_{ij} S(X_j), i = 1, 2, \ldots, M \quad (15)$$

Substitute Eqs. (13), (14) and (15) in Eq. (12)

$$\frac{dS(X_i)}{dT} = \left(\sum_{j=1}^{M} C_{ij} S(X_j)\right)^2 + S(X_i)\left(\sum_{j=1}^{M} D_{ij} S(X_j)\right), i = 1, 2, \ldots, M \quad (16)$$

Initial condition selected as follow

$$S(X, 0) = 1 - X^2, 0 < X < 1 \quad (17)$$

Boundary Conditions as follow

$$S(0,T) = a,\ X = 0, T > 0$$
$$S(1,T) = b,\ T > 0, a \text{ and } b\text{are arbitrary constant} \quad (18)$$

To solve equation number (16), we are applying equation number (5), (6), (9), (10), (17) and (18) respectively. We also used RK-4 method for solving above system of first order differential equation [13].

3 Conclusion

In this research paper, we studied an analytical approach based on Polynomial differential quadrature for solving second order one dimensional non-linear PDE. This method gives more accurate numerical solution with less number of grid points [10]. We analyze numerical result (Tables 1, 2, 3 and 4 and Figs. 1, 2, 3 and 4) which is obtained by Chebyshev-Gauss-Lobatto and uniform grid points. In the observation of result comparison, we agree with the theory of Shu given in his book [14].

Table 1 Numerical results applying chebyshev-gauss-lobatto grid points $N = 5$

Distance (X)	0	0.3083	0.4996	0.83	0.999999
Time (T)					
0	0.7464	0.7190	0.6820	0.6302	0.5526
0.1	0.6554	0.6123	0.5732	0.5260	0.4605
0.2	0.5806	0.5360	0.4965	0.4442	0.3705
0.3	0.5129	0.45806	0.4182	0.3661	0.2976
0.4	0.4576	0.4006	0.3612	0.3010	0.2287
0.5	0.4035	0.3480	0.3067	0.2391	0.1622
0.6	0.3539	0.2877	0.2502	0.1872	0.1070
0.7	0.3035	0.2513	0.2102	0.1492	0.0610
0.8	0.2644	0.2108	0.1702	0.1076	0.0304
0.9	0.2294	0.1718	0.1302	0.0901	0.0122
1	0.1979	0.1444	0.1062	0.0720	0.0092

Table 2 Numerical results applying uniform grid points $N = 5$

Distance (X) / Time (T)	0	0.25	0.5	0.75	1
0	0.7764	0.7391	0.7023	0.6503	0.5727
0.1	0.6854	0.6327	0.5931	0.5462	0.4809
0.2	0.6106	0.5666	0.5167	0.4647	0.3906
0.3	0.5529	0.4980	0.4383	0.3861	0.3178
0.4	0.4976	0.4406	0.3810	0.3219	0.2489
0.5	0.4435	0.3858	0.3269	0.2594	0.1824
0.6	0.3939	0.3377	0.2706	0.2076	0.1270
0.7	0.3535	0.2913	0.2306	0.1694	0.0811
0.8	0.3144	0.2508	0.1907	0.1278	0.0412
0.9	0.2794	0.2118	0.1505	0.1034	0.0174
1	0.2479	0.1844	0.1267	0.0721	0.0058

Table 3 Result comparison between chebyshev grid points and uniform grid points

Distance (X)	0.3		0.5		0.8		1	
Time (T)	Uniform	Chebyshev	Uniform	Chebyshev	Uniform	Chebyshev	Uniform	Chebyshev
0	0.73915	0.71905	0.70238	0.68208	0.65031	0.6302	0.57272	0.552612
0.1	0.63281	0.61241	0.59313	0.57323	0.54626	0.52602	0.48094	0.460542
0.2	0.56662	0.53602	0.51675	0.49655	0.46473	0.44424	0.39064	0.370513
0.3	0.49807	0.45807	0.43835	0.41825	0.38612	0.36611	0.31783	0.297612
0.4	0.44062	0.40062	0.38109	0.36129	0.32192	0.30102	0.24896	0.228762
0.5	0.38581	0.34807	0.32693	0.30673	0.25945	0.23915	0.18247	0.162224
0.6	0.33778	0.28778	0.27067	0.25027	0.20768	0.18728	0.12705	0.107012
0.7	0.29134	0.25134	0.23067	0.21027	0.16942	0.14922	0.08118	0.061056
0.8	0.25086	0.21086	0.19074	0.17024	0.12782	0.10762	0.04124	0.030442
0.9	0.21188	0.17188	0.15053	0.13023	0.10348	0.09018	0.01742	0.012221
1	0.18444	0.14444	0.12676	0.10626	0.07217	0.07207	0.00587	0.009268

Table 4 Comparison between chebyshev grid points and uniform grid points

Time (T)	0.2		0.4		0.6		0.8	
Distance (X)	Uniform	Chebyshev	Uniform	Chebyshev	Uniform	Chebyshev	Uniform	Chebyshev
0	0.610619	0.580619	0.497663	0.457663	0.39399	0.35399	0.314477	0.264477
0.3	0.566622	0.536022	0.440617	0.400617	0.327775	0.287775	0.250856	0.210856
0.5	0.526752	0.496552	0.39109	0.361291	0.280665	0.250265	0.200738	0.170238
0.8	0.464727	0.444243	0.321922	0.301022	0.207681	0.187281	0.127801	0.097601
1	0.390635	0.370513	0.248958	0.228762	0.127046	0.107012	0.05124	0.03044

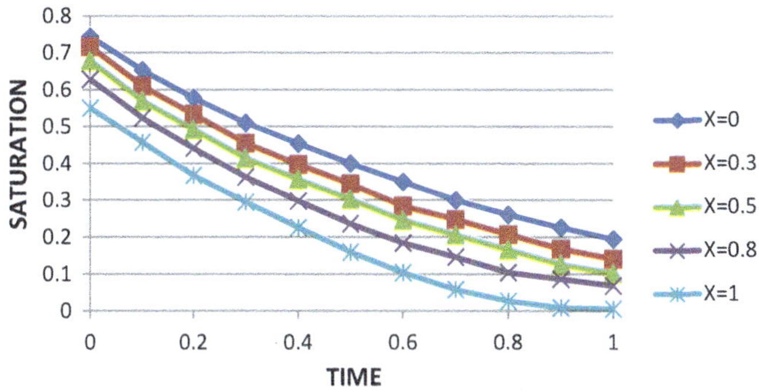

Fig. 1 Saturation versus time by chebyshev grid points

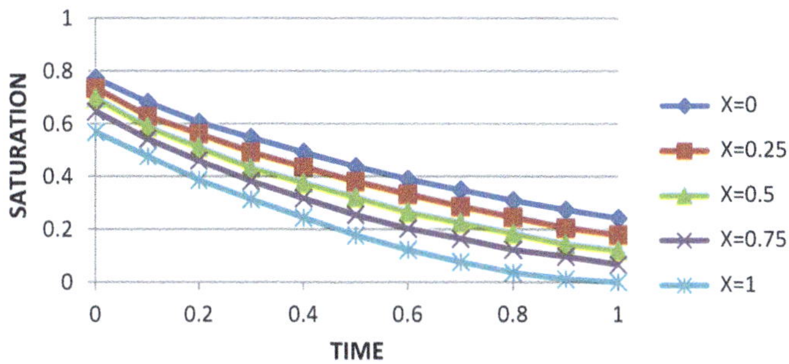

Fig. 2 Saturation versus time by uniform Grid points

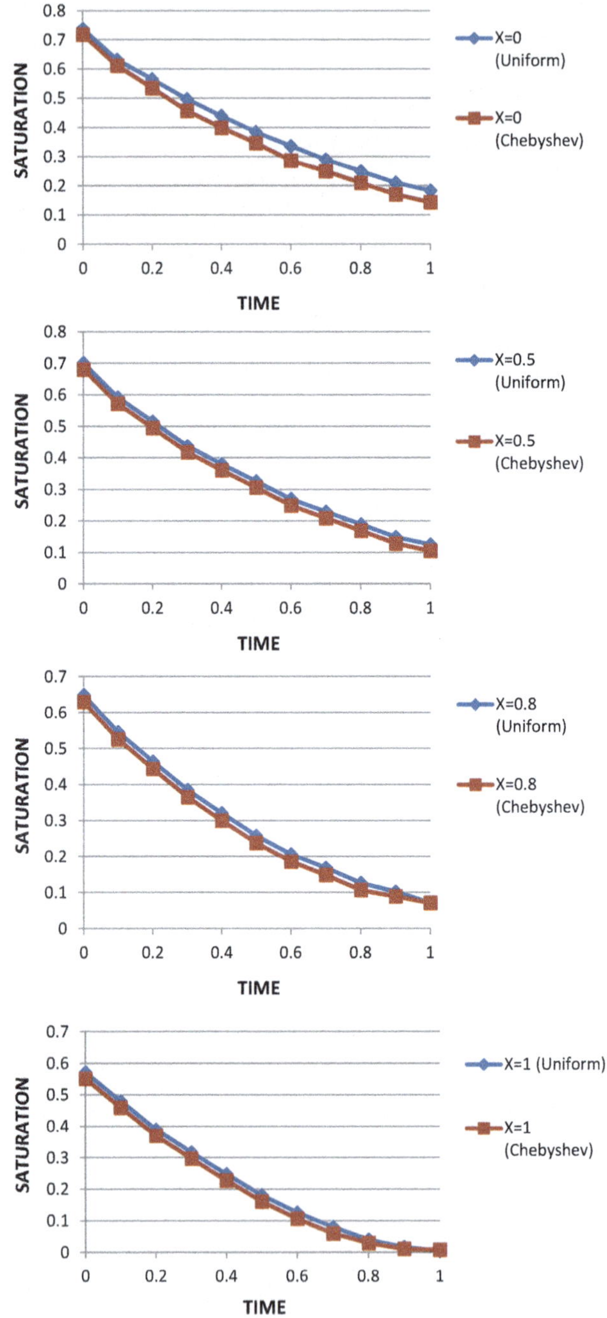

Fig. 3 Result comparisons (saturation versus time)

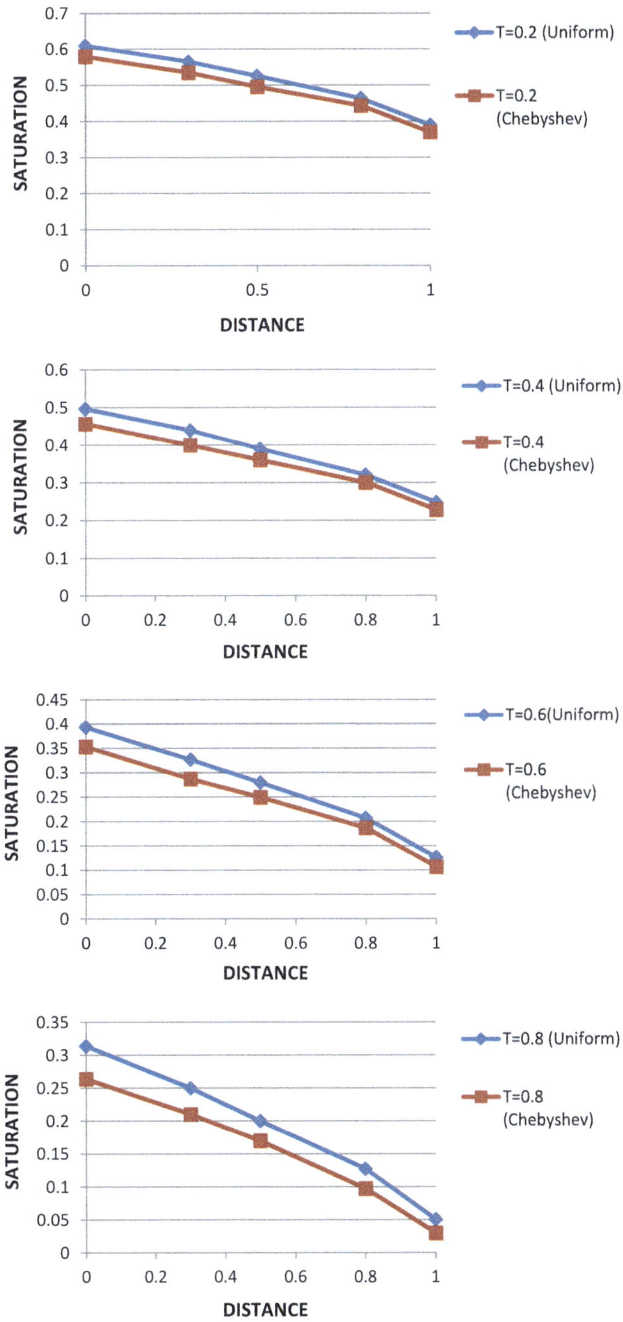

Fig. 4 Result comparisons (saturation versus distance)

References

1. Scheideggar, A.E., Johnson, E.F.: The statistical behavior of instabilities in displacement process in porous media. Can. J. Phys. **39**, 326–333 (1961)
2. Mehta, M.N., Verma, A.P.: A singular perturbation of double phase flow due to differential wettability. Ind. J. Pure Appl. Math. **8**(5), 523–526 (1977)
3. Graham, J.W., Richardson, J.G.: Theory and applications of imbibition phenomena in recovery of oil, p. 216. Trans, AIME (1959)
4. Parikh, A.K.: Generalized separable solutions of counter current imbibitions phenomenon in homogeneous porous medium in horizontal direction. Int. J. Eng. Sci. (IJES) **2**(10), 220–226 (2013)
5. Mason, G., Fischer, H., Morrow, N.R., Ruth, D.W.: Correlation for the effect of fluid viscosity on counter-current spontaneous imbibition. J. Petroleum Sci. Eng. **72**, 195–205 (2010)
6. Yadav, S., Mehta, M.N.: Mathematical model and similarity solution of countercurrent imbibition phenomenon in banded Porous matrix. Int. J. Appl. Math. Mech. **5**(5), 76–86 (2009)
7. Bellman, R.E., Casti, J.: Differential quadrature and long-term integration. J. Math. Anal. Appl. **34**, 235–238 (1971)
8. Kashef, B.G., Bellman, R.E.: Solution of the partial differential equation of the Hodgkings-Huxley model using differential quadrature. Math. Biosci. **19**, 1–8 (1974)
9. Verma, A.P., Ram, Mohan: Existence and uniqueness of similarity solutions of imbibition equation. Proc. Indian Acad. Soc. **88A**(3), 409–417 (1979)
10. Chang, Shu, Richards, B.E.: Application of generalized differential quadrature to solve two-dimensional incompressible Navier-stokes equations. Int. J. Numer. Methods Fluids **15**, 791–798 (1992)
11. Pathak, S., Singh, T.R.: A mathematical modelling of imbibition phenomenon in inclined homogeneous porous media during in oil recovery process, Perspect. Sci. (2016)
12. Parikh, A.K., Shaikh, J.K., Lakdawala, A.: Application of polynomial based differential quadrature method in double phase (oil-water) flow problem during secondary oil recovery process. Ind. J. Appl. Res. (IJAR) **9** (2019)
13. Raj, S., Pradhan, V.H.: Numerical simulation of one dimensional solute transport equation by using differential quadrature method. Int. J. Math. Comput. Appl. Res. (2013)
14. Chang, Shu: Differential quadrature method and its application in engineering, pp. 1–174. Springer, Great Britain (2000)
15. Joshi, M.S., Desai, N.B., Mehta, M.N.: An analytical solution of countercurrent imbibition phenomenon arising in fluid flow through homogeneous porous media, British J. Math. Comp. Sci. **3**(4), 478–489 (2013)

Spray Behavior Analysis of Ethanol

Shrimantini S. Patil and Milankumar R. Nandgaonkar

Abstract Ethyl alcohols have been promoted as prominent renewable alternative fuel due to its advantageous properties like high heat of vaporization and high octane number. In this study, the spray behavior of ethanol fuel was investigated using a high-pressure gasoline direct injection system. Spray penetration length was evaluated at different input conditions. The research was carried out for the injection pressure of 3–11 MPa, 1–5 ms injection duration. Spray images were captured using Olympus high-speed camera at 10,000 fps, and these images were analyzed using an image analysis software Image J. The CFD simulation was also carried out using Ansys Fluent 19.2 software tool. The results obtained with simulation were found in good agreement with the experimental results by 7% of error. The empirical correlations were developed for spray penetration length. Spray penetration length for was found 141 mm, at 11 MPa injection pressure. These results were also validated with the CFD simulation. The maximum error in empirical co-relation and experimental results were found at 9.18%.

Keywords GDI system · Spray penetration length · CFD analysis · Numerical model

1 Introduction

The stringent regulations and fossil fuel availability concern have initiated the modern gasoline direct injection technology research and use of alternative fuels. Intensive worldwide research is going on the use of ethyl alcohols as those are reproducible energy sources, which comes under the group of biofuel. Use of GDI is found advantageous than port fuel injection in terms of improved fuel economy, good control over air-fuel ratio, and improved engine performance. To enhance the combustion process, the research on befouls and their spray behavior is carried out by many researchers. The liquid spray and its behavior have included many complicated phenomena within

S. S. Patil (✉) · M. R. Nandgaonkar
College of Engineering, Pune, Maharashtra, India
e-mail: shrimantini.irwadkar@gmail.com

M. Sahni et al. (eds.), *Mathematical Modeling, Computational Intelligence Techniques and Renewable Energy*, Advances in Intelligent Systems and Computing 1287,
https://doi.org/10.1007/978-981-15-9953-8_18

it e.g.; primary spray break up length, secondary break up length, cavitation in the nozzle, etc. This research carried out to investigate the spray behavior of ethanol fuel injected using the GDI system. The spray penetration length is an important parameter as it is having a great impact on the combustion process. The numerical study carried out with Ansys Fluent 17.2 was compared with the experimental results. The empirical co-relation based on experimental results was developed for spray penetration length. The numerical and experimental study using the GDI system has been done by Li et al. [1]. It was observed that with an increase in injection pressure, the penetration length rapidly increases at the beginning of 0.2 ms, and vortex-like structure formation took place at 0.8 ms. The one-dimensional model was developed in AVL HYDSIM to check the reliability of results obtained. A single hole injector with a GDI system was used to study the spray analysis of gasoline at different fuel temperature conditions [2, 3]. With an increase in fuel temperature, there was a noticeable increase in penetration length, and the cone angle was found. With an increased temperature from 25 to 262 °C, the Sauter mean diameter (SMD) was also reduced from 10 to 1.5 μm. It was concluded from the results that there is no change in spray characteristics once the fuel temperature reaches a critical temperature. The spray evolution from normal-evaporating towards flash-boiling was investigated by Huang et al. It was concluded from the results that the same spray patterns were observed for both the fuels. Though ethanol evaporated slowly than gasoline, the rate of evaporation remained constant when the temperatures higher than 375 K.

2 Experimental Set-up and Procedure

2.1 Numerical Modeling

The numerical simulation is carried out in ANSYS Workbench using FLUENT as a solver to obtain spray characteristics i.e., spray penetration length of ethanol fuel at different initial conditions. The trends of results obtained from numerical simulation compared with the experimental results obtained. The computational domain consists of a 2D cut section of the spray chamber. The spray chamber has fuel input from the fuel injector, which is also a part of the computation domain. The turbulence, species transport, and discrete phase model (DPM) are used to carry out the numerical simulation. The realizable k-ε turbulence model with scalable wall function takes care of viscous effects within the system. The DPM injection model incorporates the breakup model. To consider the droplet's primary and secondary breakup, the KH-RT breakup model was used. The primary breakup is generally considered up to which continuous liquid ligaments break into spherical liquid drops. The secondary breakup is the phenomenon in which large diameter spherical droplets convert into small size droplets, which are of such fine size that they get evaporate into air present in the space. The fluid domain used in the current study is a 2D cut section of the actual cylindrical domain. The dimensions of the domain are 210 mm × 220 mm with a mesh size of 0.15 mm. The dependence of spray characteristics on injection pressure was found 80.60%, whereas, on injection duration, it was found 5.94%.

2.2 Testing Procedure for Spray Behavior Analysis

The injection pressure range was kept to 3–11 MPa, injection duration from 1 to 5 ms, and chamber pressure from 0.1 to 0.7 MPa. The spray behavior analysis is carried out based on spray penetration length. All the above three parameters are essential to determine the spray characteristics; the design of experimentation has done to determine the number of experimental sets. Figure 1 shows the design of experimentation carried out using MINITAB software, to determine the number of experiments. Due to experimental limitations, the spray chamber pressure was kept maximum up to 0.7 bar, and the dependence of spray characteristics on injection pressure was found 80.60%. In contrast, on injection duration, it was found 5.94%. Hence the injection pressure range available with the GDI injection system was utilized for further analysis purposes.

Figure 2 shows the schematic layout for the experimental set up of spray behavior study. The spray chamber is provided with windows for optical access. Olympus i-speed 3 high-speed camera at 5000 fps is used to capture the spray images.

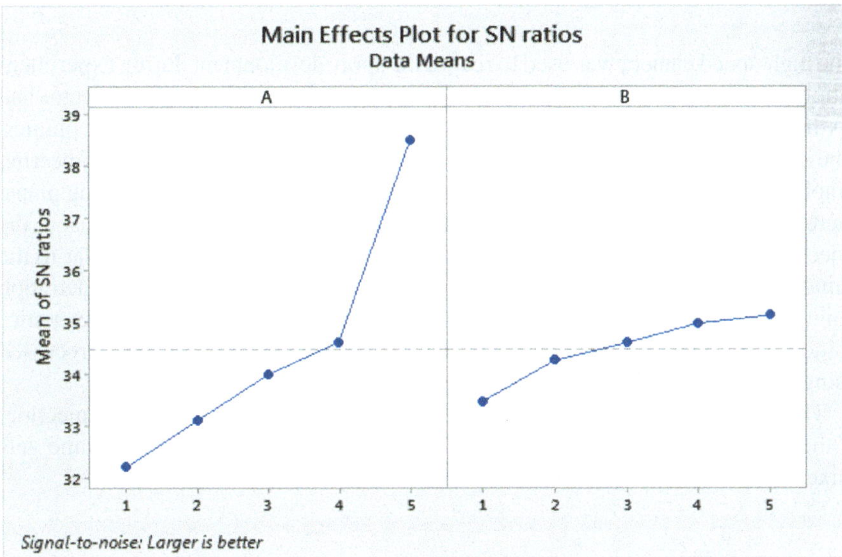

Fig. 1 **a** Effect of injection pressure on spray penetration length **b** effect of injection duration on spray penetration length

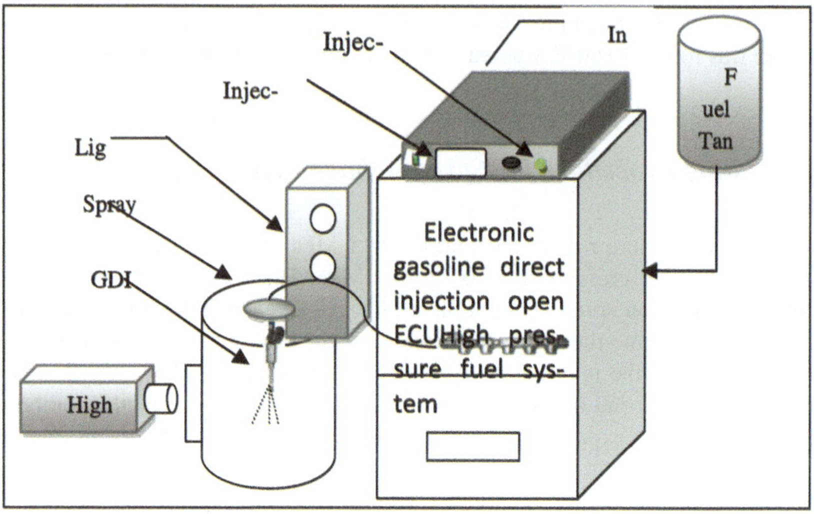

Fig. 2 Experimental setup for spray behavior study

3 Results and Discussions

3.1 Spray Behavior Study

The high-speed camera was used to record the spray development during experimentation for different injection conditions. In experimental conditions, the injector had six holes, so the images captured by high-speed camera had all six spray plumes. The spray plume is a cone with vertex at the tip of nozzle hole, and from engineering graphics, it was known that if we tilt cone in a plane perpendicular to viewing plane, there is no change in projected angle and projected length. Due to this reason, the injector was adjusted such that the rightmost spray plume was perpendicular to the camera axis leading to accurate results of spray plume angle and spray penetration length. Only the rightmost spray plume was used to find out the spray characteristics values. The spray development visualized through the camera and post-processed using the Image J software were as shown in Fig. 3.

The fully developed spray was obtained at around 2 ms after the start of injection. If the injection pressure was increased, it was observed that the spray plume gets mixes with each other, and the chaotic nature of spray occurs.

Fig. 3 Spray development of ethanol at various time instants after the start of fuel injection

3.2 *Effect of Fuel Injection Pressure, Spray Chamber Pressure and Fuel Injection Duration on Spray Penetration Length*

Figure 4a shows the effect of injection pressure and (b) effect of chamber pressure on the spray penetration length. With an increase in injection pressure, the kinetic energy of molecules increases leading to an increase in the spray penetration length. For 11 MPa injection pressure, the spray penetration length was a maximum of about 128.23 mm. The spray penetration length is important spray characteristics to ensure that spray should not splash onto the chamber walls of the engine. The trend of increase of penetration length was found linear with respect to time after the start of injection. Results indicated that the simulation gives a lower value of penetration length than experimental results. The reason behind this may be for simulation; the velocity upstream of the injector hole was zero. This assumption considers the lower total energy of fuel on the upstream side, which is going to convert into kinetic energy on the outlet through the nozzle. So, the input of velocity to simulation was less leading to a lower value of penetration length. The maximum deviation in the final value of penetration length was 1.38%.

The effect of spray chamber pressure on penetration length was given in Fig. 4b. As ambient air pressure increases, it provides higher resistance for the movement of fuel molecules into an air medium, which leads to a decrease in penetration length. For experimentation and simulation, injection pressure and injection duration kept at a constant value of 10 MPa and 2 ms, respectively. The chamber pressure is varied from 0.1 to 0.7 MPa. With an increase in chamber pressure, the low-pressure drop crested at the nozzle exit reducing the swirl motion. The results obtained were in good agreement with the literature [4]. Due to constant volume and increasing the chamber pressure of air, the temperature of air also increases. As the temperature of the surrounding air increases, fuel particles start to evaporate at a faster rate as compared to lower pressure conditions. Due to the faster evaporation of fuel particles, there is a decrease in spray penetration length. The same trend was found in the simulation and experimental results. The chamber pressure plays a vital role for spray in the engine as this pressure is pressure after compression, which is the basis for peak pressure after combustion. It was evident from results that the deviation

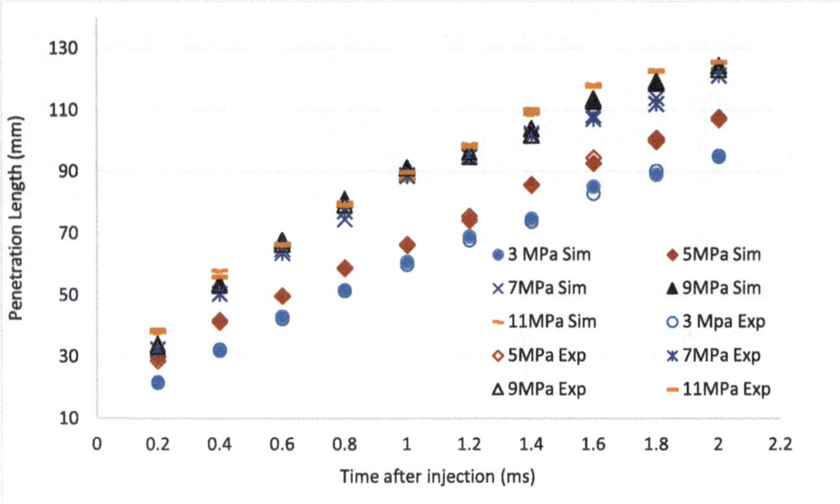

(a)

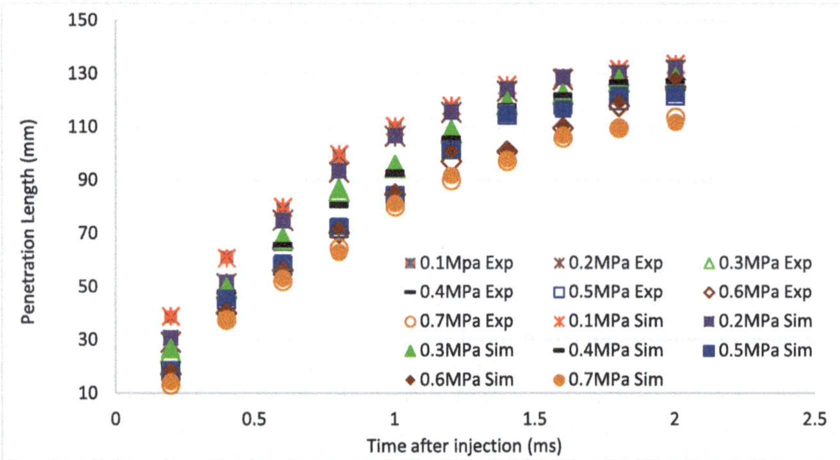

(b)

Fig. 4 **a** Effect of injection pressure **b** effect of spray chamber pressure on spray penetration length

between simulation and experimental results in the final value of penetration length with a maximum percentage error of 4.1% at a chamber pressure of 0.1 MPa.

Increased injection duration allows more amount of fuel to inject from the injection system leading to higher penetration length. Injection duration controlled by the engine speed and crank angle in practical applications [5, 6]. An increase in engine load reduces the RPM, with a requirement of an increased amount of fuel. In the GDI engine, generally, the injection duration in terms of crank angle is around 10–15° [7].

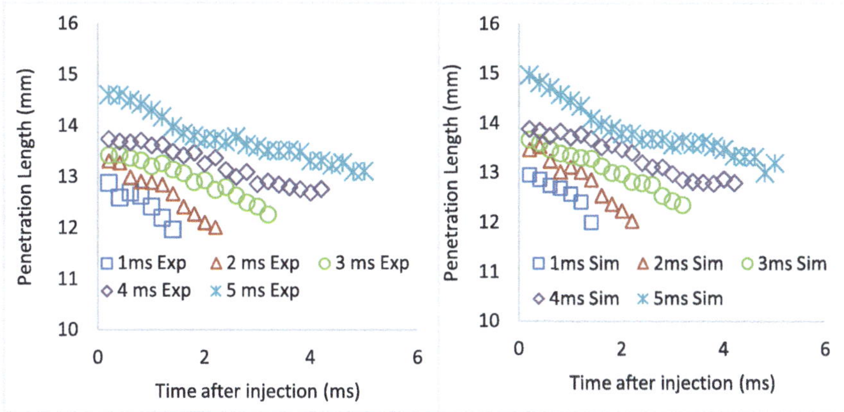

(a) Experimental results (b) Simulation results

Fig. 5 **a** Experimental results **b** simulation results of the effect of injection duration on spray penetration length

So, from the values of engine RPM and injection duration in terms of the crank angle, we can calculate what is the value of injection duration in terms of milliseconds.

The values are shown in Fig. 5a, b indicate final penetration length values for that particular injection duration. The maximum percentage error in simulation and experimental results in the final value of penetration length was 6.25%.

3.3 Empirical Co-relation

An empirical co-relation was developed based on the results obtained from experimentation.

$$L_P = a \times \left(\frac{\Delta P}{\rho}\right)^b \times t^c \tag{1}$$

where, *Lp* final spray penetration length in mm, ΔP is a pressure difference between injection pressures and chamber pressure in MPa, ρ is the density of the air inside the chamber at that particular chamber pressure condition in kg/m^3, t is the time duration of injection in milliseconds, a, b and c are constants. A similar kind of co-relation was developed for different initial conditions [7, 8]. The equation obtained from chamber pressure variation is,

$$Lp = 108.57 \times \left(\frac{\Delta P}{\rho}\right)^{0.241} \times (t)^{0.165} \tag{2}$$

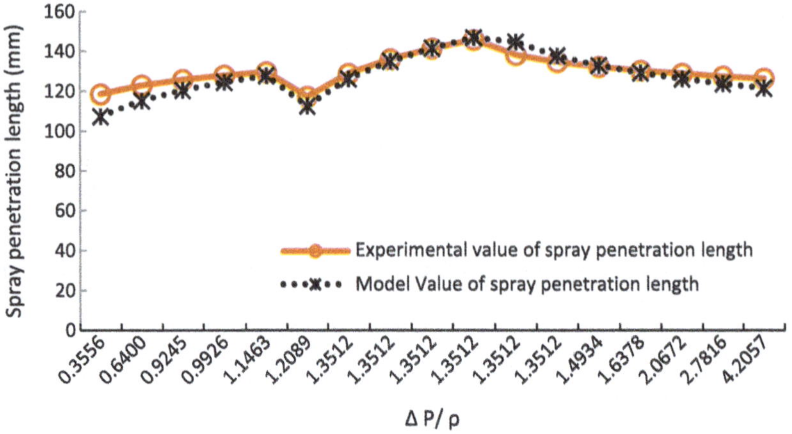

Fig. 6 Error graph in experimental and numerical results for spray penetration length

The coefficients in correlation were obtained using curve fitting. The maximum error in experimental results and results obtained from the co-relation was 8.97% at injection condition of injection pressure of 10 MPa, chamber pressure 0.4 MPa and injection duration 2 ms.

In terms of $\left(\frac{\Delta P}{\rho}\right)$ and injection duration, it was at 2.5132 and 2 ms injection duration. The co-relation developed can be used in the injection pressure range of 3–11 MPa, a chamber pressure range of 1–5 MPa and an injection duration range of 1–5 ms. This particular co-relation is for ethanol fuel since it is not having any variable which incorporates the fuel properties like viscosity, surface tension or density of the fuel (Fig. 6).

4 Conclusion

At higher injection pressure the spray evolution observed more chaotic in nature due to the high kinetic energy the spray droplets get mixed into each other.

The deviation between simulation and experimental results of spray penetration length results was found 4–6%, which indicates the model developed can be used for the study of spray characteristics of ethanol fuel.

The numerical co-relation developed using all the data points can be used to predict the value of spray penetration at any combination of pressure and injection duration within range of injection pressure 3–11 MPa, chamber pressure 0.1–0.7 MPa and injection duration 1–5 ms.

References

1. Li, B., Li, Y.: Fuel spray dynamic characteristics of GDI high-pressure injection system. Chin. J. Mech. Eng. **25**(2), 355–361 (2012)
2. Liu, Y., Pei, Y.: Spray development and droplet characteristics of high-temperature ingle-hole gasoline spray. Fuel **191**, 97–105 (2017)
3. Huang, Y., Huang, S., Huang, R., Hong, G.: Spray and evaporation characteristics of ethanol and gasoline direct injection in non-evaporating, transition and flash-boiling conditions. Energy Convers. Manag. **108**, 68–77 (2016)
4. Kale, R., Banerjee, R.: Experimental investigation on GDI spray behavior of isooctane and alcoholsat elevated pressure and temperature conditions. Fuel **236**, 1–12 (2019)
5. Mofijur, M., Rasul, M., Hassan, N.M.S.: Investigation of exhaust emissions from a stationary diesel engine fuelled with biodiesel. In: 2nd International Conference on Energy and Power, ICEP2018 (2018)
6. Zeng, W., Xu, M., Zhang, M., Zhang, Y., Cleary, D.: Atomization and vaporization for flash boiling multi-hole sprays with alcohol fuels. Fuel **95**, 287–297 (2012)
7. Song, J., Park, S.: Effect of injection strategy on the spray development process in a single-cylinder optical GDI engine. Atomization Sprays 819–836 (2015)
8. Chen, J., Li, J., Yuan, L.: Effects of inlet pressure on ignition of spray combustion. Int. J. Aerosp. Eng. Article ID 3847264 (2018)

3D Spherical—Thermal Model of Female Breast in Stages of Its Development and Different Environmental Conditions

Akshara Makrariya, Neeru Adlakha, and Shishir Kumar Shandilya

Abstract Investigators in the past proposed thermal model of breast in many—idealized conditions without considering the developed the female breast at different stages. In fact the size and the Microstructure of a female breast varies with the different stages of its development. Thus the biophysical processes affecting the heat transport also change with development of breast. Apart from this the various environmental conditions also affect the physical and physiological processes are takes place in the female breast. In this paper, a thermal analysis the three—dimensional spatial temperature in female breast during different developed stages. We have also incorporated various bio-physical processes of generation of metabolic heat generation, flow rate of blood mass and thermal conduction in the proposed model and observed the experimental results under various atmospheric temperature. We have also incorporated the observations of outer surface of the breast as per the heat loss on the surface due to radiation, evaporation, conduction and convection. A finite element method has been applied in polar, spherical coordinates to obtain the solution. The impact of different stages of development of breast and different environmental Conditions are studied on temperature profiles in skin layer regions of female breast. A significant change is observed in temperature in different atmospheric condition. The hexahedral coaxial circular sector elements have proved to be quite versatile in developing in breast yields interesting results.

Keywords Thermoregulation · Metabolic activity · Blood perfusion · Convective and adiabatic boundary conditions

---

A. Makrariya (✉) · S. K. Shandilya
VIT Bhopal University, Bhopal, India
e-mail: akshara.makrariya@vitbhopal.ac.in

N. Adlakha
SVNIT, Surat, Gujrat 395007, India

M. Sahni et al. (eds.), *Mathematical Modeling, Computational Intelligence Techniques and Renewable Energy*, Advances in Intelligent Systems and Computing 1287,
https://doi.org/10.1007/978-981-15-9953-8_19

1 Introduction

Women suffer from various breast disorders which may be benign in the initial stages and some of them if not properly diagnosed and treated may lead to the development of malignant tumors. The second largest number of death among woman takes place due to breast cancer. Also, the breast of a woman undergoes various changes in its structure shape and size it different stages of its development, due to aging and due to various benign and malignant breast disorders. The early detection of breast disorder and its treatments can prevent the development of breast cancer. Thermography is one of the useful techniques for detection of cancer and is appealing due to its non-destructive nature. Earlier experimental investigations were made by Patterson [1] to obtain temperature profiles in the human peripheral region. Some theoretical work has been carried out during the last few decades by Cooper and Trezek [2], and Chao et al. [3] and Saxena et al. [4–7] and Gurung et al. [8, 9], to study the temperature distribution in the human peripheral region under normal environmental and physiological conditions. Also, attempts have been made by Saxena and Pardasani [10, 11] to study problems of temperature distribution in the peripheral region of the human body involving abnormalities like tumors. Saxena, Pardasani, and Adlakha [12–14], developed a model to study temperature variation in human limbs for one and two—dimensional steady state cases under normal physiological and environmental conditions. Jha et al. [15–18], Naik et al. [19, 20], Agrawal et al. and Pardasani et al. [21] developed one, two and three—dimensional finite element models to study, in calcium diffusion case, calcium concentration patterns for different stages of oocyte maturation, thermal distribution in the dermal layers of human limbs with and without tumors. The thermoregulation in human head and other organs has been investigated by Khanday et al. [22–26] under cold environment. Some attempts have been made by Mittal et al. [27], Osman et al. [28–30], Sudharsan et al. [31] and Makrariya et al. [32–34] for thermal modelling of the normal and malignant tissues in female breast under different conditions. These investigators have considered female breast of fixed semi—spherical shape and structure. However, there are common benign breast changes that happen in the woman during their different stages of life. These benign changes take place in the breast due to different stages of its development and other breast disorders. These changes are affected by hormones and may get worse just before the menstrual period starts. The tissues of breast and structure begin to change according to age. Most changes in the breast due to age occurs at the time of menopause. Sometimes the change in breast occurs Due to significant distress. The investigators in the past have considered the breast of semi spherical shape without taking into account the changes in the structure, shape, and size of the breast occurring due to different stages of its development, aging and various benign and malignant disorders are the female breast. In this paper a three—dimensional finite element model is proposed to study the thermal patterns in peripheral regions of female breast under Different stages of its developments. A nude semi—spherical shape of the breast exposed to the Environment is consider in the present study. Appropriate

2 Mathematical Model

The partial differential equation for heat flow in the peripheral region of human body organs [29] is expressed as:

$$(\rho c)\frac{\partial T}{\partial t} = \nabla \cdot (K\nabla T) + m_b c_b (T_A - T) + \overline{S} \tag{1}$$

The Eq. (1) in 3D—spherical case is given by:

$$\frac{1}{r^2}\frac{\partial}{\partial r}\left(Kr^2\frac{\partial T}{\partial r}\right) + \frac{1}{r\sin\theta}\frac{\partial}{\partial\theta}\left(\frac{K\sin\theta}{r}\frac{\partial T}{\partial\theta}\right) + \frac{1}{r^2\sin^2\theta}\frac{\partial}{\partial\phi}\left(K\frac{\partial T}{\partial\phi}\right) + M(T_A - T) + S = 0 \tag{2}$$

In order to simulate a realistic condition the skin surface (r, θ, φ) of the breast is exposed to convective boundary condition as:

$$-K\frac{\partial T}{\partial r} = h(T - T_a) + LE \ at \ r = r_n, \theta \in (0, \pi) \tag{3}$$

Constant Boundary Condition

The inside surface is at T_b hence at higher and moderate atmospheric temperature inside boundary surface is:

$$T(r, \theta) = T_b \ at \ r = r_0 \tag{4}$$

The female breast is not hemispherical in shape but we can assume as hemispherical according to physiological Structure symmetric along angular directions θ and φ. Therefore we assume that the physiological properties are almost uniform along θ and φ directions [15, 19] (Fig. 1).

The female breast area is divided into the 72 elements and 128 Nodes as shown in Fig. 2.

$$I = \frac{1}{2}\int_{r_i}^{r_j}\int_{\theta_i}^{\theta_k}\int_{\phi_i}^{\phi_j}\left[\begin{array}{l}\left\{\begin{array}{l}K^{(e)}r^2\sin^2\theta\left(\frac{\partial T^{(e)}}{\partial r}\right)^2 \\ +K^{(e)}\sin^2\theta\left(\frac{\partial T^{(e)}}{\partial\theta}\right)^2 + K^{(e)}\left(\frac{\partial T^{(e)}}{\partial\phi}\right)^2\end{array}\right\} \\ +\left\{M^{(e)}\left(T_A^{(e)} - T^{(e)}\right)^2 - 2S^{(e)}T^{(e)}\right\}r^2\sin^2\theta\end{array}\right]dr\,d\theta\,d\phi$$

$$+\frac{\lambda^{(i)}}{2}\int_{\theta_j}^{\theta_l}\int_{\phi_j}^{\phi_{j'}}\left[h\left(T^{(e)} - T_a\right)^2 + 2LET^{(e)}\right]r^2\sin^2\theta\, d\theta\, d\phi \tag{5}$$

$$T = a_1^{(e)} + a_2^{(e)}r + a_3^{(e)}\theta + a_4^{(e)}\phi + a_5^{(e)}r\theta + a_6^{(e)}r\phi + a_7^{(e)}\theta\phi + a_8^{(e)}r\theta\phi \tag{6}$$

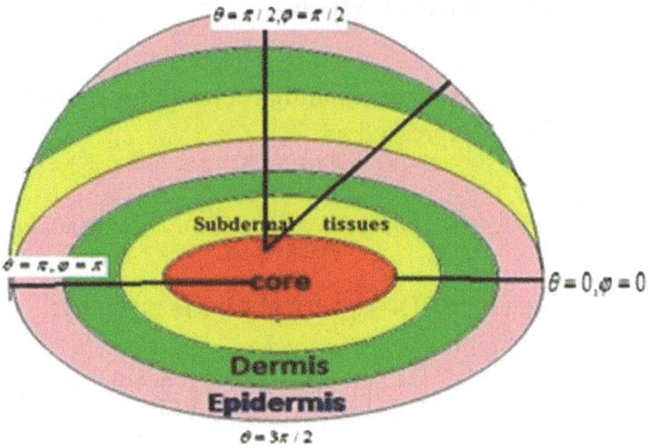

Fig. 1 The three—dimensional cross section of three peripheral layers of spherical shaped female breast dividing into three layers

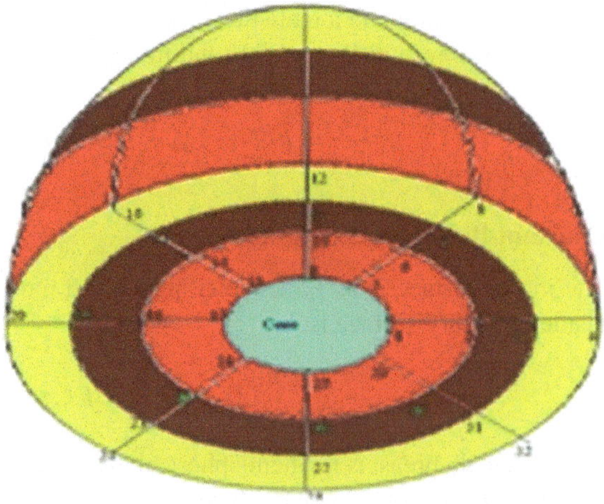

Fig. 2 FEM discretization of female breast

Equation (6) can be written as:

$$T^{(e)} = p^T a^{(e)} \tag{7}$$

where,

$$\begin{aligned} p^T &= \left[1 \; r \; \theta \; \phi \; r\theta \; r\phi \; \theta\phi \; r\theta\phi \right] \\ \left[a^{(e)} \right]^T &= \left[a_1^{(e)} \; a_2^{(e)} \; a_3^{(e)} \; a_4^{(e)} \; a_5^{(e)} \; a_6^{(e)} \; a_7^{(e)} \; a_8^{(e)} \right] \end{aligned} \tag{8}$$

The Tissue conditions are given by:

$$T^{(e)} = T_q \tag{9}$$

I is minimized on each Tissue point temperature to obtain:

$$I = \sum_{e=1}^{N} I^{(e)}$$
$$\frac{dI}{d\overline{T}} = 0 \tag{10}$$

Now from Eq. (10) we get the following set of algebraic equations regarding the Tissue temperatures

$$T_i(i = 1(1)32)$$
$$\overline{XT} = \overline{Y} \tag{11}$$

where,

$$\overline{X} = \left[X_{ij}\right]_{128\times 128}$$
$$\overline{Y} = [Y_i]_{128\times 1}, \overline{T} = [T_i]_{128\times 1}$$

Here X_{ij} and Y_i are constants.

The system of Eq. (11) is solved by Gauss elimination method. Simulation of result done by Matlab.

3 Experimental Results

The preliminary result data were obtained by [15, 23] using physiological and physical constant values as per Table 1.

The simulation of result is given by graphs and tables are given at environmental temperature and rate of evaporation [31]. The simulation was performed for at different set of element like 72, 144 and 288. We get temperature at $a = 10.9$ cm, for Stage V, phi $= \pi/2$ and theta $= \pi/2$ as 34.15760 °C for first model when no. of element is 72 and 34.15860 °C, for third model when no. of element is 288. The

Table 1 Parameter values [15, 23]

Atmospheric temperature (°C)	Rate of flow of blood mass	Generation of metabolic heat	Evaporation rate
15	0.003	0.0357	0.0
23	0.018	0.018	0.0, 0.24×10^{-3}, 0.48×10^{-3}
33	0.315	0.018	0.0, 0.24×10^{-3}, 0.48×10^{-3}, 0.72×10^{-3}

statistical analysis is [(34.1586 – 34.1576)/34.1586]*100 which works out to be 28 $\times 10^{-6}$% only. The level of significance up to 4 decimal points (0.1576/0.1586) * 100 = ~ 99.45% is the confidence level [34]. The saturation point reached at the confidence level. In this work, we have to verify that the problem be mesh sensitive [24]. The result is represented by Graph temperature distribution along the radii and circular direction with phi = $\pi/2$ for the constant boundary condition. The graphs are represented by Fig. 3, 4, 5, 6, 7, 8, 9 and 10.

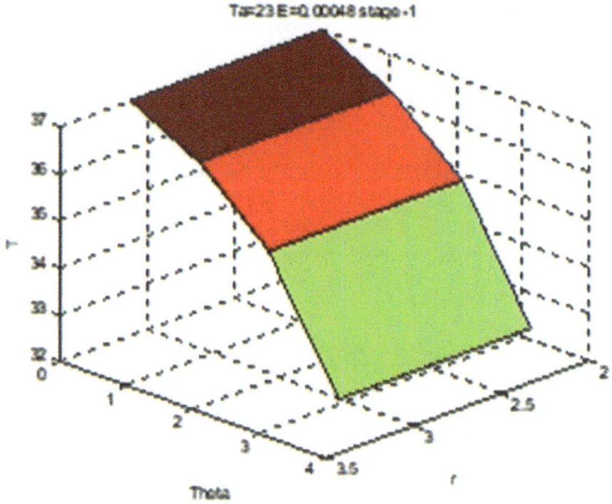

Fig. 3 Temperature Distribution in skin layer of female breast along angular and radial direction for constant boundary condition stage-II, Ta = 23 °C, E = 0.48 $\times 10^{-3}$ gm/cm^2 min, with $\theta = \pi/2$

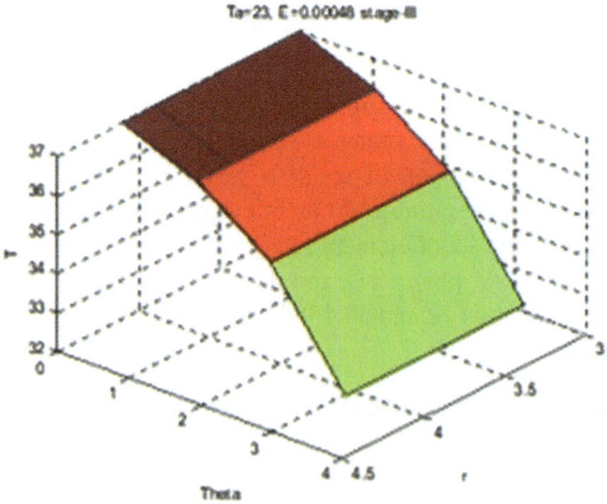

Fig. 4 Temperature distribution in skin layer of female breast along angular and radial direction for constant boundary condition stage-III, Ta = 23 °C, E = 0.48 $\times 10^{-3}$ gm/cm^2 min, with $\theta = \pi/2$

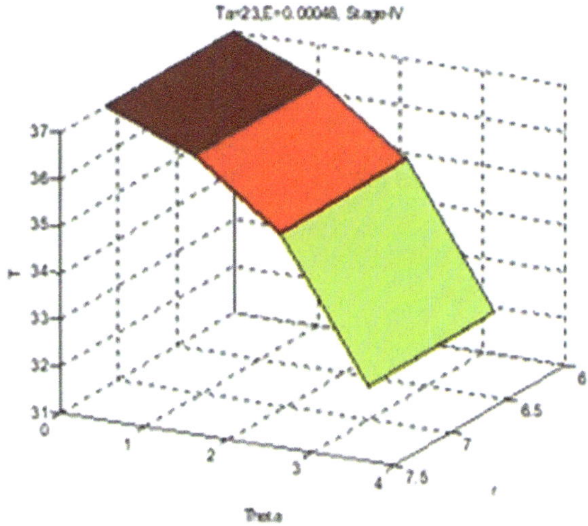

Fig. 5 Temperature distribution in skin layer of female breast on radial and angular directions at constant boundary condition stage-IV Ta = 23°C, E = 0.48×10^{-3} gm/cm^2 min, with $\theta = \pi/2$

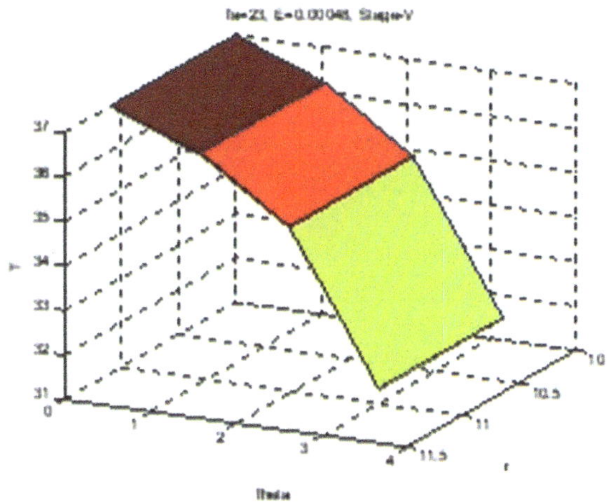

Fig. 6 Temperature distribution in skin layer of female breast on radial and angular directions at constant boundary condition stage-V Ta = 23°C, E = 0.48×10^{-3} gm/cm^2 min, with $\theta = \pi/2$

3.1 Discussion for Constant Boundary Condition

By the Fig. 3, 4, 5 and 6 we are showing that the temperature distribution at atmospheric temperature 230 °C, and rate of evaporation 0.48×10^{-3} gm/cm^2 min constant, boundary condition for various stages of female breast development. We have also noticed that the change in the slop of curve at the junctions of different skin layers. A decline in temperature at stage II is less as compared in stage V. Figure 7, 8, 9 and 10 are showing that the temperature distribution at atmospheric temperature 330C, and rate of evaporation 0.48×10^{-3} gm/cm^2 min constant, boundary condition

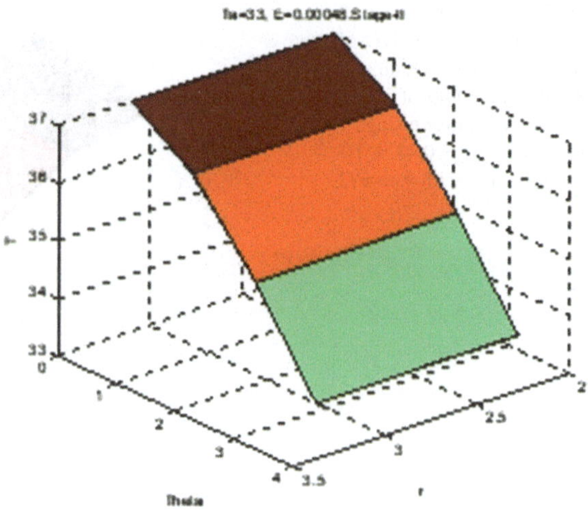

Fig. 7 Temperature Distribution in skin layer of female breast along angular and radial direction for constant boundary condition stage-II, Ta = 33 °C, E = 0.48×10^{-3} gm/cm^2 min, with $\theta = \pi/2$

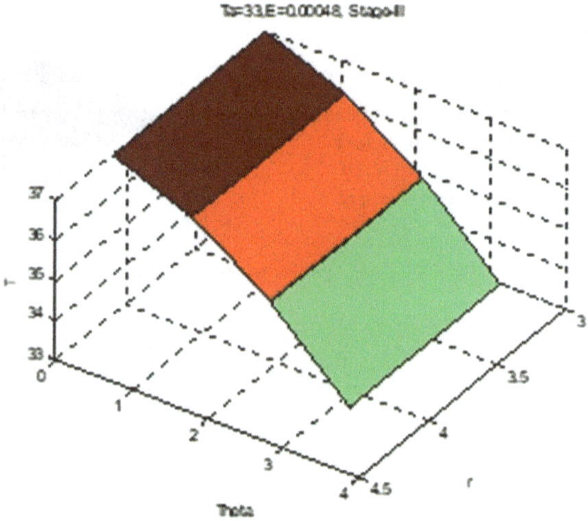

Fig. 8 Temperature Distribution in skin layer of female breast along angular and radial direction for constant boundary condition stage-III, Ta = 33°C, E = 0.48×10^{-3} gm/cm^2 min, with $\theta = \pi/2$

for the different stages of development of female breast. Again it is analysed that the temperature fall down in stage II is less as comparison to stage V.

4 Conclusion

The hexahedral coaxial circular sector elements have proved to be quite versatile in developing 3d-FEM of thermal analysis in the semi spherical shaped human breast. The developing 3d-FEM of thermal analysis in spherical shaped human breast yields

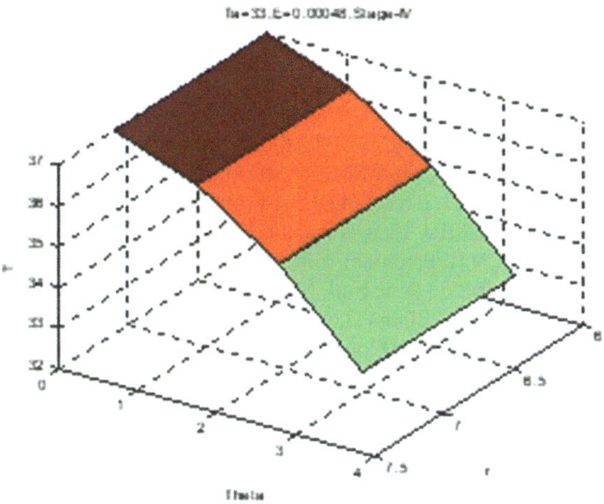

Fig. 9 Temperature Distribution in skin layer of female breast along angular and radial direction for constant boundary condition stage-IV, Ta = 33°C, E = 0.48 × 10^{-3} gm/cm^2 min, with $\theta = \pi/2$

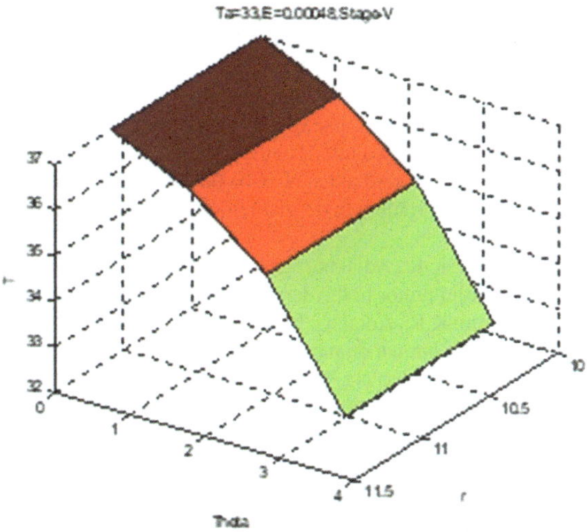

Fig. 10 Temperature Distribution in skin layer of female breast along angular and radial direction for constant boundary condition, stage-V, Ta = 33°C, E = 0.48 × 10^{-3} gm/cm^2 min, with $\theta = \pi/2$

interesting results. The simulation and analysis of temperature at development of stages with the effect of shape, size and microstructure is to be quite significant. The major reason behind this may be the variation of the surface area which is exposed to the environment. Also, this effect is highest for the higher rate of evaporation. The model can be developed further to study effect of different shapes and sizes of the breast due to the various benign disorder of breast. The results were also compared with the previous researches [16, 19].

Acknowledgements We are grateful to SERB (DST), New Delhi, India to support to support this work under National Post-Doctoral Fellowship- Scheme. Authors are also thankful to SVNIT Surat, INDIA.

References

1. Patterson, A.M.: Measurement of temperature profiles in human skin. S. Afr. J. Sci. **72**, 79 (1976)
2. Chao, K.N., Yang, W.J.: Response of skin and tissue temperature in sauna and steam baths. Bio. Mech. Symp, ASME, AMD (1), 69–71 (1975)
3. Cooper, T.E., Trezek, G.J.: A probe technique for determining the thermal conductivity of tissue. ASMEJ. Heat Trans. **94**, 133–140 (1972)
4. Saxena, V.P., Pardasani, K.R.: Effect of dermal tumor on temperature distribution in skin with variable blood flow. Bull. Math. Biol. U.S.A. (1991)
5. Saxena, V.P., Bindra, J.S.: Steady state temperature distribution in dermal regions of human body with variable blood flow perspiration and self-controlled metabolic heat generation. Indian J. Pure Appl. Math. **15**(1), 31–42 (1984)
6. Saxena, V.P., Pardasani, K.R.: Steady state radial heat flow in skin and underlying tissue of layers of spherical regions of human or animal body. Ind. J. Tech. **25**, 501–505 (1987)
7. Saxena, V.P., Pardasani, K.R., Agarwal, R.: Unsteady state heat flows in epidermis and dermis of human body. J. Indian. Acad. Sci. Math. Sci. **98**(1), 71–80 (1988)
8. Gurung, D.B.: FEM approach to one dimensional unsteady temperature distribution in human dermal parts with quadratic shape function. J. Appl. Math. Inform. **27**(1–2), 301–303 (2009)
9. Gurung, D.B.: Transient temperature distribution in human dermal part with protective layer at low atmospheric temperature. Int. J. Biomathematics **3**(4), 439–451 (2010)
10. Pardasani, K.R., Saxena, V.P.: Exact solution to temperature distribution problem in annular skin layers. Bull. Calcutta Math. Soc. 81–108 (1989)
11. K.R., Pardasani, Saxena, V.P.: Temperature distribution in skin and subcutaneous tissues with a uniformly perused tumor in the dermis. Proc. Nat. Acad. Sci. India Sec. A. **60**, 11–20 (1990)
12. Pardasani, K.R., Adlakha, N.: Coaxial circular sector elements to study two- dimensional heat distribution problem in dermal regions of human limbs. Math. Comp. Model. **22**(9), 127–140 (1995)
13. Pardasani, K.R., Adlakha, N.: Exact solution to a heat flow problem in peripheral tissue layers with a solid tumor in the dermis. Indian J. Pure Appl. Math. **22**(8), 679–687 (1991)
14. Pardasani, K.R., Adlakha, N.: Two dimensional steady state temperature distribution in annular tissues of a human or animal body. Indian J. Pure Appl. Math. **24**, 721–728 (1993)
15. Jha, A., Adlakha, N.: Two-dimensional finite element model to study unsteady state Ca^{2+} diffusion in neuron involving ER LEAK and SERCA. Int. J. Biomathematics **8**(01), 1550002 (2015)
16. Jha, A., Adlakha, N., Jha, B.K.: Finite element model to study effect of Na^{+-} Ca^{2+} exchangers and source geometry on calcium dynamics in a neuron cell. J. Mech. Med. Biol. **16**(02), 1650018 (2016)
17. Jha, B.K., Jha, A., Adlakha, N.: Finite element estimation of calcium ions in presence of NCX and buffer in astrocytes. Int. J. Pharma Med. Biol. Sci. **5**(1), 7 (2016)
18. Jha, B.K., Jha, A., Adlakha, N.: Three-dimensional finite element model to study calcium distribution in astrocytes in presence of VGCC and excess buffer. Differ. Equ. Dyn. Syst. 1–14 (2019)
19. Naik, P.A., Pardasani, K.R.: Three-dimensional finite element model to study effect of RyR calcium channel, ER leak and SERCA pump on calcium distribution in oocyte cell. Int. J. Comput. Methods **16**(01), 1850091 (2019)
20. Naik, P.A., Pardasani, K.R.: 2D finite-element analysis of calcium distribution in oocytes. Netw. Model. Anal. Health Inform. Bioinform. **7**(1), 1–11 (2018)
21. Agrawal M., Pardasani, K.R.: Finite element model to study temperature distribution in skin and deep tissues of human limbs. J. Thermal Biol. (2016)
22. Khanday, M.A., Rafiq, A.: Variational finite element method to study the absorption rate of drug at various compartments through transdermal drug delivery system. Alexandria J. Med. Elsevier **51**(3), 219–223 (2014). (Online Available from November 22)

23. Khanday, M.A., Saxena, V.P.: Finite element approach for the study of thermoregulation in human head exposed to cold environment. Proc. J. Am. Inst. Phys. **1146**, 375–385 (2009)
24. Khanday, M.A., Najar, A.: Maclaurin's series approach for the analytical solution of oxygen transport to the biological tissues through capillary bed. J. Med. Imaging Health Inform. **5**(5), 959–963 (2015)
25. Khanday, M.A., Hussain, F.: Explicit formula of finite difference method to estimate human peripheral tissue temperatures during exposure to severe cold stress. J. Thermal Biol. Elsevier **48**, 51–55(2015)
26. Khanday, M.A.: Finite element approach to study the behaviour of fluid distribution in the dermal regions of human body due to thermal Stress. J. Egypt. Math. Soc. Elsevier (2015)
27. Mittal, M., Pidaparti, R.M.: Breast tumor a simulation and parameters estimation using evolutionary algorithms. Hindvani Publishing Corporation Model. Simul. Eng. Volume Article ID 756436 (2008)
28. Osman, M.M., Afify, E.M.: Thermal modelling of the malignant female breast. J. Biomech. Eng. **110**, 269–276 (1988)
29. Osman, M.M., Afify, E.M.: Thermal modelling of the normal female breast. J. Biomech. Eng. **106**, 123–130 (1984)
30. Osman, M.M.: Effect of arterio-venous heat exchange on breast temperature profile. J. Phys. III France **4**, 435–442 (1994)
31. Sudharsan, N.M., Ng, E.Y.K., Teh, S.L.: Surface temperature distribution of a breast with and without tumor. Comput. Method Biomech. Biomed. Eng. **23**, 187–199 (1999)
32. Makrariya, A., Adlakha, N.: Two—dimensional finite element model to study temperature distribution in peripheral regions of extended spherical human organs involving uniformly perfused tumors. Int. J. Biomath. **8**(6), 1550074–15500104 (2015)
33. Makrariya, A., Pardasani, K.R.: Numerical simulation of thermal changes in tissues of woman's breast during menstrual cycle in different stages of its development. Int. J. Simul. Process Model. **14**(4), 348–359 (2019)
34. Makrariya, A., Adlakha, N.: Two-dimensional finite element model of temperature distribution in dermal tissues of extended spherical organs of a human body. Int. J. Biomath. **6**(1), 1250065-1–1250065-15 (2013)

A Novel Approach for Sentiment Analysis of Hinglish Text

Himanshu Singh Rao, Jagdish Chandra Menaria, and Satyendra Singh Chouhan

Abstract *Hinglish* is an informal language, which is written entirely in Latin script but contains informal words, phrases or even slang in a piece of writing, written both in Hindi and English. The field of sentiment analysis is vast, and there have been a lot of work done in the past couple of decades, yet sentiment analysis of *Hinglish* texts remains unexplored. Therefore, this paper presents a new approach for sentiment analysis of *Hinglish* text. The proposed approach uses algorithms such as *Stemming*, *Levenshtein* distance and *Soundex* index for preprocessing the data. Thereafter, it applies various classification models to identify the polarity of *Hinglish* text. The results of experiment show that the proposed approach is effective for sentiment analysis of *Hinglish* text.

Keywords *Hinglish* text · Sentiment analysis · *Soundex* index · *Levenshtein* distance

1 Introduction

Sentiment analysis is one of the of machine learning approaches that quantifies the inclination of human's opinions with the help of natural language processing (NLP), computational linguistics and textual analysis. Using these techniques, we can acquire and study users' information from the Internet that includes social networks and other platforms. This study evaluates the general user's assumptions toward specific items, individuals or thoughts and uncovers the relevant polarity of the data.

H. S. Rao · J. C. Menaria
College of Technology and Engineering, Udaipur 313001, India
e-mail: Singhraoh1@gmail.com

J. C. Menaria
e-mail: Jagdishcm41297@gmail.com

S. S. Chouhan (✉)
MNIT, Jaipur 302017, India
e-mail: sschouhan.cse@mnit.ac.in

M. Sahni et al. (eds.), *Mathematical Modeling, Computational Intelligence Techniques and Renewable Energy*, Advances in Intelligent Systems and Computing 1287,
https://doi.org/10.1007/978-981-15-9953-8_20

Sentiment analysis assists data scientists in various companies to gage users' opinion, to perform nuanced market research, monitor brand and product demand and understand customer experiences. As per the comprehension review given in [1], sentiment analysis is mostly applied on the English text; whereas, a few works have been done in other languages such as Arabic, Spanish and Chinese. The amount of work done on Indian languages further decreases.

In this paper, we did the sentiment analysis of one of the most used languages in Indian context, *Hinglish*, which is written entirely in Latin script but contains informal words, phrases or even slang in a piece of writing, written both in Hindi and English. For instance, consider a movie review written in Latin text but in Hindi language:

"*gaane film mein achhae lagae gae hain*"

In this example, Hindi words are written in English language. Unlike, English language, here the word's spelling and pronunciation are subjective to individual being, i.e., "*aachae*" can be written as "*acche*" by other individuals. Therefore, we cannot apply bag of word approach directly on the *Hinglish* context. Furthermore, while creating sparse matrix, the main problem is to match the same Hindi words written with different spellings, e.g., "paisa" and "pesa" both have the same meaning as "money" in English.

In this paper, to counter the abovementioned problem, we use *Levenshtein* method to find the most similar words and finally used *Soundex* function to give the words a score based on their pronunciation. If the difference between the score of new word and to be matched word is less than a threshold value, then both words are supposed to be same else different column is added in sparse matrix. A dictionary of *Hinglish* words is also used in the model to improve the accuracy of the approach.

For the classification purpose, standard classification techniques like logistic regression (LR), support vector machine (SVM), random forest (RF), Naive Bayes (NB) and decision tree (DT) are used. We calculate their performance metrics and evaluate the best model suited for *Hinglish* text classification. The contribution of this work is significant for many reasons such as it can be useful for e-commerce companies, movie productions and social media content analysis as it covers analyzing precious customer's feedback written in different languages which cannot be analyzed using traditional methods.

The rest of the paper is structured as follows. Next, we discuss the literature review in Sect. 2. In Sect. 3, we present the proposed work. Section 4 discusses the experiments and results. Finally, Sect. 5 discusses conclusions and future directions.

2 Literature Review

In this section, we present the literature review in the area of sentiment analysis using machine learning (ML) techniques.

In [1], author(s) presented an analysis of various feature selection techniques with different classifiers using term frequency–inverse document frequency (TF–IDF) feature representation. They performed total 840 number of experiments in order to identify the best classifiers for sentiment analysis in the news data and *Hinglish*-based comments written in Facebook. The experimental results indicate that the set of TF–IDF-based feature representation, radial basis function neural network and gain ratio selection is the optimal combination to identify sentiments written in the *Hinglish* text.

In [2], authors applied dictionary-based approach to analyze the polarity of review. In this work, the result is the collection of reviews among positive, negative and neutral reviews of the sentiment of the sentence.

In [3], authors applied Hindi SentiWordNet (HSWN) to identify the polarity of reviews made on Hindi movie. The objective the work was to improve the word net and to analyze the polarity of the review. For the words those do not exist in WordNet were first translated into English and then added to WordNet based on the result and hence WordNet will also be improved. To find the aggregate polarity of reviews, each word is given some score when found in HSWN and overall score is calculated to identify the polarity of the review.

In [4], authors made a lightweight stemmer for Hindi, this stemmer worked just on some basic ground rules and is valid only for Devnagri script. We used his work in Devnagri script to be applied in Latin script, thus providing us with a lightweight stemmer that works for *Hinglish* language.

In [5], author worked on analyzing sentiments of text containing both Hindi and English words. They used statistical method to find the polarity. In their work, if the frequency of positive words in the statement is higher than the negative, then the statement is considered positive or vice versa. In their approach, each word is tagged as E if word is English or H if the word is Hindi. Then correct spelling of each word is found. Sounds like "haww," "boo" and "oopps" are also considered by the model. Then, Roman Hindi is transliterated to Devanagari Hindi and WordNet is used to find review of the statement.

In [6], authors used a dictionary-based approach to classify *Hinglish* text based, they used simple preprocessing model and spell corrections and to correctly identify the incorrect spellings in the *Hinglish* text. Their approach is based on TF–IDF model in which *unigram, bigram and trigram* feature selection models are used. They also handled the negation. For sentimental analysis, they used popular classification rhythms that are used in the industry to classify *Hinglish* text.

In [7], author worked on multilingual sentiment analysis. This model used multi-input–multi-channel transfer learning with multiple feature inputs to classify offensive tweets. Their proposed model, multi-channel CNN-LSTM, has been pre-trained on English tweets using transfer learning. However, due to variations in spelling and lack of grammar rules, training of model was a complex task and thus results are inconsistent. However, their proposed model outperforms other baseline supervised classification models.

In [8], author proposes an approach, sentiment analysis of code-mixed text (SACMT), to classify sentences into their corresponding sentiment using contrastive learning. It used the shared parameters of Siamese networks to map the sentences of code-mixed and standard languages to a shared sentiment space. The results show the improvement of 7.6% accuracy in comparison with the existing approaches.

The literature survey shows that most of the work on *Hinglish* text has been done entirely using machine learning models and no proper implementation of preprocessing is done regarding *Hinglish* language. These approaches are not very effective, and they did not focus on the preprocessing part, i.e., to say that the dataset cleaning was not done efficiently solely based on *Hinglish* texts.

In light of the above works, we present a novel approach that classifies the *Hinglish* text. The proposed approach will not only work properly for *Hinglish* language but also for other languages whose dataset is limited. The detailed discussion of the proposed approach is given in Sect. 3.

3 Proposed Work

The proposed work works in three phases: (1) data acquisition, (2) data preprocessing and (3) classification. The approach uses the dictionary of stop words containing 1036 words and IIT KGP dataset of 29,437 words for *Hinglish* dictionaries to check the consistency of words. For the preprocessing, we added various methods like stemming (on Hinglish words specifically), *Levenshtein* distance and Soundex index for the words in *Hinglish* text, to match as closely as possible to the dictionaries of stop words. If the word does not meet the required criterion (selected using *Stemming*, *Levenshtein* Distance, *Soundex* Index), then the word is declared new and passed on to the feature extraction phase, else the word is removed if found in stop words, or various operations are performed on the word using stemming, *Levenshtein* or *Soundex* index. The overall approach is shown in Fig. 1. The description of each step of process is as follows.

3.1 Dataset Acquisition

Due to the lack of dataset in *Hinglish* language, a new dataset was built from scratch using Google transliteration[1] from the Hindi movie review dataset created by IIT Patna,[2] later converted into *Hinglish* movie review dataset. Total 1100 reviews were converted into *Hinglish* in which 550 were positive reviews and 550 were negative reviews. Stop words dictionary[3] was used, which contains 1036 stop words of *Hinglish* language. The dataset was created manually by transliterating all 29,374 words into *Hinglish* language using Google transliterator.

[1]https://www.google.com/inputtools/services/features/transliteration.html.

[2]https://www.iitp.ac.in/ai-nlp-ml/resources.html.

[3]https://github.com/TrigonaMinima/HinglishNLP/blob/master/data/assets/stop_Hinglish.

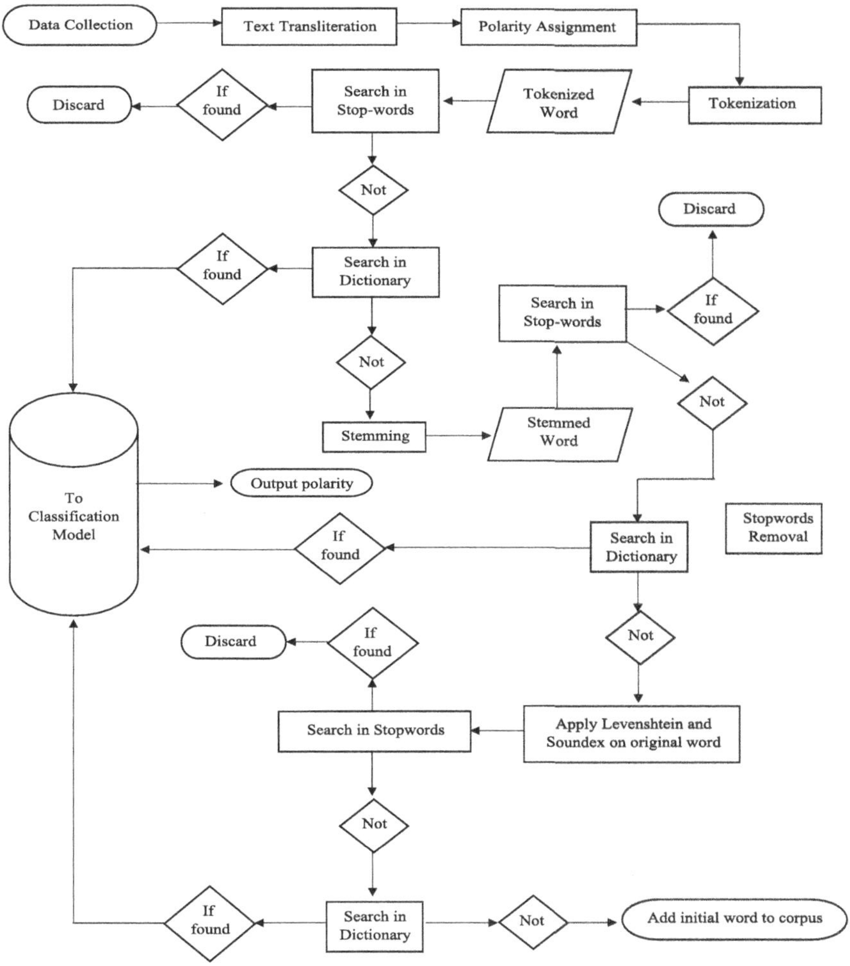

Fig. 1 Model for *Hinglish* text polarity classification

3.2 Data Preprocessing

Text preprocessing involves the following tasks.

3.2.1 Tokenization

In tokenization process, each sentence/string is broken into number of pieces such as keywords, phrases, words and symbols. These pieces are called tokens. In this process, punctuation marks and tags are also discarded. Moreover, all the letters are converted into lower case.

3.2.2 Preprocessing Model

This model is used for removing stop words and finding close relationships between words in the movie review and words in the dictionary, e.g., the word "achaaa" and "acha" have close relationship as to say they sound similar and have exact meaning. But the word "achaaa" is not in the dictionary but the word "acha" is, so to remove this ambiguity, various operations were performed on the word which are (1) stemming, (2) *Levenshtein* distance, (3) *Soundex* Index.

1. **Stemming**. Usually stemming is done on English or Hindi text only. The model contains various libraries like NLTK that performs stemming on English text and Lucene's Hindi Stemmer[4] for stemming in Hindi. The problem is that the Hindi stemmers are for Devanagari script not Roman. Therefore, implementing the same logic, we found a regex which does the same work as Hindi stemmer but in Roman script.

$$\text{Regex : re.sub}(r'(.\{2,\}?)([\text{aeiougyn}]+\$)', r'\backslash 1', \text{word}) \quad (1)$$

The above given regex operation removes all the vowels along with g, y and n from the end of the word, however leaves at least a two characters long stem. Therefore, the words like "aayenga" do not completely remove. Although this stemmer does not work perfectly for all the *Hinglish* words as *Hinglish* language is too informal and versatile, still it covers majority of *Hinglish* words. Table 1 shows the stemmed word for some inputs.

2. **Levenshtein distance**. It is a metric for measuring the difference between two sequences or in other words the *Levenshtein* distance between two words is the minimum number of single character edits (insertions, deletions or substitutions) required to change one word into the other. The Levenshtein distance between two strings a, b (of length $|a|$ and $|b|$, respectively) given by $\text{lev}_{a,b}(|a|, |b|)$ where

$$\text{lev}_{a,b}(x, y) = \left\{ \begin{matrix} \max(x, y) & \text{if } \min(x, y) = 0 \\ \min \left\{ \begin{matrix} \text{lev}_{a,b}(x-1, y) + 1 \\ \text{lev}_{a,b}(x, y+1) + 1 \\ \text{lev}_{a,b}(x-1, y-1) + 1_{(a_x \neq b_y)} \end{matrix} \right\} & \text{otherwise} \end{matrix} \right\} \quad (2)$$

Table 2 shows the *Levenshtein* distance between two *Hinglish* words.

3. **Soundex Index**. It is a phonetic algorithm for indexing names by sound, as pronounced in Latin script. The goal is for homophones to be encoded to the same representation so that they can be matched despite minor differences in spelling. Since similar sounding words are mostly used in *Hinglish*, we have taken a review and converted each word into its Soundex index and then the Soundex index was compared to the Soundex index of the words in the dictionaries. Thus,

[4]http://hitesh.in/2012/stemming-transliterated-hindi/.

Table 1 Stemming table

Input word	Stemmed word
dosti	Dost
doston	dost
boliye	bol
bola	bol
jana	ja
jaenge	ja

Table 2 Levenshtein distance between two words

Word 1	Word 2	Levenshtein distance
Kahaanee	Kahani	3
Paisa	Pesa	2
Ishq	Ishk	1
Paani	Paani	0

Table 3 Soundex index of various words and their absolute difference

Soundex index of word in dictionary	Soundex index of word in movie review	Absolute Soundex difference
Kahani—K500	Kahaanee—K500	0
Anupam—A515	Anupaa—A510	5
Kiska—K200	Kisne—K250	50
Drishya—D620	Drishti—D623	3
Himanshu—H552	Himanshi—H551	1
Ameer—A560	Ameen—A550	10
Jaa—J000	Jaunga—J520	520
Abhi—A100	Aaabhi—A100	0

Soundex becomes a very powerful tool while comparing words that have similar sounds in the dictionary. Table 3 shows the Soundex index of various *Hinglish* words and their absolute differences.

Combining all these operations on *Hinglish* words, it becomes a very efficient approach while comparing *Hinglish* words both in their meaning and in their phonetic transcription. Using these operations, each word is modified accordingly using stemming, *Levenshtein* or Soundex Index and is thus matched more precisely with the dictionaries.

3.2.3 Example (Data Preprocessing)

Consider the sentence in the movie reviews:

1. "*Kahaanee ke abhaav me unake vyaktigat prayaas kaa prabhaav kam ho jaataa hai.*"

2. "*film bhee thodee lambee ho gaee hai*"

The sentence 1 will be tokenized into

{*"kahaanee", "ke", "abhaav", "me", "unake", "vyaktigat", "prayaas", "kaa", "prabhaav", "kam", "ho", "jaataa", "hai"*}

Special characters will also be removed, and all letters are converted into lower case.

Thereafter, as the per the preprocessing shown in Fig. 1, the stop words found in the stop words dictionary were *"ke", "me", "ho", "hai"*.
"kaa"—after applying stemming the word changed to *"ka"*.

after applying *Levenshtein* {"Kahaanee"—"kahani"—3 (Levenshtein distance)} and Soundex {"Kahaanee"-K500—"Kahani"-K500—0 (Absolute difference)} the nearest word for "kahaanee" was found to be "kahani", therefore the word "kahaanee" is replaced with "kahani".

Last corpus created is {*kahani abhaav unake vyaktigat prayaas prabhaav kam jaataa*}

This review is classified as negative review by classification model.

The sentence 2 will be tokenized into

{"film", "bhee", "thodee", "lambee", "ho", "gaee", "hai"}

Stemming—"bhee"—"bhi"
Levenshtein {"lambee"—"lambi"—1 (*Levenshtein* Distance)}
"thodee"—"thodi"—1 (Levenshtein Distance)
Soundex "gaee"-K200—"gae"-K200—0 (Absolute difference)
Stop words {"bhi", "ho", "hai"}

Last corpus created is {film thodi lambi gae} This review is classified as negative review by classification model.

Feature Extraction Each word is then fed into bag of words model, and sparse matrix of corpus is created for each review. The 2000 words with max frequencies were stored in sparse matrix and then fed to classification model. Tokenization accuracy is improved because special characters will be removed and all letters are converted into lower case, and stop word removal is done later in preprocessing model as shown in Fig. 1. The dataset contains 1336 negative and 1339 positive polarity words (Table 4).

3.3 *Classification*

After preprocessing, bag of words (BoW) model is used. BoW is the frequency of a word occurs in the collection of sentences. In this model, occurrence of each word or the word count is used as feature. Bag of words model feature selection plays a crucial role in classifying. It tells us the how the words will be used as the features while classifying a review or text. There are three feature selection models: unigram, bigram and N gram; in this paper, unigram feature selection model is used.

Unigram Model It takes individual words present in a sentence as features, and the whole sentence is divided into individual words.
Example: "mujhe toh film boht acchi lagi"
Feature Set: {mujhe, toh, film, boht, acchi, lagi} (preprocessing not done)

These words are then fed into vectorizer that counts the occurrences of each word and transforms words into features based on their frequencies. Next, some classification algorithms such as Naive Bayes (NB), decision tree (DT), support vector machine (SVM), logistic regression (LR) and random forest (RF) classifiers were used to identify the polarity of the sentence.

4 Result and Discussion

The proposed approach is implemented using Python 3.7 using machine learning and natural language processing libraries such as NLTK and scikit-learn. After pre-processing and feature extraction, the dataset containing 1100 samples of positive and negative polarity were divided into training and testing test.

The classification methods with their default parameters' setting were used. These default parameters are presented in scikit-learn library. However, some parameters such as the number of features, trees and depth were customized with respect to the volume of the data and limitations associated with the computing resources. The performance of the classifiers was examined based on the accuracy and the following performance metrics:

- **Precision**. It is the ratio of true positive observations to the total positive observations (predicted). High precision is directly related to the low false positive rate.

$$\text{Precision} = \text{True Positives(TP)}/\text{True Positives(TP)} + \text{False Positives(FP)} \quad (3)$$

- **Recall (sensitivity)**. It is the ratio of true positive observations (predicated) to the all observations in the actual class. It is different from precision in the fact that it is a quantity measure to check how many relevant results are being returned by the algorithm while precision checks from the selected items, how many of them were relevant.

Table 4 Positive–negative polarity words

Positive words	Negative words
Achaa	kharab
pyaar	abhaav
badhiya	galat
shaandar	jhagde
khubsoorat	tanaav
romanchak	todfod

Table 5 Experimental results

Performance metrics in %				
Classifiers	Accuracy	Precision	Recall	F1
SVM	70.9	71.79	69.55	70.31
Naive Bayes	59.09	58.815	56.08	57.41
Decision Tree	64.54	59.4	59.31	59.35
Logistic Regression	70	70.24	68.53	69.37
Random Forest	66.36	66.71	64.45	65.56

$$\text{Recall} = \text{True Positives(TP)}/\text{True Positives(TP)} + \text{False Negatives(TP)} \quad (4)$$

- **F1-score**. It is the weighted average of precision and recall. Therefore, this score considers both false positives and false negatives.

$$\text{F1} = 2 * \frac{\text{precision.recall}}{\text{precision} + \text{recall}} \quad (5)$$

Table 5 shows the results obtained of the proposed approach on the abovementioned performance metrics.

From the results, following observation is made:

- On the movie review dataset, SVM classifier performs better (accuracy—70.90%, precision—71.79%, recall—69.55% and F1-score—70.31%) in comparison with that of other classifiers.
- Naive Bayes classifier is least stable method on proposed approach as the values of average accuracy, precision and F1-score are very low in comparison with other classifiers.
- The performance of decision tree classifier is fine but the difference between accuracy and other measures is larger.
- Logistic regression performs as good as SVM with accuracy of 70%, precision—70.24%, recall—69.55% and F1-score of 70.3%.

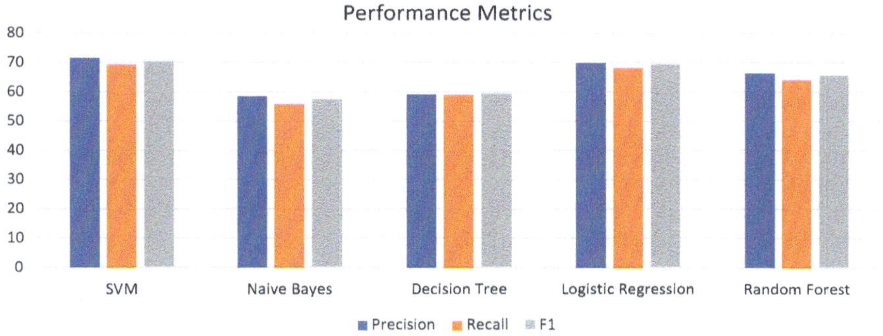

Fig. 2 Graphical representation of various performance metrics

Figure 2 graphically illustrates that the average values of classification accuracy, precision, recall and F1-score of logistic regression (LR) and support vector machine (SVM) are similar, and support vector machines have achieved higher average classification accuracy, precision, recall and F1-score results in comparison with Naive Bayes, decision tree, logistic regression and random forest but the difference is not statistically significant, except in the case of Naive Bayes and random forest.

5 Conclusions and Future Directions

This paper presented an approach for the sentimental analysis of Hindi text written in English, i.e., *Hinglish*. We also built the dataset and dictionaries from scratch. The dataset is open sourced for further research work in the *Hinglish* language. We implemented various algorithms like stemming, *Levenshtein* and Soundex that efficiently preprocess data unlike used in the past approaches. The processed data was fed into various classifiers using unigram bag of Words model.

The comparison of various classifiers including Naive Bayes, decision tree, support vector machine, logistic regression and random forest is also presented in the paper. Out of these classifiers, support vector machine did the best job while classifying movie reviews and pointing out the polarity of reviews with accuracy—70.90%, precision—71.79%, recall—69.55% and F1-score—70.31%. In future work, the proposed approach can be improved by using a larger dataset, different feature selection models like bigram, Ngram and hybrid classification models.

References

1. Kumar, R., Vadlamani, R.: Sentiment classification of Hinglish text. In: Proceedings 3rd International Conference on Recent Advances in Information Technology (RAIT), pp. 641–664 (2016)
2. Richa, S., Shweta, N., Rekha, J.: Polarity detection movie reviews in Hindi language. arXiv preprint arXiv:1409.3942 (2014)

3. Pooja, P., Sharvari, G.: A framework for sentiment analysis in Hindi using HSWN. Int. J. Comput. Appl. **119**(19), 23–26 (2009)
4. Ananthakrishnan, R., Durgesh, D.R.: A lightweight stemmer for Hindi. In: Proceedings of EACL, pp. 1–6 (2003)
5. Shashank, S., Srinivas, P.Y.K.L., Rakesh, C.B.: Sentiment analysis of code-mix script. In: Proceedings of the International Conference on Computing and Network Communications (CoCoNet), pp. 530–534 (2015)
6. Harpreet, K., Veenu, M., Nidhi, K.: Dictionary based sentiment analysis of Hinglish text. Int. J. Adv. Res. Comput. Sci. **8**(5), 816–822 (2017)
7. Puneet, M., Ramit, S., Meghna, A., Rajiv, S.: Did you offend me? Classification of offensive tweets in Hinglish language. In: Proceedings of the 2nd Workshop on Abusive Language Online (ALW2), pp. 138–148 (2018)
8. Nurendra, C., Rajat, S., Ishita, B., Manish, S.: Sentiment analysis of code-mixed languages leveraging resource rich languages. arXiv preprint arXiv:1804.00806 (2018)

Evolution of Sea Ice Thickness Over Various Seas of the Arctic Region for the Years 2012–13 and 2018–19

Dency V. Panicker, Bhasha Vachharajani, and D. Ram Rajak

Abstract Sea ice is formed when ocean water freezes at a temperature of −1.8 °C. This ice formed over the ocean plays a crucial role in the global climate system. Most of the solar radiation which is incident on sea ice is reflected, and thus, it acts as the earth's polar refrigerator by limiting the heat absorption. However, it is found over the decades that sea ice is declining dramatically, and this can cause various negative impacts. Firstly, the warming up of the poles disrupts the earth's overall heat flow, and secondly, it alters the wind pattern by pushing more ice towards the Atlantic. The study aims at extracting the rate of growth and attainment of sea ice over various seas of the Arctic region. The seas mainly considered for the purpose of study are East Siberian, Kara and Barents. Using sea ice thickness data from NSIDC, for the year 2012–13 and 2018–19, various episodes of sea ice thickness over the region are distinctly identified and compared with each other. It is found that the fluctuation in sea ice thickness is different for each sea mainly due to its topography. Despite the cool nature of the Arctic, it is also found that unexpected warmth comes beneath the Arctic Ocean because of the high heat capacity of water, which in turn serves as the causative factor for mild climates in the coastal regions. On the contrary, land has a lower heat capacity, so it gets heated up quickly during the day and cools down with dawn. Not only the heat capacity factor but also other parameters such as pressure, temperature, wind, humidity and its related precipitation also contribute to the variation in sea ice thickness.

D. V. Panicker (✉)
Department of Science, School of Technology, Pandit Deendayal Petroleum University, Gandhinagar, Gujarat, India
e-mail: Dency.Panicker@sot.pdpu.ac.in

B. Vachharajani
Department of Mathematics, School of Technology, Pandit Deendayal Petroleum University, Gandhinagar, Gujarat, India
e-mail: bhasha.vachharajani@sot.pdpu.ac.in

D. Ram Rajak
Cryosphere Sciences Division, GHCAG/EPSA, Space Applications Centre—ISRO, Ahmedabad, Gujarat, India
e-mail: rajakdr@sac.isro.gov.in

M. Sahni et al. (eds.), *Mathematical Modeling, Computational Intelligence Techniques and Renewable Energy*, Advances in Intelligent Systems and Computing 1287,
https://doi.org/10.1007/978-981-15-9953-8_21

Keywords Sea ice thickness · East Siberian Sea · Kara Sea · Barents Sea · NSIDC

1 Introduction

In recent years, a decreasing trend is seen to prevail in Arctic sea ice as ice extent and volume have dropped. It is known that several parameters influence the formation of sea ice. Due to the difference between air temperature and water temperature of the sea, the air plays its role like an energy absorbent. Two types of ice are formed depending on the state of the water, calm and rough. Moreover, the latitude also influences the formation of sea ice because the pycnoclines are found in higher depths whereas deep layer ice (with varying density) is found in lower depths. With the formation of the pack ice, its thickness continues to increase. Initially, the ice grows thicker at its bottom by the influence of freezing water. Later, the heat reaches the cold air moving across the ice sheet. Ice grows by a scale proportional to the temperature gradient between the air and the water. However, the snow cover acts as an insulator by influencing the heat transfer. To scale the growth rate, 8–10 cm of accumulation is found in the initial 24 h, and then, the rate gradually comes down as ice thickens. During one season, ice can be 1–1.5 m thick. In the Arctic sea at the north pole, the perennial ice is 3–5 m thick.

The Kara Sea is one of the Siberian shelf seas covering an area of 926,000 km^2. It ranges up to 81 °N; the eastern border is shared with the Laptev Sea, which is at 100 °E [9]. It is shallow, with an average depth of 417 ft and maximum depth of 2034 ft. Bottom of the sea, at its centre, is comparatively flat. It is deeper up to 100 m in its western region, with several small depressions divided by ridges of variable heights [8]. Frontal structures mixing and alteration procedures explain the hydrography of the sea. This leads to penetration of saline and warm waters of the North Atlantic from the boundaries of the west, which in turn results into excessive river runoff from the east Atlantic water coming from the Barents Sea to the north of Kara Sea through the channel between Novaya Zemlya and Franz Josef Land. Whereas in the south, water in the Barents Sea from Atlantic spread eastward through the Kara Strait [7]. This causes the relatively warm Atlantic water to enter Kara Sea through the St. Anna and Voroninm troughs, causing the temperature of the water to rise at 50–100 m depth, gaining a maximum value in the range of 1.0–1.50 °C leading to the melt of sea ice over the region [1]. The usual months for the formation of ice in these areas vary from March to April in the central Barents Sea, September in the distant north of Kara Sea to mid-November in the south Kara Sea and Pechora Sea (south-eastern Barents Sea) [4]. Melting of the sea ice initiates during end of April in the peripheral ice sector of the Barents Sea. However, by the time June ends, the central Barents Sea and the Pechora Sea frequently become completely ice-free. Over Kara Sea, melting initiates in May, and the melting continues steadily through July and August. Kara Sea remains ice-free for almost one month during July–August. Here, the ice season persists for almost half a year to nine months depending on its seasonal conditions and location. The East Siberian Sea is the broadest amongst all the Arctic Ocean shelf seas, having an area of 895,103 km^2, but as the average depth is only 52 m, it has the second minimum volume, when compared to the Chukchi Sea. The wind patterns impact the motion of sea ice to a larger extent, as sea ice has the general tendency of moving in the direction of wind [3].

An exception to this behaviour is found during winters in the zone closer to shore when stationary land fast ice is existing [2]. Studies have reported a profound loss of sea ice cover in the Arctic Ocean during summer for past years; in fact, this has made the East Siberian Sea almost ice-free during summer. Earlier, it had a large area covered with sea ice [6]. Due to the formation of ice and surface cooling, sea water temperature is normally close to the freezing point over the entire water column during winter. Whereas, during summers, the temperature attains positive values near the regions, closer to the surface in ice-free areas. The lowermost layer may also be affected by the interferences of warmer Atlantic water approaching from the continental slope during upsurge environments [1].

2 Data Sets and Methodology

For the current study, we have used sea ice thickness (hereafter referred to as SIT), confining the study to the Arctic region. Our aim at studying SIT was to analyse how various seas behave during the winters. Here, we have considered the evolution of sea ice over various seas of Arctic region, that is, right from the freeze onset to its maximum attainment. SIT data have been taken from CryoSat-2 [5] for the months of October–April for the years 2012–13 and 2018–19. This dataset comprises of daily averaged Arctic sea ice thickness, with freeboard and roughness of ice surface, derived from the Synthetic Aperture Interferometric Radar Altimeter (SIRAL). Similarly, the shape files for Kara, Barents and East Siberian Sea are extracted from www.marineregions.org.

Daily data for the years 2012–13 and 2018–19 are collected using fttp, from the above-mentioned site. Further, the script is run on Python-Spyder for the extraction of SIT data. While plotting it spatially, the undefined values (denoted by 'NaN') have been removed from the dataset. However, the latitude and longitude are kept confined to the seas of the Arctic region as the current study involves the evolution of SIT. Furthermore, SIT is spatially plotted using a basemap, and the values are imported in excel file to create various plots required to analyse these features. Generated contour plots are studied daily for the entire span of the years 2012–13 and 2018–19. The rate of growth of SIT and the maximum attainment of SIT for each sea is calculated using the formula:

$$\mathrm{GSIT} = (\mathrm{SIT}_{n-1} - \mathrm{SIT}_n) * 100/\mathrm{SIT}_{n-1} \tag{1}$$

where

GSIT Growth rate of SIT,

SIT_n SIT on a given day,
SIT_{n-1} SIT on the previous day.

To study the evolution of SIT, we have considered particular dates on which the sea ice begins to form and then rises to attain maximum. These dates are found to be different for different seas, owing to the topography and the climatic condition of the seas. Since we have considered the growing seasons for the years 2012–13 and 2018–19, we are able to highlight how the growth is affected with passage of time. At same dates of different years, sea ice exhibits distinct behaviour; the same has been discussed in the next section. To study the evolution of SIT, we have considered particular dates on which the sea ice begins to form and then rises to attain maximum. These dates are found to be different for different seas, owing to the topography and the climatic condition of the seas. Since we have considered the growing seasons for the years 2012–13 and 2018–19, we are able to highlight how the growth is affected with passage of time. At same dates of different years, sea ice exhibits distinct behaviour; the same has been discussed in the next section.

3 Results and Discussion

In Kara sea, on 26th December 2012 the SIT is found over higher latitudes, after which it is seen to be expanding towards lower latitude. By 12th February 2013, we find the entire sea is spatially covered with SIT of about 1.25 m. However, SIT of about 1.25 m starts to evolve over the north-east corner of the sea by 22nd February 2013. This trend is continued, and finally, by 4th March 2013, the entire sea is found to be covered with SIT of approximately 1.5 m with small patches of 1 m ice at higher latitude (Fig. 1).

For the year 2018, it is found that on 12th December, SIT starts to develop on higher latitudes. Gradually, it is found that by 28th January 2019, the entire sea gets covered with SIT of 1 m. Soon by 12th February 2019, towards the higher latitude, the sea attains a maximum thickness of 1.75 m. Eventually, on 22nd February 2019, the entire sea gains thickness of 1.765 m of sea ice. On 27th February 2019, SIT of 3 m is found to develop towards the north-east corner of the sea, and this trend is seen to exist till 4th March 2019.

Similar to Kara, sea ice over Barents starts to appear at higher latitudes on 11th January 2013. Additionally, a small patch of 1.5 m is simultaneously seen to be developing at lower latitudes towards the east. By 12th February 2013, SIT of approximately 1 m is found completely over higher latitudes ranging from 78 °E to 82 °E. By 22nd February 2013, SIT of almost 1.5 m starts to form over the north-west corners of the sea. Soon this feature is observed over the north-east corners on 4th March 2013. By 14th March 2013, the sea has SIT of only 1.5 m, which means that all mass of ice which was 1.5 m thick has now grown up to 1.75 m. Further, the SIT intensifies to almost 2 m by 28th April 2013 (Figs. 2 and 3; Table 1).

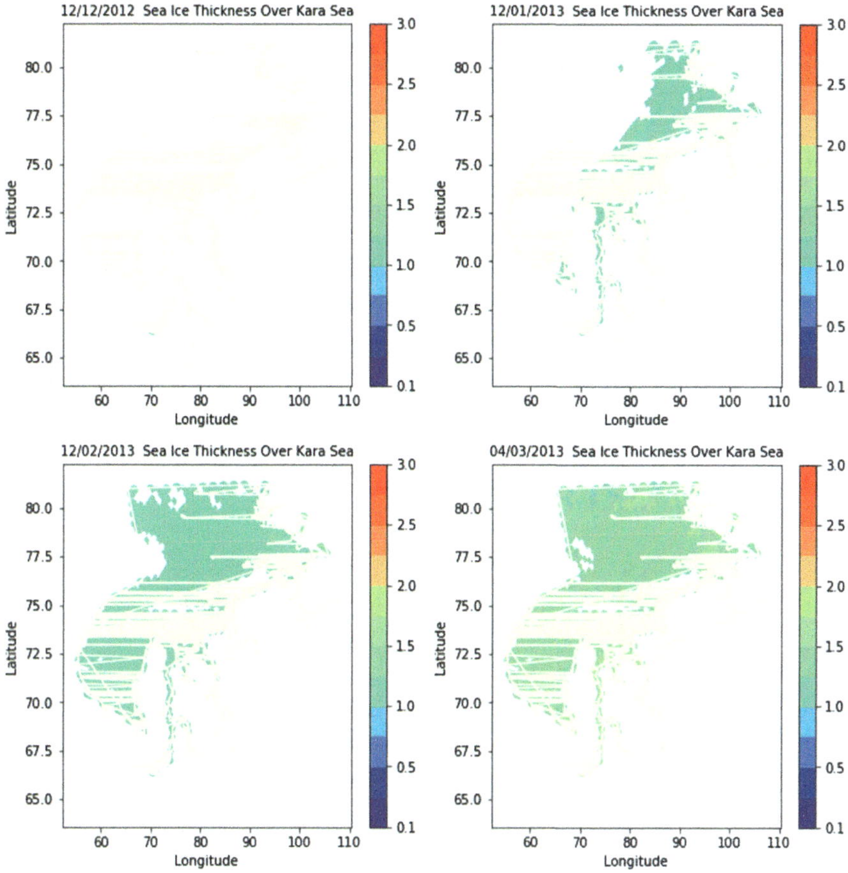

Fig. 1 Evolution of SIT over Kara Sea (2012–13)

On 11th January 2019, SIT of 1.25 m starts to appear towards the higher latitude of the sea which is prolonged till 12th February 2019. In another 10 days, i.e. by 22nd February 2019, SIT of 1.25 m is completely replaced with SIT of 1.75 m, and the entire northern portion of the sea is covered with the same thickness. However, by 4th March 2019, there seems to be thinning of ice by 0.5 m from north-east to north-west. Again, the entire sea thickens with SIT of 2 m by 28th April 2019 (Fig. 4).

Unlike Barents, in East Siberian, sea ice develops along the circumference of the sea which is clearly seen in Fig. 5. By 27th November 2012, the entire sea is covered with sea ice showing a distinct feature. Stretching from north-west to north-east, the sea has SIT of 1.25 m whereas from south-west to south-east, it is 1.75 m which is clearly shown in the plots. On 7th December 2012, the sea ice again thickens to attain SIT of 3 m. This trend is seen to prevail till 20th December 2012. However, on 12th January 2013, no portion of the sea is observed to have SIT of 3 m instead the dominant SIT ranges from 1.5 to 2 m. Thus, the East Siberian Sea stands apart

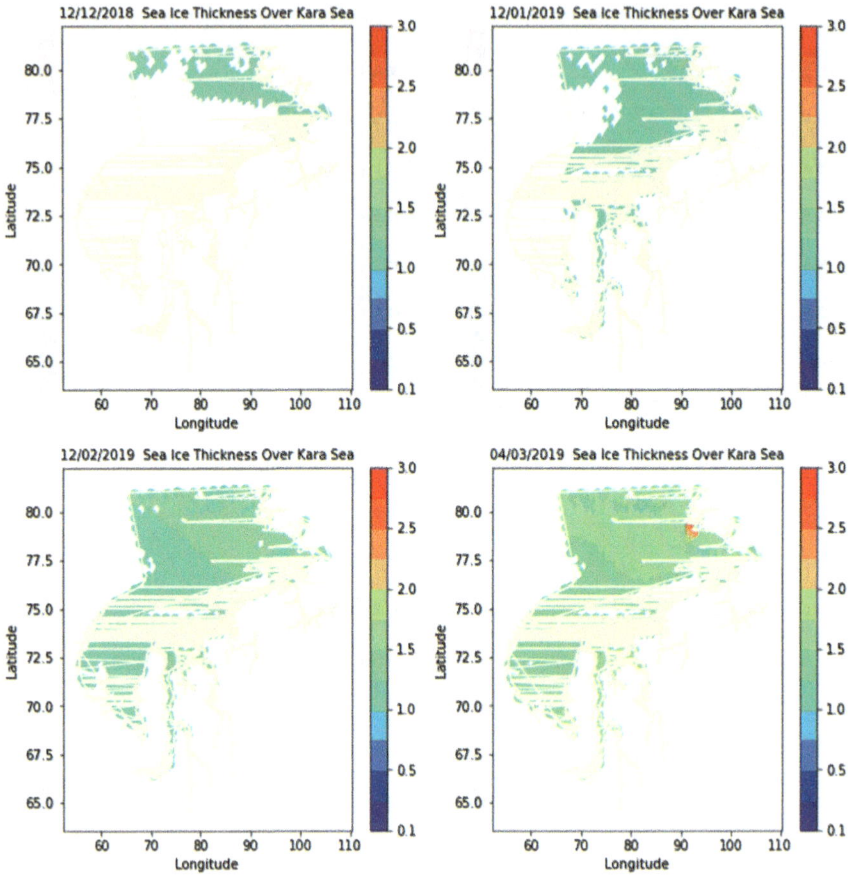

Fig. 2 Evolution of SIT over Kara Sea (2018–19)

in early loss of sea ice, i.e. the SIT peaks up to 3 m in December and then decreases by 1 m in January (Table 2).

Over the central latitudes of East Siberian Sea, on 11th November 2018, SIT of 2 m started to evolve along with SIT of 1.5 m. For the subsequent days, it is found to spread out to its remaining sides. By 27th November 2018, almost the entire sea is covered with SIT of 1.25 m and 1.5 m. However, on 7th December 2018, the west part of the sea gains thickness of 1.75 m with small patches of SIT with 3 m, whereas the east portion still has SIT of 1.25 m. Later, the sea ice over the entire sea grows to attain SIT of 1.75 m (Fig. 6).

The thickening of sea ice over Kara Sea has advanced for the year 2018–19 compared to 2012–13 as shown in Fig. 7a. However, for the year 2012–13, it was found that SIT has melted away after achieving its maximum thickness. This decline in the year 2012–13 for the month of December has clearly affected its rate profile as greater fluctuation is seen during that month (Table 3).

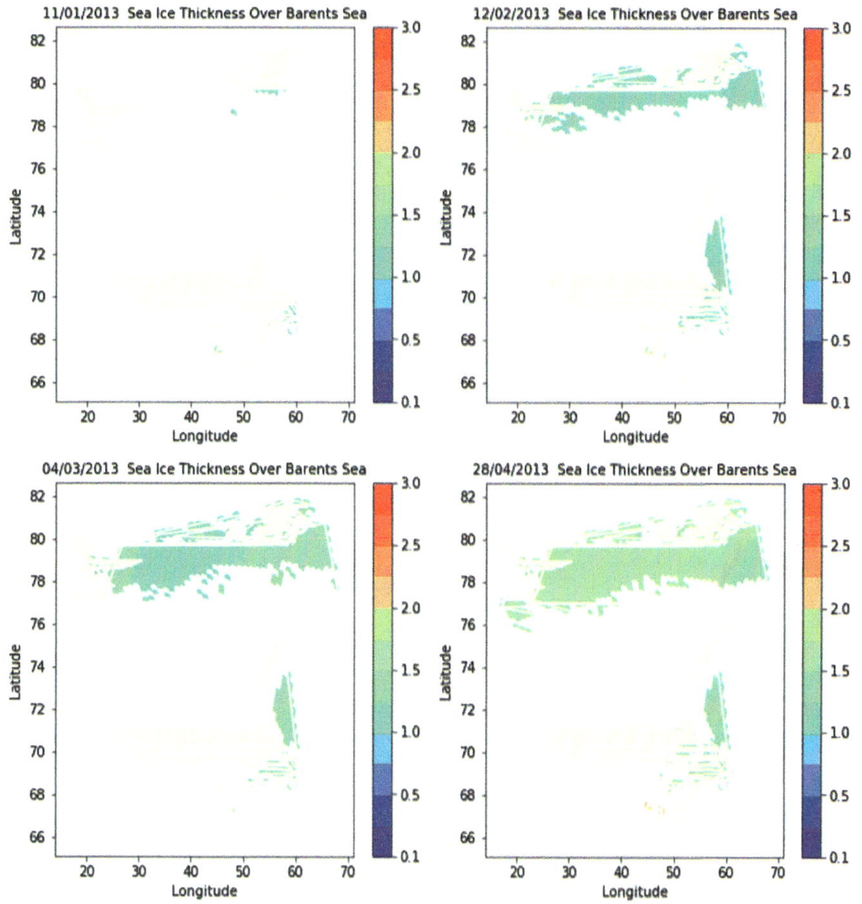

Fig. 3 Evolution of SIT over Barents Sea (2012–13)

Table 1 Difference in the evolution of SIT over Kara Sea for 2012–13 and 2018–19

Dates	2012–13	2018–19
12th December	No sea ice	Sea ice starts to evolve
12th February	SIT of 1.25 m only	SIT of 1.25 m at lower latitude and 1.75 m at higher latitude
22nd February	Entire sea is covered with 1.25 m SIT	Entire sea is covered with SIT of 1.75 m SIT
27th February	SIT of both 1.25 m and 1.75 m appears	SIT of 3 m starts to appear at higher latitudes

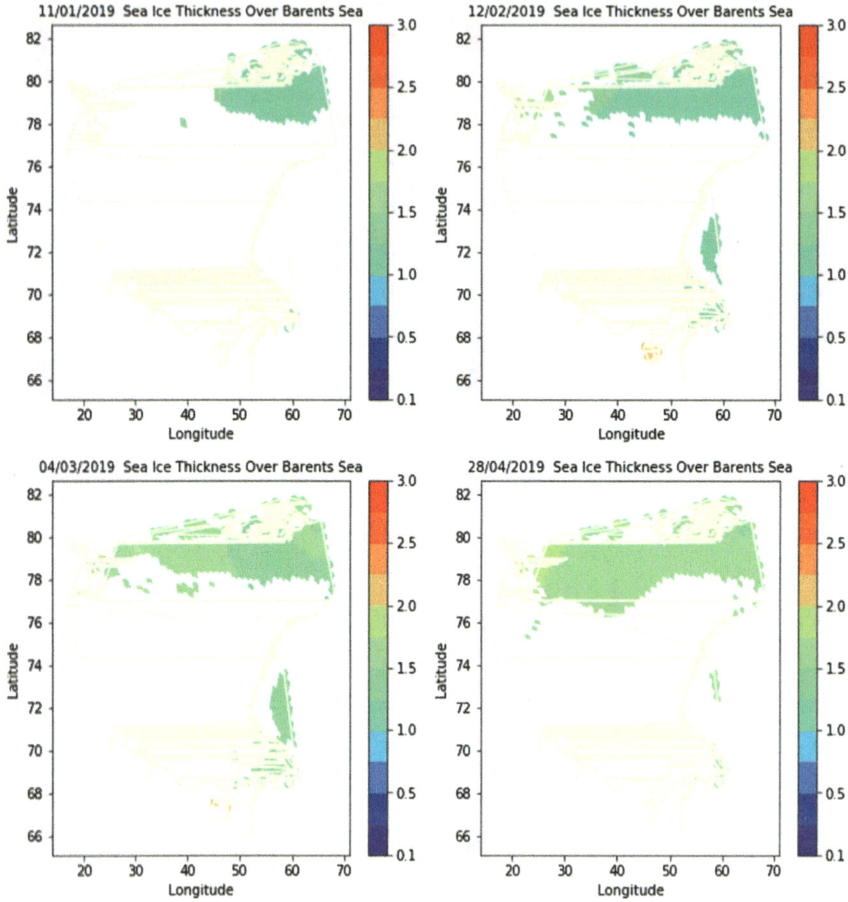

Fig. 4 Evolution of SIT over Barents Sea (2018–19)

Like Kara Sea, over Barents SIT have started to grow way before in 2018–19 compared to 2012–13. However, after the month of February, similar patterns of SIT are observed for both the years. Compared to Kara and East Siberian Seas, the growth of SIT over Barents is much delayed for both the years; this lag might be due to the topography of the sea (Fig. 8).

Unlike Barents Sea, the growth of SIT has started in the month of October for both the years. The rate profile in Fig. 9b shows a tremendous fluctuation in the month of October and November as the SIT growth was not gradual since there were frequent ups and downs.

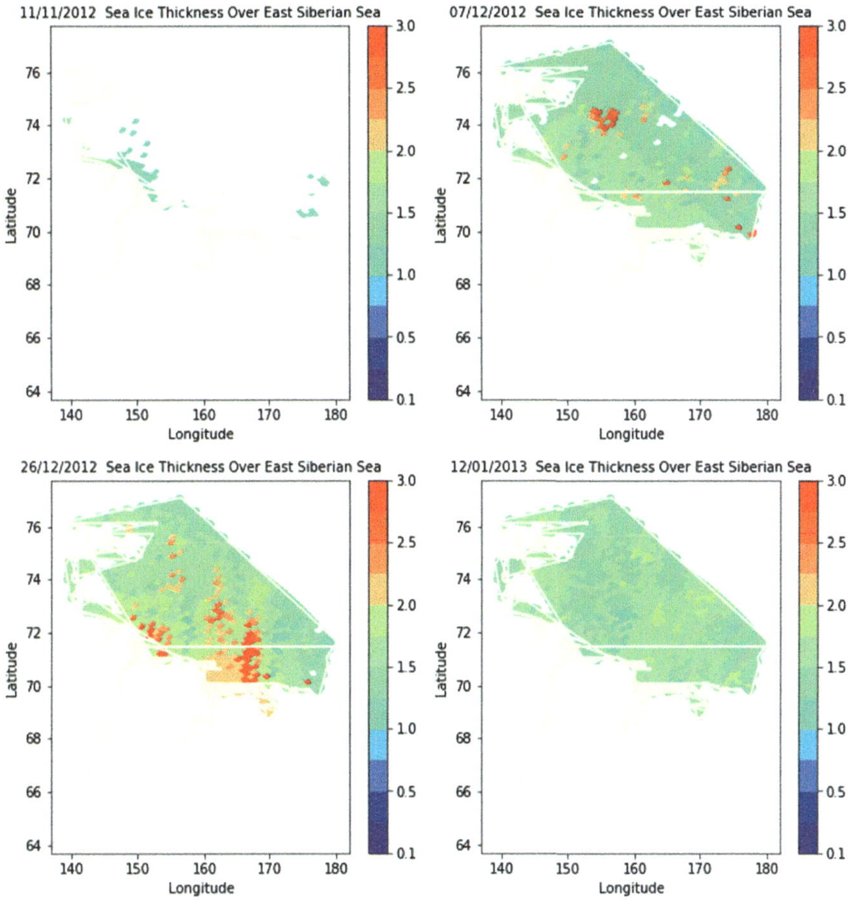

Fig. 5 Evolution of SIT over East Siberian Sea (2012–13)

Table 2 Difference in the evolution of SIT over Barents Sea for 2012–13 and 2018–19

Dates	2012–13	2018–19
11th January	No much of sea ice	Sea ice towards north-east corners of the sea
22nd February	1.25 m of SIT spreads across the north	1.25 m of SIT thickens to become 1.75 m, and it spreads across the north
14th March	Slight appearance of SIT with 0.75 m	Transition of SIT from 1.75 to 2 m

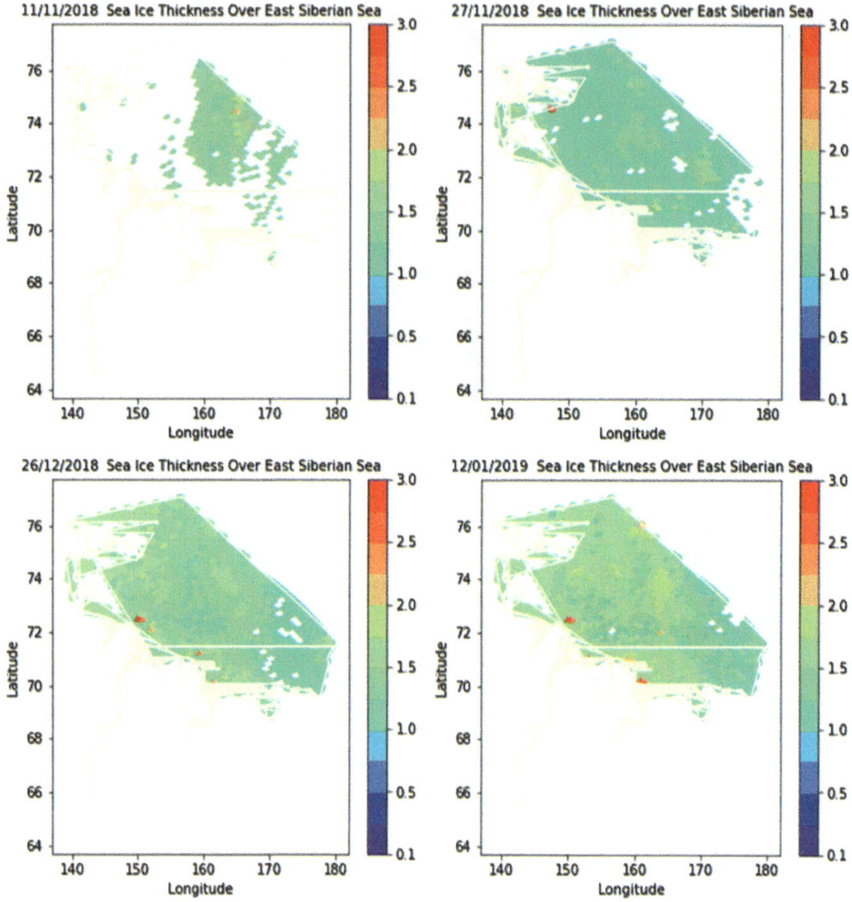

Fig. 6 Evolution of SIT over East Siberian Sea (2018–19)

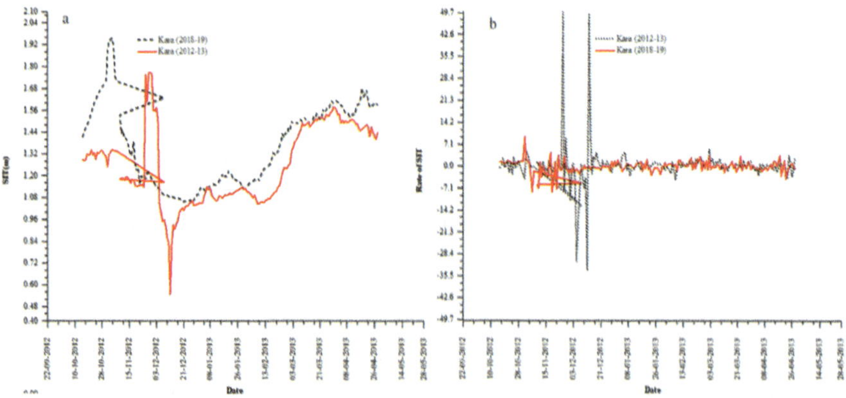

Fig. 7 **a** Daily growth of SIT and **b** rate profile over Kara Sea for the year 2012–13 and 2018–19

Table 3 Difference in the evolution of SIT over East Siberian Sea for 2012–13 and 2018–19

Dates	2012–13	2018–19
11th November	Evolution of sea ice from the circumference of the sea (1.25 m)	Evolution of sea ice from the central latitudes (1.75 m)
22nd November	Entire sea is covered with SIT of 1.25 m except the central regions	Entire sea is covered mostly with SIT of 1.75 m except for the two sides of the sea
7th December	100% area of the sea is covered with SIT mostly of 1.75 m and very few areas with 3 m. Evolution of 3 m SIT starts	Some portions of the sea don't have sea ice. A transition of SIT from 1.25 m to 1.75 m is found from lower to higher latitudes
26th December	SIT thickens to 3 m at lower latitudes	No such trend is found

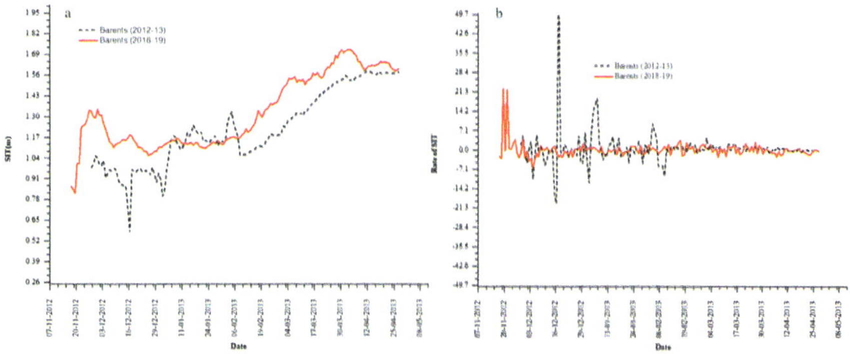

Fig. 8 **a** Daily growth of SIT and **b** rate profile of SIT over Barents Sea for the year 2012–13 and 2018–19

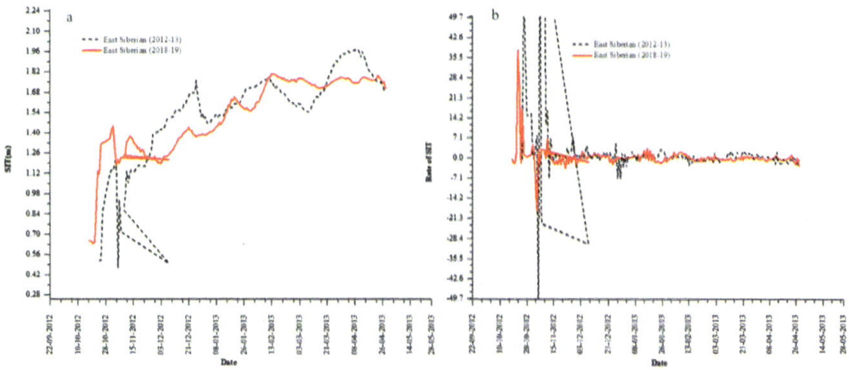

Fig. 9 **a** Daily growth of SIT and **b** rate profile of SIT over East Siberian Sea for the year 2012–13 and 2018–19

4 Conclusions

From the study, it is found that Kara, Barents and East Siberian Sea even being part of the Arctic region behave differently in terms of SIT, which is due to the topography. Kara and East Siberian Sea show early growth for SIT (October) for the year 2018–19, whereas it shows a decline in SIT for the year 2012–13. The rate of SIT is highly fluctuating for the year 2012–13 for all the three seas. Whereas for the year 2018–19, it was found that only during the growing phase, the fluctuations were prominent; later on, they attain a steady state.

References

1. Dmitrienko, I.A., Kirillov, S.A., Tremblay, L.B., Bauch, D., Hölemann, J.A., Krumpen, T., Kassens, H., Wegner, C., Heinemann, G., Schröder, D.: Impact of the arctic ocean atlantic water layer on siberian shelf hydrography. J. Geophys. Res. Oceans. **115**(C8) (2010)
2. Holt, B., Martin, S.: The effect of a storm on the 1992 summer sea ice cover of the Beaufort, Chuk- chi, and East Siberian seas. J. Geophys. Res. **106**(C1), 10171032 (2001)
3. Jakobsson, M.: Hypsometry and volume of the Arctic Ocean and its constituent seas. Geochem. Geophys. Geosyst. **3**(5) (2002). https://doi.org/10.1029/2001gc000302
4. Johannessen, O.M., Alexandrov, V.Y., Frolov, I.Y., Sandven, S., Pettersson, L.H., Bobylev, L.P., Kloster, K. et al.: Remote sensing of sea ice in the Northern Sea route: studies and applications. Springer–Praxis, Chichester, UK (2007)
5. Kurtz, N., Harbeck, J.: CryoSat-2 Level-4 Sea Ice Elevation, Freeboard, and Thickness, Version 1. Boulder, Colorado USA. NASA National Snow and Ice Data Center Distributed Active Archive Center (2017). https://doi.org/10.5067/96JO0KIFDAS8
6. Kwok, R., Cunningham, G.F., Wensnahan, M., Rigor, I., Yi, D.: Thinning and volume loss of the Arctic Ocean sea ice cover: 2003–2008. J. Geophys. Res. **114**, C07005 (2009). https://doi.org/10.1029/2009JC005312
7. Makkaveev, P.N., Polukhin, A.A., Kostyleva, A.V., Protsenko, E.A., Stepanova, S.V., Yakubov, S.K.: Hydrochemical features of the Kara Sea water structure in the summer 2015. Oceanology **57**(1), 57–66 (2017)
8. Pavlov, V., Pfirman, S.: Hydrographic structure and variability of the Kara Sea: implications for pollutant distribution. Deep Sea Res. **42**(6), 1369–1390 (1995)
9. Pivovarov, S., Schlitzer, R., Novikhin, A.: River run-off influence on the water mass formation in the Kara Sea. In: Ruediger S (ed) Siberian River Runoff in the Kara Sea: characterization quantification variability and environmental significance. Proc. Mar. Sci. **6**, 9–27 (2003)

Einstein's Cluster Demonstrating a Stable Relativistic Model for Strange Star SAX J1808.4-3658

R. Goti, S. Shah, and D. M. Pandya

Abstract A new class of solution of Einstein's field equations for super-dense stars is discussed on the ground of Einstein's cluster. We chose a particular metric assumption in the form of Finch-Skea ansatz. This mathematical model has gone under various stringent conditional testing for the fitment of physical viability. For the fulfilment of this verification, we chose a well-known compact star candidate SAX J1808.4-3658 whose observational mass and radii are discussed in detail in Gangopadhayay T. et al. (2013). The salient feature of this model is the increasing nature of tangential pressure from centre to the boundary. The regularity conditions ($\rho \geq 0$, $p_r \geq 0$, $p_t \geq 0$) and 'weak ($\rho \geq p_t, \rho \geq p_r$) and strong energy conditions ($\rho - p_r - 2p_t \geq 0$)' are well behaved. The stability condition becomes peculiar here and is required to be given special attention as suggested by Abreu H. et al. (2007). The model shows a viable behaviour in the interval suggested for cracking condition therein. Thus, we found that this model describes the large class of self-gravitating objects with vanishing radial pressure throughout the distribution.

Keywords Einstein's cluster · Cracking condition · Self-gravitating objects · Radial and tangential pressure

---

R. Goti · S. Shah
Department of Physics, Pandit Deendayal Petroleum University, Raisan, Gandhinagar, Gujarat 382007, India
e-mail: rid181198@gmail.com

S. Shah
e-mail: theshlokguy@yahoo.com

D. M. Pandya (✉)
Department of Mathematics, Pandit Deendayal Petroleum University, Raisan, Gandhinagar, Gujarat 382007, India
e-mail: dishantpandya777@gmail.com

M. Sahni et al. (eds.), *Mathematical Modeling, Computational Intelligence Techniques and Renewable Energy*, Advances in Intelligent Systems and Computing 1287,
https://doi.org/10.1007/978-981-15-9953-8_22

1 Introduction

Einstein's cluster was conceptually established in 1939 to describe the motion of stationary gravitating particles under the combined gravitational field. The spacetime is spherically symmetric, and the radial pressure of the complex system would be vanishing because of centrifugally supported systems. These stars move in the circular orbits with different inclined angles and create Einstein's shell. Different researchers [1–4] have studied extensively about the cluster system. The effects of radial pressure on this dynamic system can increase the anisotropic nature much higher because of $p_r = 0$. The exact category of solution for Einstein's field equations was studied by Herrera and Ponce de León [5] with consideration of anisotropic fluid sphere. The further generalization of field equations for a self-gravitating system using killing vectors of motion was studied Maartens and Maharaj [6]. New approach to a pressureless dust had been studied by Florides [1] and derived new interior uniform density solution because it was found that pressureless dust can't sustain in equilibrium. After Florides' [1] solutions satisfied Einstein's cluster, a modified approach to the problem was established by Zel'dovich and Polnarev [7]. The stability of Einstein's cluster was studied by Gilbert [3]. Eventually, the charged model of Einstein's cluster and spin fluid concept without pressure has been considered by Banerjee and Som [8] and Bedran and Som [9]. Modified gravity attracted the attention of Einstein's cluster modeling (Singh et al. [10]). Einstein's cluster model for weakly interacting massive particles (WIMPs) was proposed by Böhmer and Harko [11], Lake [12] that generates a spherically symmetric gravitational field. The dynamic stability of Einstein's clusters under radial and non-radial perturbations was discussed by Böhmer and Harko [11]. An anisotropic compact stellar model becomes meritless if it has instability due to the various physical parameters. The existence of a stability criterion for relativistic anisotropic neutron stars was shown by Lemaître [13]. Chan et al. [14] demonstrated how small anisotropies lead to a sudden change in the progress of self-gravitating systems. The concept of cracking or overturning was given by (Herrera et al. [15]). It was used to understand the behaviour of fluid beyond the equilibrium point (hydrostatic equilibrium). Herrera et al. [15] and Lemaître [16] found out that deviation from equilibrium is due to perturbations in both local anisotropy and matter density. The state of cracking, overturning, expansion or collapse is determined through integration of the Einstein's field equations. The potential stability and instability of anisotropic matter configurations over specified regions are shown by Abreu et al. [17].

In this paper, we have used the metric potential given by Sharma and Ratanpal [18], to describe the anisotropic fluid composition in self-gravitating systems—Einstein's clusters. The results and calculations have been verified for the relativistic star SAX J1808.3-3658, whose mass and radius are known (Gangopadhyay et al. [19]). The article is ordered in the following way: Section 2 renders information about basic field equations. Section 3 demonstrates how Einstein's cluster concept is used to describe the behaviour of a strange star. In succeeding Sect. 4, we have compared our interior metric with Schwarzschild's exterior solution. Sections 5 and 6 are devoted to the explanation of the physical validity of our model and graphical analysis with conclusion in Sect. 7.

2 Basic Field Equations

The static pseudo spheroidal spacetime element of a star is described in spherical coordinates (t, r, θ, ϕ) as [5]:

$$ds^2 = e^{\nu(r)}dt^2 - e^{\lambda(r)}dr^2 - r^2\left(d\theta^2 + \sin^2\theta d\phi^2\right) \tag{1}$$

where λ and ν are dependent on r.

The corresponding Einstein's field equations with consideration of anisotropy are [5]:

$$8\pi\rho = \frac{1 - e^{-\lambda}}{r^2} + \frac{e^{-\lambda}\lambda'}{r} \tag{2}$$

$$8\pi p_r = \frac{e^{-\lambda} - 1}{r^2} + \frac{e^{-\lambda}\nu'}{r} \tag{3}$$

$$8\pi p_t = e^{-\lambda}\left(\frac{\nu''}{2} + \frac{\nu'^2}{4} - \frac{\nu'\lambda'}{4} + \frac{\nu' - \lambda'}{2r}\right) \tag{4}$$

Here, $'$ denotes derivative with respect to r.

3 Einstein's Cluster Model

We assume Finch-Skea ansatz [20],

$$e^{\lambda} = 1 + \frac{r^2}{R^2} \tag{5}$$

where 'R' is curvature constant. The prescribed relation between e^{λ} and mass of a star $m(r)$ is given by,

$$e^{-\lambda} = 1 - \frac{2m(r)}{r} \tag{6}$$

Thus, using Eqs. (5) and (6), mass distribution of a star is represented as,

$$m(r) = \frac{r^3}{2\left(r^2 + R^2\right)} \tag{7}$$

Here, $m(r)$ is increasing and regular at centre $(m(r)_{r=0} = 0)$ of the star. Since Einstein's cluster has zero radial pressure ($p_r = 0$), Eq. (3) leads to

$$e^{\lambda} = 1 + r\nu' \tag{8}$$

Using Eq. (5) in Eq. (8), ν is given by,

$$\nu = \frac{r^2}{2R^2} + C \tag{9}$$

where 'C' is an arbitrary constant. Putting Eqs. (5) and (9) into field Eqs. (2–4), the final expressions would be,

$$\rho = \frac{r^2 + 3R^2}{8\pi\left(r^2 + R^2\right)^2} \tag{10}$$

$$p_r = 0 \tag{11}$$

$$\Delta = p_t = \frac{r^4 + 3r^2R^2}{32\pi R^2\left(r^2 + R^2\right)^2} \tag{12}$$

Since radial pressure p_r is zero throughout the star, the anisotropy $\Delta = p_t - p_r$ is same as tangential pressure p_t.

4 Matching Conditions

The obtained interior solution of model is matched with Schwarzschild's exterior solution given by,

$$\mathrm{d}s^2 = \left(1 - \frac{2M}{r}\right)\mathrm{d}t^2 - \left(1 - \frac{2M}{r}\right)^{-1}\mathrm{d}r^2 - r^2\left(\mathrm{d}\theta^2 + \sin^2\theta\,\mathrm{d}\phi^2\right) \tag{13}$$

where $r = a$ is radius of the star. Now, using matching conditions of Eqs. (1) and (13), we obtain mass-radius relationship as follows:

$$R = \frac{a\sqrt{a - 2M}}{\sqrt{2M}} \tag{14}$$

We have considered the star SAX J1808.4-3658 with mass $M = 0.9M_{\odot}$ where $M_{\odot}$ is solar mass, and $a = 7.951$ km is radius of the star (Gangopadhyay et al. [19]). Hence, the corresponding curvature parameter $R = 11.172$ km and $C = -0.663$ is obtained from matching of Eqs. (1) and (13).

5 Validation of Physical Parameters

To validate our model, we consider the stability conditions of Abreu et al. [17], Abbas et al. [21], Böhmer and Harko [22], Buchdahl [23].

5.1 Regularity of Metric Potentials

Putting $r = 0$ in our assumed metric, we get, $e^{\lambda} = 1$ and $e^{\nu} = e^{C}$, which are positive constants. Also, derivatives of Eqs. (5) and (9) give,

$$(e^{\nu})' = \left(e^{\frac{r^2}{2R^2}+C}\right).\frac{r}{R^2} \tag{15}$$

$$\left(e^{\lambda}\right)' = \frac{2r}{R^2} \tag{16}$$

Clearly, we can see from Eqs. (15) and (16) that $(e^{\nu})'_{r=0} = 0$ and $(e^{\lambda})'_{r=0} = 0$, which implies that metric coefficients of space and time are not singular at $r = 0$.

5.2 Radial Pressure at Boundary

Since Einstein's cluster has interesting behaviour that radial pressure vanishes all over the system, pressure at boundary obviously becomes zero ($p_{r(r=a)} = 0$).

5.3 Energy Conditions

(i) $\rho - p_r - 2p_t \geq 0$ **(strong energy condition)**

The representation of the SEC (strong energy condition) is given by:

$$\rho - p_r - 2p_t = -\frac{r^4 + r^2R^2 - 6R^4}{16\pi R^2\left(r^2 + R^2\right)^2} \tag{17}$$

Table 1 Conditions at centre and boundary of the star

Conditions	Centre ($r = 0$)	Boundary ($r = a$)
$\rho - p_r - 2p_t$	722.4724	277.8425
$\rho - p_t$	722.4724	324.9577
M/a	$0.1690 \leq \frac{4}{9}$ (mass-radius condition from Buchdahl [23])	
$v_t^2 - v_r^2$	−0.1500	−0.1132
Z_s	0.0000	0.2274

The positive nature of $\rho - p_r - 2p_t$ is shown in Fig. 4 and Table 1.

(ii) $\rho \geq p_t$ **(weak energy condition)**

Weak energy condition (WEC) states that $\rho - p_t \geq 0$, and the equation obtained is:

$$\rho - p_t = \frac{-r^4 + r^2R^2 + 12R^4}{32\pi R^2\left(r^2 + R^2\right)^2} \tag{18}$$

For the considered strange star SAX J1808.4-3658, putting the values of R and a (radius of the star) (Gangopadhyay et al. [19]) in strong and weak energy condition expressions, we get distinct positive values at $r = 0$ and $r = a$, which is one of the necessary bounds to justify our model's validation. The positive nature of $\rho - p_t$ is shown in Fig. 3 and Table 1.

5.4 *Central Anisotropy*

The difference of transverse and radial pressure must be null at the centre of the star. That means the anisotropic pressure values at point $r = 0$ are same. For our case, $p_r = 0$. Thus, putting $r = 0$ in Eq. (12), we get $\Delta_{r=0} = 0$, where Δ is anisotropy of the star. Thus, our model satisfies this condition which is shown in Fig. 2.

5.5 *Stability Conditions*

(i) $-1 \leq v_t^2 - v_r^2 \leq 0$ **(cracking condition)**

From Abreu et al. [17], it is found that anisotropic nature can induce cracking or collapsing of self-gravitating system. For a potentially stable system, $v_t^2 - v_r^2$ should be between −1 and 0. Sound speed should be considered for this stability condition. Here, v_t^2 is tangential sound speed, and v_r^2 is radial sound speed. They are expressed as,

$$v_t^2 = \frac{dp_t}{d\rho} = -\frac{3}{20} + \frac{2r^2}{5(r^2 + 5R^2)} \tag{19}$$

$$v_r^2 = \frac{dp_r}{d\rho} = 0 \tag{20}$$

Difference of Eqs. (19) and (20) would give,

$$v_t^2 - v_r^2 = -\frac{3}{20} + \frac{2r^2}{5(r^2 + 5R^2)} \tag{21}$$

From Fig. 5, we found that $v_t^2 - v_r^2$ lies between -1 and 0. Thus, our Finch-Skea ansatz [20] satisfies the cracking condition. Also from Table 1, it is shown that $v_t^2 - v_r^2$ stays within the range of 0 and -1. It is necessary to mention that, associated with this criterion, one of the extreme matter structure mentioned here ($p_t \neq 0$, $p_r = 0$) is always potentially stable for cracking, and the other one ($p_t = 0$, $p_r \neq 0$) becomes potentially unstable.

(ii) **Adiabatic index** $\Gamma_r = \frac{\rho + p_r}{p_r} \frac{dp_r}{d\rho} \geq \frac{4}{3}$ (evident from $p_r = 0$)
(iii) **Redshift *Z***

The surface redshift should be finite between range of $0 \leq Z_s \leq 5$, as derived by Böhmer and Harko [22]. The expression of surface redshift is given by,

$$Z_s = \left(\sqrt{e^{\lambda}} - 1\right)_{r=a} = -1 + \sqrt{1 + \frac{a^2}{R^2}} \tag{22}$$

In Table 1, we find the redshift (Z_s) at centre and boundary of the star, and it remains below 5. Here, Z_s is redshift at surface $r = a$.

Also, the gravitational redshift is given by,

$$Z_g = \left(\sqrt{e^{-\upsilon}} - 1\right)_{r=a} = -1 + e^{\left(-\frac{r^2}{4R^2} + 0.3315\right)} \tag{23}$$

The representation of expression $Z_g + 1$ is given in Fig. 6, and it is decreasing which is well-behaved function of gravitational redshift.

6 Physical Analysis

To scrutinize the agreement of our proposed model with observationally obtained results, we have assumed Finch-Skea ansatz $1 + \frac{r^2}{R^2}$ [20] and have considered the strange star SAX J1808.4-3658.

The calculated physical parameters are given in Table 1, and their variations against r have been plotted. Variation of matter density ρ against radial parameter

r is shown in Fig. 1, which can be seen as a monotonically decreasing function. Tangential pressure p_t and the anisotropic nature of the star composition Δ have been plotted in Fig. 2. An increasing trend of p_t against r can be seen, which is a noteworthy feature of our model. Δ shows the same trend as shown by p_t because of $p_r = 0$; for $0 < r \leq a$. Δ being positive and $\Delta_{r=0} = 0$ (Fig. 2), one can infer that anisotropic force is directed outwards, which acts as a counter-phenomenon against gravitational collapse, hence, making our model well behaved.

Figures 3 and 4 depict weak energy and strong energy conditions, respectively, which have positive values for $0 < r \leq a$. The cracking condition is given in Fig. 5. It can be noticed that $v_t^2 - v_r^2$ lies between -1 and 0, which makes our model potentially stable. Figure 6 shows variation of gravitational redshift Z against r, which is a decreasing function. More is the compactness of a star; more is the redshift.

Fig. 1 Density profile

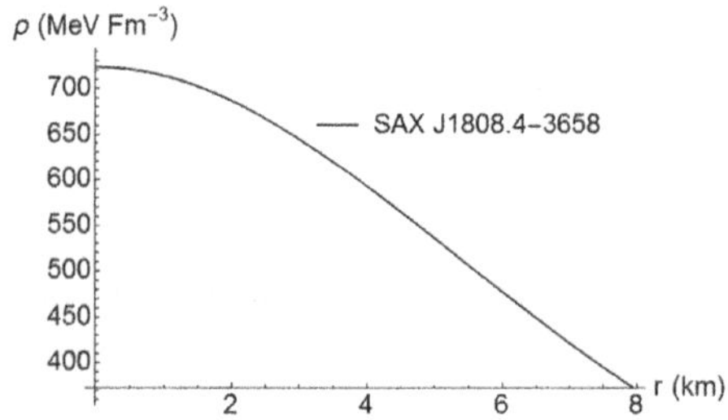

Fig. 2 Tangential pressure, Anisotropy profile

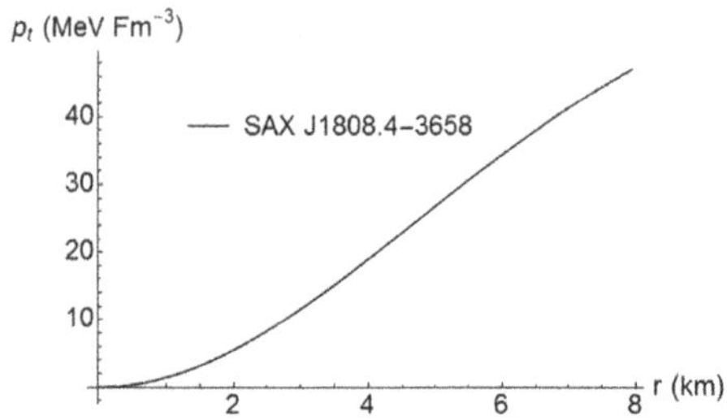

Fig. 3 Weak energy condition profile

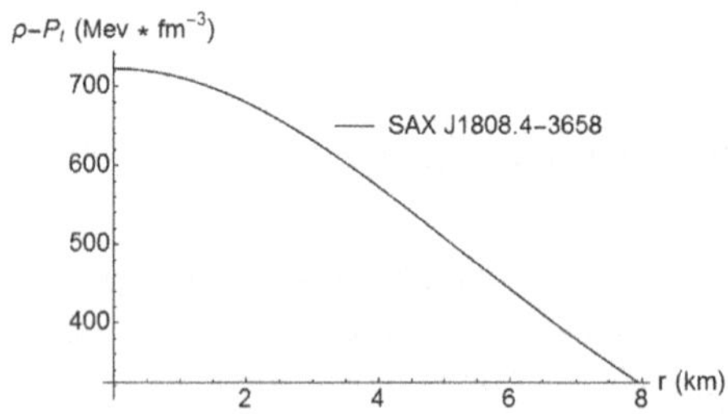

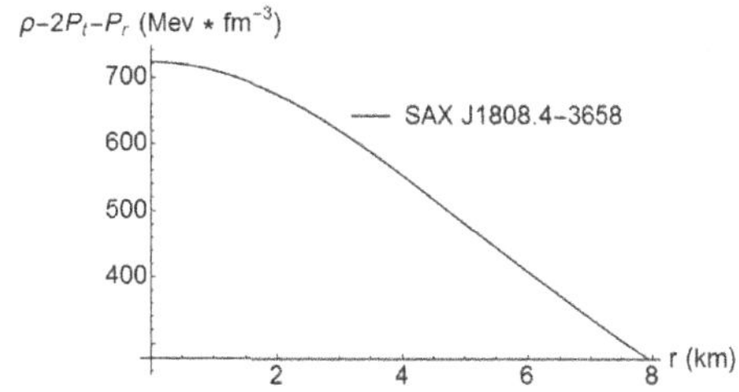

Fig. 4 Strong energy condition profile

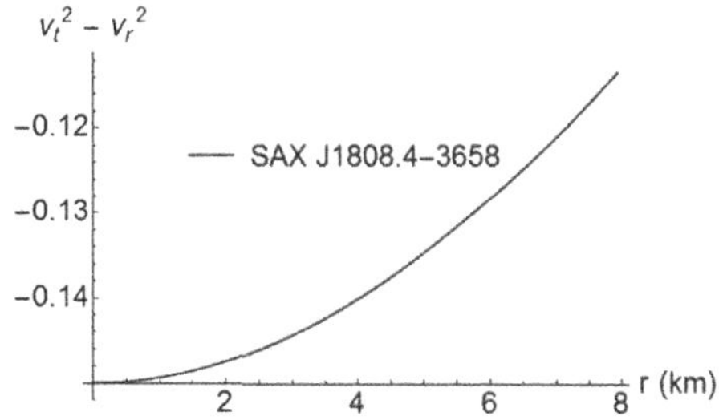

Fig. 5 Cracking condition profile

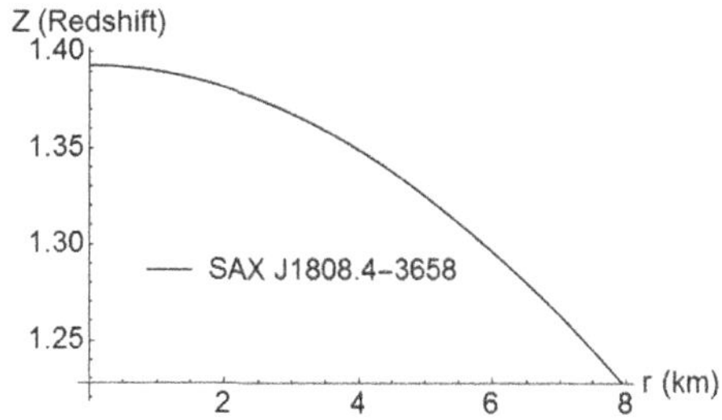

Fig. 6 Gravitational redshift $Z = Z_g + 1$

7 Conclusion

Using various stability conditions, energy conditions and mathematical bound parameters, we obtained solution of Einstein's field equations on the background of spacetime representing anisotropic compact star SAX J1808.4-3658. The spherically symmetric spacetime of Einstein's cluster with Finch-Skea ansatz [20] is crucial and at the same time well-behaved assumption of metric potential in this paper. Generated anisotropic spacetime outlines plausibility of various physical parameters like density, using the fact that radial pressure vanishes through the cluster was a great challenge, and at the same time, it is a salient feature of this work too. The compactness of star decreases with radius and anisotropy goes tremendously high compared to the anisotropic stars with non-vanishing radial pressure. The strong and weak energy conditions are well behaved within the permissible range. Another notable feature of this work is that the cracking condition is satisfied within the bounds for tangential pressure, anisotropy, redshift, etc. to achieve a potentially stable model of gravitating star. Moreover, surface redshift and adiabatic index lie within the range of stability. Considering all these parameters, we conclude that the present model has generated a new class of potentially stable stellar configurations on the background of spacetime with Einstein cluster. Noteworthy aspect of this work is that, it generates the compatible model with observational data of the large class of pulsars. The mimicking of compact star for Einstein cluster has provided a possible model to describe the nature of matter state. Future scope may lie in considering different spacetimes and ansatz which may possibly generate interesting results.

References

1. Florides, P.S.: A new interior Swarzchild solution. Proc. R. Soc. A R. Soc. **337**, 529–535 (1974). (Great Britain)
2. Zapolsky, H.S.: Can the redshifts of quasi-stellar objects BE gravitational? Astrophys. J. **153**, 163–169 (1968)
3. Gilbert, C.: The stability of a spherically symmetric cluster of stars describing circular orbits. Mon. Not. R. Astron. Soc. **114**, 628–634 (1954)
4. Comer, G.L., Katz, J.: Thick Einstein's Shells and Their Mechanical Stability, vol. 10. Classical Quantum Gravity, IOP Publishing, UK, pp 1751–1765 (1993)
5. Herrera, L., Ponce de León, J.: Isotropic and anisotropic charged spheres admitting a one-parameter group of conformal motions. J. Math. Phys. **26**, 2302–2307 (1985)
6. Maartens, R., Maharaj, S.D.: Collisionfree gases in spatially homogeneous spacetimes. J. Math. Phy. **26**, 2869–2880 (1985). (AIP Publishing)
7. Zel'dovich, Y.B., Polnarev, A.G.: Radiation of gravitational waves by a cluster of super dense stars. Sov. Astron. **18**, 17–23 (1974)
8. Banerjee, A., Som, M.M.: Einstein cluster of charged particles. Prog. Theor. Phys. **65**, 1281–1289 (1981)
9. Bedran, M.L., Som, M.M.: Conformally flat solution of a static dust sphere in Einstein-Cartan theory. J. Phys. A **15**, 3539–3542 (1982)
10. Singh, K.N., Rahaman, F., Banerjee, A.: Einstein's cluster mimicking compact star in the teleparallel equivalent of general relativity. Phys. Rev. D **100**, 084023–1-16 (2019)
11. Böhmer, C.G., Harko, T.: On Einstein clusters as galactic dark matter haloes. Mon. Not. R. Astron. Soc. **379**, 393–398 (2007)
12. Lake, K.: Galactic halos are Einstein clusters of WIMPs. arXiv:gr-qc/0607057 (2006)
13. Lemaître, G.A.: The expanding universe. Monthly Not. R. Astron. Soc. **91**, 490–501 (1933)
14. Chan, R., Herrera, L., Santos, N.O.: Dynamical instability for radiating anisotropic collapse. Mon. Not. R. Astron. **265**, 533–544 (1993)
15. Herrera, L., et al.: Spherically symmetric dissipative anisotropic fluids: a general study. Phys. Rev. D **69**, 084026-1-12 (2004)
16. Lemaître, G.A.: The expanding universe. Gen. Relativ. Gravit. **29**, 641–680 (1997)
17. Abreu, H., Hernandez, H., Nunez, L.A.: Sound speeds, cracking and the stability of self-gravitating anisotropic compact objects. Class. Quantum Gravity **24**, 4631–4646 (2007)
18. Sharma, R., Ratanpal, B.S.: Relativistic stellar model admitting a quadratic equation of state. Int. J. Mod. Phys. D **22**, 1350074-1-15 (2013)
19. Gangopadhyay, T., et al.: Strange star equation of state fits the refined mass measurement of 12 pulsars and predicts their radii. Mon. Not. R. Astron. Soc. **431**, 3216–3221 (2013)
20. Finch, M.R., Skea, J.E.F.: A realistic stellar model based on an ansatz of Duorah and Ray. Class. Quantum Gravity **06**, 467–476 (1989)
21. Abbas, G., Momeni, D., Aamir Ali, M., Myrzakulov, R., Qaisar, S.: Anisotropic compact stars in f(G) gravity. Astrophys. Space Sci. **357**, 1–11 (2015)

22. Böhmer, C.G., Harko, T.: Bounds on the basic physical parameters for anisotropic compact general relativistic objects. Class. Quantum Gravity **23**, 6479–6491 (2006)
23. Buchdahl, H.A.: Regular general relativistic charged fluid spheres. Acta Phys. Pol. **10**, 673–685 (1979)

A Mathematical Model to Study the Role of Buffer and ER Flux on Calcium Distribution in Nerve Cells

Hardik Joshi and Brajesh Kumar Jha

Abstract Ca^{2+} signalling is an essential process for initiating and performing numerous cellular activities of nerve cells. Ca^{2+} flux continuously modulates the Ca^{2+} signalling process by monitoring free Ca^{2+} ion in the cytosol. In the view of the above, a dynamic fractional reaction-diffusion model associated with the evolution of the neuronal disorder is developed. The mathematical model includes Ca^{2+} diffusion through a synapse, Ca^{2+} buffering near the plasma membrane, and Ca^{2+} flux between the endoplasmic reticulum (ER) and cytosol. The purpose of this study is to investigate the impact of Ca^{2+} flux on Ca^{2+} distribution in neurones. A simulation is performed in MATLAB with and without different Ca^{2+} flux to portray the profile of Ca^{2+} ions in nerve cells. The obtained result based on this study gives the new site as the role of fractional calculus in the modelling of Ca^{2+} signalling in nerve cells.

Keywords Calcium · Calcium flux · Fractional calculus · Fractional integral transform

1 Introduction

Ca^{2+} is a second messenger in the nervous system and plays a very important role in signal transduction. The process of Ca^{2+} signalling depends on various parameters like buffers, ER, mitochondria, voltage-gated Ca^{2+} channel, etc. [1]. Buffer is essential for the nervous system throughout life [2]. An adequate amount of the buffer is required to make a biological process alive. ER is a source of Ca^{2+} flux that modifies the cellular activities control by the Ca^{2+} signalling process. Thus, buffer and ER are actively involved in the Ca^{2+} signalling process to control the adequate amount

H. Joshi (✉) · B. K. Jha
Department of Mathematics, School of Technology, Pandit Deendayal Petroleum University, Gandhinagar, Gujarat, India
e-mail: hardik.joshi8185@gmail.com

B. K. Jha
e-mail: brajeshjha2881@gmail.com

M. Sahni et al. (eds.), *Mathematical Modeling, Computational Intelligence Techniques and Renewable Energy*, Advances in Intelligent Systems and Computing 1287,
https://doi.org/10.1007/978-981-15-9953-8_23

of calcium concentration ([Ca^{2+}]). The alteration or dysfunction in this leads to cell death or shown the early symptom of the neuronal disorder [3].

In the last decay, a researcher has found the impact of buffer on a variety of cells like astrocytes, myocytes, fibrocytes, oocytes, neurones, etc. [4–25]. In most of the cases, researchers have used the integer-order derivative to study the role of various parameters on cytosolic Ca^{2+} profile. Jha et al. have calculated the effect of a buffer, voltage-gated Ca^{2+} channel and ER on astrocytes cells [6, 9, 10, 12, 23]. Pathak and Adlakha have calculated the effect of Ca^{2+} flux on myocytes cells [5, 20]. Naik and Pardasani have calculated the effect of a buffer, ER flux on oocytes cells [16, 24, 25]. Panday and Pardasani have studied the effect of buffer and ER flux on oocytes cells [11, 21]. Jha and Adlakha have calculated the effect of sodium Ca^{2+} exchanger, ER flux on neurone cells [8, 17, 18]. Dave et al. have calculated the effect of ER flux on nerve cells [4, 15]. Joshi and Jha have calculated the effect of a buffer, voltage-dependent Ca^{2+} channel and ryanodine receptors on nerve cells by using a fractional-order model [7, 13, 14, 19, 22]. Most of the researchers have calculated the effect of different parameters of Ca^{2+} tools kit on various cells by using the integer-order model.

This work aims to study the effect of EGTA and ER flux on Ca^{2+} distribution by using a fractional-order model. The mathematical formulation and development of the model are given in the next section.

2 Mathematical Formulation

To develop a mathematical model first, we have defined the basic definition and its integral transforms which are used to solve a model [26].

1. Caputo fractional derivative

$$ {}_0^C D_t^{\alpha} h(t) = \begin{cases} \frac{1}{\Gamma(n-\alpha)} \int\limits_0^t (t-\tau)^{n-\alpha-1} \frac{\mathrm{d}^n h(\tau)}{\mathrm{d}\tau^n} \mathrm{d}\tau, & n-1 < \alpha < n \\ \frac{\mathrm{d}^n h(t)}{\mathrm{d}t^n} & , \alpha = n \in Z^+ \end{cases} \tag{1} $$

2. Mittage-Leffler function

$$ E_\alpha(p) = \sum_{m=0}^{\infty} \frac{p^m}{\Gamma(\alpha m + 1)}, \; R(\alpha) > 0, \quad \alpha, p \in Ç \tag{2} $$

3. Wright function

$$ \phi(\alpha, \beta; r) = \sum_{m=0}^{\infty} \frac{r^m}{\Gamma(\alpha m + \beta) \cdot m!}, \quad \alpha > -1, \beta \in Ç \tag{3} $$

4. Mainardi function

$$M_p(r) = \sum_{m=0}^{\infty} \frac{(-1)^m r^m}{\Gamma(-pm + (1-p)) \cdot m!}, \quad 0 < p < 1 \tag{4}$$

5. Laplace transform of Caputo fractional derivative

$$L\{{}_0^C D_t^\alpha h(t)\} = s^\alpha H(s) - \sum_{m=0}^{n-1} s^{\alpha-m-1} h^{(m)}(0), \quad n-1 < \alpha \leq n \tag{5}$$

6. Fourier transform of Caputo fractional derivative

$$F\left({}_{-\infty}^{C} D_x^\beta h(x)\right) = (-i\upsilon)^\beta F(\upsilon) \tag{6}$$

2.1 Mathematical Model

We have developed the fractional mathematical model of Ca^{2+} distribution in nerve cells. The law of mass conservation in the form of a differential equation is [27]

$$\frac{\partial C}{\partial t} - \frac{\partial M}{\partial x} = h(x, t, C) \tag{7}$$

where C is the $[Ca^{2+}]$; h is the reaction rate of Ca^{2+} buffering in neurone; and M is the rate of change of C represented by

$$M(x, t) = -D_{\mathrm{Ca}} \frac{\partial C}{\partial x} \tag{8}$$

The model incorporates the reaction between free Ca^{2+} ion and buffer inside the cytosol [9, 18, 24].

$$\left[\mathrm{Ca}^{2+}\right] + [B] \underset{k^-}{\overset{k^+}{\leftrightarrow}} [\mathrm{Ca}B] \tag{9}$$

where $[B]$ is a free buffer, and $[\mathrm{Ca}B]$ is Ca^{2+} bound buffer.

$$\frac{\partial}{\partial t}\left[\mathrm{Ca}^{2+}\right] = D_{\mathrm{Ca}} \cdot \nabla^2 \left[\mathrm{Ca}^{2+}\right] + R_j \tag{10}$$

$$\frac{\partial}{\partial t}[B] = D_B \cdot \nabla^2 [B] + R_j \tag{11}$$

$$\frac{\partial}{\partial t}[\text{Ca}B] = D_{\text{Ca}B} \cdot \nabla^2 [\text{Ca}B] - R_j \tag{12}$$

where $R_j = -k^+[B][\text{Ca}^{2+}] + k^-[\text{Ca}B]$.

Using Eqs. (7)–(12), the model is framed as

$$\begin{aligned} {}_0^C D_t^\alpha [\text{Ca}^{2+}](x,t) &= D_{\text{ca}} \cdot {}_{-\infty}^{\ C} D_x^\beta [\text{Ca}^{2+}](x,t) - u D_x [\text{Ca}^{2+}](x,t) \\ &\quad - k^+[B]_\infty \{[\text{Ca}^{2+}] - [\text{Ca}^{2+}]_\infty\}(x,t) + J_{\text{Leak}} - J_{\text{Pump}} + J_{\text{Chan}} \end{aligned} \tag{13}$$

where ${}_0^C D_t^\alpha [Ca^{2+}]$ and ${}_{-\infty}^{\ C} D_x^\beta [Ca^{2+}]$ are the fractional derivative along time and space, respectively, [26]. J_{Leak}, J_{Chan}, and J_{Pump} are the Ca^{2+} flux given by [1, 6, 23]

$$\begin{aligned} J_{\text{Leak}} &= D_L c_1 \\ &= \frac{D_L}{C_v}\{(C^* - (1 + C_v)[\text{Ca}^{2+}])\} \end{aligned} \tag{14}$$

where $c_1 = [\text{Ca}^{2+}]_{\text{ER}} - [\text{Ca}^{2+}]$, J_{Leak} is the flux passed through ER to the cytosol, and D_L is the Ca^{2+} leak flux; C_v is the ratio of ER and the cytosol volume and $C^* = C_v [\text{Ca}^{2+}]_{\text{ER}} + [\text{Ca}^{2+}]$.

$$\begin{aligned} J_{\text{Chan}} &= P \cdot D_C \cdot c_1 \\ &= \frac{D_C}{C_v}\{(C^* - (1 + C_v)[\text{Ca}^{2+}])\} \end{aligned} \tag{15}$$

where J_{Chan} is the channel probability that directly proportional to total numbers of open channels, and D_C is the channel conductance, and P is the probability of maximum Ca^{2+} uptake.

$$J_{\text{Pump}} = P_R^{\max} \cdot \left([\text{Ca}^{2+}]^2 \Big/ \left([\text{Ca}^{2+}]^2 + (K_R^M)^2 \right) \right) \tag{16}$$

Using Eqs. (13)–(16), the fractional diffusion equation for Ca^{2+} signalling becomes as

$$\begin{aligned} {}_0^C D_t^\alpha C(x,t) &= D_{\text{ca}} {}_{-\infty}^{\ C} D_x^\beta C(x,t) - u D_x C(x,t) - k_j^+ [B]_\infty (C - C_\infty) \\ &\quad + \frac{D_L}{C_v}(C^* - C(1 + C_v)) + \frac{D_C}{C_v}(C^* - C(1 + C_v)) - P\frac{C^2}{C^2 + K^2} \end{aligned} \tag{17}$$

where $C = [\text{Ca}^{2+}]$. To handle the non-linearity occurs in Eq. (17), we consider two cases when $K \gg C$ and $K \ll C$.

The initial and boundary conditions according to the physiology of nerve cells are

$$\left[\mathrm{Ca}^{2+}\right](x,t) = g(x); \quad t = 0,\ \left[\mathrm{Ca}^{2+}\right](x,t) = 0; \quad x \to \pm\infty \tag{18}$$

For simplicity, Eq. (17) is written as

$${}_0^C D_t^\alpha C(x,t) = D_{Ca} \cdot {}_{-\infty}^{C} D_x^\beta C(x,t) - u D_x C(x,t) - aC(x,t) + b \tag{19}$$

Applying the integral transform on Eq. (19), we get the solution as

$$G(x,t) = \frac{1}{2\pi} \int_{-\infty}^{\infty} \mathrm{e}^{-ikx} E_\alpha \theta \mathrm{d}k + \frac{bt^\alpha}{2\pi} \int_{-\infty}^{\infty} \mathrm{e}^{-ikx} E_{\alpha,\alpha+1} \theta \mathrm{d}k \tag{20}$$

where $\theta = \left\{iuk + D_{\mathrm{Ca}}(-ik)^\beta - a\right\} \cdot t^\alpha$ and $G(x,t)$ are the Green's function.

The solution is found to time domain and space domain separately in a form of special function. The solution of the advection-diffusion equation according to time is

$$G_{\alpha,2}(x,t) = \frac{1}{\pi\sqrt{2D_{Ca}t^\alpha}} \int_0^\infty \mathrm{e}^{-\frac{x^2}{4D_{Ca}t^\alpha k} - at^\alpha k} k^{-\frac{1}{2}} M_\alpha(k)\theta_1 \mathrm{d}k$$
$$+ \frac{bt^\alpha}{\pi\sqrt{2D_{Ca}t^\alpha}} \int_0^\infty \mathrm{e}^{-\frac{x^2 t^{-\alpha}}{4kD_{Ca}} - at^\alpha k} \phi(-\alpha, 1; -k) k^{-\frac{1}{2}} \theta_1 \mathrm{d}k \tag{21}$$

where $\theta_1 = \frac{sin(x - kut^\alpha)}{(x - kut^\alpha)}$.

The solution of the advection-diffusion equation according to space is

$$G_{1,\beta}(x,t) = \frac{\mathrm{e}^{-at}}{\theta_3}\theta_2 + \frac{bt}{\theta_3\{\ln\theta_2 - at\}} \times \left[\left\{\frac{\mathrm{e}^{-at}}{\theta_3}\theta_2\right\} - 1\right] \tag{22}$$

where $\theta_2 = S\left(\frac{x-u}{(D_{\mathrm{Ca}_\alpha} t)^{1/\beta}} / \beta, 1, 1, u; 1\right)$ is a Levy stable distribution [28] and $\theta_3 = (D_{\mathrm{Ca}_\alpha} t)^{1/\beta}$.

3 Results and Discussion

To portray the results, we have used the numerical values which are mention in Table 1. Figures 1 and 2 express the spatial $[\mathrm{Ca}^{2+}]$ distribution in the existence of buffer and ER flux. The mass of buffer concentration taken into account is 30 μM and 70 μM in Figs. 1 and 2, respectively. The results are obtained for different values

Table 1 Values of parameters [6, 9, 23, 24]

Symbol	Parameter	Value
D_{Ca}	Diffusion coefficient	200–300 μm^2/s
$[Ca^{2+}]_\infty$	Background $[Ca^{2+}]$	0.1 μM
$[Ca^{2+}]_{ER}$	ER $[Ca^{2+}]$	500 μM
$[B]_\infty$	Buffer concentration	30–150 μM
k^+	Buffer association rate constant	1.5 μM^{-1} s^{-1}
k^-	Buffer dissociation rate constant	1 s^{-1}
C_v	V_{ER}/V_{cyt}	0.185
P	Maximum uptake of Ca^{2+}	0.9
D_L	Leak flux of Ca^{2+}	0.11 s^{-1}
D_C	Channel conductance	6 s^{-1}

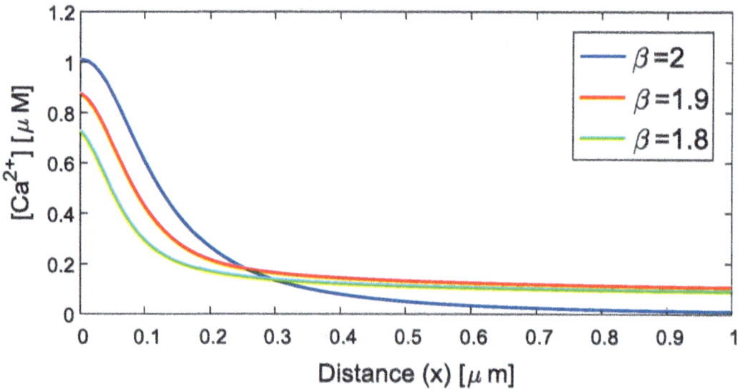

Fig. 1 Spatial distribution of $[Ca^{2+}]$ in nerve cells for a low amount of buffer $[B] = 30$ μM

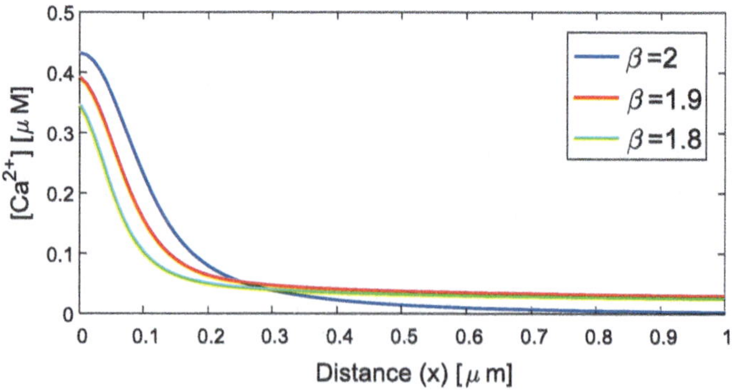

Fig. 2 Spatial distribution of $[Ca^{2+}]$ in nerve cells for a high amount of buffer $[B] = 70$ μM

of β at $\alpha = 1$. The time taken into consideration in both the figure is $t = 0.1$ s. By comparing Figs. 1 and 2, we have observed that the higher level of EGTA buffers lowers down the amount of $[Ca^{2+}]$ in the nerve cells. Physiologically is happened due to the mass of EGTA that reacts with Ca^{2+} in cytosol and produces a Ca^{2+} bound buffer.

Figures 3 and 4 express the temporal $[Ca^{2+}]$ distribution in the existence of buffer and ER flux. The mass of buffer concentration taken into account in Figs. 3 and 4 is 30 μM and 70 μM, respectively. The results are obtained for various values of α at $\beta = 2$. Both the results are obtained for $x = 0.01$ μm. The same physiological effect of EGTA buffers has been observed in the temporal profile of $[Ca^{2+}]$ distribution in the nerve cells. It is observed that the profile of $[Ca^{2+}]$ attains the peak near the source than achieved the steady-state level.

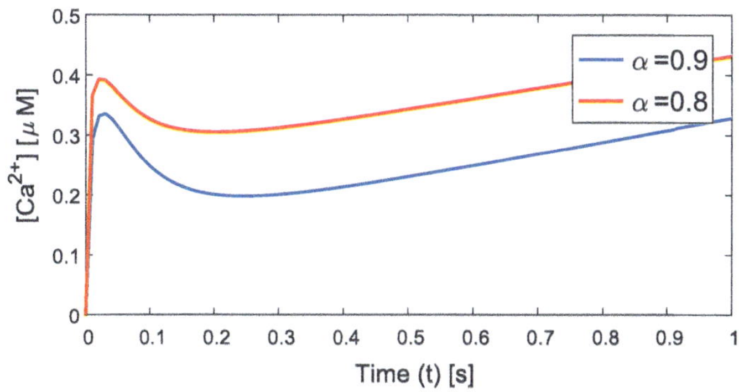

Fig. 3 Temporal distribution of $[Ca^{2+}]$ in nerve cells for a low amount of buffer $[B] = 30$ μM

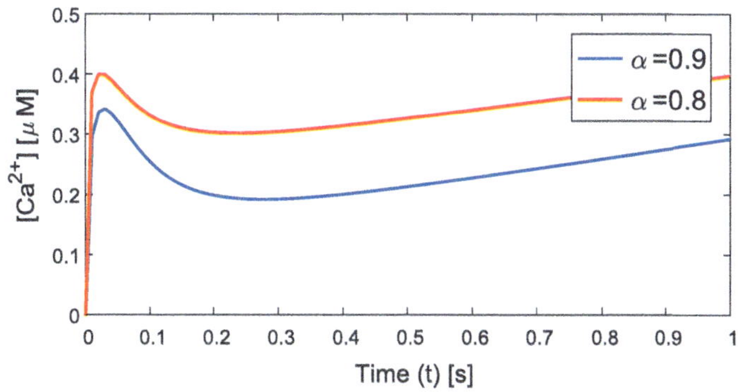

Fig. 4 Temporal distribution of $[Ca^{2+}]$ in nerve cells for a high amount of buffer $[B] = 70$ μM

4 Conclusion

In this study, a fractional-order model is developed to visualize the effects of EGTA and ER flux on Ca^{2+} distribution in nerve cells. It is observed that to maintain the adequate level of $[Ca^{2+}]$ in the cytosol EGTA and ER flux plays an important role. Also, it is observed that ER flux is more efficient in the existence of a low quantity of EGTA. Furthermore, we have observed that the fractional-order model provides a well understanding of a complex biomathematical problem.

Acknowledgements We are very grateful to the reviewers for their constructive suggestions that make updations in the research article.

References

1. Keener, J., Sneyd, J.: Mathematical Physiology. Interdisciplinary Applied Mathematics. Springer, US (2009)
2. Fall, C.: Computational Cell Biology. Interdisciplinary Applied Mathematics. Springer, Berlin (2002)
3. Carafoli, E., Brini, M.: Calcium signalling and disease. Springer Science and Business Media, Netherlands (2007)
4. Dave, D.D., Jha, B.K.: Delineation of calcium diffusion in alzheimeric brain. J. Mech. Med. Biol. **18**(2), 1–15 (2018)
5. Pathk, K., Adlakha, N.: Finite element model to study two dimensional unsteady state calcium distribution in cardiac myocytes. Alexandria J. Med. **52**(3), 261–268 (2016)
6. Jha, B.K., Jha, A.: Two dimensional finite volume model to study the effect of ER on cytosolic calcium distribution in astrocytes. Procedia Comput. Sci. **46**, 1285–1293 (2015)
7. Jha, B.K., Joshi, H.: A Fractional Mathematical Model to Study the Effect of Buffer and Endoplasmic Reticulum on Cytosolic Calcium Concentration in Nerve Cells. Fractional Calculus in Medical and Health Science. CRC Press/Taylor & Francis Group CRC Press, Boca Raton, FL, pp. 211–227 (2020)
8. Jha, A., Adlakha, N.: Analytical solution of two dimensional unsteady state problem of calcium diffusion in a Neuron cell. J. Med. Imaging Heal. Inf. **4**(4), 547–553 (2014)
9. Jha, B.K., Adlakha, N., Mehta, M.N.: Two-dimensional finite element model to study calcium distribution in astrocytes in presence of excess buffer. Int. J. Biomath. **7**(3), 1–11 (2014)
10. Jha, B.K., Adlakha, N., Mehta, M.N.: Two-dimensional finite element model to study calcium distribution in astrocytes in presence of VGCC and excess buffer. Int. J. Model. Simulation, Sci. Comput. **4**(2), 1250030 (2013)
11. Panday, S., Pardasani, K.R.: Finite element model to study the mechanics of calcium regulation in oocyte. J. Mech. Med. Biol. **14**(2), 1–16 (2014)
12. Jha, B.K., Jha, A., Adlakha, N.: Three-dimensional finite element model to study calcium distribution in astrocytes in presence of VGCC and excess buffer. Differ. Equ. Dyn. Syst. **28**, 603–616 (2020)
13. Joshi, H., Jha, B.K.: Fractional-order mathematical model for calcium distribution in nerve cells. Comput. Appl. Math. **39**, 56 (2020)
14. Jha, B.K., Joshi, H., Dave, D.D.: Portraying the effect of calcium-binding proteins on cytosolic calcium concentration distribution fractionally in nerve cells. Interdiscip. Sci. **10**(4), 674–685 (2018)

15. Dave, D.D., Jha, B.K.: Analytically depicting the calcium diffusion for Alzheimer's affected cell. Int. J. Biomath. **11**(7), 1850088 (2018)
16. Naik, P.A., Pardasani, K.R.: Three-dimensional finite element model to study effect of RyR calcium channel, ER leak and SERCA pump on calcium distribution in Oocyte cell. Int. J. Comput. Methods **15**(3), 1850091 (2019)
17. Jha, A., Adlakha, N., Jha, B.K.: Finite element model to study effect of $na^{+} - ca^{2+}$ exchangers and source geometry on calcium dynamics in a neuron cell. J. Mech. Med. Biol. **16**(2), 1650018 (2016)
18. Jha, A., Adlakha, N.: Two-dimensional finite element model to study unsteady state Ca^{2+} diffusion in neuron involving ER LEAK and SERCA. Int. J. Biomath. **8**(1), 1550002 (2015)
19. Joshi, H., Jha, B.K.: Fractional reaction diffusion model for Parkinson's disease. In: D. Pandian, X. Fernando, Z. Baig, F. Shi (eds) International Conference on ISMAC in Computational Vision and Bio-Engineering, vol. 30. LNCVB, Springer, Berlin, pp. 1739–1748 (2018)
20. Pathak, K., Adlakha, N.: Finite element model to study calcium signalling in cardiac myocytes involving pump, leak and excess buffer. J. Med. Imaging Heal. Informatics. **5**(4), 683–688 (2015)
21. Panday, S., Pardasani, K.R.: Finite element model to study effect of buffers along with leak from er on cytosolic Ca^{2+} distribution in oocyte. J. Med. Imaging Heal. Informatics. **4**(5), 1–8 (2013)
22. Joshi, H., Jha, B.K.: Fractionally delineate the neuroprotective function of calbindin-28 k in Parkinson's disease. Int. J. Biomath. **11**(8), 1850103 (2018)
23. Jha, B.K., Adlakha, N., Mehta, M.N.: Finite volume model to study the effect of ER on cytosolic calcium distribution in astrocytes. J. Comput. **3**, 74–80 (2011)
24. Naik, P.A., Pardasani, K.R.: One dimensional finite element model to study calcium distribution in oocytes in presence of VGCC, RyR and buffers. J. Med. Imaging Heal. Informatics. **5**, 471–476 (2015)
25. Naik, P.A., Pardasani, K.R.: Finite element model to study calcium distribution in oocytes involving voltage gated Ca2+ channel, ryanodine receptor and buffers. Alexandria J. Med. **52**, 43–49 (2016)
26. Podlubny, I.: Fractional differential equations. Academic Press, New York (1999)
27. Crank, J.: The Mathematics of Diffusion. Oxford University Press, London (1975)
28. Borak, S., Härdle, W., Weron, R.: Stable Distributions. Statistical Tools for Finance and Insurance, pp. 21–44. Springer, Berlin (2005)

Pattern Dynamics of Prey–Predator Model with Swarm Behavior via Turing Instability and Amplitude Equation

Teekam Singh, Shivam, Mukesh Kumar, and Vrince Vimal

Abstract In this paper, we propose and analyze the spatial prey–predator system with swarm behavior. Firstly, we examine the possibility and uniqueness of the coexistence equilibrium solutions of the model. We obtain Turing instability for spatial prey–predator model. By the use of weakly nonlinear analysis near the threshold value of parameter for patterns, we obtain the amplitude equation. Finally, we present numerical outcomes to justify the numerical results. Some interesting patterns show the spatial dynamical complexity of prey–predator model.

Keywords Prey–predator model · Swarm behavior · Amplitude equations · Pattern formation · Turing instability

1 Introduction

Social interaction between the individual is an integral part of life to survive on the earth for every species. In particular, swarm behavior within a population is a widespread and important phenomenon in ecological world [1–5]. There is a truth that living in groups is more reliable than living in an individual. This behavior is observed

T. Singh · Shivam (✉) · M. Kumar
Department of Mathematics, Graphic Era Deemed to be University, Dehradun 248002, India
e-mail: shivam.ma.geu@gmail.com

T. Singh
e-mail: tsingh@ma.iitr.ac.in; tsingh@gehu.ac.in

M. Kumar
e-mail: mukeshkumar.maths@geu.ac.in

T. Singh
Department of Mathematics, Graphic Era Hill University, Dehradun 248002, India

V. Vimal
Department of Computer Science & Engineering, Graphic Era Hill University, Dehradun 248002, India
e-mail: vvimal@ec.iitr.ac.in

M. Sahni et al. (eds.), *Mathematical Modeling, Computational Intelligence Techniques and Renewable Energy*, Advances in Intelligent Systems and Computing 1287,
https://doi.org/10.1007/978-981-15-9953-8_24

in many organisms on land or in the ocean for getting food, protection from predator and natural disaster and so on. It is good for them for their existence. Predation is the most important type of interaction that affects the population dynamics of all the species living on earth or in the ocean [6, 7]. As a result, prey–predator model has long been and will sustain to be one of the principal concerns in both ecosystem and mathematical biology [8–10].

Most of the mathematical systems in population ecology assumed the homogeneous distribution of living organisms in their habitat. The interaction systems are well known to ecologist and have been subject of much recent theoretical development with mathematical approaches like continuum partial differential equations or reaction–diffusion systems. Reaction refers the population interaction, often of the predator prey kind and diffusion to population redistribution of interacting species. One of the most intriguing predictions from reaction–diffusion concepts is that an interaction among communication and species interaction may govern to spatial pattern formation in the uniform habitat. The idea of Turing patterns was proposed by Turing [11] to show simple interaction of species can be responsible for patchiness even in perfectly uniform environment. Idea of Turing pattern formation in spatiotemporal system was first applied by Levin and Segel [12] to justify ecological pattern formulation. The basic idea of "diffusion driven instability (Turing instability)" was also applied in an ecological context by Segel and Jackson [13].

The present article is structured as follows. Section 2 describes the reaction-diffusion partial differential equations of the model with swarm behavior in prey with quadric death rate and positive starting population distributions. Furthermore, we analyze the equilibrium points, local stability of temporal model and obtain the Turing patterns condition in Sect. 3. Section 4 discusses the amplitude equations for Turing instability with Neumann condition. Section 5 carries the computer simulations to show that there is a variation of distinct patterns formation with respect to distinct predators death rate and diffusion coefficients.

2 Mathematical Model

Incorporating spatial aspects of prey and predator into the primal prey–predator model with swarm functional response [1, 8], we acquire the spatial prey–predator system as

$$\begin{aligned}\frac{\partial \Phi}{\partial t} &= \alpha\Phi\left(1-\frac{\Phi}{N}\right) - \frac{\beta\sqrt{\Phi}\Psi}{1+t_h\beta\sqrt{\Phi}} + k_1\nabla^2\Phi,\\ \frac{\partial \Psi}{\partial t} &= \frac{\gamma\beta\sqrt{\Phi}\Psi}{1+t_h\beta\sqrt{\Phi}} - \delta\Psi + k_2\nabla^2\Psi,\end{aligned} \tag{1}$$

where Φ and Ψ are the population density of the species (Φ: Prey and Ψ: Predator). α is the increasing rate of prey. Here, N is holding capacity of the system. The parameter δ is the predator's death rate, and γ is the transforming rate of species (prey to predator). β and t_h are the finding capacity of predator for food and mean dealing time of predator. The positive k_1 and k_2 are the diffusions for prey and predator, respectively.

Introducing the following non-dimensionalized variables in model (1)

$$\phi = \frac{\Phi}{N}, \quad \psi = \frac{\beta\Psi}{\alpha\sqrt{N}}, \quad \tau = \alpha t, \quad u = x\frac{\alpha}{k_2}, \quad v = y\frac{\alpha}{k_2},$$
$$\eta = \frac{\delta}{\alpha}, \quad \sigma = t_h\beta\sqrt{N}, \quad \rho = \gamma\frac{\beta\sqrt{N}}{\alpha}, \quad \kappa = \frac{k_2}{k_1}.$$

We get the modified model as in the line of the following references [5]; we choose the quadric rate of death for predator. Then the system (1):

$$\begin{aligned}\frac{\partial\phi}{\partial t} &= \phi(1-\phi) - \frac{\sqrt{\phi}\psi}{1+\sigma\sqrt{\phi}} + \nabla^2\phi,\\ \frac{\partial\psi}{\partial t} &= \frac{\rho\sqrt{\phi}\psi}{1+\sigma\sqrt{\phi}} - \eta\psi^2 + \kappa\nabla^2\psi,\end{aligned} \tag{2}$$

where κ is the proportion of diffusions of species densities for model (2). $\nabla^2\left(= \frac{\partial^2}{\partial u^2} + \frac{\partial^2}{\partial v^2}\right)$ is Laplacian in 2-dimensional area. For spatial patterns of model (2), we use the following initial species densities

$$\begin{aligned}&\phi(u, v, 0) > 0, \quad \psi(u, v, 0) > 0, \quad (u, v) \in [0, a] \times [0, a],\\ &\frac{\partial\phi}{\partial\xi} = \frac{\partial\psi}{\partial\xi} = 0,\end{aligned} \tag{3}$$

where a denotes the size of the domain in the direction of (ψ, ψ). ξ is outward unit normal on the boundary. Conditions (3) indicate that no species leave the system.

3 Equilibrium and Local Stability Analysis

Without diffusion, the equilibria of the model (2) are as follows

$$\begin{aligned}\phi(1-\phi) - \frac{\sqrt{\phi}\psi}{1+\sigma\sqrt{\phi}} &= 0,\\ \frac{\rho\sqrt{\phi}\psi}{1+\sigma\sqrt{\phi}} - \eta\psi^2 &= 0.\end{aligned}$$

Obviously (0, 0), (1, 0), $(\phi^*,\ \psi^*)$ are the equilibria, where $\psi^* = \frac{\rho}{\eta}\left(\frac{\sqrt{\phi^*}}{1+\sigma\sqrt{\phi^*}}\right)$ and ϕ^* is the root of

$$\eta\sigma^2\phi^{*2} - \eta(\sigma^2 - 1)\phi^* + 2\eta\sigma\sqrt{\phi^*}(\phi^* - 1) + (\rho - \sigma) = 0.$$

The Jacobian about the equilibria $(\phi^*,\ \psi^*)$

$$\begin{bmatrix} 1 - 2\phi - \frac{\psi}{2\sqrt{\phi}(1+\sigma\sqrt{\phi})^2} & -\frac{\sqrt{\phi}}{1+\sigma\sqrt{\phi}} \\ \frac{\rho\psi}{2\sqrt{\phi}(1+\sigma\sqrt{\phi})^2} & \frac{\rho\sqrt{\phi}}{1+\sigma\sqrt{\phi}} - 2\eta\psi \end{bmatrix}_{(\phi^*,\ \psi^*)}.$$

(i) At (1, 0), the Jacobian is

$$\begin{bmatrix} -1 & -\frac{1}{1+\sigma} \\ 0 & \frac{\rho}{1+\sigma} \end{bmatrix},$$

whose characteristic values are −1 and $\frac{\rho}{1+\sigma}$. Hence, the model is not stable (since $\frac{\rho}{1+\sigma}$ is constantly greater then zero).

(ii) At $(\phi^*,\ \psi^*)$, the Jacobian is (because $\psi^* = \frac{\rho}{\eta}\left(\frac{\sqrt{\phi^*}}{1+\sigma\sqrt{\phi^*}}\right)$

$$\begin{bmatrix} 1 - 2\phi^* - \frac{\rho}{2\eta(1+\sigma\sqrt{\phi^*})^3} & -\frac{\sqrt{\phi^*}}{1+\sigma\sqrt{\phi^*}} \\ \frac{\rho^2}{2\eta(1+\sigma\sqrt{\phi^*})^3} & \frac{\rho\sqrt{\phi^*}}{1+\sigma\phi^*} \end{bmatrix},$$

and then the corresponding characteristic equation is

$$\mu^2 + L_1\mu + L_2 = 0,$$

where

$$L_1 = \frac{\rho+2\eta(1+\sigma\sqrt{\phi^*})^2\left(2\phi^*-1+(\rho-\sigma)\sqrt{\phi^*}+2\sigma\phi^{*\frac{3}{2}}\right)}{2\eta(1+\sigma\sqrt{\phi^*})^3} \quad \text{and}$$

$$L_2 = \frac{\rho\sqrt{\phi^*}\left(\rho+\eta(2\phi^*-1)(1+\sigma\sqrt{\phi^*})^3\right)}{\eta(1+\sigma\sqrt{\phi^*})^4}.$$

Model (2) will be locally stable about $(\phi^*,\ \psi^*)$ if $L_i > 0$ $(i = 1, 2)$, by Routh-Hurwitz principle.

With diffusion, linearizing the model (2) about nontrivial equilibria $(\phi^*,\ \psi^*)$, we acquire the Jacobian as

$$\begin{bmatrix} 1 - 2\phi^* - \frac{\rho}{2\eta(1+\sigma\sqrt{\phi^*})^3} - k^2 & -\frac{\sqrt{\phi^*}}{1+\sigma\sqrt{\phi^*}} \\ \frac{\rho^2}{2\eta(1+\sigma\sqrt{\phi^*})^3} & \frac{\rho\sqrt{\phi^*}}{1+\sigma\phi^*} - \kappa k^2 \end{bmatrix}. \tag{4}$$

The condition for spatial pattern formation is

$$\frac{2\eta(1-2\phi^*)(1+\sigma\sqrt{\phi^*})^3-\rho}{2\eta(1+\sigma\sqrt{\phi^*})^3}-\frac{\kappa\rho\sqrt{\phi^*}}{1+\sigma\sqrt{\phi^*}}$$
$$-\sqrt{\frac{4\kappa\left(\rho\sqrt{\phi^*}(\rho+\eta(2\phi^*-1)(1+\sigma\sqrt{\phi^*})^3)\right)}{\eta(1+\sigma\sqrt{\phi^*})^4}}>0.$$

4 Amplitude Equation

Close to the instability threshold, the equilibria are not stable near to the emergence $\rho=\rho_T$, the eigenvalue related with the crucial modes is near to zero, and these modes are gradually changing, whereas the modes slack hastily which are off-critical; therefore, sole disruptions with k nearby k_T require considering. The spatial dynamics may be hence diminished to the dynamics of dynamic dormant modes. We will obtain the equations of amplitude by the method of multiple scales [5]. Turing pattern formation is thus nicely discussed by a system of dynamic resonant couples of modes $(-k_j, k_j)$ $(j=1,2,3)$ forming angle of $\frac{2\pi}{3}$ and $|k_j|=k_T$.

For obtaining the equations of amplitude, we should linearized the model (2) at equilibria (ϕ^*, ψ^*) and then the solution of model (2), at close to onset $\rho=\rho_T$ can be expanded as

$$\Phi=\Phi_s+\sum_{j=1}^{3}\Phi_0[A_j\exp(r\cdot ik_j)+\bar{A}_j\exp(r\cdot -ik_j)], \tag{5}$$

$$\Phi^0=\sum_{j=1}^{3}\Phi_0[A_j\exp(r\cdot ik_j)+\bar{A}_j\exp(r\cdot -ik_j)], \tag{6}$$

where ϕ_s represent thee uniform steady state and $\phi_0=[\frac{a_{11}^*\kappa+a_{22}^*}{2a_{21}^*\kappa},\ 1]^T$ $(\kappa=\frac{k_2}{k_1})$ is the characteristic vector of linearized operator. A_j and $\bar{A}_j$ (conjugate of A_j) are the amplitude related with modes k_j and $-k_j$, respectively. By the method of multiple scale, spatiotemporal evolution of the amplitudes A_j by the analysis of symmetries is given as

$$\rho_0\frac{\partial A_1}{\partial\tau}=\mu A_1+h\bar{A}_2\bar{A}_3-[g_1|A_1|^2+g_2(|A_2|^2+|A_3|^2)]A_1, \tag{7}$$

$$\rho_0\frac{\partial A_2}{\partial\tau}=\mu A_2+h\bar{A}_1\bar{A}_3-[g_1|A_2|^2+g_2(|A_1|^2+|A_3|^2)]A_2, \tag{8}$$

$$\rho_0\frac{\partial A_3}{\partial\tau}=\mu A_3+h\bar{A}_1\bar{A}_2-[g_1|A_3|^2+g_2(|A_1|^2+|A_2|^2)]A_3, \tag{9}$$

where ρ_0 is the relaxation time and μ is the normalized distance to onset. Now we will acquire the precise but complicated expression of the coefficient ρ_0, h, g_1, g_2. The model (2) can be written in the form:

$$\frac{\partial \Phi}{\partial \tau} = B\Phi + C \tag{10}$$

$$\text{where } \Phi = (\phi, \psi)^T, \quad B = \begin{bmatrix} a_{11} + \nabla^2 & a_{12} \\ a_{21} & a_{22} + \kappa \nabla^2 \end{bmatrix}.$$

Close to onset $\rho = \rho_T$, then the expansion of bifurcation parameter(s) as follows

$$\rho_T - \rho = \epsilon \rho_1 + \epsilon^2 \rho_2 + \epsilon^3 \rho_3 + O(\epsilon^4), \tag{11}$$

where $|\epsilon| \ll 1$. Expanding the variable Φ and C to the small parameter ϵ, we have

$$\Phi = \begin{bmatrix} \phi \\ \psi \end{bmatrix} + \epsilon \begin{bmatrix} \phi_1 \\ \psi_1 \end{bmatrix} + \epsilon^2 \begin{bmatrix} \phi_2 \\ \psi_2 \end{bmatrix} + \epsilon^3 \begin{bmatrix} \phi_3 \\ \psi_3 \end{bmatrix} + O(\epsilon^4) \tag{12}$$

$$C = \epsilon^2 h^2 + \epsilon^3 h^3 + O(\epsilon^4) \tag{13}$$

where h^2 and h^3 are the second and third order of ϵ. The linear terms B may be expressed as

$$B = B_T + (\rho_T - \rho)M, \tag{14}$$

where

$$B_T = \begin{bmatrix} a^*{}_{11} + \nabla^2 & a^*{}_{12} \\ a^*{}_{21} & a^*{}_{22} + \kappa \nabla^2 \end{bmatrix}, \quad M = \begin{bmatrix} b_{11} & b_{12} \\ b_{21} & b_{22} \end{bmatrix}.$$

Now we need separate the time scale for model (2) where each time scale T_i can be considered as an independent variable. By the chain rule, the derivative should convert as

$$\frac{\partial}{\partial t} = \frac{\partial}{\partial T_0} + \epsilon \frac{\partial}{\partial T_1} + \epsilon^2 \frac{\partial}{\partial T_2} + O(\epsilon^3) \tag{15}$$

As the variable amplitude A is varying very slowly, therefore $\frac{\partial}{\partial T_0} = 0$. Then Eq. (15) reduces to:

$$\frac{\partial}{\partial t} = \epsilon \frac{\partial}{\partial T_1} + \epsilon^2 \frac{\partial}{\partial T_2} + O(\epsilon^3) \tag{16}$$

Substituting Eqs. (11)–(15) into Eq. (10). According to different order of ϵ, expanding Eq. (10) as follows:

For the first order of ϵ:

$$B_T \begin{bmatrix} \phi_1 \\ \psi_1 \end{bmatrix} = 0 \tag{17}$$

The second order of ϵ:

$$B_T \begin{bmatrix} \phi_2 \\ \psi_2 \end{bmatrix} = \frac{\partial}{\partial T_1} \begin{bmatrix} \phi_1 \\ \psi_1 \end{bmatrix} - \rho_1 M \begin{bmatrix} \phi_1 \\ \psi_1 \end{bmatrix} - h_2 \tag{18}$$

The third order of ϵ:

$$B_T \begin{bmatrix} \phi_3 \\ \psi_3 \end{bmatrix} = \frac{\partial}{\partial T_1} \begin{bmatrix} \phi_2 \\ \psi_2 \end{bmatrix} = \frac{\partial}{\partial T_2} \begin{bmatrix} \phi_1 \\ \psi_1 \end{bmatrix} - \rho_1 M \begin{bmatrix} \phi_2 \\ \psi_2 \end{bmatrix} - \rho_2 M \begin{bmatrix} \phi_1 \\ \psi_1 \end{bmatrix} - h_3 \tag{19}$$

From Eq. (17), near to the starting B_T is the operator of the system, $(\phi_1, \psi_1)^T$ is the linear span of characteristic vector corresponding to characteristic value 0. Solving the first order of ϵ, we have

$$\begin{bmatrix} \phi_1 \\ \psi_1 \end{bmatrix} = \begin{bmatrix} \theta \\ 1 \end{bmatrix} [W_1 \exp(r \cdot ik_1) + W_2 \exp(r \cdot ik_2) + W_3 \exp(r \cdot ik_3)] + \text{complex conjugate} \tag{20}$$

when the system is under first perturbation, $\exp(r \cdot ik_j)$ mode amplitude is W_j, $|k_j| = k^*{}_T$ and this amplitude is determined by the higher order terms.

By the stipulation of Fredholm solvability, the right side of Eq. (18) must be orthogonal with zero eigenvector of $B^*{}_T$ to make certain the solution of this equation. The zero eigenvectors of $B^*{}_T$ are

$$\begin{bmatrix} 1 \\ \varphi \end{bmatrix} \exp(r \cdot -ik_j) + \text{complex conjugate}. \tag{21}$$

For second order ϵ, we have

$$B_T \begin{bmatrix} \phi_2 \\ \psi_2 \end{bmatrix} = \frac{\partial}{\partial T_1} \begin{bmatrix} \phi_1 \\ \psi_1 \end{bmatrix} - \rho_1 \begin{bmatrix} b_{11}\phi_1 & b_{12}\psi_1 \\ b_{21}\phi_1 & b_{22}\psi_1 \end{bmatrix} -$$
$$\begin{bmatrix} \left(\frac{\rho(3\phi^*\sigma^2+4\sigma\sqrt{\phi^*}+1)}{8\eta\phi^*(1+\sigma\sqrt{\phi^*})^5} - 1\right){\phi_1}^2 - \left(\frac{1}{2\sqrt{\phi^*}(1+\sigma\sqrt{\phi^*})^2}\right)\phi_1\psi_1 \\ -\left(\frac{\rho^2(3\phi^*\sigma^2+4\sigma\sqrt{\phi^*}+1)}{8\eta\phi^*(1+\sigma\sqrt{\phi^*})^5}\right){\phi_1}^2 + \left(\frac{\rho}{2\sqrt{\phi^*}(1+\sigma\sqrt{\phi^*})^2}\right)\phi_1\psi_1 - \left(\eta\right){\psi_1}^2 \end{bmatrix} = \begin{bmatrix} F_\phi \\ F_\psi \end{bmatrix} \tag{22}$$

The stipulation of orthogonality is

$$[1, \varphi] \begin{bmatrix} F^i{}_\phi \\ F^i{}_\psi \end{bmatrix} = 0 \tag{23}$$

where $F^i{}_\phi$ and $F^i{}_\psi$, the coefficient corresponding to $\exp(r \cdot ik_j)$. Applying stipulation of orthogonality, we have

$$
\begin{aligned}
(\theta+\varphi)\frac{\partial W_1}{\partial T_1} &= \rho_1[\theta b_{11}+b_{12}+\varphi(\theta b_{21}+b_{22})]W_1 - 2(h_1+\varphi h_2)\bar{W}_2\bar{W}_3\\
(\theta+\varphi)\frac{\partial W_2}{\partial T_1} &= \rho_1[\theta b_{11}+b_{12}+\varphi(\theta b_{21}+b_{22})]W_2 - 2(h_1+\varphi h_2)\bar{W}_2\bar{W}_3\\
(\theta+\varphi)\frac{\partial W_3}{\partial T_1} &= \rho_1[\theta b_{11}+b_{12}+\varphi(\theta b_{21}+b_{22})]W_3 - 2(h_1+\varphi h_2)\bar{W}_1\bar{W}_2
\end{aligned}
$$

Solving Eq. (22), we have

$$
\begin{aligned}
\begin{bmatrix}\phi_2\\ \psi_2\end{bmatrix} &= \begin{bmatrix}\Phi_0\\ \Psi_0\end{bmatrix} + \sum_{j=1}^{3}\begin{bmatrix}\Phi_j\\ \Psi_j\end{bmatrix}\exp(r\cdot ik_j) + \sum_{j=1}^{3}\begin{bmatrix}\Phi_{jj}\\ \Psi_{jj}\end{bmatrix}\exp(r\cdot ik_j)\\
&\quad + \begin{bmatrix}\Phi_{12}\\ \Psi_{12}\end{bmatrix}\exp[r\cdot i(k_1-k_2)] + \begin{bmatrix}\Phi_{23}\\ \Psi_{23}\end{bmatrix}\exp[r\cdot i(k_2-k_3)]\\
&\quad + \begin{bmatrix}\Phi_{31}\\ \Psi_{31}\end{bmatrix}\exp[r\cdot i(k_3-k_1)] + \text{complex cojugate.}
\end{aligned} \tag{24}
$$

Coefficients of above equation are obtain by solving the equations about exp(0), $\exp(r\cdot ik_j)$, $\exp(r\cdot 2ik_j)$ and $\exp(r\cdot i(k_j-k_k))$. Using above method, we have

$$
\begin{aligned}
\begin{bmatrix}\Phi_0\\ \Psi_0\end{bmatrix} &= \begin{bmatrix}\phi_0\\ \psi_0\end{bmatrix}(|W_1|^2+|W_2|^2+|W_3|^2),\quad \Phi_j = f\Psi_j\\
\begin{bmatrix}\Phi_{jj}\\ \Psi_{jj}\end{bmatrix} &= \begin{bmatrix}\phi_{11}\\ \psi_{11}\end{bmatrix}W_j^{\,2},\quad \begin{bmatrix}\Phi_{jk}\\ \Psi_{jk}\end{bmatrix} = \begin{bmatrix}\phi^*\\ \psi^*\end{bmatrix}W_j\bar{W}_k.
\end{aligned}
$$

From Eq. (19), we have

$$
\begin{aligned}
B_T\begin{bmatrix}\phi_3\\ \psi_3\end{bmatrix} &= \frac{\partial}{\partial T_1}\begin{bmatrix}\phi_2\\ \psi_2\end{bmatrix} = \frac{\partial}{\partial T_2}\begin{bmatrix}\phi_1\\ \psi_1\end{bmatrix} - \rho_1\begin{bmatrix}b_{11}\phi_2+b_{12}\psi_2\\ b_{21}\phi_2+b_{22}\psi_2\end{bmatrix} - \rho_2\begin{bmatrix}b_{11}\phi_1+b_{12}\psi_1\\ b_{21}\phi_1+b_{22}\psi_1\end{bmatrix}\\
&- \begin{bmatrix}
\left(\frac{\rho(3\phi^*\sigma^2+4\sigma\sqrt{\phi^*}+1)}{8\eta\phi^*(1+\sigma\sqrt{\phi^*})^5}-1\right)\phi_1\phi_2 - \left(\frac{1}{2\sqrt{\phi^*}(1+\sigma\sqrt{\phi^*})^2}\right)(\phi_1\psi_2+\phi_2\psi_1)\\
-\left(\frac{3\rho\sqrt{\phi^*}(5\phi^{*2}\sigma^3+9\phi^*\sigma^2\sqrt{\phi^*}+5\phi^*\sigma+\sqrt{\phi^*})}{48\eta\phi^{*3}(1+\sigma\sqrt{\phi^*})^6}\right)\phi_1^{\,3} + \left(\frac{3\phi^*\sigma^2+4\sigma\sqrt{\phi^*}+1}{8\phi^*\sqrt{\phi^*}(1+\sigma\sqrt{\phi^*})^4}\right)\phi_1^{\,2}\psi_1\\
-\left(\frac{\rho^2(3\phi^*\sigma^2+4\sigma\sqrt{\phi^*}+1)}{8\eta\phi^*(1+\sigma\sqrt{\phi^*})^5}\right)\phi_1\phi_2 + \left(\frac{\rho}{2\sqrt{\phi^*}(1+\sigma\sqrt{\phi^*})^2}\right)(\phi_1\psi_2+\phi_2\psi_1) - \left(\eta\right)\psi^2\\
+\left(\frac{3\rho^2\sqrt{\phi^*}(5\phi^{*2}\sigma^3+9\phi^*\sigma^2\sqrt{\phi^*}+5\phi^*\sigma+\sqrt{\phi^*})}{48\eta\phi^{*3}(1+\sigma\sqrt{\phi^*})^6}\right)\phi_1^{\,3} - \left(\frac{\rho(3\phi^*\sigma^2+4\sigma\sqrt{\phi^*}+1)}{8\phi^*\sqrt{\phi^*}(1+\sigma\sqrt{\phi^*})^4}\right)\phi_1^{\,2}\psi_1
\end{bmatrix}
\end{aligned}
$$

By the condition of Fredholm solubility, we get

$$
\begin{aligned}
(\theta+\varphi)\left(\frac{\partial Y_1}{\partial T_1}+\frac{\partial W_1}{\partial T_2}\right) &= [\theta b_{11}+b_{12}+\varphi(\theta m_{21}+m_{22})](\rho_1 Y_1+\rho_2 W_1)\\
&\quad + H(\bar{Y}_2\bar{W}_3+\bar{Y}_3\bar{W}_2) - [G_1|W_1|^2+G_2(|W_2|^2+G_3|W_3|^2)]W_1
\end{aligned}
$$

The remaining other equations can be acquired by the conversion of the subscript of W. W_i can be expanded as

$$W_i = \epsilon W_i + \epsilon^2 V_i + O(\epsilon^3). \tag{25}$$

We obtained the equations of amplitude using Eqs. (25) and (15)

$$\rho_0 \frac{\partial A_1}{\partial t} = \mu A_1 + h\bar{A}_2\bar{A}_3 - [g_1|A_1|^2 + g_2(|A_2|^2 + |A_3|^2)]A_1.$$

The other two equation can be obtained by transformation of the subscript W_i.

5 Numerical Simulation

In this section, we discuss the emergence of patterns for a certain set of parameter values and diffusion coefficients of model system (2) that illustrates the stable steady state becomes unstable due to diffusion. For example, we consider following parameter values

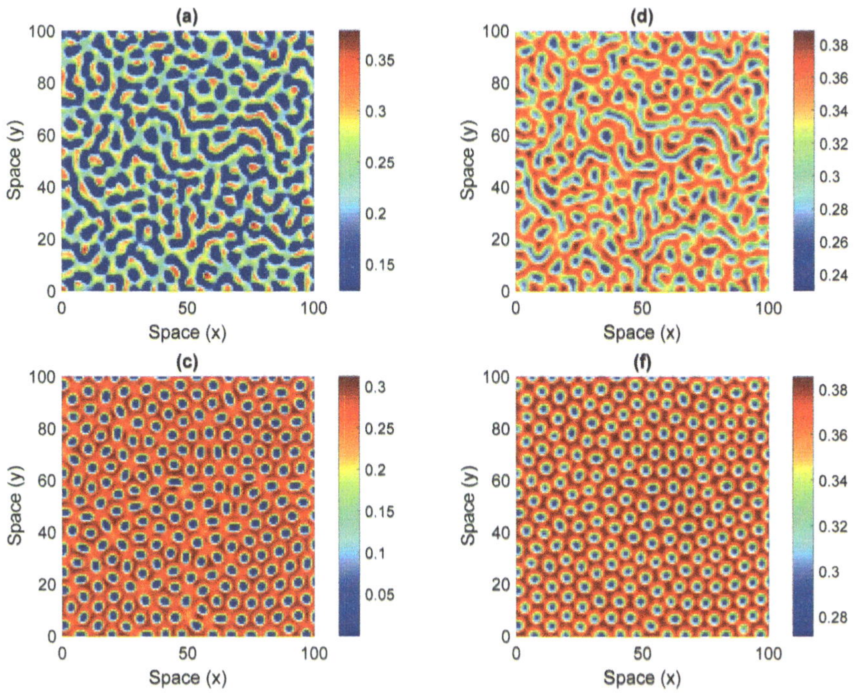

Fig. 1 Spatial distribution of the species (left: prey) and (right: predator) at time $\tau = 2500$

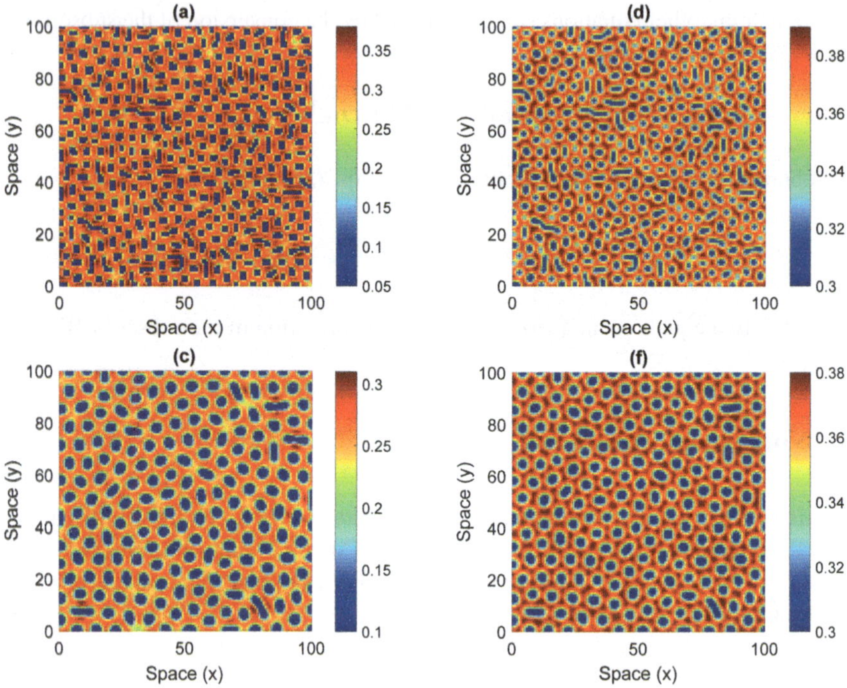

Fig. 2 Spatial distribution of the species (left: prey) and (right: predator) at time $\tau = 2500$

$$\sigma = 0.01, \quad \rho = 0.8, \quad k_1 = 8, \quad k_2 = 100, \quad \kappa = \frac{k_2}{k_1} = \frac{100}{8} \tag{26}$$

the initial distribution are $\phi(u_i, v_j, 0) = 0.21 + \nu_1 \xi_{ij}$ and $\psi(u_i, v_j, 0) = 0.36 + \nu_2 \omega_{ij}$ where ν_1 and ν_2 are small reals and ξ_{ij} and ω_{ij} are uncorrelated white Gaussian noise.

Numerical results are obtained by finite difference method that reveals spatial complexity of the model (2). From Figs. 1 and 2, we observe that pattern of spatial distribution depend upon death rate of predators. We obtain pattern for time $\tau = 2500$ and choose diffusion coefficients by $k_1 = 8$, $k_2 = 100$. We see blue and red spot patterns where blue spot shows patches of low population density and red spot shows patches of high population density. These patterns are well known as spot patterns in literature [4, 5, 9].

6 Conclusion

In this paper, we have incorporated the swarm type of functional response for the interaction between the species. We have obtained the uniform steady state and derived the conditions that confirms the stability of steady states. It is observed that Turing instability may occur in the system and it depends upon death rate and diffusion coefficients. We perform numerical simulation to illustrate hidden realities of the proposed system. We have obtained various patterns such as spot, stripes and mixed in two dimensional space. Various patterns reveal the spatial complexity of reaction-diffusion systems. This study reveals that decreasing κ, the spatial pattern converts from mixed to patches; this mean that the species populations are flocking together. The outcomes and analysis in this article may enhance the research study of spatial pattern formation in the prey–predator system with swarm behavior and may describe numerous realm.

References

1. Ajraldi, V., Pittavino, M., Venturino, E.: Modeling herd behavior in population systems. Nonlinear Anal. Real World Appl. **12**(4), 2319–2338 (2011)
2. Kooi, B.W., Venturino, E.: Ecoepidemic predator–prey model with feeding satiation, prey herd behavior and abandoned infected prey. Math. Biosci. **274**(1), 58–72 (2016)
3. Matia, S.N., Alam, S.: Prey–predator dynamics under herd behavior of prey. Univ. J. Appl. Math. **1**(4), 251–257 (2013)
4. Singh, T., Banerjee, S.: Spatiotemporal model of a predator–prey system with herd behavior and quadratic mortality. Int. J. Bifurc. Chaos. **29**(4), 1950049 (2019)
5. Yuan, S., Xu, C., Zhang, T.: Spatial dynamics in a predator–prey model with herd behavior. Chaos **23**(3), 033102 (2013)
6. Ma, X., Shao, Y., Wang, Z., Luo, M., Fang, X., Ju, Z.: An impulsive two-stage predator-prey model with stage-structure and square root functional responses. Math. Comput. Simul. **119**(1), 91–107 (2016)
7. Xu, C., Yuan, S., Zhang, T.: Global dynamics of a predator–prey model with defense mechanism for prey. Appl. Math. Lett. **62**(1), 42–48 (2016)
8. Liu, X., Zhang, T., Meng, X., Zhang, T.: Turing–Hopf bifurcations in a predator–prey model with herd behavior, quadratic mortality and prey-taxis. Physica A **496**(1), 446–460 (2018)
9. Tang, X., Song, Y.: Bifurcation analysis and Turing instability in a diffusive predator–prey model with herd behavior and hyperbolic mortality. Chaos Soliton Fract. **81**(1), 303–314 (2015)
10. Belvisi, S., Venturino, E.: An ecoepidemic model with diseased predators and prey group defense. Simul. Model. Pract. Theory **34**(1), 144–155 (2013)
11. Turing, A.M.: The chemical basis of morphogenesis. Philos. Trans. R. Soc. Lond. B: Biol. Sci. **237**(641), 37–72 (1952)
12. Levin, S.A., Segel, L.A.: Hypothesis for origin of planktonic patchiness. Nature **259**(5545), 659 (1976)
13. Segel, L.A., Jackson, J.L.: Dissipative structure: an explanation and an ecological example. J. Theor. Biol. **37**(3), 545–559 (1972)

Unsteady Magnetohydrodynamic Flow of Two Immiscible Fluids Through a Pipe in Presence of Heat Transfer

Ankush Raje and M. Devakar

Abstract The intent of the current work is to study the time-dependent MHD flow of two immiscible Newtonian and micropolar fluids in a porous pipe in presence of heat transfer. The numerical solutions for the flow under consideration are obtained using Crank–Nicolson approach. The classical no-slip condition at the boundary of the pipe is used along with the appropriate interface conditions at the interface, to solve the initial boundary value problem governing the flow. The results for fluid velocities, microrotation and temperatures with varying physical parameters are displayed through graphs and are discussed as well. The Nusselt number is also studied for the considered flow, and the results are tabulated.

Keywords Micropolar fluid · Immiscible fluids · Circular pipe · Heat transfer · Unsteady flow · MHD · Porous medium

1 Introduction

Among the assortment of non-Newtonian liquid models, one conspicuous model called micropolar fluid model was coined by Eringen [1] in mid sixties. The micropolar fluid model has the capacity to depict the unpredictable liquids like creature blood, polymeric liquids, colloidal suspensions, slurries and so forth, whose singular particles are of various shape, which can expand and shrink. A free vector accounting for the revolution of the fluid particles at the small scale level is present in micropolar fluids. This vector is called microrotation vector, and it is combined with fluid

A. Raje (✉)
Department of Mathematics, School of Technology, Pandit Deendayal Petroleum University, Gandhinagar 382007, India
e-mail: ankush.mth@gmail.com

M. Devakar
Department of Mathematics, Visvesvaraya National Institute of Technology, Nagpur 440010, India

M. Sahni et al. (eds.), *Mathematical Modeling, Computational Intelligence Techniques and Renewable Energy*, Advances in Intelligent Systems and Computing 1287,
https://doi.org/10.1007/978-981-15-9953-8_25

velocity vector in the modeling of micropolar fluid flows. A definite record of the theory of micropolar liquids can be alluded from [2–4].

The magnetohydrodynamic (MHD) flows gained significant consideration in the research arena due to appropriateness in numerous fields of technology and science. In the treatment of cancer, magnetic drug targeting is generally utilized technique for the conveyance of medicines to influenced organs. The fluid flow in permeable medium is likewise of significant for many fields of engineering. To mention a few, oil recuperation process, filtration of liquids, stream of water in waterway bed, and so forth. Furthermore, the studies on the effect of heat transfer and magnetic field on immiscible fluid flows through porous medium are of prime importance in biomechanics and medicine [5].

Flow through pipes has picked up the consideration taking into account its relevance in plentiful fields. Specifically, the two-fluid flows through circular pipe discover application in blood through veins [6, 7]. Blood, when moves through arteries, can be demonstrated as immiscible liquid stream with suspension of erythrocytes and granulocytes in the inward (center) district and plasma layer in the external (fringe) one [8, 9]. The fluid in the center locale, which has suspended particles in it as erythrocytes and granulocytes, carries on as non-Newtonian liquid, while the liquid in fringe layer takes after Newtonian liquid. Apart from the applications in blood rheology, immiscible liquid courses through pipes have significance in chemical industries, petroleum industries, particularly in enhanced oil recovery (EOR). Some recent studies on the flow through pipes can be referred from [10, 11] and the references therein.

In this paper, the magnetohydrodynamic flow of immiscible fluids (Newtonian and micropolar) through a porous pipe alongside heat transfer is thought of, which is not concentrated till now in literature. The micropolar liquid occupies the core region of the pipe and the peripheral region of the pipe is filled up with Newtonian liquid. The governing problem is solved for fluid flow and heat transfer profiles.

2 Mathematical Formulation of the Problem

We consider a laminar, time-dependent and axi-symmetric flow of two immiscible fluids in a pipe of radius R_0. The pipe is filled with homogeneous porous medium having permeability k^*. The core region ($r = 0$ to $r = R$) is occupied with micropolar fluid, whereas the peripheral region ($r = R$ to $r = R_0$) is assumed to be filled with Newtonian fluid. Further, the fluids in both regions are conducting having electrical conductivity σ, and a transverse magnetic field of influence H_0 is applied perpendicular to the pipe as shown in Fig. 1.

Owing to the considered unsteady flow, initially, fluids and pipe are stationary. At time $t > 0$, pressure gradient is applied for the generation of the flow in positive Z-direction. The fluid velocity and temperature profiles are assumed to be in the form as described in [10].

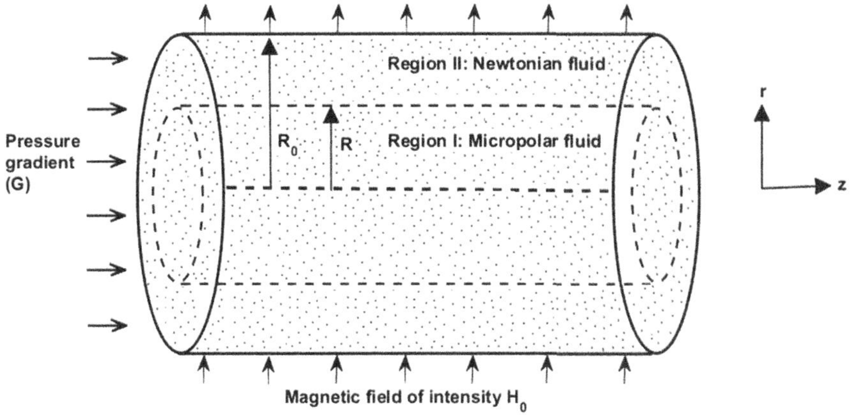

Fig. 1 Geometrical setup of the flow

Under these assumptions, the considered unsteady two-fluid flow is governed by the following equations [10, 11]. It has to be noted that, the non-dimensional form of the governing equations is directly presented here for the sake of brevity.

Region-I: Micropolar fluid region ($0 \leq r \leq 1$)

$$\frac{\partial w_1}{\partial t} = G + \frac{(1+n_1)}{\mathrm{Re}}\frac{1}{r}\frac{\partial}{\partial r}\left(r\frac{\partial w_1}{\partial r}\right) + \frac{n_1}{\mathrm{Re}}\frac{1}{r}\frac{\partial}{\partial r}(rb) - \frac{\left(M^2 + \frac{1}{\mathrm{Da}}\right)}{\mathrm{Re}}w_1, \tag{1}$$

$$\frac{\partial b}{\partial t} = \frac{\left(1 + \frac{n_1}{2}\right)}{\mathrm{Re}}\frac{\partial}{\partial r}\left[\frac{1}{r}\frac{\partial}{\partial r}(rb)\right] - \frac{n_1}{\mathrm{Re}}\left[\frac{\partial w_1}{\partial r} + 2b\right], \tag{2}$$

$$\begin{aligned}\frac{\partial T_1}{\partial t} = & \frac{1}{\mathrm{Re}P_R}\frac{1}{r}\frac{\partial}{\partial r}\left(r\frac{\partial T_1}{\partial r}\right) + \frac{B_R}{\mathrm{Re}P_R}\left\{\left(\frac{\partial w_1}{\partial r}\right)^2 + n_1\left[\frac{\partial w_1}{\partial r} + 2b\right]^2\right. \\ & \left. -2\delta_1\frac{b}{r}\frac{\mathrm{d}b}{\mathrm{d}r} + \delta_2\left[\left(\frac{\partial b}{\partial r}\right)^2 + \frac{b^2}{r^2}\right]\right\},\end{aligned} \tag{3}$$

Region-II: Newtonian fluid region ($1 < r \leq s$)

$$\frac{\partial w_2}{\partial t} = \frac{G}{m_2} + \frac{m_1}{\mathrm{Re}m_2}\frac{1}{r}\frac{\partial}{\partial r}\left(r\frac{\partial w_2}{\partial r}\right) - \frac{\left(M^2 + \frac{1}{\mathrm{Da}}\right)}{\mathrm{Re}m_2}w_2, \tag{4}$$

$$\frac{\partial T_2}{\partial t} = \frac{K}{\mathrm{Re}P_R m_2 C_p}\frac{1}{r}\frac{\partial}{\partial r}\left(r\frac{\partial T_2}{\partial r}\right) + \frac{B_R m_1}{\mathrm{Re}P_R m_2 C_p}\left(\frac{\partial w_2}{\partial r}\right)^2, \tag{5}$$

where, $s = \frac{R_0}{R} \geq 1$, $n_1 = \frac{\kappa}{\mu_1}$ is micropolarity parameter, $\mathrm{Re} = \frac{\rho_1 W R}{\mu_1}$ is Reynolds number, $G = -\frac{\partial p}{\partial z}$ is pressure gradient, $M = \sqrt{\frac{\sigma {H_0}^2 R^2}{\mu_1}}$ is Hartmann number, $\mathrm{Da} = \frac{k^*}{R^2}$ is Darcy number, $P_R = \frac{\mu_1 C_{p_1}}{K_1}$ is the Prandtl number, $B_R = \frac{\mu_1 W^2}{K_1 \Delta T}$ is the Brinkman number, $\delta_1 = \frac{\gamma}{R^2 \mu_1}$ and $\delta_2 = \frac{\beta}{R^2 \mu_1}$, $m_1 = \frac{\mu_2}{\mu_1}$ is the ratio of viscosities, $m_2 = \frac{\rho_2}{\rho_1}$ is the ratio of densities, $C_p = \frac{C_{p_2}}{C_{p_1}}$ is the ratio of specific heats, $K = \frac{K_2}{K_1}$ is the ratio of thermal conductivities.

The non-dimensional conditions to be satisfied are,
Initial conditions:

$$w_1(r, 0) = 0 \quad \text{for } 0 \leq r \leq 1, \tag{6}$$

$$w_2(r, 0) = 0 \quad \text{for } 1 < y \leq s, \tag{7}$$

$$b(r, 0) = 0 \quad \text{for } 0 \leq r \leq 1, \tag{8}$$

$$T_1(r, 0) = T_0 \quad \text{for } 0 \leq r \leq 1, \tag{9}$$

$$T_2(r, 0) = T_0 \quad \text{for } 1 < y \leq s. \tag{10}$$

Interface and boundary conditions: For $t > 0$,

$$\frac{\partial w_1}{\partial r} = 0 \quad \text{at } r = 0, \tag{11}$$

$$\frac{\partial b}{\partial r} = 0 \quad \text{at } r = 0, \tag{12}$$

$$w_2(r, t) = 0 \quad \text{at } r = s, \tag{13}$$

$$w_1(r, t) = w_2(r, t) \quad \text{at } r = 1, \tag{14}$$

$$b(r, t) = -\frac{1}{2} \frac{\partial w_1}{\partial r} \quad \text{at } r = 1, \tag{15}$$

$$\left(1 + \frac{n_1}{2}\right) \frac{\partial w_1}{\partial r} = m_1 \frac{\partial w_2}{\partial r} \quad \text{at } r = 1, \tag{16}$$

$$\frac{\partial T_1}{\partial r} = 0 \quad \text{at } r = 0, \tag{17}$$

$$T_2(r, t) = 0 \quad \text{at } r = s, \tag{18}$$

$$T_1(r, t) = T_2(r, t) \quad \text{at } r = 1, \tag{19}$$

$$\frac{\partial T_1}{\partial r} = K\frac{\partial T_2}{\partial r} \quad \text{at } r = 1, \tag{20}$$

where, $T_0 = \frac{T_0{}^* - T_w}{T_\infty - T_w}$.

3 Numerical Solution

The system (1)–(20) is coupled, non-linear and unsteady; it is troublesome to find exact solution. It can be identified that the partial differential equations (1), (2) and (4) are coupled in w_1, w_2 and b, but are decoupled from the temperatures T_1 and T_2; therefore, once w_1, w_2 and b are obtained by solving the PDEs (1), (2) and (4), the temperature fields T_1 and T_2 can be acquired afterward, by solving the PDEs (3) and (5).

3.1 *Velocity and Microrotation*

Discretizing the computational domain and employing the Crank–Nicolson methodology for Eqs. (1), (2) and (4), we get the finite difference scheme for each time level j.

Upon imposition of boundary conditions, the resultant linear system of order $(3m + 1)$ is expressed as,

$$ZX_{j+1} = \bar{b} + YX_j, \tag{21}$$

where Y and Z are the sparse square matrices and $\bar{b}$ is the known vector and X_j is the solution vector for each time level j.

The components of the solution at zeroth time level, X_0, are given through the initial conditions. The system (21), for each time level j, is solved to obtain the fluid velocities and microrotation, numerically.

3.2 *Temperature*

After obtaining the fluid velocities and microrotation from Sect. 3.1, the goal of current sub-section is to solve, numerically, the differential equations (3) and (5) for obtaining temperature fields. Following the approach similar to that of Sect. 3.1, using the same approach for Eqs. (3) and (5), after considerable simplifications, the finite difference equations for temperatures are obtained. This system, in matrix form, is the tri-diagonal system of linear equations. This system is solved to get the heat transfer characteristics.

The Nusselt number is calculated at the boundary as,

$$(\mathrm{Nu})_{r=s} = -\left(\frac{\partial T_2}{\partial r}\right)_{r=s}.$$

A finite difference approximation has been used to compute Nusselt number, and its variation with relevant flow parameters is presented through Table 2.

4 Discussion of Results

The intent of this section is to display and interpret the effects of relevant parameters on the fluid flow and temperature. A grid independence test for flow and heat transfer profiles is conducted to obtain a solution which is independent of the choice of the grid. The numerical results for fluid velocities throughout the pipe, keeping all the fluid parameters fixed ($n_1 = 0.5$, $m_1 = 0.5$, $m_2 = 0.5$, $\mathrm{Re} = 1$, $G = 5$, $M = 0.2$, $\mathrm{Da} = 0.2$), are presented in Table 1.

Figure 2 displays the fluid velocity profiles for different times. It has been observed that the velocities of fluid in micropolar and Newtonian fluid regions are increasing as time progresses and eventually attaining the steady state for higher values of time. Further, it has been contemplated that the values of microrotational velocity and fluid temperatures are also growing with time and finally entering into steady state. The fluid velocities are declining with that of micropolarity parameter (see Fig. 3). However, the decrease in Newtonian fluid region is slow as compared to that of micropolar fluid region. The micropolarity parameter $n_1 = \frac{\kappa}{\mu_1}$ is there for micropolar fluid region only, but, at interface, due to continuity of fluid velocities as well as shear stresses, we have this little change in region-II too.

Table 1 Fluid velocity values versus radius of pipe r

r	Fluid velocity	r	Fluid velocity
0.0	2.8235	1.0	2.8277
0.1	2.8239	1.1	2.8118
0.2	2.8253	1.2	2.7387
0.3	2.8276	1.3	2.6086
0.4	2.8303	1.4	2.4208
0.5	2.8332	1.5	2.1742
0.6	2.8357	1.6	1.8673
0.7	2.8372	1.7	1.4984
0.8	2.8369	1.8	1.0656
0.9	2.8341	1.9	0.5667
1.0	2.8277	2.0	0.0

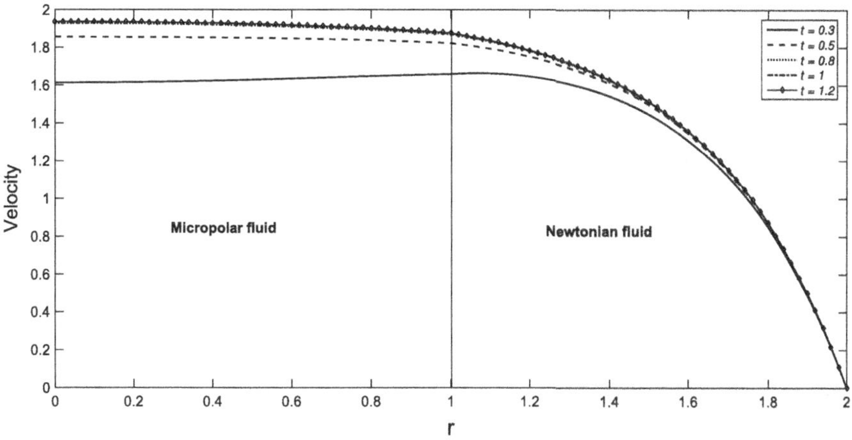

Fig. 2 Impact of time on fluid velocities

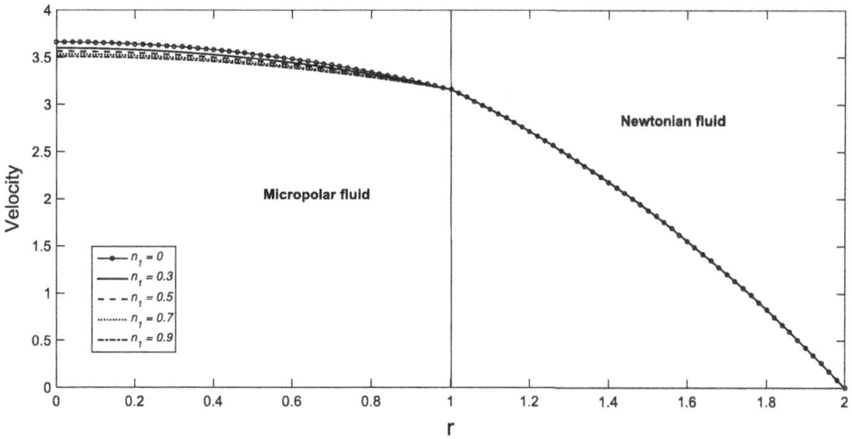

Fig. 3 Impact of micropolarity parameter n_1 on fluid velocities

The fluid flow profiles are found to be increasing with the Darcy number and decreasing with Hartmann number. From Figs. 4 and 5, it is clear that fluid velocities show exactly counterproductive effect with respect to Hartmann number and Darcy number. Fluid temperatures are also increasing with increasing of porosity parameter and decreasing with magnetic field intensity parameter. From Fig. 6, it is clear that microrotation is getting lower with the increase of micropolarity parameter. Fluid temperature profiles when plotted against the radius of pipe, are perceived that, with increasing values of Brinkmann number, fluid temperature in both the regions is increasing (see Fig. 7). Also, one can note from Fig. 8 that Prandtl number has a decreasing effect on fluid temperature profiles.

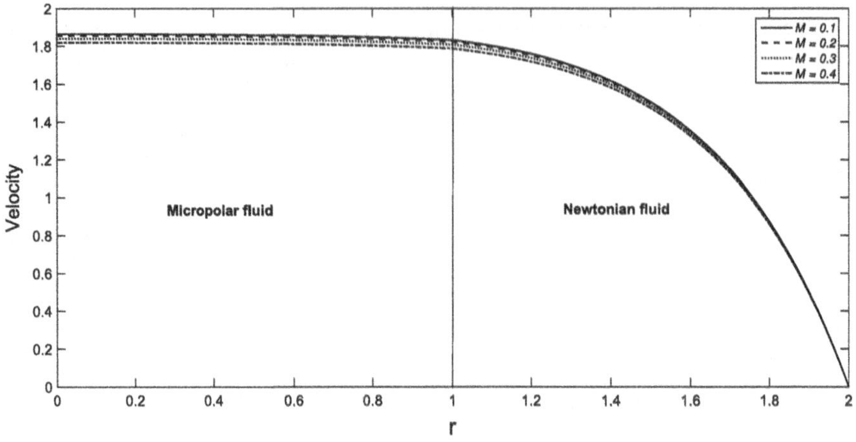

Fig. 4 Impact of Hartmann number M on fluid velocities

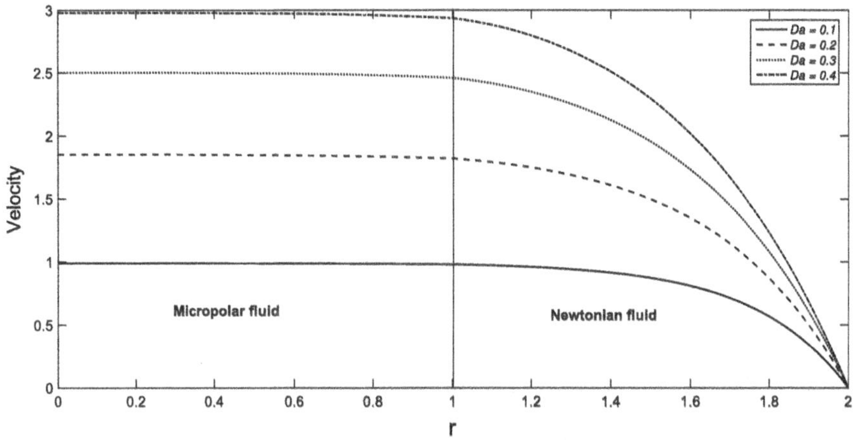

Fig. 5 Impact of Darcy number Da on fluid velocities

Table 2 displays the numerical values of Nusselt number for various fluid flow parameters. With growing values of micropolarity parameter, the Nusselt number increases. The porosity parameter renders the increasing impact on Nusselt number, whereas the magnetic parameter reduces the same. Nusselt number at the boundary is showing a tendency to increase with time and gets settled for higher values of time.

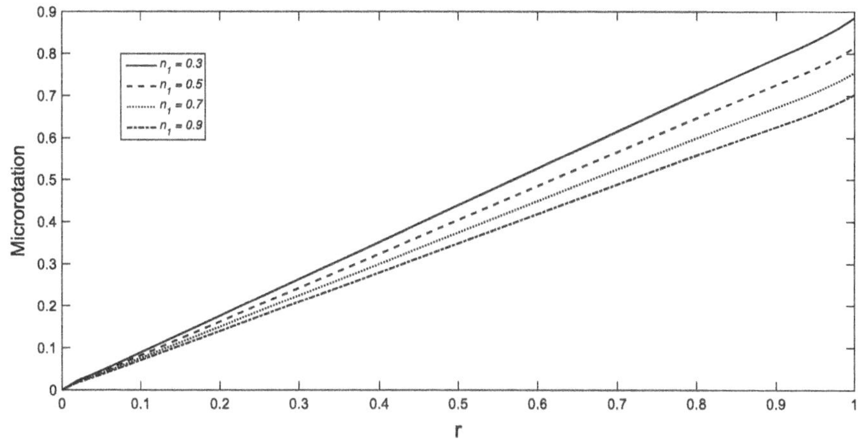

Fig. 6 Impact of micropolarity parameter n_1 on microrotation

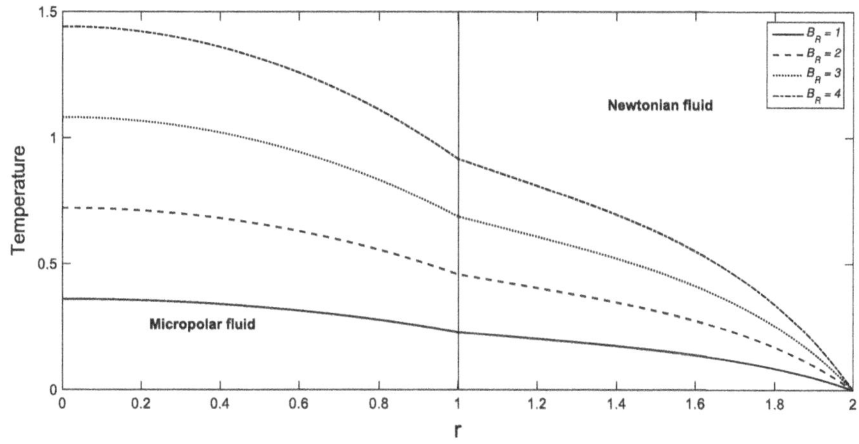

Fig. 7 Impact of Brinkmann number B_R on fluid temperatures

5 Conclusions

The main outcomes presented two-fluid flow through pipe are summarized as follows:

- The presence of magnetic effects decrease the fluid velocities and temperatures in both regions of the flow.
- The effect of uniform porous medium is promoting the fluid velocity profiles.
- The flow and temperature profiles are increasing with time and becomes steady, eventually.

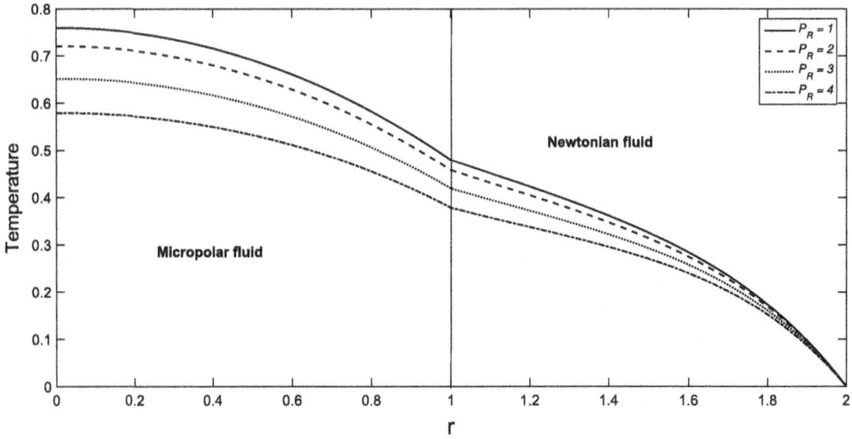

Fig. 8 Impact of Prandtl number P_R on fluid temperatures

Table 2 Nusselt number for various values of fluid parameters

n_1	Nu	Da	Nu	M	Nu	t	Nu
0.3	7.4402	0.1	7.3642	0.1	9.5987	0.5	9.0922
0.5	7.3980	0.2	9.4483	0.2	9.4483	2	13.4289
0.7	7.2684	0.3	10.3275	0.3	9.2050	2.5	13.4390
0.9	7.1036	0.3	10.8101	0.4	8.8789	2.6	13.4395

- The numerical values of Nusselt number falls down with rising micropolarity parameter.

References

1. Eringen, A.C.: Theory of micropolar fluids. J. Appl. Math. Mech. **16**(1), 1–16 (1966)
2. Stokes, V.K.: Theories of Fluids with Microstructure. An Introduction. Springer, New York (1984)
3. Eringen, A.C.: Microcontinuum Field Theories I: Foundations and Solids. Springer, New York (1999)
4. Eringen, A.C.: Microcontinuum Field Theories II: Fluent Media. Springer, New York (2001)
5. Umavathi, J.C., Chamkha, A.J., Mateen, A., Mudhaf, A.A.: Unsteady two-fluid flow and heat transfer in a horizontal channel. Heat Mass Transfer **42**, 81–90 (2005)
6. Bugliarello, G., Sevilla, J.: Velocity distribution and other characteristics of steady and pulsatile blood flow in fine glass tubes. Biorheology **7**, 85–107 (1970)
7. Shukla, J.B., Parihar, R.S., Gupta, S.P.: Biorheological aspects of blood flow through artery with mild stenosis: effects of peripheral layer. Biorheology **17**, 403–410 (1980)
8. Chaturani, P., Samy, R.P.: A study of non-Newtonian aspects of blood flow through stenosed arteries and its applications in arterial diseases. Biorheology **22**(6), 521–531 (1985)

9. Haldar, K., Anderson, H.I.: Two-layered model of blood flow through stenosed arteries. Acta Mech. **117**, 221–228 (1996)
10. Devakar, M., Raje, A.: Modelling and analysis of the unsteady flow and heat transfer of immiscible micropolar and Newtonian fluids through a pipe of circular cross section. J. Braz. Soc. Mech. Sci. Eng. **40**, 325 (2018)
11. Raje, A., Devakar, M.: MHD flow and heat transfer of immiscible micropolar and Newtonian fluids through a pipe: a numerical approach. Numer. Heat Transfer Fluid Flow: Select Proc. NHTFF **3**, 55–64 (2018)

A Computational Model to Study the Effect of Amyloid Beta on Calcium Dynamics

Hemlata Jethanandani and Amrita Jha

Abstract Calcium signaling is critically important for various cellular activities to maintain homeostasis. The accumulation of Alzheimer disease in the form $A\beta$ interferes with calcium levels. In the existing model, the study of intracellular calcium for the effect of amyloid beta has been done. In the present study, we investigate the effect of amyloid beta proteins on various physiological parameters like inositol triphosphate IP3 receptors, ryanodine receptor (RyR) on intracellular calcium dynamics. In the proposed study, we construct a mathematical model of intracellular calcium dynamics with effect of amyloid beta through ryanodine receptor is developed. We predict that as the doses of receptors gradually influence the level of calcium. The bifurcation analysis is done for solving the proposed mathematical model and results simulated on XPPAUT. The results of present study are compared with the existing mathematical model that shows the effect for RYR reaction rate to manage Ca^{2+} signals.

Keywords Amyloid beta · Ryanodine receptor · Bifurcation · Calcium channels

1 Introduction

In most of cellular mechanism, calcium (Ca^{2+}) is unique second messenger that controls various physiological cellular activities in proper form. The endoplasmic reticulum and receptor have been the center of attention in the study of intracellular calcium dynamics [1, 2].

Alzheimer's disease is a neurological disorder which effect memory, thinking skills and then the ability to carry out the simplest tasks. The main component of the Alzheimer's disease is the amyloid beta found in the brains [3, 4]. The results

---

H. Jethanandani · A. Jha (✉)
Department of Mathematics, ISHLS, Indus University, Ahmedabad, India
e-mail: amritajha.gd@indusuni.ac.in

H. Jethanandani
e-mail: hemlatajethanandani.rs@indusuni.ac.in

M. Sahni et al. (eds.), *Mathematical Modeling, Computational Intelligence Techniques and Renewable Energy*, Advances in Intelligent Systems and Computing 1287,
https://doi.org/10.1007/978-981-15-9953-8_26

of amyloid beta transform the abnormal intracellular Ca^{2+} levels. Accumulation of amyloid beta peptide $A\beta$ to increase the intracellular Ca^{2+} levels [5]. Intracellular accumulation of $A\beta$ to bring about the calcium release from internal store endoplasmic reticulum (ER). The accumulation of $A\beta$ has been linked alteration of Ca^{2+} signaling within nerve cells. The description of mathematical model with different case under which resulting $A\beta$ to increase the level of calcium signaling [6].

Synaptic failure and neuronal death are happened due to slow accumulation of $A\beta$ peptides that changes Ca^{2+} signaling process. The receptors IP_3R, CICR play vital role for the production of calcium signaling. The $A\beta$ interprets for the production of IP_3R through RyR and plasma membrane. We develop the mathematical model to express the biological process under which $A\beta$ impacts on calcium signaling. The accumulation of $A\beta$ is effected in various ways in physiological process.

In the literature survey, there are lot of work that have been done for the effect of Alzheimer's disease on calcium dynamics. Demuro et al. [7] have shown the effect of intracellular Ca^{2+} liberation induced by injections of $A\beta$ oligomers into oocytes. Hao et al. [8] have developed a mathematical model with effect of Alzheimer's disease for various physiological cells. Dave and Jha [9] have been studied the modeling of calcium homeostatic in the presence of protein and VGCC for Alzheimer's cell. Dave et al. [4] have shown the effect of Alzheimer's disease in the form 3D model in the calcium signaling.

The present study is discussed that aberrant calcium immerges in the presence of $A\beta$. There are various physiological changes in the form of Ca^{2+} oscillations within an AD environment have presented to understand the mechanisms. $A\beta$ effect on Ca^{2+} flux through many physiological parameters like channels, pumps, leak, etc., have been studied and solutions of model provide the impact of $A\beta$ on Ca^{2+} over various temporal scales [3]. Amyloid betas peptides which are involved in Alzheimer's disease (AD) that causes levels of intracellular calcium (Ca^{2+}) by drastically affecting on Ca^{2+} levels in the nerve cell. Due to disruptions, the cell functions and survival are the effects on cell [4, 5, 6, 9, 10, 11].

2 Mathematical Formulation

To show the effect of $A\beta$ on Ca^{2+} of cell, we develop a simple entire cell Ca^{2+} model of Ca^{2+} dynamics and the calcium signaling. The model of Ca^{2+} dynamics using the transition in and out of the cytosol [5, 11, 12].

The calcium combination of free Ca^{2+} ion in cytosol is represented as c, and rate of change of calcium in form of flux is given as follows:

$$\frac{\mathrm{d}c}{\mathrm{d}t} = J_{\text{in}} - J_{\text{out}} \tag{1}$$

We suppose the ER and cytoplasm together exist at the same point. Simplified whole cell model is used to review the impact of $A\beta$ on Ca^{2+} dynamics. We consider

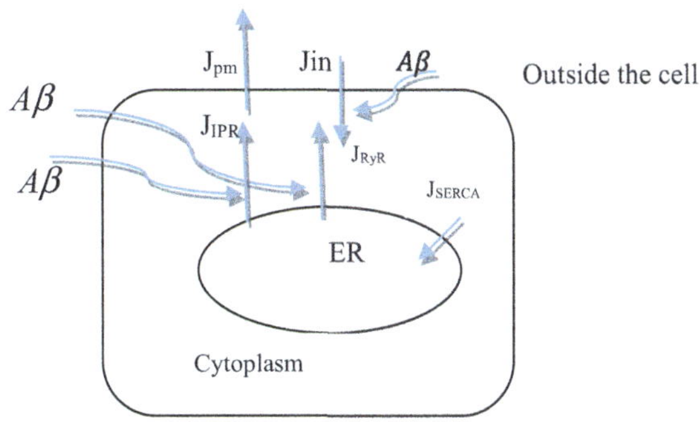

Fig. 1 Two pool model diagram

that intracellular Ca^{2+} fluxes are corresponding to biophysical parameters RYRs, IP_3, sink membrane J_{in}. The out fluxes are modeled using a sarco/endoplasmic reticulum Ca^{2+} ATPase(SERCA) pump Jserca and a plasma membrane pump J_{pm}. A diagram of various types fluxes and interaction of $A\beta$ in the form of model is shown below [3, 4, 5, 6, 7, 9, 10, 11, 13, 14, 15, 16, 17, 18, 19]:

Figure 1 affected fluxes by $A\beta$ in the formulation of mathematical model using various physiological processes in a connection of Ca^{2+} model.

The mathematical model of Ca^{2+} in a connection of various physiological processes is represented as follows.

$$\frac{d}{dt}[Ca^{2+}] = J_{IPR} + J_{RYR} - J_{SERCA} + J_{in} - J_{pm} \tag{2}$$

$$\frac{d}{dt}[Ca^{2+}]_{ER} = -\gamma(J_{IPR} + J_{RYR} - J_{SERCA}) \tag{3}$$

where $[Ca^{2+}]$ is the concentration of Ca^{2+} in cytoplasm and $[Ca^{2+}]_{ER}$ is the concentration of the ER and γ is the ratio of cytoplasm volume to the ER volume [16].

2.1 IP_3 Receptor Model

Inositol 1,4,5-triphosphate receptor is a intracellular channel which has a major role in controlling Ca^{2+} levels in nerve cells. The IP_3 transforms many vital physiological functions, to release Ca^{2+} from ER through IP_3 receptors

$$J_{IPR} = (k_f P_o + J_{er})([Ca^{2+}]_{ER} - [Ca^{2+}]) \tag{4}$$

where k_f directs the concentration of the IP_3 receptor, J_{er} is a release from the ER to the cytosol, and P_0 is the open probability of the IPR. The leakage expression is significant to changeability in the ATPase flux at equilibrium. [19].

Table 1 Physiological parameters [5]

k_f	$0.98\ s^{-1}$	γ	5.4
RyR			
k_1	$0.013\ s^{-1}$	k_d	$0.13\ \mu M$
k_2	$0.18\ s^{-1}$	k_a	0.75
SERCA			
K_1	0.0001	K_2	0.007 s
K_3	$0.06\ \mu M^{-1}$ s	K_4	$0.0014\ \mu M^{-1} s$
K_5	$0.007\ \mu M^{-2}$ s		
Transport			
a_1	$0.003\ \mu M s^{-1}$	k_β	$1\ s^{-1}$
a_2	$0.02\ s^{-1}$	m	4
V_{pm}	$2.8\ \mu M s^{-1}$	K_{pm}	$0.425\ \mu M$

2.2 RYR Model

Ca^{2+} release from intracellular stores through ryanodine receptors is second principal way. RYR have similar structure and functions with IP3 receptors. Ca^{2+} can activate IP3 receptors and increase the Ca^{2+} flux. The flux through RYR is given by:

$$J_{\text{RYR}} = K_3\left(\left[\text{Ca}^{2+}\right]_{\text{ER}} - \left[\text{Ca}^{2+}\right]\right) \tag{5}$$

Ca^{2+} triggers Ca^{2+}-induced Ca^{2+} release (CICR) from the endoplasmic reticulum through ryanodine receptors. The rate constant K_3 is describe as

$$K_3 = k_1 + \frac{k_2\left[\text{Ca}^{2+}\right]^3}{k_d^3 + \left[\text{Ca}^{2+}\right]^3} \tag{6}$$

where k_1, k_2 and k_d are parameters, and values of parameters of nonlinear model are given in Table 1 [20, 21].

2.3 Leak, Membrane Pump and SERCA

We include only IP_3 concentration is given by membrane leak J_{in} in linear form. Linear increase confirms steady-state Ca^{2+} combination which depends on p.

$$J_{\text{in}} = a_1 + a_2 p \tag{7}$$

where a_1 and a_2 are parameters. The plasma membrane pump in Hill equation is given

$$J_{\mathrm{pm}} = \frac{V_{\mathrm{pm}}[\mathrm{Ca}^{2+}]^2}{K_{\mathrm{pm}}^2 + [\mathrm{Ca}^{2+}]^2} \tag{8}$$

To express the impact of the SERCA pump here, we use four state bidirectional Markov model of the form

$$J_{\mathrm{SERCA}} = \frac{[\mathrm{Ca}^{2+}] - k_1[\mathrm{Ca}^{2+}]_{\mathrm{ER}}}{k_2 + k_3[\mathrm{Ca}^{2+}] + k_4[\mathrm{Ca}^{2+}]_{\mathrm{ER}} + k_5[\mathrm{Ca}^{2+}][\mathrm{Ca}^{2+}]_{\mathrm{ER}}} \tag{9}$$

2.4 *Amyloid Beta Hypothesis*

The consequence of $A\beta$ on existence pump, channels and exchangers remains extensively unrevealed. Despite, to view the $A\beta$ influence, we have taken some provided studies [7]. To incorporate the effect of $A\beta$ in the model to include $k_\beta a^m$ in J_{in}

$$J_{\mathrm{in}} = a_1 + a_2 p + k_\beta a^m \tag{10}$$

where m denotes cooperatively coefficients and k_b is a constant of speed. Furthermore $A\beta$ can increase RYR channel open probability by disrupt RYR-regulated Ca^{2+} signal. To alter the RYR channel term to assume the effect of $A\beta$.

$$K_3 = k_1 + \frac{k_2[\mathrm{Ca}^{2+}]^3}{(k_d + k_\alpha a)^3 + [\mathrm{Ca}^{2+}]^3} \tag{11}$$

where ka is a positive and denotes carrying out the intensity of the effects of $A\beta$.

2.5 *Calcium Model Formulation with Amyloid Beta*

The model for the fluxes by putting the different exchangers, channels and pumps that are important for control of cytosolic Ca^{2+} concentration in the presence of $A\beta$ is a function of time which takes the form

$$\begin{aligned}\frac{\mathrm{d}}{\mathrm{d}t}[\mathrm{Ca}^{2+}] = {} & (k_f P_o + J_{\mathrm{er}})([\mathrm{Ca}^{2+}]_{\mathrm{ER}} - [\mathrm{Ca}^{2+}]) + K_3([\mathrm{Ca}^{2+}]_{\mathrm{ER}} - [\mathrm{Ca}^{2+}]) \\ & - \frac{[\mathrm{Ca}^{2+}] - k_1[\mathrm{Ca}^{2+}]_{\mathrm{ER}}}{k_2 + k_3[\mathrm{Ca}^{2+}] + k_4[\mathrm{Ca}^{2+}]_{\mathrm{ER}} + k_5[\mathrm{Ca}^{2+}][\mathrm{Ca}^{2+}]_{\mathrm{ER}}} + a_1\end{aligned}$$

$$+ a_2 p + k_\beta a^m - \frac{V_{\text{pm}}[\text{Ca}^{2+}]^2}{K_{\text{pm}}^2 + [\text{Ca}^{2+}]^2} \tag{12}$$

$$\frac{\text{d}}{\text{d}t}[\text{Ca}^{2+}]_{\text{ER}} = -\gamma\Big[(k_f P_o + J_{\text{er}})([\text{Ca}^{2+}]_{\text{ER}} - [\text{Ca}^{2+}]) + K_3([\text{Ca}^{2+}]_{\text{ER}} - [\text{Ca}^{2+}]) - \frac{[\text{Ca}^{2+}] - k_1[\text{Ca}^{2+}]_{\text{ER}}}{k_2 + k_3[\text{Ca}^{2+}] + k_4[\text{Ca}^{2+}]_{\text{ER}} + k_5[\text{Ca}^{2+}][\text{Ca}^{2+}]_{\text{ER}}}\Big] \tag{13}$$

In the present study, we assume $A\beta$ create plasma membrane $A\beta$ pores which subsequently increases membrane permeability and have used the XPPAUT to show the effects of $A\beta$ in membrane in flux. Also included the RYR effects in presence of $A\beta$. Firstly, the study has been done for IP_3 as IP_3 triggers Ca^{2+} release from IP3 receptors. Subsequently, it releases activities from RYR and CICR. In the present study, the analysis in the presence of membrane potential and then incorporating the presence of $A\beta$ in absence of membrane potential. Further to include other $A\beta$ parameters k_a and k_2 to show the effect of presence of $A\beta$ [5, 11, 17, 18, 19, 20, 21, 22, 23, 24]. We determine numerically the amplitude oscillation of the system as changing the membrane potential parameters and $A\beta$ parameters. The stability analysis of differential equation has been done in AUTO extension of XPPAUT. The results have been simulated using AUTO software to investigate the bifurcation analysis of the systems of ordinary differential equations. A Hopf bifurcation is a critical point where a system's stability switches and a periodic solution arises [5]. Initial conditions are set at $[Ca^{2+}]$=0.05, $[Ca^{2+}]_{ER} = 10$.

3 Results and Discussion

The influence of $A\beta$ is deviating impact on promotions a compound of various cellular processes. We divided the given model by tracking solutions in terms of result analysis. The results have shown the impact of different cellular process like IP3, RYR in terms of calcium dynamics with or without impact of. Lastly, we consider model solving for different states of $A\beta$ when IP3 combination is steady by incorporating the effect of membrane potentials. The results have been compared with model of amyloid beta on calcium.

3.1 IP3 Influence on Model Dynamics

$A\beta$ represents as 'a' ($a = 0$). Figure 2a, b shows the concentration oscillations of Ca^{2+} with $p = 5$ and $p = 10$, respectively. Figure 2c illustrates the steady state as increases p, Fig. 2d shows corresponding nullcline graph, and Fig. 2e is bifurcation

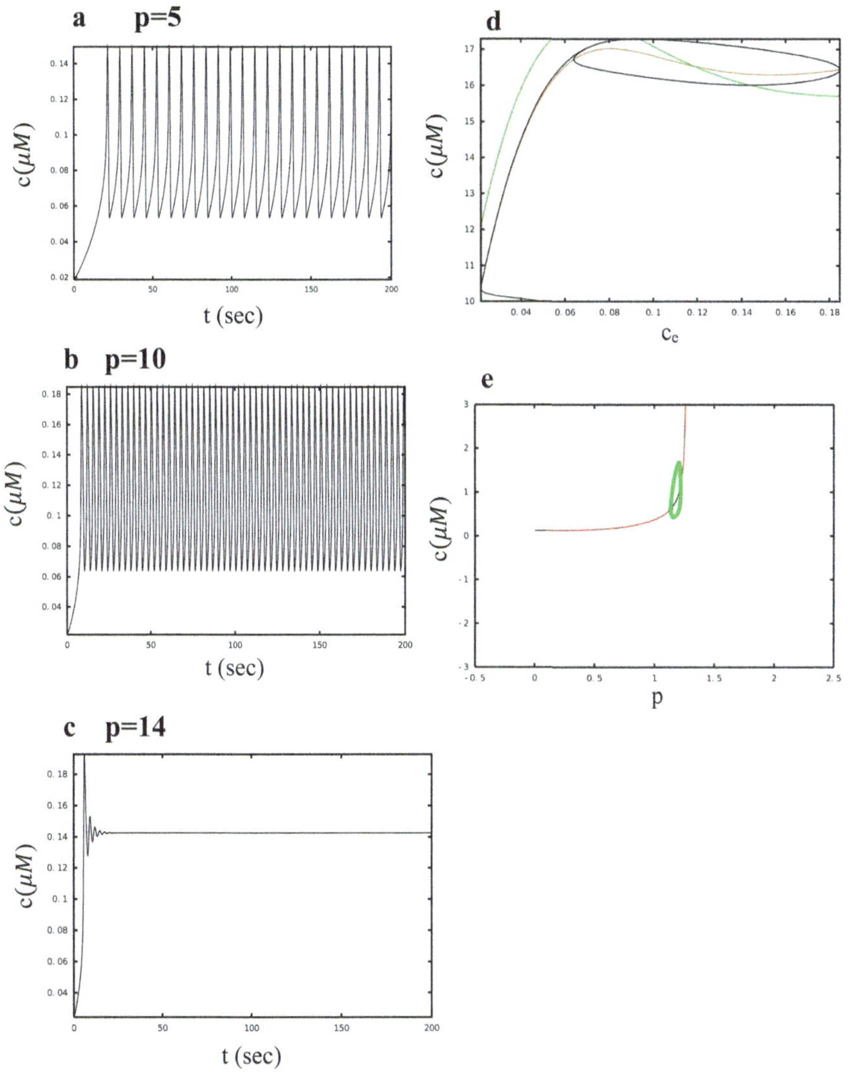

Fig. 2 Calcium dynamics for fixed IP_3 levels

graph between concentration of Ca^{2+} and parameter p that gives the periodic limit cycles.

In Fig. 2a, b, there are Ca^{2+} oscillations as simulation of the model with $a = 0$, the oscillations increased with $p = 5$ to $p = 10$. Figure 2c shows the oscillation converges to steady state with the value of $p = 14$. Under such conditions, the Ca^{2+}-nullcline of cytoplasmic and endoplasmic concentration given in Fig. 2d, nullcline intersect in one point. Figure 2a, b shows the calcium dynamics for constant IP_3 levels. Properties of calcium model are when IP_3 concentrations are fixed, no $A\beta$ is present ($a = 0$). It is observed that in both graphs, frequency of oscillations increases when p increases. After certain level of p, frequency of oscillation decreases and goes steady state at $p = 13.5$. Hopf bifurcation occurs due to that limit cycle which is given in Fig. 2d [20, 21, 22, 23].

3.2 The Presence of Amyloid Beta

In Fig. 3a–c the absence of IP_3, the oscillation of amplitude of Ca^{2+} has shown with different values of Fig. 3a, d which shows the phase space and nullcline. Figure 3e shows the bifurcation diagram with a as parameter. Here c concentration of Ca^{2+}.

Figure 3 shows the effect of $A\beta$ on equilibrium state level of calcium dynamics in the absence of IP_3. In both graphs when we increase the value of amyloid beta so the level of calcium increases. The graphs show the effect of amyloid beta, stable calcium oscillation for $a = 1.2$ and 1.22, and the steady level rapidly becomes abstruse as the proportion of $A\beta$ expands toward $a = 1.15$.

3.3 Two Parameter Analysis

Here C represents concentration of Ca^{2+}. Two sets of parameters choice of a and p are shown in Fig. 4a, b. Figure 4a shows Ca^{2+} oscillations with high point amplitude around 1 corresponding to $a = 1.15$ and $p = 10$. Figure 4b shows the high point of concentration around 1.5 corresponding to $a = 1.2$ and $p = 5$.

In Fig. 4, the influence of a and p is on calcium oscillation amplitudes. Figure 4a shows calcium oscillation with peak amplitude around 0.9 corresponding to $a = 1.15$ and $p = 10$. Figure 4b shows calcium oscillation with high point amplitude closer to 1.5 relating to $a = 1.2$ and $p = 5$.

3.4 Calcium Signaling Through the Ryanodine Receptor

Here C represents concentration of Ca^{2+}. Figure 5a shows stable periodic oscillations $a = 1.15, p = 10$, and Fig. 5b shows stable periodic oscillations when $a = 1.15, p = 10, k_a = 0.7, k_2 = 0.5$. Figure 5c shows with increasing values of k_a and k_2, the oscillation turns to study state. Figure 5d is the nullcline for $a = 1.15, p = 10, k_a = 0.7, k_2 = 0.5$. Figure 5e shows bifurcation diagram with parameter k_a.

To find the effects of RYR, we first fixed the a and p and to change the values of parameters k_a and k_2 to show the effects of $A\beta$. Figure 5a, b shows to increase the values of parameters k_a and k_2, the peak level of concentration of Ca^{2+} becomes higher. After the certain level, the graph approaches to steady state. Figure 5c shows

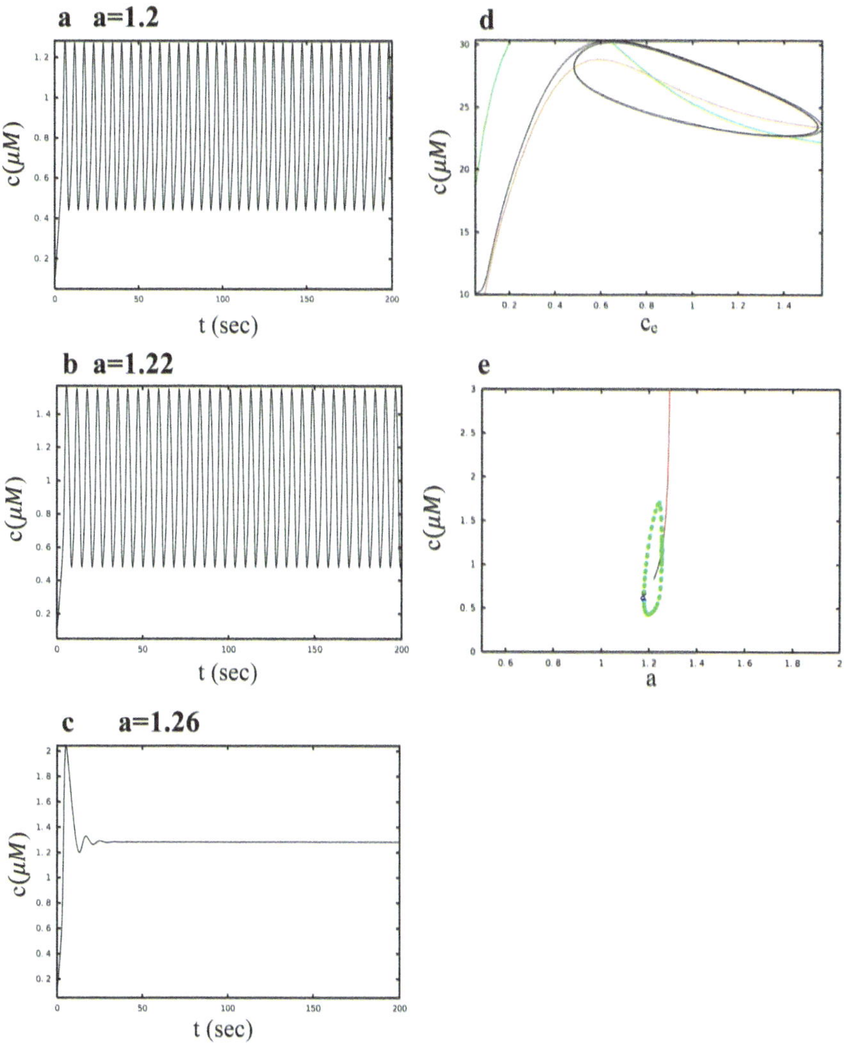

Fig. 3 Effect of $A\beta$ on equilibrium state levels

corresponding nullcline. The limit cycles for Hopf points are given in Fig. 5d, e, respectively.

4 Conclusion

We have shown the effect of calcium signals in a entire cell model supporting the effects of $A\beta$. It is observed that $A\beta$ places a vital role in the cognitive decline

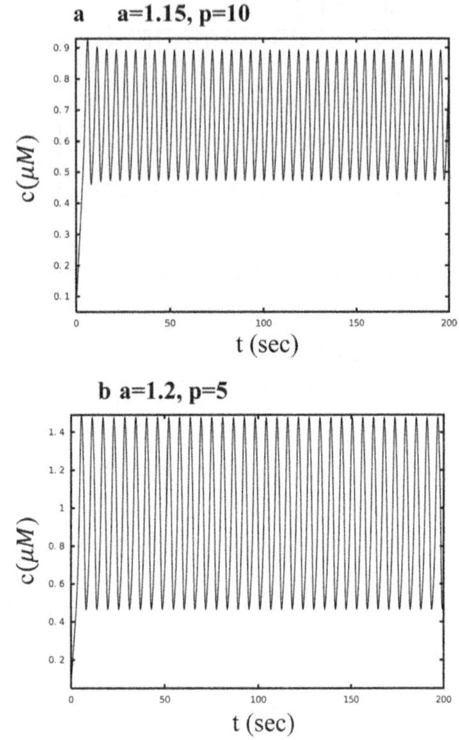

Fig. 4 Sequence of a and p on Ca^{2+} oscillation amplitudes

(abnormal functioning of brain) by direct synaptic transmission. The effect of $A\beta$ on various cellular activities increases homeostasis calcium levels. The model can be useful for better understanding the influence of $A\beta$ on calcium flux through individual physiological parameters such as RYR, IP_3 and plasma membrane [25].

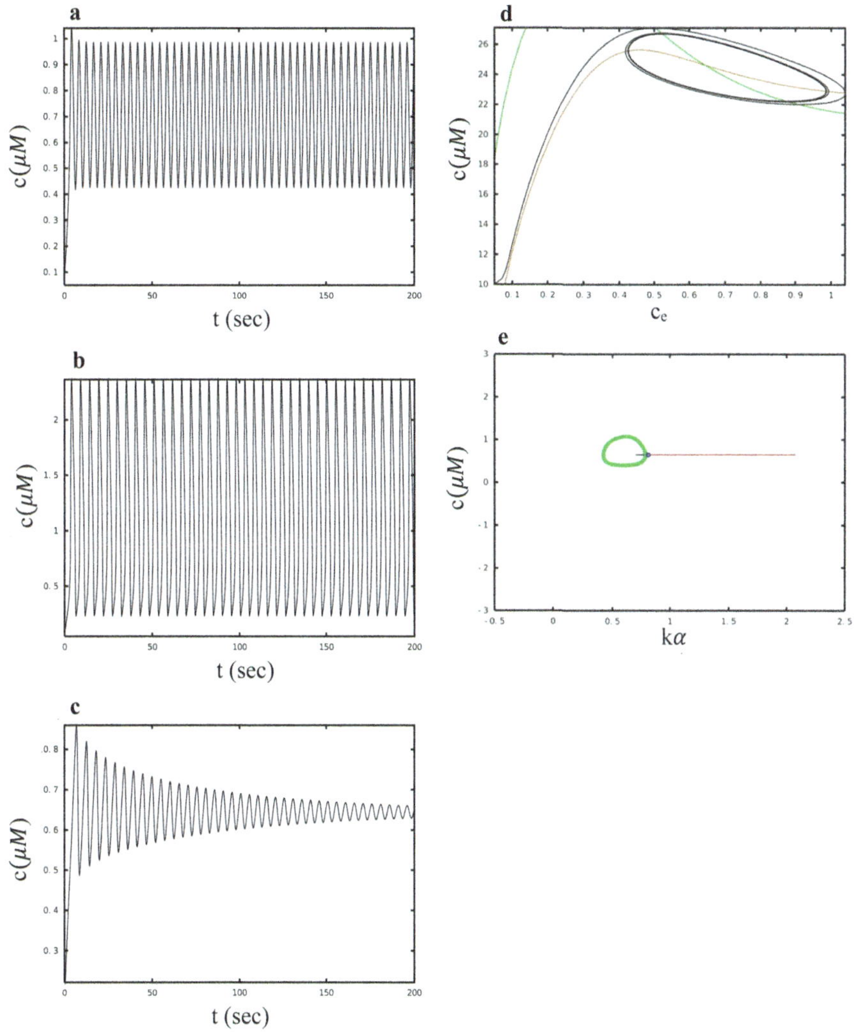

Fig. 5 Change the maximal reaction rate of the RyR

References

1. Sneyd, J., Dufour, J.F.: A dynamic model of the type-2 inositol trisphosphate receptor. Proc. Natl. Acad. Sci. U. S. A. **99**(4), 2398–2403 (2002)
2. Sneyd, J., Tsaneva-Atanasova, K., Bruce, J.I., Straub, S.V., Giovannucci, D.R., Yule, D.I.: A model of calcium waves in pancreatic and parotid acinar cells. Biophys. J. **85**(3), 1392–1405 (2003)
3. Jha, B.K., Dave, D.D.: Approximation of calcium diffusion in Alzheimeric cell. J. Multiscale Model. **11**(2), 2050001-17 (2019)

4. Costa, R.O., Lacor, P.N., Ferreira, I.L., Resende, R., Auberson, Y.P., Klein, W.L., et al.: Endoplasmic reticulum stress occurs downstream of GluN2B subunit of N-methyl-d-aspartate receptor in mature hippocampal cultures treated with amyloid-beta oligomers. Aging Cell **11**(5), 823–833 (2012)
5. Latulippe, J., Lotito, D., Murby, D.: A mathematical model for the effects of amyloid beta on intracellular calcium. PLoS One **13**(8), e0202503 (2018)
6. Dave, D.D., Jha, B.K.: Modelling the alterations in calcium homoeostasis in the presence of protein and VGCC for Alzheimeric cell. Soft Comput. Theories Appl. 181–189 (2018)
7. Demuro, A., Parker, I.: Cytotoxicity of intracellular AB_{42} amyloid oligomers involves Ca^{2+} release from the endoplasmic reticulum by stimulated production of inositol triphosphate. J. Neurosci. 3824–3833 (2013)
8. Hao, W., Friedman, A.: Mathematical model on Alzheimer's disease. BMC Syst. Biol. **10**(108) (2016)
9. Dave, D.D., Jha, B.K.: 3D mathematical modelling of calcium signalig in Alzheimer disease. Netw. Model. Anal. Health Inf. Bioinform. **9**(1), 1 (2020)
10. Dupont, G., Falcke, M., Kirk, V., Sneyd, J.: Models of Calcium Signaling, Interdisciplinary Applied Mathematics, vol. 4. Springer International Publishing, Switzerland (2016)
11. Fall, C.P., Marland, E.S., Wagner, J.M., Tyson, J.J.: Computational Cell Biology. Interdisciplinary Applied Mathematics, vol. 20. Springer, Berlin (2002)
12. Li, Y.X., Rinzel, J.: Equations for InsP3 receptor-mediated $[Ca^{2+}]$i oscillations derived from a detailed kinetic model: a Hodgkin-Huxley like formalism. J. Theor. Biol. **166**(4), 461–473 (1994)
13. Jha, A., Adlakha, N., Jha, B.K.: Finite element model to study the effect of $Na^+ - Ca^{2+}$ exchangers and source geometry on calcium dynamics in a neuron cell. J. Mech. Med. Biol. **16**(2), 1–22 (2015)
14. Berridge, M.J., Bootman, M.D., Roderick, H.L.: Calcium signalling: dynamics, homeostasis and remodelling. Nat. Rev. Mol. Cell Biol. **4**(7), 517–529 (2003)
15. Berridge, M.J.: Inositol trisphosphate and calcium signalling. Nature **361**(6410), 315–325 (1993)
16. Berridge, M.J.: The inositol trisphosphate/calcium signaling pathway in health and disease. Physiol. Rev. **96**(4), 1261–1296 (2016)
17. Jha, A., Adlakha, N.: Analytical solution of two dimensional unsteady state problem of calcium diffusion in a neuron. J. Med. Imaging Health Inf. **4**(4), 541–553 (2014)
18. Jha, A., Adlakha, N.: Finite element model to study the effect of exogenous buffer on calcium dynamics in dendrite spines. Int. J. Model. Simul. Sci. Comput. **5**(12), 1–12 (2014)
19. Keener, J., Sneyd, J.: Mathematical Physiology I Cellular Physiology. Interdisciplinary Applied Mathematics, vol. 2/I, 2nd edn. Springer, Berlin (2009)
20. De Pitta, M., Volman, V., Levine, H., Ben-Jacob, E.: Multimodal encoding in a simplified model of intracellular calcium signaling. Cogn. Proc. **10**(Suppl 1), 55–70 (2009)
21. Pitta, M., Goldberg, M., Volman, V., Berry, H., Ben-Jacob, E.: Glutamate regulation of calcium and IP3 oscillating and pulsating dynamics in astrocytes. J. Biol. Phys. **35**, 383–411 (2009)
22. De Pittà, M., Volman, V., Levine, H., Pioggia, G., De Rossi, D., Ben-Jacob, Eshel: Coexistence of amplitude and frequency modulations in intracellular calcium dynamics. Phys. Rev. E **77**, 030903(R) (2008)
23. Paula-Lima, A.C., Adasme, T., Sanmartin, C., SEBOLLELA, A., Hetz Ccarrasco, M.A., et al.: Amyloid beta peptide oligomers stimulate RyR- mediated Ca^{2+} release inducing mitochondrial fragmentation in hippocampal neurons and prevent RyR-mediated dendritic spine remodeling produced by BDNF. Antioxid. Redox Signaling. **14**(7), 1209–23 (2011)
24. Ermentrout, B.: Simulating, analyzing, and animating *dynamical systems* a guide to xppaut for researchers and students. In: SIAM, 1st edn (2002)
25. Tewari, V., Tewari, S., Pardasani, K.R.: A Model to Study the Effect of Excess buffers and Na^+ ions on Ca^{2+} diffusion in Neuron cell. Int. J. Bioeng. Life Sci. 5(4) (2011)

Hybrid User Clustering-Based Travel Planning System for Personalized Point of Interest Recommendation

Logesh Ravi, V. Subramaniyaswamy, V. Vijayakumar, Rutvij H. Jhaveri, and Jigarkumar Shah

Abstract In the recent times, the massive amount of user-generated data acquired from Internet has become the main source for recommendation generation process in various real-time personalization problems. Among various types of recommender systems, collaborative filtering-based approaches are found to be more effective in generating better recommendations. The recommendation models that are based on this collaborative filtering approach are used to predict items highly similar to the interest of an active target user. Thus, a new hybrid user clustering-based travel recommender system (HUCTRS) is proposed by integrating multiple swarm intelligence algorithms for better clustering. The proposed HUCTRS is experimentally assessed on the large-scale datasets to demonstrate its performance efficiency. The results obtained also proved the potential of proposed HUCTRS over traditional approaches by means of improved user satisfaction.

Keywords Recommender systems · Personalization · Point of Interest · Clustering · Prediction · E-Tourism

L. Ravi
Department of Computer Science and Engineering, Vel Tech Rangarajan Dr. Sagunthala R&D Institute of Science and Technology, Chennai, India
e-mail: logeshphd@gmail.com

V. Subramaniyaswamy (✉)
School of Computing, SASTRA Deemed University, Thanjavur, India
e-mail: vsubramaniyaswamy@gmail.com

V. Vijayakumar
School of Computer Science and Engineering, University of New South Wales, Kensington, Australia
e-mail: vijayakumar.varadarajan@gmail.com

R. H. Jhaveri · J. Shah
School of Technology, Pandit Deendayal Petroleum University, Gandhinagar, India

M. Sahni et al. (eds.), *Mathematical Modeling, Computational Intelligence Techniques and Renewable Energy*, Advances in Intelligent Systems and Computing 1287,
https://doi.org/10.1007/978-981-15-9953-8_27

1 Introduction

The rapid advancements of internet technologies have drastically increased the information overload and personalization concern, since modern users are provided with plenty of choices [1–3]. The recommender system provides an excellent solution for personalization problem in various application domains. To suggest the list of relevant recommendations to the active user, the recommender system strongly studies the association between the item and target user from existing information like historical purchase record, browser footprint and evaluation information. Some of the commercial applications of recommender system include art, e-commerce, social events, etc. Considerably, personalized recommendation increases the probability of suggesting more interested items when compared to generic recommendations. Further collaborative filtering method helps to provide personalized recommendations by increasing the prediction accuracy [4–6].

Most of the commercial applications are based on collaborative filtering recommender system (CFRS) and gained a tremendous amount of popularity [7, 8]. In CFRS, the relevant item for the active user is suggested on the basis of ratings given by same user for other equivalent items. In general, CFRS uses an item-based or user-based model for guessing ratings. In user-based model, collaborative filtering studies the behavior of target user and predicts the rating for unknown items based on the ratings of identical users [9–15]. Accordingly to boost the accuracy of prediction ratings, clustering techniques were exploited and thereby generate more relevant recommendations [16–18]. Based on the similarity measures, clustering techniques group the user into a single or multiple classes of similar users [19, 20]. In CFRS, many clustering methods such as expectation maximization, K-means clustering, evolutionary clustering, fuzzy c-means clustering, etc., were used to generate personalized recommendations [21–24].

The key applications of clustering techniques in data mining are pattern recognition, image processing, web mining, market research, document classification and spatial data analysis. Though the user clustering-based recommender system provides better results, the information processing becomes more complex. Besides, the conventional clustering algorithms have limitation to provide optimal solution for large-scale datasets [25–28]. Hence, the utilization of such traditional clustering techniques fails to generate efficient recommendations. To overcome the drawbacks of traditional approaches, bio-inspired intelligent clustering algorithms [29–35] have been introduced to generate optimal recommendations in real-time recommender systems. The bio-inspired intelligent clustering algorithms implemented in different application areas such as decision support system, market research, spatial data analysis and pattern recognition have already proved their improved performance efficiency [36–38].

The recommender system generates the recommendations based on the feedback of active target user on previously consumed items. The explicit user ratings together with implicit user behavior are utilized to generate target user's personalized recommendation list [39, 40]. After predicting the ratings for the point of interest (POI), recommender system organizes and ranks the POIs. Finally, the items with high prediction ratings were presented at the top, and first n items were suggested as relevant recommendation to the target user. To enhance the recommendations, hybridization is embraced by combining multiple recent technologies with the traditional recommendation approaches. As an efficient decision-making approach, recommender system was adopted in the field of travel and e-tourism [41–44]. Personalized travel recommendation is a challenging task compared to other recommender problems. Such challenge in the domain of e-tourism can be resolved by adopting user's preferences, opinion mining and contextual features in the recommendation prediction process. Therefore, a new hybrid user clustering-based travel recommender system is proposed using swarm intelligence algorithms through a clustering ensemble approach. To review the prediction accuracy of proposed recommender system, experiments were carried out on large-scale real-time Yelp and TripAdvisor datasets. The experimental result demonstrates the outperformance of the proposed recommender system with more relevant travel recommendations to target users.

2 Hybrid User Clustering-Based Travel Recommender System

A new hybrid user clustering model is designed by integrating three different swarm intelligence algorithms to address the limitations of stand-alone models which are used in the collaborative filtering-based recommender systems. To achieve better POI recommendations, we have exploited P-SSO [45], DPSO [46] and HPSO [47] through an ensemble approach to generate user clusters [25]. The HUCTRS generates a final list with n POIs which are highly relevant to the preference of active target user. It is predicted by utilizing the data acquired from location-based social network (LBSN). The proposed HUCTRS consists of clustering, rating prediction and recommendation generation phase. The structure of the proposed HUCTRS is shown in Fig. 1. Initially, the LBSN data is used to generate clusters through swarm intelligence algorithms. Clusters are formed by training 70% of the dataset, and the remaining 30% are used as testing data. Then on the basis of generated clusters, the active user is mapped to the highly identical user cluster determined through a neighborhood search to predict ratings by DML-PCC metric [48]. Based on the ratings, n POIs listed on the top are suggested to the target user.

The main objective of using ensemble model is to exploit the important features of multiple clustering algorithms and thereby boost the performance efficiency of the recommendations. The user clustering is performed iteratively with different number of clusters to produce better clusters. From the clustering results of the individual smarm intelligence algorithms, similarity matrix is generated for the further processing. Consensus similarity matrix is constructed from the results of individual

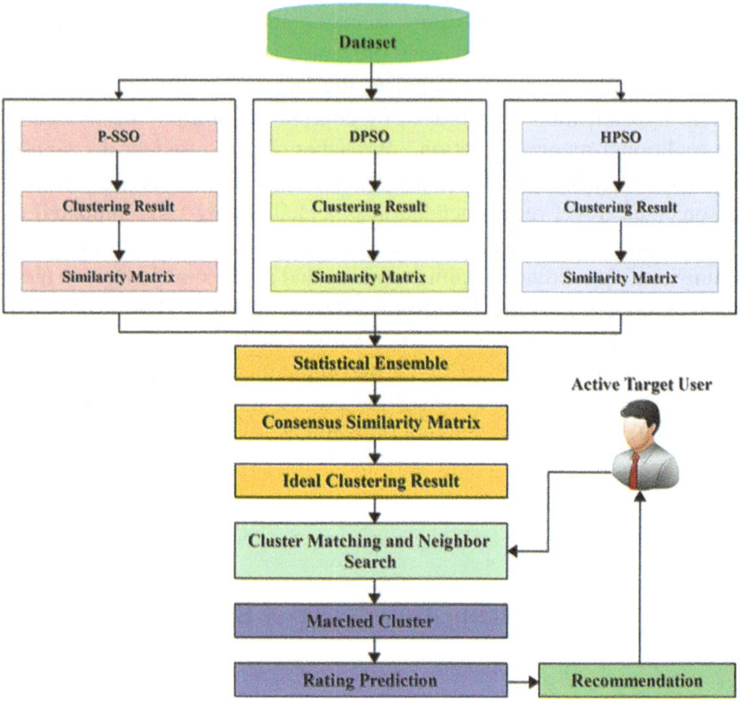

Fig. 1 Proposed system architecture

similarity matrix. From the resultant matrix, ideal user clusters are produced and active user is mapped to the highly relevant cluster.

2.1 Parallel Social Spider Optimization

The swarm intelligent algorithm introduced on the basis of social behavior of the spider in the colony is the social spider optimization (SSO) [49]. The position updating procedure of the SSO is modified with a parallel algorithm to form parallel social spider optimization (P-SSO) [45]. The P-SSO has been proven to be computationally faster than the SSO by nearly ten times. The existing experimental results have clearly presented the clustering capabilities of P-SSO over traditional clustering approaches.

2.2 Density-Based Particle Swarm Optimization

Particle swarm optimization is widely used to address the limitations of clustering algorithms in various domains. The PSO is designed in a better way to withstand premature convergence, and also it requires tuning of learning coefficients to produce better clusters. The popularity-based PSO approach is further hybridized with the kernel density estimation technique as density-based particle swarm optimization (DPSO) to address the premature convergence problem in clustering [46].

2.3 Hierarchical Particle Swarm Optimization

The hybridization of the traditional PSO has been done in a pace to meet the demands of the better solutions for various engineering problems. Especially to address the clustering problem, the PSO algorithm is combined with the hierarchical agglomerative data clustering technique as hierarchical particle swarm optimization (HPSO) to improve the clustering accuracy [47].

2.4 Dynamic Multi-Level Pearson Correlation Coefficient

The considerable concern of the CFRS is the accuracy of the resultant recommendations. Though the traditional cosine and PCC metrics are capable of providing an improved recommendation, yet there is a space that requires improvement in the process. To progress the quality of recommendation process, the traditional PCC is modified as dynamic multi-level Pearson correlation coefficient (DML-PCC) in an efficient way to address the requirement of active user [48].

3 Experimental Evaluation

The recommendation efficiency of the proposed HUCTRS has evaluated experimentally on the large-scale real-time TripAdvisor and Yelp datasets. The experiments were carried out on the PC running with 64-bit Windows 7 OS and Intel Core i7-5500U clocked at 3.00 GHz and 16 GB of memory. The comparative analyses were performed between experimental results of HUCTRS and other relevant recommendation methods. For evaluation process, three metrics such as coverage, F-measure and root mean square error (RMSE) were utilized to analyze the recommendation performance of HUCTRS.

3.1 Root Mean Square Error (RMSE)

RMSE is the evaluation metric applied to estimate the error in the resultant recommendations. In general, RMSE is calculated as follows:

$$\mathrm{RMSE} = \sqrt{\frac{\sum_{\mathrm{user,point}} \left(\mathrm{ActualRating}_{\mathrm{user,point}} - \mathrm{PredictRating}_{\mathrm{user,point}}\right)^2}{\mathrm{TotalRatings}_{\mathrm{tested}}}} \quad (1)$$

where *user* is the active target user, *point* is the point of interest (PoI), and actual $\mathrm{rating}_{\mathrm{user,point}}$ is the rating actually provided by the active target user to the particular PoI. Similarly, predict $\mathrm{rating}_{\mathrm{user,point}}$ is the rating calculated by the recommendation approach for the same user and PoI.

3.2 Coverage

An evaluation metric coverage is utilized to determine the percentage of ratings predicted for overall user-POI pairs by the recommendation model.

$$\mathrm{Coverage} = \frac{\text{Number of Ratings}_{\mathrm{Predicted}}}{\text{Number of Ratings}_{\mathrm{Tested}}} \quad (2)$$

3.3 F-*Measure*

Another metric F-measure is utilized to evaluate the resultant point of interest recommendations. To compute F-measure, coverage and precision are required. Precision is calculated as follows:

$$\mathrm{Precision} = 1 - \frac{\mathrm{RMSE}}{4} \quad (3)$$

F-measure is determined using the resultant precision and coverage values as follows:

$$F\text{ - Measure} = 2 \times \frac{\mathrm{Precision} \times \mathrm{Coverage}}{\mathrm{Precision} + \mathrm{Coverage}} \quad (4)$$

3.4 Discussions

The experiments were carried out on large-scale real-time TripAdvisor and Yelp datasets to demonstrate the potential capabilities of the proposed HUCTRS. The experiments were also conducted with the existing baseline approaches [50, 51], namely LBPARS, LBCFRS, LFARS, UPARS, MPCBRS and PBCFRS to make experimental analyses. Figures 2, 3 and 4 present the comparative graph of different recommendation algorithms applied on TripAdvisor and Yelp datasets in terms of coverage, RMSE and *F*-measure values. The effectiveness of the recommended poIs generated by the proposed model is evaluated on the basis of user's previous travel history. Our proposed HUCTRS makes recommendations based on the learning process made on the previously visited venues of the user.

When the recommendation approach is capable of generating more number of location recommendations, the developed approach is considered to be more efficient and effective. The result depicts the enhanced performance of the proposed HUCTRS over existing methods on both TripAdvisor and Yelp datasets. Following the proposed

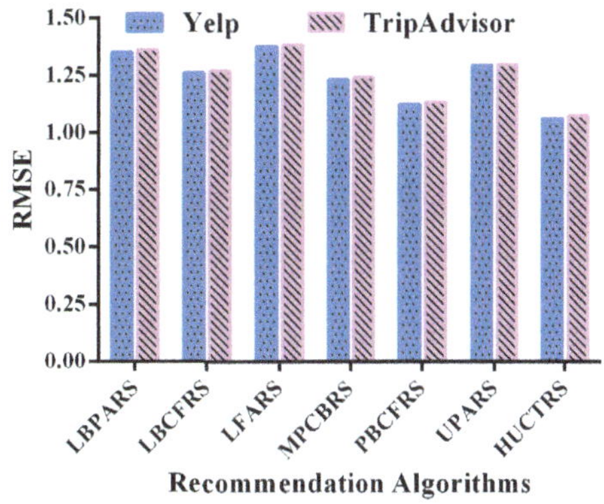

Fig. 2 Comparison of RMSE for various recommendation algorithms

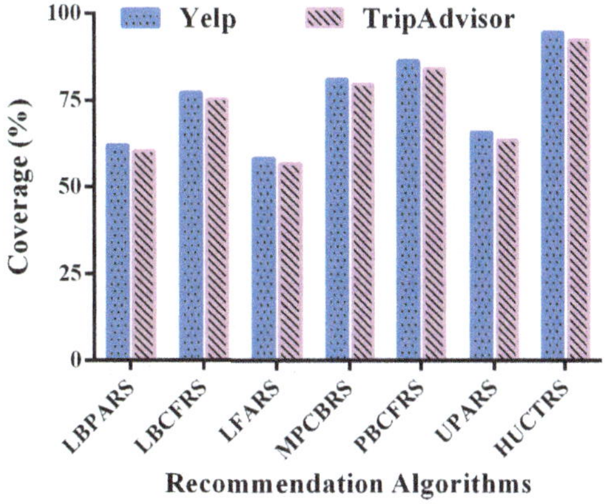

Fig. 3 Comparison of resultant coverage with other recommendation algorithms

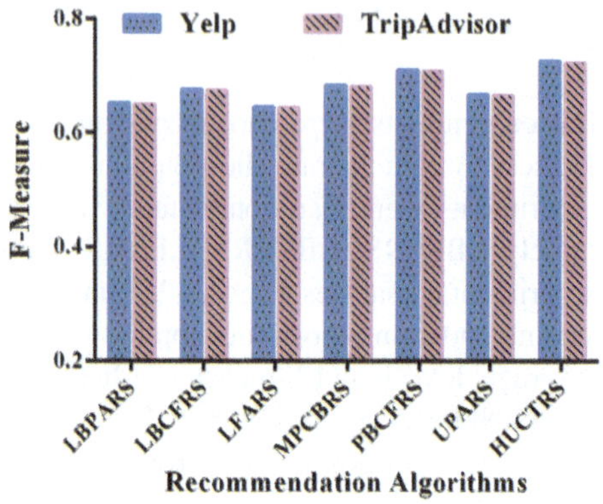

Fig. 4 Comparison of resultant *F*-measure with other recommendation algorithms

HUCTRS, the PBCFRS recommendation approach has been performed competitively. Compared to all recommendation approaches considered for the experimental evaluation, the LFARS is found to be the least performer on both Yelp and TripAdvisor datasets.

The proposed HUCTRS approach is found to be a better performer due to the global optimization provided by the selected swarm intelligence algorithms. To mention about the RMSE, the proposed HUCTRS has the lesser error rate between the predicted and actual ratings. The ratings prediction function of the proposed HUCTRS utilizes the advantageous feature of DML-PCC to determine ratings for the unknown items. The HUCTRS approach is designed to be location-aware, and the spatial attributes of the POIs play a crucial role in generating final recommendation list. From the overall analysis, it is noticeable that HUCTRS generates user satisfiable recommendations in an efficient and effective manner.

4 Conclusion

Recommender systems are developed to assist users in selecting interesting and relevant services by addressing the information overload problem. In the field of e-tourism, personalization is the research challenge of recommender system in recent times. To address the challenge, we have developed a HUCTRS based on swarm intelligence algorithms. Our proposed hybrid recommendation approach exploits the advantageous features of P-SSO, DPSO and HPSO algorithms through an ensemble model. The developed recommendation approach predicts ratings through an enhanced DML-PCC metric to make recommendations. The proposed HUCTRS outperforms the baseline approaches on both TripAdvisor and Yelp datasets. The results obtained from experiment were assessed by means of standard evaluation metrics, namely RMSE, *F*-measure and coverage. The HUCTRS provides better POI recommendations due to the tendency of providing global optimization by ensemble swarm intelligent algorithms. The utilization of social network data significantly reduces the interactions between the user and the system. The proposed HUCTRS is designed in a way to be a proficient support tool for the active user. In future, we intend to develop the recommender system with multi-agent technology to accumulate user-generated information from various resources.

Acknowledgements The authors gratefully acknowledge the Science and Engineering Research Board (SERB), Department of Science and Technology, India, for the financial support through Mathematical Research Impact Centric Support (MATRICS) scheme (MTR/2019/000542). The authors also acknowledge SASTRA Deemed University, Thanjavur, for extending infrastructural support to carry out this research work.

References

1. Goldberg, D., Nichols, D., Oki, B.M., Terry, D.: Using collaborative filtering to weave an information tapestry. Commun. ACM **35**, 61–70 (1992)
2. Resnick, P., Iacovou, N., Suchak, M., Bergstrom, P., Riedl, J.: Grouplens: An open architecture for collaborative filtering of netnews. In: Proceedings of the 1994 ACM Conference on Computer Supported Cooperative Work, pp. 175–186 (1994)
3. Adomavicius, G., Tuzhilin, A.: Toward the next generation of recommendation system: a survey of the state-of-the-art and possible extensions. IEEE Trans. Knowl. Data Eng. **17**, 734–749 (2005)
4. Subramaniyaswamy, V., Logesh, R.: Adaptive KNN based recommender system through mining of user preferences. Wirel. Pers. Commun. **97**, 2229–2247 (2017)
5. Logesh, R., Subramaniyaswamy, V.: A reliable point of interest recommendation based on trust relevancy between users. Wirel. Pers. Commun. **97**, 2751–2780 (2017)
6. Subramaniyaswamy, V., Logesh, R., Abejith, M., Umasankar, S., Umamakeswari, A.: Sentiment analysis of tweets for estimating criticality and security of events. J. Organ. End User Comput. **29**, 51–71 (2017)
7. Deshpande, M., Karypis, G.: Item-based top-N recommendation algorithms. ACM Trans. Inf. Syst. **22**, 143–177 (2004)
8. Basu, C., Hirsh, H., Cohen, W.: Recommendation as classification: using social and content-based information in recommendation. In: Proceedings of the 1994 ACM Conference on Computer Supported Cooperative Work, pp. 714–720 (1998)
9. Burke, R.: Hybrid recommender systems: survey and experiments. User Model. User-Adapt. Interact. **12**, 331–370 (2002)
10. Koren, Y.: The bellkor solution to the netflix grand prize. Available http://www.netflixprize.com/assets/GrandPrize2009_BPC_BellKor.pdf (2009)
11. Pilaszy, I., Tikk, D.: Recommending new movies: even a few ratings are more valuable than metadata. In: Proceedings of the third ACM Conference on Recommender systems, pp. 93–100 (2009)
12. Abreu, P.H., Silva, D.C., Almeida, F., Mendes-Moreira, J.: Improving a simulated soccer team's performance through a memory-based collaborative filtering approach. Appl. Soft Comput. J. **23**, 180–193 (2014)
13. Abreu, P.H., Silva, D.C., Portela, J., Mendes-Moreira, J., Reis, L.P.: Using modelbased collaborative filtering techniques to recommend the expected best strategy to defeat a simulated soccer opponent. Intell. Data Anal. **18**, 973–991 (2014)
14. Taneja, A., Arora, A.: Modeling user preferences using neural networks and tensor factorization model. Int. J. Inf. Manage. **45**, 132–148 (2019)

15. Taneja, A., Arora, A.: Recommendation research trends: review, approaches and open issues. Int. J. Web Eng. Technol. **13**(2), 123–186 (2018)
16. Ericson, K., Pallickara, S.: On the performance of high dimensional data clustering and classification algorithms. Future Gener. Comput. Syst. **29**, 1024–1034 (2012)
17. Ungar, L.H., Foster, D.P.: Clustering methods for collaborative filtering. Paper presented at the AAAI workshop on recommendation systems (1998)
18. Bilge, Polat, H.: A comparison of clustering-based privacy-preserving collaborative filtering schemes. Appl. Soft Comput. **13**, 2478–2489 (2013)
19. Logesh, R., Subramaniyaswamy, V.: Learning recency and inferring associations in location based social network for emotion induced point-of-interest recommendation. J. Inf. Sci. Eng. **33**, 1629–1647 (2017)
20. Arunkumar, S., Subramaniyaswamy, V., Devika, R., Logesh, R.: Generating visually meaningful encrypted image using image splitting technique. Int. J. Mech. Eng. Technol. **8**, 361–368 (2017)
21. Cheng, L., Wang, H.: A fuzzy recommender system based on the integration of subjective preferences and objective information. Appl. Soft Comput. **18**, 290–301 (2014)
22. Tsai, C.F., Hung, C.: Cluster ensembles in collaborative filtering recommendation. Appl. Soft Comput. **12**, 1417–1425 (2012)
23. Pal, N., Pal, K., Keller, J., Bezdek, J.: A possibilistic fuzzy c-means clustering algorithm. IEEE Trans. Fuzzy Syst. **13**, 517–530 (2005)
24. Zhang, L., Pedrycz, W., Lu, W., Liu, X., Zhang, L.: An interval weighed fuzzy cmeans clustering by genetically guided alternating optimization. Expert Syst. Appl. **41**, 5960–5971 (2014)
25. Logesh, R., Subramaniyaswamy, V., Vijayakumar, V., Gao, X.-Z., Indragandhi, V.: A hybrid quantum-induced swarm intelligence clustering for the urban trip recommendation in smart city. Future Gener. Comput. Syst. (in Press)
26. Subramaniyaswamy, V., Logesh, R., Chandrashekhar, M., Challa, A., Vijayakumar, V.: A personalised movie recommendation system based on collaborative filtering. Int. J. High Perform. Comput. Networking **10**, 54–63 (2017)
27. Senthilselvan, N., Udaya Sree, N., Medini, T., Subhakari Mounika, G., Subramaniyaswamy, V., Sivaramakrishnan, N., Logesh, R.: Keyword-aware recommender system based on user demographic attributes. Int. J. Mech. Eng. Technol. **8**, 1466–1476 (2017)
28. Logesh, R., Subramaniyaswamy, V., Malathi, D., Senthilselvan, N., Sasikumar, A., Saravanan, P., Manikandan, G.: Dynamic particle swarm optimization for personalized recommender system based on electroencephalography feedback. Biomed. Res. **28**, 5646–5650 (2017). (India)
29. Abraham, A., Guo, H., Liu, H.: Swarm intelligence: foundations, perspectives and applications. Swarm Intelligent Systems. Springer, Berlin, pp. 3–25(2006)
30. Alam, S., Dobbie, G., Koh, Y.S., Riddle, P., Rehman, S.U.: Research on particle swarm optimization based clustering: a systematic review of literature and techniques. Swarm Evol. Comput. **17**, 1–13 (2014)
31. Li, C., Zhou, J., Kou, P., Xiao, J.: A novel chaotic particle swarm optimization based fuzzy clustering algorithm. Neurocomputing **83**, 98–109 (2012)
32. Chen, S., Xu, Z., Tang, Y.: A hybrid clustering algorithm based on fuzzy cmeans and improved particle swarm optimization. Arab. J. Sci. Eng. **39**, 8875–8887 (2014)
33. Zhang, Y., Xiong, X., Zhang, Q.: An improved self-adaptive PSO algorithm with detection function for multimodal function optimization problems. Math. Probl, Eng (2013)
34. Izakian, H., Abraham, A.: Fuzzy c-means and fuzzy swarm for fuzzy clustering problem. Expert Syst. Appl. **38**, 1835–1838 (2011)
35. Fang, W., Sun, J., Xie, Z.-P., Xu, W.-B.: Convergence analysis of quantum-behaved particle swarm optimization algorithm and study on its control parameter. Wuli Xuebao/Acta Phys. Sin. **59**, 3686–3694 (2010)
36. Ravi, L., Vairavasundaram, S.: A Collaborative location based travel recommendation system through enhanced rating prediction for the group of users. Comput. Intell. Neurosci. 1291358 (2016)

37. Subramaniyaswamy, V., Vijayakumar, V., Logesh, R., Indragandhi, V.: Intelligent travel recommendation system by mining attributes from community contributed photos. Procedia Comput. Sci. **50**, 447–455 (2015)
38. Subramaniyaswamy, V., Logesh, R., Vijayakumar, V., Indragandhi, V.: Automated message filtering system in online social network. Procedia Comput. Sci. **50**, 466–475 (2015)
39. Vairavasundaram, S., Varadharajan, V., Vairavasundaram, I., Ravi, L.: Data mining-based tag recommendation system: an overview. Wiley Interdiscip. Rev. Data Min. Knowl. Discov. **5**, 87–112 (2015)
40. Subramaniyaswamy, V., Vijayakumar, V., Logesh, R., Indragandhi, V.: Unstructured data analysis on big data using map reduce. Procedia Comput. Sci. **50**, 456–465 (2015)
41. Han, Y., Guan, H., Duan, J.: Tour route multiobjective optimization design based on the tourist satisfaction. Discrete Dyn. Nat. Soc. (2014). https://doi.org/10.1155/2014/603494
42. Brilhante, I.R., Macedo, J.A., Nardini, F.M., Perego, R., Renso, C.: On planning sightseeing tours with tripbuilder. Inform. Process. Lett. **51**, 1–15 (2015)
43. Kurata, Y., Hara, T.: CT-Planner4: Toward a more user-friendly interactive day-tour planner. In: Proceedings of the 2014 International Conference on Information and Communication Technologies in Tourism, pp. 73–86 (2014)
44. Hasuike, T., Katagiri, H., Tsubaki, H., Tsuda, H.: Route planning problem with groups of sightseeing sites classified by tourist's sensitivity under time expanded network. In Proceedings of the 2014 IEEE International Conference on Systems, Man and Cybernetics, pp. 188–193 (2014).
45. Shukla, U.P., Nanda, S.J.: Parallel social spider clustering algorithm for high dimensional datasets. Eng. Appl. Artif. Intell. **56**, 75–90 (2016)
46. Alswaitti, M., Albughdadi, M., Isa, N.A.M.: Density-based particle swarm optimization algorithm for data clustering. Expert Syst. Appl. **91**, 170–186 (2018)
47. Wai, E.N.C., Tsai, P.W., Pan, J.S.: Hierarchical PSO clustering on MapReduce for scalable privacy preservation in big data. In: International Conference on Genetic and Evolutionary Computing. Springer International Publishing, Berlin, pp. 36–44 (2016)
48. Polatidis, N., Georgiadis, C.K.: A dynamic multi-level collaborative filtering method for improved recommendations. Comput. Stan. Interfaces **51**, 14–21 (2017)
49. Cuevas, E., Cienfuegos, M., Zaldívar, D., Pérez-Cisneros, M.: A swarm optimization algorithm inspired in the behavior of the social-spider. Expert Syst. Appl. **40**, 6374–6384 (2013)
50. Bao, J., Zheng, Y., Mokbel, M.F.: Location-based and preference-aware recommendation using sparse geo-social networking data. In: Proceedings of the 20th International Conference on Advances in Geographic Information Systems, pp. 199–208 (2012)
51. Zhu, X., Hao, R., Chi, H., Du, X.: Personalized location recommendations with local feature awareness. In: Proceedings of IEEE Global Communications Conference, GLOBECOM 2016, Art. No. 7842140 (2017)

Finite Element Technique to Study Calcium Distribution in Alzheimer's Disease

Devanshi D. Dave and Brajesh Kumar Jha

Abstract The study of calcium diffusion in Alzheimer's disease (AD) is an active field of research. This study is carried out using various methods and techniques. In the present paper, a 2D unsteady-state computational model has been constructed to study the physiological phenomena of calcium diffusion. Calcium is known as the second messenger having plethora of functions to perform and maintain. The alteration may lead to devastating neurological disorders. Here, the role of calcium diffusion is studied in AD. In order to know their effect on cytosolic calcium concentration in regular and diseased brain, physiological parameters such as calcium-binding buffers and sodium calcium exchanger have been taken into consideration. Estimation of the domain using discrete elements is implemented to achieve the results. Apposite boundary conditions are employed which matches well with the physiological condition. The graphs are plotted in MATLAB. The obtained graphs clearly demonstrate the significance of buffers and NCX on the cytosolic calcium in AD.

Keywords Calcium concentration · Buffers · Sodium calcium exchanger · Alzheimer's disease · Finite element estimation

1 Introduction

Over the last few years, the study of calciumopathy is increasing significantly suggesting the importance of normal calcium homeostasis. It plays important roles in almost all kinds of cells but the role of calcium in brain is immense and important. Plethora of functions is carried out by cell calcium. Calcium is basically known as second messenger. Hence, it has key role in signaling cascades [1]. The influx and efflux of calcium via certain channels and pumps and other calcium sequestering organelles lead to various cellular functions [1]. The normal cytosolic calcium level

D. D. Dave (✉) · B. K. Jha
Department of Mathematics, School of Technology, Pandit Deendayal Petroleum University, Gujarat, India
e-mail: ddave1822@gmail.com

M. Sahni et al. (eds.), *Mathematical Modeling, Computational Intelligence Techniques and Renewable Energy*, Advances in Intelligent Systems and Computing 1287,
https://doi.org/10.1007/978-981-15-9953-8_28

helps in sustaining several; if not all; neuronal functions. It has been found that almost 95–98% of the calcium is situated in the nucleus of the cell and only the remaining calcium is found in a free state. But this free calcium plays a critical role for determining the cell fate. From this free calcium located in the cytoplasm, most amount of calcium gets buffered with the help of calcium-binding proteins and buffers. Rest of the calcium concentration gradient is processed and managed by calcium channels, several exchangers and pumps, intracellular entities like mitochondria and endoplasmic reticulum [2]. Hence, on the whole, it works as a syncytium and maintains the cytosolic cell calcium.

The alterations in the intracellular calcium lead to several devastating neuronal disorders like Alzheimer's disease, epilepsy, Parkinson's disease, etc. [3] Out of all, Alzheimer's disease is experienced by billions of people around the globe and is the utmost prevailing form of dementia. In it, mostly hippocampal area is damaged. The hallmark causes of Alzheimer's disease are amyloid beta plaques and neurofibrillary tangle [4]. Over and above these two hallmarks, research suggests that there are several other factors which may contribute in prevailing this dementia. Out of that, the development and research of calciumopathy and calcium imaging techniques have enabled and opened a new door in knowing the insights about this dementia. In this article, the aim is to study the calcium diffusion in normal and Alzheimeric cells in the presence of physiological parameters like buffers and sodium calcium exchanger. Buffers are present at the periphery of the cell. Calcium after entering the cell reacts with cytosolic buffers and results into calcium bound buffers. In Alzheimeric condition, the amount of buffer is significantly decreased which consequently increases the cell calcium. Sodium calcium exchanger exchanges the sodium to calcium in 3:1 ratio from the plasma membrane. It plays an important role in maintaining cell calcium [5]. The modification and alteration in the normal functioning of buffer and NCX lead to hike in cytoplasmic calcium level which results into toxicity to the cell.

The rendered toxicity disables the normal signaling cascades of calcium. Also, this toxicity leads to the cell death phenomenon which may further cause hindrance in formation and maintenance of memory in the brain [4]. Thus, the impact of these parameters of the calcium toolkit is immerse at the cytosolic level. Hence, we have considered these parameters to delineate the physiology of calcium diffusion in neurons using mathematical model.

Literature survey suggests that the authors have used finite difference and finite element method to study the impact of buffers, sodium calcium exchanger and other parameters on cytosolic calcium concentration level in various cells [6, 7]. Tewari and Pardasani have used finite difference method to know the possible effects of the sodium flux on the cytosolic calcium concentration distribution [8]. In the year, 2012, Tewari and Pardasani have mathematically modeled the phenomenon of exchange of sodium calcium at cytosolic level [9]. Their results show the oscillatory behavior of the calcium concentration distribution. In 2015, Jha et al. have employed finite element method to portray the impact of the sodium calcium exchanger and the calcium influx on the cytosolic calcium level in a neuronal cell [10]. In 2016, Jha et al. have estimated the calcium diffusion in astrocytes along with buffers and sodium calcium exchanger [11]. Also, Jha and Dave have used FEM to explain the level of calcium concentration for normal and Alzheimeric cells [12].

Thus, on the basis of the literature survey done, it has been found that no efforts were made to study calcium diffusion in view of Alzheimer's disease in the presence of buffers and sodium calcium exchanger using irregular neuronal cells. Hence, in view of the physiology described and the literature survey, we have the mathematical model of calcium diffusion in the next section having appropriate boundary conditions. Further, the finite element technique and the results obtained are discussed.

2 Mathematical Formulation

The mathematical formulation of the problem depends on the formulation of the calcium buffering and sodium calcium exchange phenomena. Thus, they are formulated and discussed in the subsections hereafter.

2.1 Calcium Buffering

Fundamentally, the calcium buffering phenomenon can be mathematically stated as follows [13, 14]:

$$\left[\mathrm{Ca}^{2+}\right]+\left[B_j\right]=\left[\mathrm{Ca}B_j\right] \tag{1}$$

where $[\mathrm{Ca}^{2+}]$ is the calcium concentration, $[B]$ is the buffer concentration, and $[\mathrm{Ca}B]$ is the calcium bound buffers.

Now, it can further be stated in the form of PDEs using Fick's law as follows [15]:

$$\frac{\partial\left[\mathrm{Ca}^{2+}\right]}{\partial t}=D_{Ca}\left(\frac{\partial^2\left[\mathrm{Ca}^{2+}\right]}{\partial x^2}+\frac{\partial^2\left[\mathrm{Ca}^{2+}\right]}{\partial y^2}\right)+\sum_j R_j \tag{2}$$

$$\frac{\partial[B_j]}{\partial t}=D_B\left(\frac{\partial^2[B_j]}{\partial x^2}+\frac{\partial^2[B_j]}{\partial y^2}\right)+R_j \tag{3}$$

$$\frac{\partial\left[\mathrm{Ca}B_j\right]}{\partial t}=D_{\mathrm{Ca}B}\left(\frac{\partial^2\left[\mathrm{Ca}B_j\right]}{\partial x^2}+\frac{\partial^2\left[\mathrm{Ca}B_j\right]}{\partial y^2}\right)-R_j \tag{4}$$

where

$$R_j=-k^+\left[B_j\right]\left[\mathrm{Ca}^{2+}\right]+k^-\left[\mathrm{Ca}B_j\right] \tag{5}$$

2.2 Sodium Calcium Exchanger

The flux through sodium calcium exchanger is mathematically modeled as follows [11, 16]:

$$\sigma_{\mathrm{NCX}} = Ca_0\left(\frac{\mathrm{Na}_i}{\mathrm{Na}_0}\right)^3 \times \exp\left(\frac{\mathrm{FV}_m}{\mathrm{RT}}\right) \tag{6}$$

where Ca_o is the extracellular calcium, Na_i and Na_o are the intracellular and extracellular sodium concentrations. F is the Faraday's constant, V_m is the voltage, R is the gas constant, and T is the temperature.

Thus, on the basis of these formulations, the mathematical problem of two-dimensional calcium concentration distribution in neuronal cell along with these parameters is stated as follows:

$$D_{\mathrm{Ca}}\left(\frac{\partial^2 C}{\partial x^2} + \frac{\partial^2 C}{\partial y^2}\right) - k_j^+[B]_\infty\,(\mathrm{C} - C_\infty) = \frac{\partial C}{\partial t} \tag{7}$$

The correct boundary conditions are described in accordance with brain physiology as follows:

$$-D_{\mathrm{Ca}}\frac{\partial\left[\mathrm{Ca}^{2+}\right]}{\partial\eta} = \sigma_{\mathrm{Ca}} - \sigma_{\mathrm{NCX}} \tag{8}$$

where η is perpendicular to the surface.

$$\left[\mathrm{Ca}^{2+}\right] = 0.1\ \mu\mathrm{M},\ \text{ for } t > 0 \text{ and } x, y \to \infty \tag{9}$$

3 Finite Element Estimation

The method of finite elements is commonly used to address various physical and physiological problems. It beautifully deals with the irregular and complex domains. The method gives the approximate solution to the problems. In this section, we have defined the domain of the problem, i.e., the geometry of the neuron cell (hippocampal neuron) as shown in Fig. 1. It includes of dendrites, soma and axon terminals.

Further, we have shown the calcium flux in the neuron using single and multiple boundaries. Figure 2 shows the single and multiple influxes of calcium and the exchange of calcium occurring via sodium calcium exchanger.

To obtain the results, the estimated geometry is discretized into 614 triangles and 504 nodes. Discretizing the domain leads to approximate solution of calcium diffusion taking place in hippocampal neuron. Figure 3 shows the discretization of

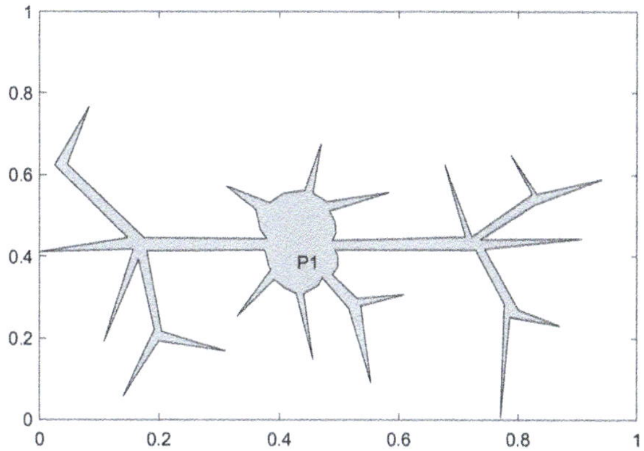

Fig. 1 Estimated geometry of a hippocampal neuron

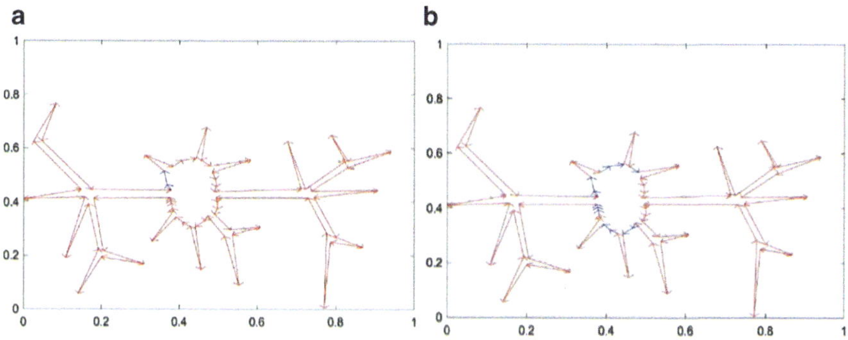

Fig. 2 Calcium influx in a hippocampal neuron using **a** single and **b** multiple boundaries

Fig. 3 Discretization of the domain of hippocampal neuron

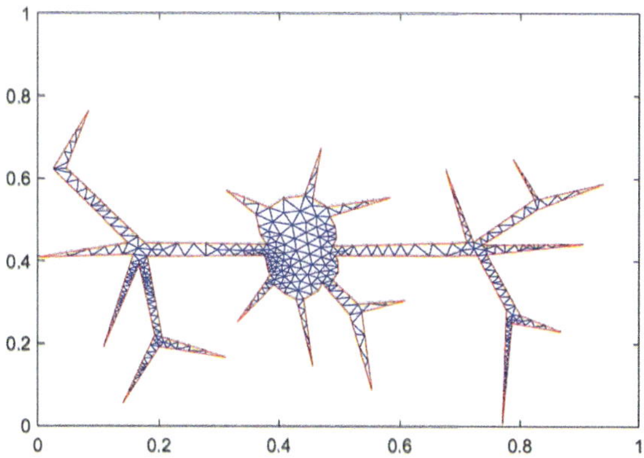

the domain of typical hippocampal neuron.

Thus, the estimated calcium diffusion in normal and Alzheimeric neuronal cells in the presence of buffer and NCX is shown in the next section.

4 Results and Discussion

Table 1 displays the values of the biophysical parameters used to achieve the results.

Figure 4 shows the calcium diffusion in normal and Alzheimeric cells. Here, single

Table 1 Values of biophysical parameters [8, 10]

	Biophysical parameter	Values
D	Diffusion coefficient	200–300 $\mu m^{-1}\ s^{-1}$
k^{+} Calmodulin	Buffer association rate	250 $\mu m^{-1}\ s^{-1}$
$[B]_{\infty}$	Buffer concentration	1–100 $\mu m^{-1}\ s^{-1}$
$[Ca^{2+}]_{\infty}$	Background calcium conc.	0.1 μM
$[Na^{+}]_{out}$	Extracellular conc. of Na^{+}	145 mM
$[Na^{+}]_{in}$	Intracellular conc. of Na^{+}	12 mM
R	Gas constant	8.314 J
T	Temperature	20 °C
F	Faradays constant	94,687
Z	Calcium valency	2

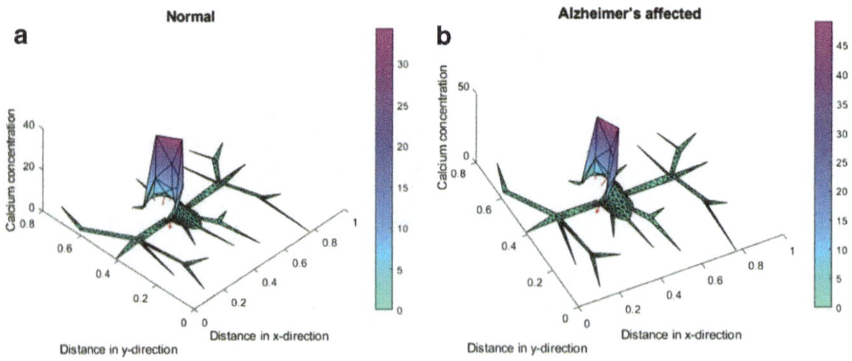

Fig. 4 Calcium diffusion in **a** normal and **b** Alzheimeric cells having sole flux

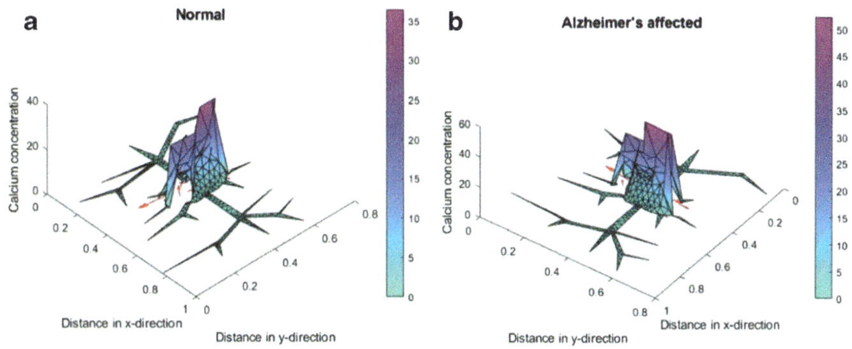

Fig. 5 Calcium diffusion in **a** normal and **b** Alzheimer's affected cell having multiple fluxes

flux is considered to estimate the calcium diffusion. The endogenous buffer Calmodulin has its significant role in hippocampal area and is reduced to almost 30% in Alzheimer's disease. The decrease in buffer leads to increase in calcium level. Moreover, it has been found that the NCX activities are highly increased in Alzheimeric neuronal cells. The combined impact of the decreased buffer and increased sodium calcium exchanger activities is seen from Fig. 4b. In normal condition, the cytosolic calcium concentration level is around 35, whereas in Alzheimeric condition it is found to be around 50.

Similarly, Fig. 5 shows the distribution of calcium for several fluxes. We have measured it for several influxes to gain a better understanding of the mechanism behind calcium diffusion. The values of the parameters are same as that in Fig. 4. Similar pattern of the calcium diffusion is seen for normal and Alzheimeric cells. It has been observed that there is high level of calcium concentration due to low amount of buffer and high NCX activities. Due to this high calcium concentration level, the cell gets toxic and causes hindrance in memory formation, further leading to Alzheimeric symptoms.

5 Conclusion

In this article, the study of spatial spread of calcium concentration is estimated in normal and diseased Alzheimeric cells in the presence of a buffer and sodium calcium exchanger. It is found that the cells with decreased level of Calmodulin and increased activities of NCX have a greater level of calcium concentration which leads to cell toxicity. This toxicity in the normal cell is dominant in the Alzheimeric disorder and could lead to cell death. The effect of buffer Calmodulin is found to be significant in Alzheimeric cells which can be clearly seen on the concentration level of cytosolic calcium. This research is useful in understanding the functions of buffer and NCX in the prevalent dementia for cognitive scientists and biologists. Further, more physiological parameters can be added to consider its effect on the distribution of cytosolic calcium.

References

1. Kawamoto, E.M., Vivar, C., Camandola, S.: Physiology and pathology of calcium signaling in the brain. Front Pharmacol. **3**, 1–17 (2012)
2. Magi, S., Castaldo, P., Macrì, M.L., Maiolino, M., Matteucci, A., Bastioli, G.: Intracellular calcium dysregulation: implications for Alzheimer's disease. Biomed. Res. Int. **2016**, 1–14 (2016)
3. Carafoli, E., Brini, M. (eds.) Calcium Signalling and Disease. Springer, Berlin (2007)
4. Turkington, C., Mitchell, D.: The Encyclopedia of Alzheimer's Disease, 2nd edn. Facts On File: An imprint of Infobase Publishing, p. 321 (2010)
5. Blaustein, M.P., Lederer, W.J.: Sodium/calcium exchange: its physiological implications. Physiol. Rev. **79**(3), 763–854 (1999)
6. Naik, P.A., Pardasani, K.R.: Finite element model to study calcium distribution in oocytes involving voltage gated Ca^{2+} channel, ryanodine receptor and buffers. Alexandria J. Med. [Internet] **52**(1), 43–49 (2016)
7. Naik, P.A., Pardasani, K.R.: Two dimensional finite element model to study calcium distribution in oocytes. J. Multiscale Model. **6**(1), 1450002-1–1450002-15 (2015)
8. Tewari, S., Pardasani, K.R.: Finite difference model to study the effects of Na^{+} influx on cytosolic Ca^{2+} diffusion. World Acad. Sci. Eng. Technol. **5**, 670–675 (2008)
9. Tewari, S.G., Pardasani, K.R.: Modeling effect of sodium pump on calcium oscilations in neuron cells. J. c Model. **4**(3), 1–16 (2012)
10. Jha, A., Adlakha, N., Jha, B.: Finite element model to study effect of Na^{+}–Ca^{2+} exchangers and source geometry on calcium dynamics in a neuron cell. J. Mech. Med. Biol. **16**(2), 1–22 (2015)
11. Jha, B.K., Jha, A.: Two dimensional finite element estimation of calcium ions in presence of NCX and buffers in Astrocytes. Bol. Soc. Parana. Mat. **36**(1), 151–160 (2018)
12. Jha, B.K., Dave, D.D.: Approximation of calcium diffusion in Alzheimeric cell Brajesh. J Multiscale Model. 1–21(2020)
13. Keener, J., Sneyd, J.: Mathematical Physiology, 2nd edn. Springer US, p. 1067 (2009)
14. Smith, G.D.: Analytical steady-state solution to the rapid buffering approximation near an open Ca^{2+} channel. Biophys. J. [Internet]. **71**, 3064–3072 (1996). Available from http://dx.doi.org/10.1016/S0006-3495(96)79500-0
15. Crank, J.: The Mathematics of Diffusion, 2nd edn. Clarendon Press Oxford, p. 421 (1975)
16. Panday, S., Pardasani, K.R.: Finite element model to study effect of advection diffusion and Na +/Ca2 + exchanger on $^{Ca2+}$ distribution in oocytes. J. Med. Imaging Heal Inf. **3**(3), 374–379 (2013)

The Interval Estimation of the Shapley Value for Partially Defined Cooperative Games by Computer Simulations

Satoshi Masuya

Abstract In this paper, we investigate the Shapley value for partially defined cooperative games by computer simulations using the theory of the statistical estimation. The population of the Shapley values is assumed to follow the normal distribution. However, the mean and the standard deviation of the population are not known. For each unknown coalition, random numbers which follow the uniform distributions are given between the lower bound and the upper bound of each unknown coalitional worth. Finally, using the formula of the interval estimation of the mean of the population, we obtain the mean of the population of the Shapley value as the interval with 95% of the confidence coefficients.

Keywords Partially defined cooperative game · Shapley value · Interval estimation

1 Introduction

In the classical approach to cooperative games it is assumed that the worth of every coalition is known. However, in the real world problems there may be situations in which the amount of information is limited and consequently the worths of some coalitions are unknown. The games corresponding to those problems are called partially defined cooperative games (PDGs in short).

Partially defined cooperative games were first studied by Willson [6]. Willson [6] proposed and characterized an extension of the Shapley value [5] for partially defined cooperative games. This extended Shapley value coincides with the ordinary Shapley value of a complete game (we say that a game is complete if the worths of all the coalitions are known). In this complete game the coalitions whose worth were known in the original game maintain the same worth, but otherwise they are assigned

S. Masuya (✉)
Faculty of Business Administration, Daito Bunka University,
1-9-1, Takashimadaira Itabashi-ku, Tokyo 175-8571, Japan
e-mail: masuya@ic.daito.ac.jp

M. Sahni et al. (eds.), *Mathematical Modeling, Computational Intelligence Techniques and Renewable Energy*, Advances in Intelligent Systems and Computing 1287,
https://doi.org/10.1007/978-981-15-9953-8_29

a worth zero, that seems to be not well justified. After that, Housman [3] continued the study of Willson [6]. Housman [3] characterized the generalized Shapley value by Willson [6].

Recently, Masuya and Inuiguchi [4], and Albizuri et al. [1] considered the Shapley values for PDGs. Masuya and Inuiguchi [4] defined complete games called the lower and the upper games for the given PDGs. Assuming that the PDG is superadditive, they showed that the lower game is superadditive, but the upper game need not be superadditive. On the other hand, Albizuri et al. [1] proposed an extension of the Shapley value for general partially defined cooperative games by following the Harsanyi's approach. That is, it is assumed that each coalition guarantees certain payments, called the Harsanyi dividends [2], to its members. They assumed that coalitions whose worths are not known are assigned a dividend equal to zero. The final payoff will be the sum of these dividends. They axiomatized this value using four axioms.

Although the Shapley value for general PDGs proposed by [1] has been succeeded for an axiomatization, generally, theoretical investigations of a solution defined on PDGs are very complicated. Then we focus on using computational simulations to analyze the Shapley value for PDGs.

In this paper, we investigate the Shapley value for PDGs by computer simulations using the theory of the statistical estimation. First, the population of the Shapley values is assumed to follow the normal distribution. However, the mean and the standard deviation of the population are not known. For each unknown coalition, random numbers which follow the uniform distributions are given between the lower bound and the upper bound of each unknown coalitional worth.

For obtained complete game v, we check whether v is superadditive or not. If v is superadditive, we calculate the Shapley value for v. If v is not superadditive, we obtain the next random number.

Iterating this processing, obtained Shapley values can be interpreted as the samples of the population of the Shapley value. Finally, using the formula of the interval estimation of the mean of the population, we can obtain the mean of the population of the Shapley value as the interval with the confidence coefficients such as 95 or 99%.

This paper is organized as follows. In Sect. 2, we present the definition of PDGs, and the lower and upper games associated with them and the superadditivity of PDGs. In Sect. 3, we consider the Shapley value for PDGs by computer simulations using the theory of the statistical estimation. In Sect. 4, we investigate how to obtain good interval estimation of the Shapley value (this means the width of the interval is small and so on). In Sect. 5, concluding remarks and future directions are given.

2 Partially Defined Cooperative Games

Let $N = \{1, 2, \ldots, n\}$ be the set of players and $v\colon 2^N \to \mathbb{R}$ such that $v(\emptyset) = 0$. A classical cooperative game, i.e., a coalitional game with transferable utility (a TU game) is characterized by a pair (N, v). A set $S \subseteq N$ is regarded as a coalition of players and function value $v(S) \in \mathbb{R}$ shows a collective payoff that players in S can gain by forming coalition S.

A cooperative game (N, v) is said to be superadditive if and only if

$$v(S \cup T) \geq v(S) + v(T), \ \forall S, \ T \subseteq N \text{ such that } S \cap T = \emptyset. \tag{1}$$

The superadditivity is a natural property for giving players incentives to form a bigger coalition.

As the representative one-point solution for cooperative games, the Shapley value is well-known. The Shapley value is defined as follows:

$$\phi_i(v) = \sum_{\substack{S \ni i \\ S \subseteq N}} \frac{(|S|-1)!(n-|S|)!}{n!}(v(S) - v(S \setminus i)), \ \forall i \in N, \tag{2}$$

where ϕ_i is the i-th component of ϕ and $|S|$ is the cardinality of the set S.

Throughout this paper, we assume that the underlying complete game is set to be superadditive and that all players know this fact although coalitional worths are unknown. Therefore, forming a bigger coalition is advantageous for players.

A PDG can be characterized by a set of players $N = \{1, 2, \ldots, n\}$, a set of coalitions whose worths are known, say $\mathcal{K} \subseteq 2^N$, and a function $v: \mathcal{K} \to \mathbb{R}$, where we basically assume that $\emptyset \in \mathcal{K}$ and $v(\emptyset) = 0$. We assume that worths of singleton and the grand coalitions are at least known, i.e., $\{i\} \in \mathcal{K}$, $i = 1, 2, \ldots, n$ and $N \in \mathcal{K}$. Because of the superadditivity of the underlying complete game, $v(\{i\}) \geq 0$, $i = 1, 2, \ldots, n$ are assumed.

A PDG v is assumed to be superadditive, i.e.,

$$v(S) \geq \sum_{i=1}^{s} v(T_i), \ \forall S, T_i \in \mathcal{K}, \ i = 1, 2, \ldots, s \text{ such that } \bigcup_{i=1,2,\ldots,s} T_i = S$$
$$\text{and } T_i, i = 1, 2, \ldots, s \text{ are disjoint.} \tag{3}$$

As defined above, a triple $(N, \mathcal{K}, v)$ can identify a PDG. When we consider only games under fixed N and $\mathcal{K}$, PDG $(N, \mathcal{K}, v)$ is simply written as v. The set of all the superadditive PDGs is denoted Γ.

Associated with a given PDG $(N, \mathcal{K}, v)$, we may define two complete games $(N, \underline{v})$ and $(N, \overline{v})$:

$$\underline{v}(S) = \max_{\substack{T_i \in \mathcal{K}, \ i=1,2,\ldots,s \\ \cup_i T_i \subseteq S, T_i \text{ are disjoint}}} \sum_{i=1}^{s} v(T_i), \tag{4}$$

$$\overline{v}(S) = \min_{\hat{S} \in \mathcal{K}, \ \hat{S} \supseteq S} \left(v(\hat{S}) - \underline{v}(\hat{S} \setminus S) \right) \tag{5}$$

A complete game (N, w) such that $w(T) = v(T)$, $\forall T \in \mathcal{K}$ is called a complete extension of $(N, \mathcal{K}, v)$, or simply a complete extension of v.

Theorem 1 *([4]) Let $(N, \mathcal{K}, v)$ be a PDG, and $(N, \underline{v})$ and $(N, \overline{v})$ the complete extensions defined by (4) and (5). For an arbitrary superadditive complete extension (N, v) of $(N, \mathcal{K}, v)$, we obtain*

$$\underline{v}(S) \leq v(S), \ \forall S \subseteq N, \tag{6}$$

$$\overline{v}(S) \geq v(S), \ \forall S \subseteq N. \tag{7}$$

3 Computer Simulations to Obtain the Interval Estimations of the Shapley Value

In this section, we obtain the confidence interval of the mean of the Shapley value such that the confidence coefficient is 95% using the theory of statistical estimation. Let $(N, \mathcal{K}, v)$ be a superadditive PDG satisfying $\mathcal{K} \supseteq \{\{1\}, \ldots, \{n\}, N\}$. The method of simulations is as follows:

1. The population of the Shapley values is assumed to follow the normal distribution. The mean and the standard deviation of the population are not known. For each unknown coalition, random numbers which follow the uniform distributions are given between the lower bound $\underline{v}$ and the upper bound $\overline{v}$ of each unknown coalitional worth.
2. For obtained complete game w, we check whether w is superadditive or not. [4] shows that lower game $\underline{v}$ is superadditive for all $v \in \Gamma$ while upper game $\overline{v}$ is not necessarily superadditive. Hence, obtained complete game w would be checked whether w is superadditive or not. If w is superadditive, we calculate the Shapley value for w. If w is not superadditive, we do not use w and obtain the next random number.
3. Iterating processing 2, obtained Shapley values can be interpreted as the samples of the population of the Shapley value. Finally, using the formula of the interval estimation of the mean of the population, we can obtain the mean of the population of the Shapley value as the interval with 95% of the confidence coefficient. The formula of the mean of the population is represented as follows:

$$\left[m - 1.96\frac{S}{\sqrt{n}}, \ m + 1.96\frac{S}{\sqrt{n}}\right], \tag{8}$$

   where m is the mean of samples, n is the number of samples and S is the sample standard deviation. Note that the Eq. (8) can be used when the number of samples is large (Concretely, larger than around 30) and the standard deviation of the population is not known.

From the formula of the estimation, if the standard deviation grows n-fold, the width of the interval also grows n-fold. Hence, it is necessary to obtain a small standard deviation to have a good estimation.

Table 1 Coalitional worths which are used to define 5-person PDGs

$v(1) = 10$	$v(124) = 26$
$v(2) = 8$	$v(125) = 25$
$v(3) = 6$	$v(134) = 22$
$v(4) = 4$	$v(135) = 20$
$v(5) = 2$	$v(145) = 20$
$v(12) = 20$	$v(234) = 22$
$v(13) = 16$	$v(235) = 20$
$v(14) = 15$	$v(245) = 19$
$v(15) = 14$	$v(345) = 16$
$v(23) = 15$	$v(1234) = 37$
$v(24) = 16$	$v(1235) = 35$
$v(25) = 10$	$v(1245) = 34$
$v(34) = 10$	$v(1345) = 28$
$v(35) = 9$	$v(2345) = 35$
$v(45) = 9$	$v(12345) = 60$
$v(123) = 28$	

Now we perform the simulation following the method described above. In this paper, the simulation is performed using a 5-person PDG v. First, each coalitional worth of PDG v is taken from Table 1 described below.

Further, as the first step to investigate the Shapley value for PDGs using computer simulations , we assume the following.

Assumption 1 For two players $i \in N, j \in N$, we consider a coalition $S \ni i$ such that $S \not\ni j$ and $S \subseteq N$. Further, let S' be a coalition interchanging i and j in S. That is, we consider $S' = S \cup \{j\} \backslash \{i\}$. Then, if $S \in \mathcal{K}$, $S' \in \mathcal{K}$ holds.

The set of known coalitions satisfying the assumption described above can be represented by $M \subseteq \{0, 1, \ldots, n\}$ satisfying $M \supseteq \{0, 1, n\}$. That is, the elements of M represent the cardinalities of known coalitions. In this research, the possible set of known coalitions is each subset of $\{0, 1, \ldots, n\}$ including $\{0, 1, n\}$. Then the number of possible sets of known coalitions is $2^{n-2} - 1$ (if the complete game is not included). Therefore, the number of possible sets of known coalitions is seven when the number of players is five.

We performed simulations following the method described above for all the possible sets of known coalitions. The number of samples from the population is the number of computation times. The number of computation times are 50 for each case in this paper.

4 Observations on the Results

In this section, we will observe the outcomes of the simulation.

First, we observe the computation times which are shown in Table 2 from Case 1 to 7. The computation time of Case 1 is the longest. Needless to say, this result was obtained since only the least number of worths was given in the simulation. However, the reason is not only the least number of worths but also the difficulty of obtaining a superadditive game in Case 1. In Case 1, since only the worths of singleton and the grand coalitions are given to the PDG, the difference between the lower bound and the upper bound becomes large. Then the condition of superadditivity is not obtained easily since the upper game is not necessarily superadditive although the lower game is always superadditive [4]. Also, those reasons made the standard deviations and the width of confidence intervals very large. Next to Case 1, computation times of Case 2 and 5 are long. These outcomes were obtained because of the same reason of Case 1. However, the standard deviations of Case 5 are very small although those of Case 2 are large. These results represent the decrease of upper bound is more effective than the increase of the lower bound in this example (Tables 3, 4, 5, 6, 7, 8 and 9).

In contrast to these cases, the computation times of the other cases (Case 3, 4, 6 and 7) are short. This is because almost all the coalitional worths are known in all the cases other than Case 4. Moreover, the standard deviations are very small in the cases in which $M \ni 4$ holds. The computation time of Case 4 is short although it has not a small number of unknown coalitions. It seems that the difference between the lower bound and the upper bound of each coalitional worth became similar since we took the middle point of the cardinalities of coalitions.

The purpose of this research is how to obtain a good interval estimation of the Shapley value (this means the width of the interval is small) using the small number of known coalitions and the short computation time. However, there is a trade-off

Table 2 The computation time of each case [sec]

	Case 1	Case 2	Case 3	Case 4	Case 5	Case 6	Case 7
M	$\{0, 1, 5\}$	$\{0, 1, 2, 5\}$	$\{0, 1, 2, 3, 5\}$	$\{0, 1, 3, 5\}$	$\{0, 1, 4, 5\}$	$\{0, 1, 2, 4, 5\}$	$\{0, 1, 3, 4, 5\}$
Time	480	71	18	18	68	16	16

Table 3 The mean of the Shapley values, the SD and the confidence interval of Case 1 ($M = \{0, 1, 5\}$)

	Mean	SD	Interval
Player 1	16.26	2.65	[15.53, 16.99]
Player 2	13.55	2.89	[12.75, 14.35]
Player 3	12.20	2.80	[11.42, 12.98]
Player 4	9.37	2.48	[8.68, 10.06]
Player 5	8.61	2.15	[8.01, 9.21]

Table 4 The mean of the Shapley values, the SD and the confidence interval of Case 2 ($M = \{0, 1, 2, 5\}$)

	Mean	SD	Interval
Player 1	15.92	2.05	[15.35, 16.49]
Player 2	14.36	2.16	[13.76, 14.96]
Player 3	11.44	2.15	[10.84, 12.04]
Player 4	10.08	1.70	[9.61, 10.55]
Player 5	8.21	2.16	[7.61, 8.81]

Table 5 The mean of the Shapley values, the SD and the confidence interval of Case 3 ($M = \{0, 1, 2, 3, 5\}$)

	Mean	SD	Interval
Player 1	15.37	1.75	[14.88, 15.86]
Player 2	14.41	1.63	[13.96, 14.86]
Player 3	11.33	1.52	[10.91, 11.75]
Player 4	10.18	1.59	[9.74, 10.62]
Player 5	8.71	1.58	[8.27, 9.15]

Table 6 The mean of the Shapley values, the SD and the confidence interval of Case 4 ($M = \{0, 1, 3, 5\}$)

	Mean	SD	Interval
Player 1	15.61	1.71	[15.14, 16.08]
Player 2	14.52	1.64	[14.07, 14.97]
Player 3	11.66	1.67	[11.20, 12.12]
Player 4	9.83	1.50	[9.41, 10.25]
Player 5	8.38	1.80	[7.88, 8.88]

Table 7 The mean of the Shapley values, the SD and the confidence interval of Case 5 ($M = \{0, 1, 4, 5\}$)

	Mean	SD	Interval
Player 1	14.64	0.39	[14.53, 14.75]
Player 2	15.25	0.40	[15.14, 15.36]
Player 3	11.78	0.38	[11.67, 11.89]
Player 4	10.13	0.35	[10.03, 10.23]
Player 5	8.20	0.43	[8.08, 8.32]

Table 8 The mean of the Shapley values, the SD and the confidence interval of Case 6 ($M = \{0, 1, 2, 4, 5\}$)

	Mean	SD	Interval
Player 1	14.48	0.21	[14.42, 14.54]
Player 2	15.42	0.16	[15.38, 15.46]
Player 3	11.46	0.21	[11.40, 11.52]
Player 4	10.45	0.22	[10.39, 10.51]
Player 5	8.18	0.22	[8.12, 8.24]

Table 9 The mean of the Shapley values, the SD and the confidence interval of Case 7 ($M = \{0, 1, 3, 4, 5\}$)

	Mean	SD	Interval
Player 1	14.48	0.09	[14.46, 14.50]
Player 2	15.37	0.13	[15.33, 15.41]
Player 3	11.64	0.08	[11.62, 11.66]
Player 4	10.21	0.12	[10.18, 10.24]
Player 5	8.30	0.12	[8.27, 8.33]

Table 10 The evaluations of each criterion given to each computational result

	C1	C2	C3
Case 1	2.89	25	480
Case 2	2.16	15	71
Case 3	1.75	5	18
Case 4	1.8	15	18
Case 5	0.43	20	68
Case 6	0.22	10	16
Case 7	0.12	10	16

among these factors. Hence, we give three criteria for evaluating the wellness of the solution as follows:

C1. The standard deviations of the solution of each player are small.

C2. The number of known coalitions is small.

C3. The computation time is short.

For each case (from Case 1 to 7), let the evaluation of C1 be the maximal standard deviation among players of each case, that of C2 the number of unknown coalitions, and let the evaluation of C3 be the computation time of each case. Then the evaluations of each criterion are given as Table 10.

Although the evaluations of C1 for Case 5, Case 6 and Case 7 are very small, it will change dependent on coalitional worths which are given to define a PDG. That is, the difference between the worth of the grand coalition and the coalitional worths whose cardinalities are four is large in this example. Therefore, if all the criteria are of equal importance to obtain the Shapley value using the proposed method, it seems that Case 4 is reasonable.

Finally, the method to obtain random numbers which were used to obtain a superadditive PDG is described. Each unknown coalitional worth was obtained within the range from $\underline{v}$ to $\overline{v}$ using the random number for PDG v. In this method, if first random number which is given to some coalitional worth does not satisfy the superadditivity of the game, the number is discarded and the next random number is obtained. Then the game is checked to satisfy the superadditivity or not again. This procedure is continued.

However, in this simulation, the superadditive complete game was not be obtained indefinitely using the method above in particular Case 1 and 2. Then we have changed the method a little. The modified method is that if first random number does not satisfy the superadditivity of the game, the possible upper bound is changed to the upper bound—1. From this modification, the superadditive game was managed to be obtained. If the possible upper bound is decreased more than one at once, the superadditive game will be obtained faster. However, this may abandon the possible coalitional worth which satisfies the superadditivity.

5 Conclusion and Future Research

In this paper, we investigated the Shapley value for PDGs by computer simulations using the theory of the statistical estimation. First, the population of the Shapley values is assumed to follow the normal distribution although the mean and the standard deviation of the population are not known. For each unknown coalition, random numbers are given between the lower bound and the upper bound of each unknown coalitional worth.

Iterating this processing, the samples of the population of the Shapley value were taken. Finally, using the formula of the interval estimation of the mean of the population, we obtained the mean of the population of the Shapley value as the interval with 95% of the confidence coefficient.

Furthermore, to evaluate the wellness of the solution, we showed three criteria which show wellness of it and chose one solution when all the criteria are of equal importance.

As future research, we should investigate the results of simulations for general PDGs. In general PDGs, the standard deviations of players may be biased. This is one problem. Next, we should consider the efficient method to obtain random numbers which were used to obtain a superadditive game. This will lead to decreasing the time of computing the Shapley value. Finally, it will be interesting to consider other solution concepts such as the nucleolus and the core in this framework.

References

1. Albizuri, M.J., Masuya, S., Zarzuelo, J.M.: An extension of the shapley value for partially defined cooperative games. In: The 29th International Conference on Game Theory (2018)
2. Harsanyi, J.C.: A simplified bargaining model for the n-person cooperative game. Int. Econ. Rev. **4**, 194–220 (1963)
3. Housman, D.: Linear and symmetric allocation methods for partially defined cooperative games. Int. J. Game Theor. **30**, 377–404 (2001)
4. Masuya, S., Inuiguchi, M.: A fundamental study for partially defined cooperative games. Fuzzy Optim. Decis. Making **15**, 281–306 (2016)
5. Shapley, L.S.: A value for n-person games. In: Kuhn, H.W., Tucker, A.W. (eds.) Contributions to the Theory of Games 2, pp. 307–317. Princeton, Princeton University Press (1953)
6. Willson, S.J.: A value for partially defined cooperative games. Int. J. Game Theor. **21**, 371–384 (1993)

Applications in Energy and Business Sector

Techno-Economic Feasibility Study of a Hybrid Renewable Energy System for a Remote Rural Area of Karnataka, India

M. Ramesh and R. P. Saini

Abstract Global environmental issues combined with rapid developments in renewable energy technologies offer new possibilities for harnessing and using renewable energy resources in various ways. Since renewable energy resources are associated with the intermittency and irregularity, the system reliability can be enhanced by integrating different renewable energy resources together with a diesel generator. Renewable energy based system is designed for an un-electrified area located in India. Techno-economic viability analysis is carried out using HOMER Pro* software tool. The proposed study is intended to provide uninterrupted power supply. The available resources in the study area include biomass, solar, hydro, and wind. The peak demand for the area is estimated at 43.33 kW. PV/Hydro/Battery hybrid system is found to be the optimal feasible configuration. Further, a sensitivity study is carried out to assess the uncertainty in relation to the COE and NPC of the proposed hybrid system.

Keywords HRES · COE · NPC · Sensitivity analysis

1 Introduction

In remote rural communities, the need for energy-efficient and reliable energy supply is becoming a driving force for research about the stand-alone Hybrid Renewable Energy Systems (HRES) [1]. These HRESs are considered to be the most efficient and cost-effective way of electrifying off-grid communities, where extension of utility grid is not possible [2]. Renewable energy resources, being intermittent and irregular, do not follow load demand. Therefore, batteries are required to avoid wasting excess energy [3].

M. Ramesh (✉) · R. P. Saini
Indian Institute of Technology Roorkee, Roorkee, India
e-mail: mramesh@ah.iitr.ac.in

R. P. Saini
e-mail: saini.rajeshwer@gmail.com

M. Sahni et al. (eds.), *Mathematical Modeling, Computational Intelligence Techniques and Renewable Energy*, Advances in Intelligent Systems and Computing 1287,
https://doi.org/10.1007/978-981-15-9953-8_30

In addition, diesel generators are used in such systems when there is a shortage of renewable energy while meeting peak load requirements [4]. Das and Zaman [5] proposed PV/DG HRES for a rural area located in the rural region of Bangladesh using HOMER Pro. Azimoh et al. [6] presented a study on optimization of PV/WT/Hydro/DG/BT HRES for rural electrification. Sawleet al. [7] modeled the HRES with six different objectives to evaluate the system performance. Razmjoo et al. [8] proposed a study related to techno-economic aspects of PV/wind/DG HRES of an area located in the remote.

Olatomiwa et al. [9] conducted a techno-economic exercise on the HRES for health care facilities in a rural area. Duman and Guler [10] presented a study on a PV LED HRES for off-grid street lighting. Fodhil et al. [11] discussed performance and sensitivity analysis of HRES with the help of HOMER tool. Baseer et al. [12] presented a viability study of a HRES for a residential community. Javed et al. [13] presented a study of PV/Wind/BT hybrid system designed to operate in a remote area. In this study, a performance study of an off-grid HRES is presented for a cluster of un-electrified villages located in the remote regions of Chikmagalur district of Karnataka state, which is situated in India. This is accomplished with the help of Hybrid Optimization of Multiple Energy Resources (HOMER Pro*).

2 Demand and Resource Assessment

In this section, details of study area estimation of load demand and resource assessment for the proposed area has been presented as follows:

2.1 Study Area

Bannur, Aldur, and Donagudige villages of Chikmagalur district of Karnataka are located in the south of India, in Malnad region at 75.45° east longitude and 13.35° north latitude. Nearly 61 households with 273 populations are scattered in the dense forest, hence they are un-electrified.

2.2 Estimated Load Demand

The proposed study area has 427.44 kWh/day as the average daily load demand. The values of average load demand and peak load demand are 17.81 kW and 43.33 kW, respectively. The hourly load data of three seasons viz: summer, rainy, and winter is given in Fig. 1.

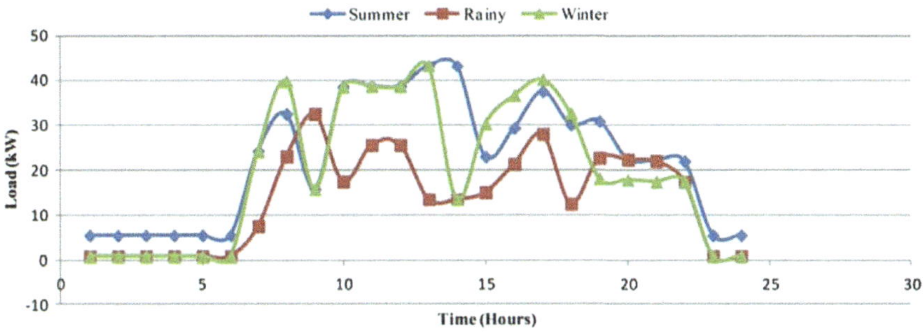

Fig. 1 Season wise daily load curves

2.3 Resource Assessment

The study area has a significant amount of potential for solar radiation. The annual average PV potential of the proposed location is 5.5 kWh/m^2/day. The solar insolation data along with the clearness index of solar potential is given in Fig. 2. The study area has an average wind speed of 2.83 m/s. Months of February and May receive maximum and minimum wind speeds respectively as shown in Fig. 3. The available biomass is about 0.001139 tons/day. The monthly rainfall is depicted in Fig. 4.

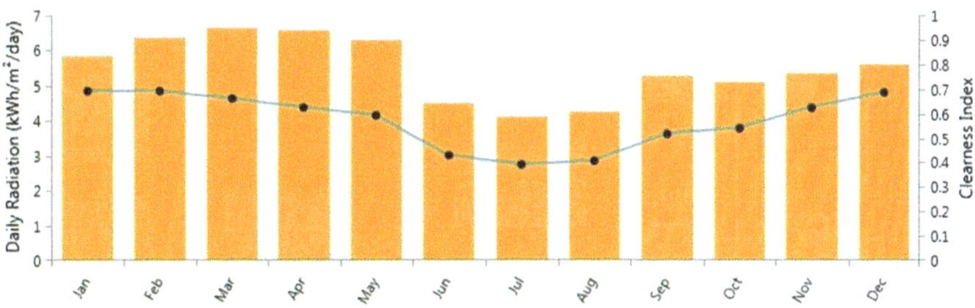

Fig. 2 Solar insolation data

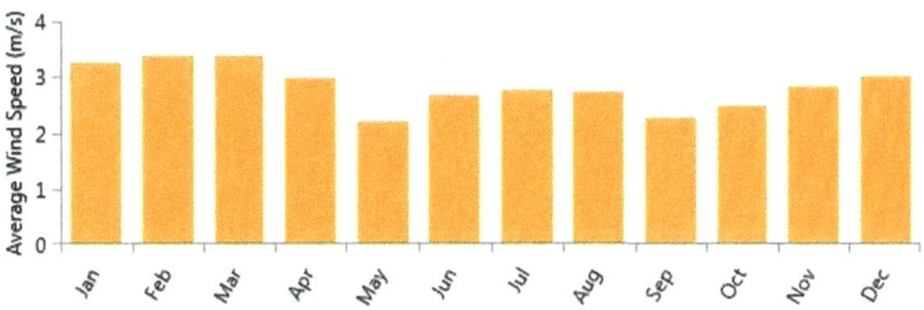

Fig. 3 Month wise average wind speed

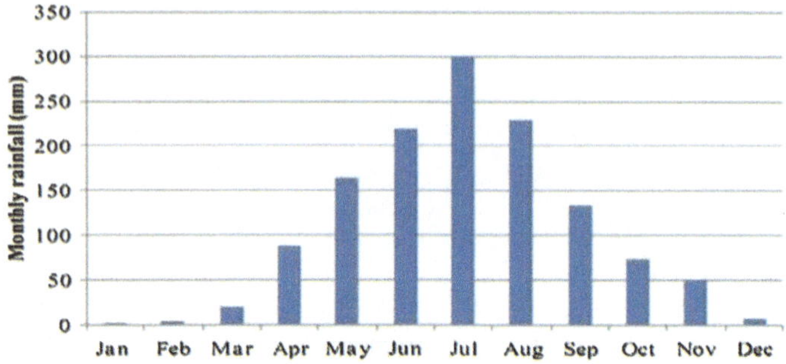

Fig. 4 Month wise rainfall received at the study area

3 Modeling and Cost of the HRES Components

The proposed HRES has been modeled using HOMER software tool. Figure 5a shows the configuration of the HRES components developed by the HOMER. Figure 5b shows the configurations of a HRES which consists of solar panel, wind turbine, hydro power biomass gasifier, diesel generator, battery storage, and bidirectional converter. Table 1 shows the several techno-economic factors used in the HRES such as initial capital cost (CC), replacement cost (RC), and O&M cost (O&M).

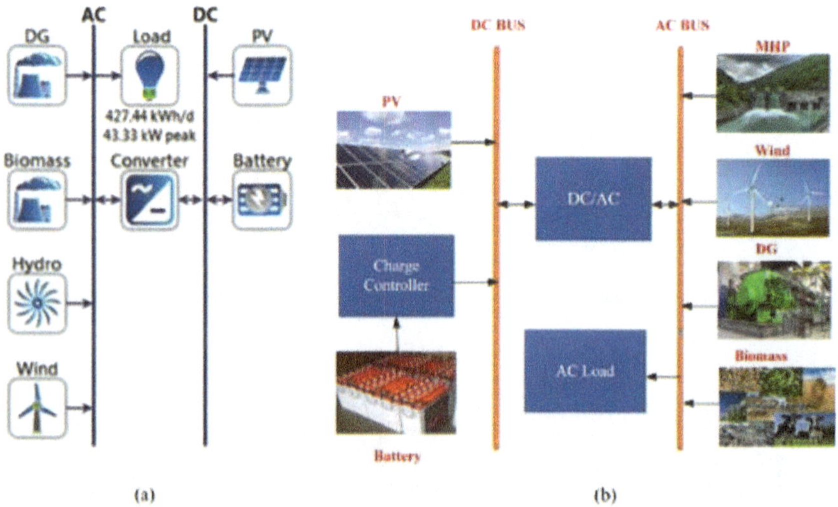

Fig. 5 Configuration of HRES: **a** HOMER Pro. **b** Schematic diagram

Table 1 Techno-economic parameters used in the HRES

Component	CC	RC	O&M	Life span	Other specifications
PV Panel (1 kW) [14]	630	0	10	25 Y	Derating factor—80%
Wind (1 kW) [14]	300	300	20	20 Y	Hub height—50 m
Hydro (1 kW) [6]	1300	0	100	30 Y	Pipe load loss—2%
Biomass (1 kW) [7]	300	300	2	15,000 h	Min. load ratio—25%
DG (1 kW) [2]	220	200	0.03	15,000 h	Min. load ratio—25%
Battery (1 kWh) [5]	167	67	1.67	10 Y	Throughput—800
Bi-directional converter (1 kW) [14]	300	300	3	15 Y	Efficiency—95%

4 Results and Discussion

Optimization of the proposed HRES is carried out by different models by using HOMER Pro*. Configurations have been considered on the basis of economic, technical, and environmental aspects which are summarized and presented in Table 2. Configurations and cost parameters of the proposed HRES are discussed as follows:

4.1 Optimal Feasible Configurations

Out of the feasible optimal configurations, six different HRES configurations are presented in Fig. 6 which include PV/Hydro/BT (Case1), PV/Wind/Hydro/BT (Case 2), PV/DG/Hydro/BT (Case 3), PV/Wind/DG/Hydro/BT (Case 4), PV/DG/BM/Hydro/BT (Case 5) and PV/Wind/DG/BM/Hydro/BT (Case 6). Further, out of six configurations considered the sharing of power demand by the most

Table 2 Summary of performance results of HRES

Description	Case-1	Case-2	Case-3	Case-4	Case-5	Case-6
PV panel (kW)	138	138	123	123	120	119
Wind (nos)	–	2	–	1	–	1
Hydro (kW)	8.83	8.83	8.83	8.83	8.83	8.83
Biomass (kW)	–	–	–	–	5	5
DG (kW)	–	–	48	48	48	48
Battery (nos)	254	254	278	278	278	277
Converter (kW)	43.3	43	47.7	47.7	47.7	47.7

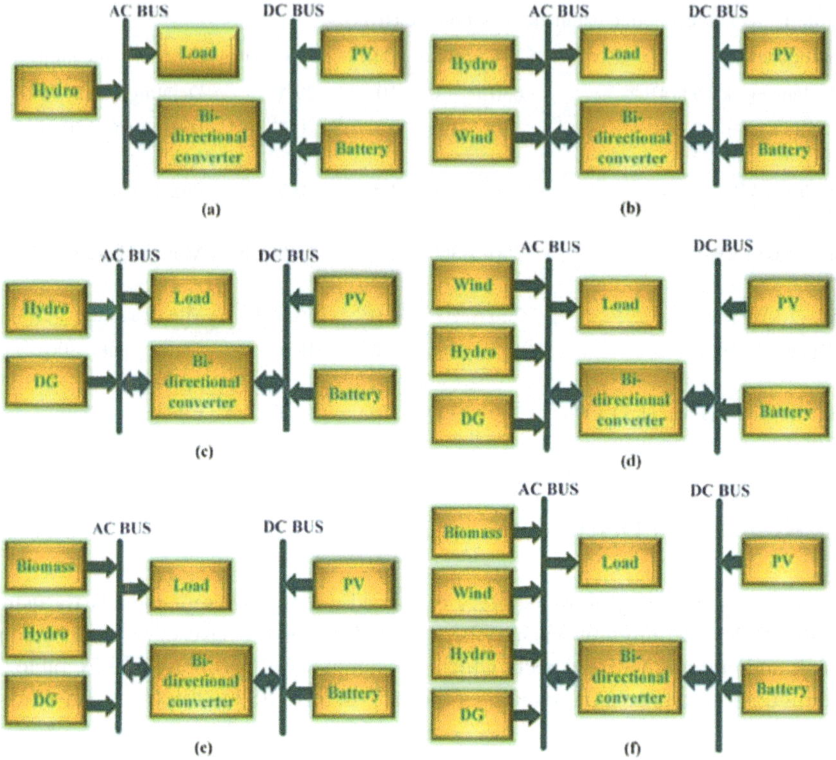

Fig. 6 Optimal feasible energy system configurations **a** PV/Hydro/BT, **b** PV/Wind/Hydro/BT, **c** PV/DG/Hydro/BT, **d** PV/Wind/DG/Hydro/BT, **e** PV/DG/Biomass/Hydro/BT, **f** PV/Wind/DG/Biomass/Hydro/BT

optimal system components PV/Hydro/BT system is 13.6 kW, 1.77 kW, and 35 kWh, respectively.

4.2 *Optimal Operating Costs*

The cost comparisons for different configurations are shown in Fig. 7. Out of these configurations of the HRES, the most feasible system is selected based on operational costs. From Fig. 8, it is seen that the optimal COE and NPC are obtained as 0.133 $/kWh and $23,060 respectively. The initial cost, replacement cost, and O&M cost are $17,227, $2866.44, and $546.34, respectively. Since no DG is connected in this configuration, no fuel cost is required.

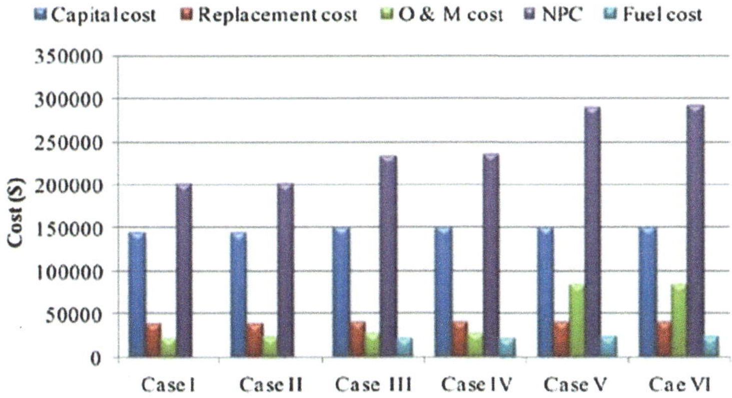

Fig. 7 Cost comparisons among different configurations

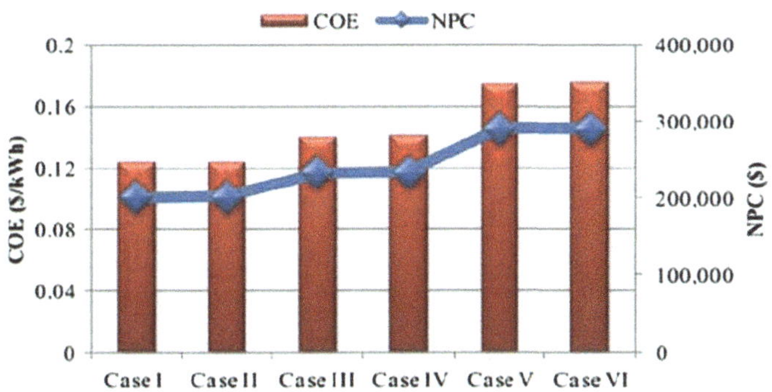

Fig. 8 Comparison between COE and NPC

4.3 *Sensitivity Analysis*

A sensitivity study is exercised to assess the uncertainty of the PV/Wind/Hydro/DG/Biomass/BT HRES with respect to COE and NPC. The parameters considered for this exercise include discount rate, wind speed, design flow rate, component cost such as PV panel, battery and the fuel cost ±20% of the input parameter value has been varied. From Fig. 9, it is seen that there is no effect on COE by varying wind speed since its capacity is very small (1 kW) as compared to that of the hybrid system. The discount rate and cost of PV and battery variability affect the HRES economics significantly. Discharge of water, fuel cost, and wind speed variations does not have a significant effect on the unit cost of energy and also NPC as shown in Fig. 10. Variation of ±20% in the cost of PV panel and storage lead to variation in the value of NPC by ±6%. Whereas, the discount rate has more effect than the PV cost and battery cost.

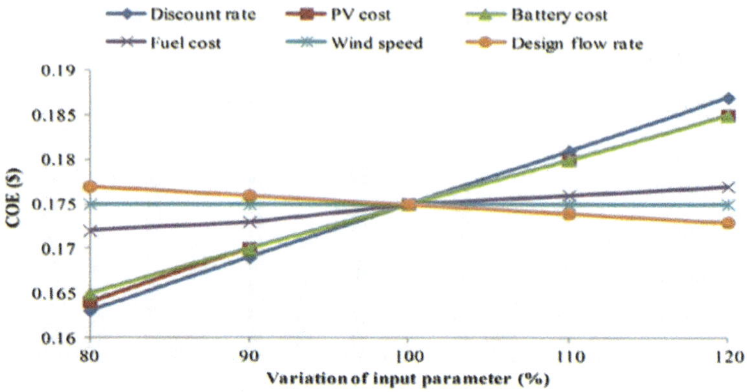

Fig. 9 Effect of various input parameters on COE

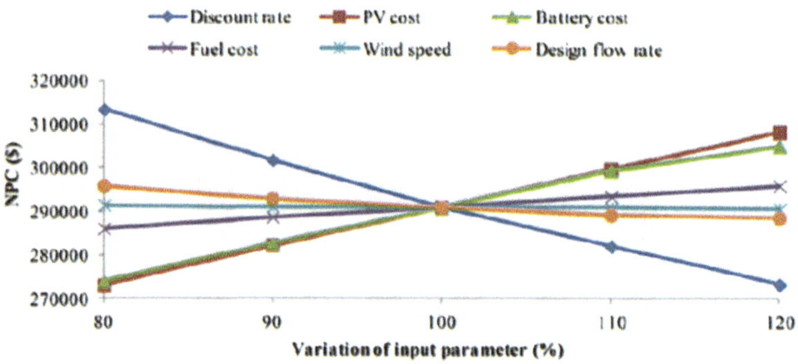

Fig. 10 Effect of variation in input parameters on NPC

5 Conclusions

This study has been intended to assess the performance of PV/Wind/Hydro/Biomass/DG/Battery based hybrid system for a cluster of villages located in the Chikmagalur district in Karnataka state, India. To evaluate the HRES performance, six different feasible configurations have been considered. Out of these configurations, PV/Hydro/Battery HRES offers the most optimal feasible configuration. The operational costs such as the COE and NPC are obtained as 0.123 $/kWh and $200,897 respectively. In addition, a sensitivity study is evaluated to predict the HRES performance on varying variable input parameters and it is found that HRES is significantly affected by the variability in the component cost of PV panel and storage and also by the discount rate.

References

1. Ahmadi, S., Abdi, S.: Application of the hybrid big bang-big crunch algorithm for optimal sizing of a stand-alone hybrid PV/wind/battery system. Sol. Energy **134**, 366–374 (2016)
2. Hossain, M., Mekhilef, S., Olatomiwa, L.: Performance evaluation of a stand-alone PV-wind-diesel-battery hybrid system feasible for a large resort center in South China Sea. Sustain. Cities Soc. **28**, 358–366 (2017)
3. Dagdougui, H., Minciardi, R., Ouammi, A., Robba, M., Sacile, R.: A dynamic decision model for the real-time control of hybrid renewable energy production systems. IEEE Syst. J. **4**, 323–333 (2010)
4. Yousefi, M., Kim, J.H., Hooshyar, D., Yousefi, M., Sahari, K.S.M., Ahmad, R.B.: A practical multi-objective design approach for optimum exhaust heat recovery from hybrid stand-alone PV-diesel power systems. Energy Convers. Manag. **142**, 559–573 (2017)
5. Das, B.K., Zaman, F.: Performance analysis of a PV/diesel hybrid system for a remote area in Bangladesh: effects of dispatch strategies, batteries. Energy Generator Sel. Energy **169**, 263–276 (2018)
6. Leonard, C., Klintenberg, P., Wallin, F., Karlsson, B., Mbohwa, C.: Electricity for development: mini-grid solution for rural electrification in South Africa. Energy Convers. Manag. **110**, 268–277 (2016)
7. Sawle, Y., Gupta, S.C., Bohre, A.K.: Socio-techno-economic design of hybrid renewable energy system using optimization techniques. Renew. Energy **119**, 459–472 (2018)
8. Razmjoo, A., Shirmohammadi, R., Davarpanah, A., Pourfayaz, F.: Stand-alone hybrid energy systems for remote area power generation. Energy Reports **5**, 231–241 (2019)
9. Olatomiwa, L., Mekhilef, S., Huda, A.S.N., Ohunakin, O.S.: Economic evaluation of hybrid energy systems for rural electrification in six geo-political zones of Nigeria. Renew. Energy **83**, 435–446 (2015)
10. Duman, A.C., Güler, Ö.: Techno-economic analysis of off -grid photovoltaic LED road lighting systems: a case study for northern, central and southern regions of Turkey. Build. Environ. **156**, 89–98 (2019)
11. Fodhil, F., Hamidat, A., Nadjemi, O.: Potential, optimization and sensitivity analysis of photovoltaic-diesel-battery hybrid energy system for rural electrification in Algeria. Energy **169**, 613–624 (2019)
12. Baseer, M.A., Alqahtani, A., Rehman, S.: Techno-economic design and evaluation of hybrid energy systems for residential communities: case study of Jubail industrial city. J. Clean. Prod. **237**, 117806 (2019)
13. Javed, M.S., Song, A., Ma, T.: Techno-economic assessment of a stand-alone hybrid solar-wind-battery system for a remote island using genetic algorithm. Energy **176**, 704–717 (2019)
14. Adefarati, T., Bansal, R.C.: Reliability, economic and environmental analysis of a microgrid system in the presence of renewable energy resources. Appl. Energy **236**, 1089–1114 (2019)

LPF-BPF Fundamental Current Extractor Based Shunt Active Filtering with Grid Tied PV System

Arpitkumar J. Patel and Amit V. Sant

Abstract Grid tied photovoltaic (PV) systems comprise of a voltage source inverter interfacing PV panels with the grid. These PV systems, which are redundant at night times, can be employed for shunt active power filtering during night times. Thus, harmonic current mitigation and reactive power compensation can be achieved during night times without effecting any change in power structure. For shunt active filtering, fundamental active current estimator is highly essential. This paper proposes a fundamental active current estimator for 3-phase shunt active filter (SAF) operation of grid tied PV system during night times. The estimation algorithm comprises of a low pass filter (LPF) and a band pass filter (BPF). The 3-phase load currents are transformed to stationary reference frame (SRF) and then 2-phase currents are individually processed through the proposed estimation algorithm for extracting the peak value of fundamental active components of load current. Further, with the help of unit vector templates (UVT), the instantaneous fundamental active components of the 2-phase quantities are determined, which can be transformed into 3-phase quantities with inverse Clarke transformation. Drawing of these currents from the grid can result in mitigation of current harmonics in grid. Moreover, as the phase angle of these currents drawn from the grid corresponds to UVT, the reactive power compensation is ensured. Hence, unity power factor operation is observed at the point of common coupling. The use of Clarke transformation results in reduced computational burden as only two quantities, instead of three, are to be processed. From the simulation studies, it is clear that the operation of 3-phase grid tied PV system as SAF, with the reported estimation algorithm, during night time results in reactive power compensation and elimination of harmonics from the grid currents.

Keywords Power quality · Grid tied photovoltaic system · Shunt active filter

---

A. J. Patel (✉) · A. V. Sant
School of Technology, Pandit Deendayal Petroleum University, Gandhinagar, Gujarat, India
e-mail: arpitk.pmtee18@sot.pdpu.ac.in

M. Sahni et al. (eds.), *Mathematical Modeling, Computational Intelligence Techniques and Renewable Energy*, Advances in Intelligent Systems and Computing 1287,
https://doi.org/10.1007/978-981-15-9953-8_31

1 Introduction

The demand for electrical energy is growing by leaps and bounds. This combined with the energy security and environmental concerns, and exhaustion of conventional energy sources has led to increased attention towards solar energy sector [1, 2]. Moreover, with the rapid progressions in photovoltaic (PV) and power electronics technology, the open and plentifully accessible solar energy is progressively finding more usage in electric power generation [2, 3]. From the point of view of reliability of electric supply and no need for battery back-up, the grid tied PV systems are largely preferred [4]. For grid synchronization and to control power flow from PV panel into the grid, power converters are employed to ensure regulated power can be delivered at day time. A number of literature have reported different configurations and control of grid tied PV system [2, 5–7]. Throughout the day time, based on the irradiance and temperature levels, the grid interfaced PV system transfers electric power of varying magnitudes to the grid. However, at night times the PV system cannot generate and deliver active power into the grid. This is a major limitation for this system.

The power structure of a grid-connected PV system consists of PV array, acting as source, interfaced with the grid by means of voltage source inverter (VSI) with capacitor connected at *dc*-link and power filter. Except for the PV array, the power structure of a shunt active filter (SAF) is the same as that of the PV system interfaced with the grid. Li et al have proposed a PV system connected with the grid that injects active power into the grid during daytimes, and acts as a SAF by mitigating current harmonics and compensating for the reactive power at night times [8]. Similarly, a grid tied PV system that provides quality enhancement at low irradiation levels is reported in [9].

With the increasing nonlinear loads, the current harmonics and resulting power quality issues are a major concern. Current harmonics result in reduced equipment efficiency, failure of control and protection system, increased energy bills, network congestion, poor voltage regulation, and increased line losses. SAF is a custom power device that caters to the mitigation of power quality issues arising from distorted load currents and reactive power demand [3, 10–12]. The use of SAF ensures the currents drawn from the grid are sinusoidal while maintaining a unity power factor (UPF). [3, 10–12]. Due to this, the issues of power quality degradation on account of current harmonics and poor power factor are eliminated.

For the control of SAF, fundamental active current estimation is of prime significance. Gohil and Sant [13] proposes low pass and band pass filter based fundamental current estimation algorithm for a single-phase distributed static compensator. However, this technique is not explored for 3-phase SAF. This paper reports low pass filter (LPF) and band pass filter (BPF) based algorithm for fundamental current estimation for a 3-phase grid tied PV system for implementing shunt active filtering during night times. In this algorithm, the 3-phase load currents undergo Clarke transformation to determine the 2-phase sinusoidal quantities in stationary reference frame. Thus derived quantities in stationary reference frame are separately processed through LPF-BPF extractor for the determination of individual fundamental in-phase and quadrature components. The peak amplitude of the fundamental component of load current is determined based on the obtained in-phase and quadrature components. Using the unit vector templates (UVT), the instantaneous fundamental component of load currents is determined in stationary reference frame. The two fundamental components in stationary reference frame are converted into 3-phase reference source currents. If the source currents match these estimated reference values, then source currents will be sinusoidal and in phase with the respective grid voltages resulting in UPF operation of the load. Thus, shunt active filtering can be implemented with the grid connected PV systems at the night time without the need for any change in the power structure.

2 Shunt Active Filtering with Grid Tied PV System

Figure 1 shows the schematic diagram of a 3-phase grid connected PV system, where v_{ga}-v_{gb}-v_{gc} are the grid voltages, L_s and R_s are the inductance and resistance of the line, i_{ga}-i_{gb}-i_{gc} are the grid currents, i_{la}-i_{lb}-i_{lc} are the currents drawn by the load, i_{fia}-i_{fib}-i_{fic} are the currents injected by the grid connected VSI, C_{dc} is the dc-link capacitor and *vdc* is the dc-link voltage. The grid connected PV system is made up of PV array feeding active power into the grid at the point of common coupling in controlled manner with the help of VSI. *Cdc* is connected across the dc terminals of the VSI, whereas coupling inductors are connected between the 3-phase output terminals of VSI and grid. At day time, switch *sw* is closed and VSI regulates active power flow from PV array to the grid. The PV array operation at maximum power point is also ensured by the VSI. During night time, when the PV array no longer generates electric power, *sw* is opened and VSI implements shunt active filtering. When the VSI is operated as SAF, *dc*-link voltage is regulated with the *dc* voltage control loop. SAF caters to the harmonic current components and reactive power demand of the load, thus necessitating the grid to supply only the fundamental active current. For control of SAF, fundamental active current estimator and phase-locked loop (PLL) are integral elements, which essentially determine the fundamental active component of load current. With the help of thus determined fundamental active component of load current, the combined harmonic and reactive components of load current are computed which acts as reference value of currents injected by the SAF for power quality enhancement. Often, hysteresis current controller is employed to generate the gate pulses for VSI.

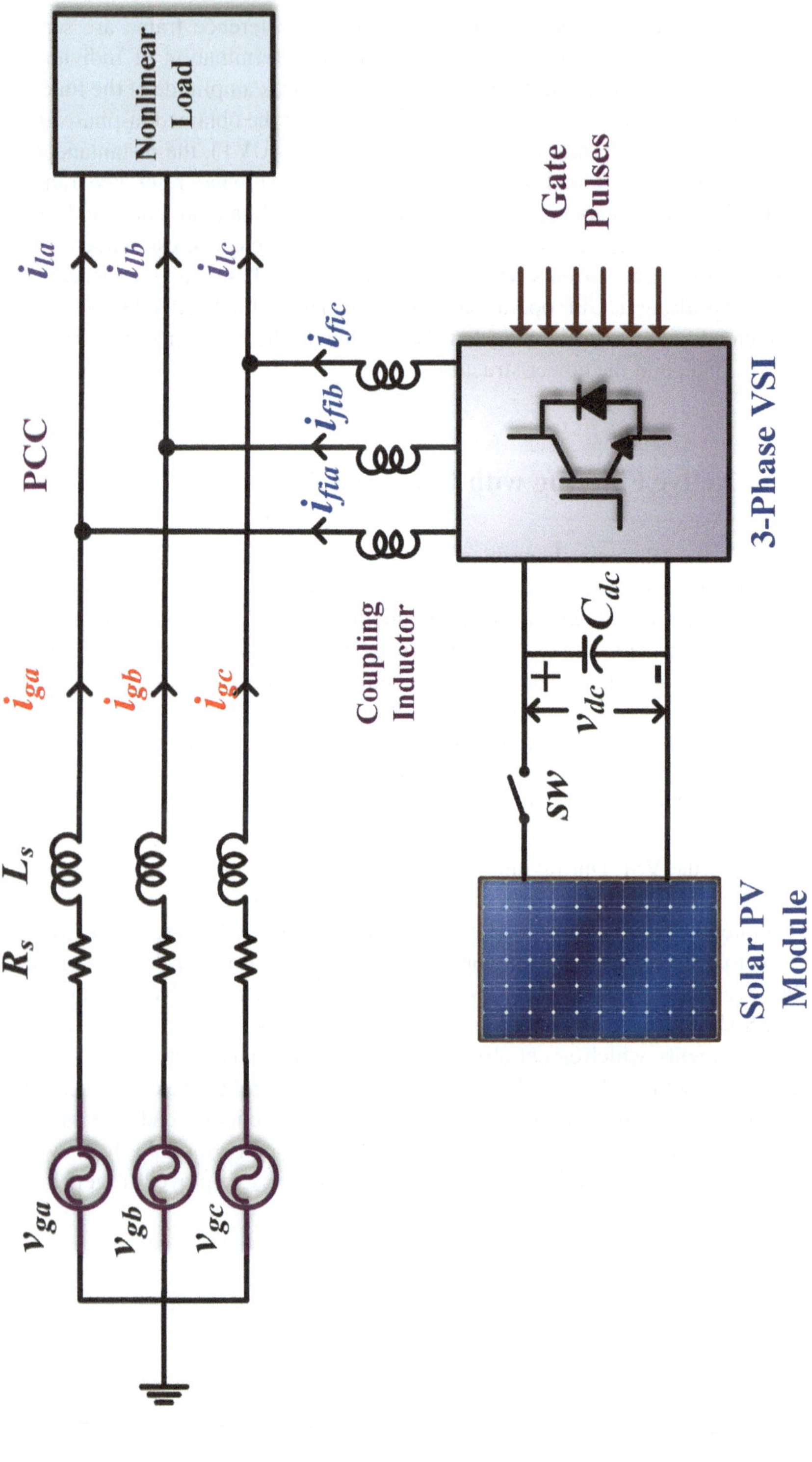

Fig. 1 Block diagram of a 3-phase grid tied PV system with SAF capability

3 3-Phase LPF-BPF Fundamental Active Current Estimator

3.1 1-Phase LPF-BPF Fundamental Active Current Estimator

The schematic diagram of fundamental active current estimator using LPF-BPF algorithm reported in [13] for a 1-phase system, is shown in Fig. 2. The load current, i_{lx}, where x represents phase a, b or c; is processed by the BPF to generate the in-phase fundamental component, i_{lxf}, represented in S-domain as in (1), [13], where $G_1(s)$ is the transfer function of the II order BPF, ω_c is the centre frequency, and P is the pass band. In the next step, the fundamental quadrature component, i_{lxfq}, of i_{lxf} is determined by processing it through an LPF. Equation (2) represents i_{lxfq} in S-domain, [13], where k is the gain, τ is the coefficient of damping and ω_o is the cut-off frequency. In Eqs. (1) and (2), it is to be noted that $\omega_o = \omega_c = 2\pi \times f$, where, f is the grid frequency, and $k = 1/\sqrt{2}$.

$$I_{lxf}(s) = G_1(s) I_{lx}(s) = \left(\frac{P \cdot s}{s^2 + B \cdot s + \omega_c^2} \right) I_{lx}(s) \tag{1}$$

$$I_{lxfq}(s) = G_2(s) I_{lxf}(s) = \left(\frac{k \cdot \omega_o^2}{s^2 + 2\tau \cdot \omega_o \cdot s + \omega_o^2} \right) I_{lxf}(s) \tag{2}$$

$i_{lxf}(t)$ is a sinusoidal quantity with frequency equal to f, while $i_{lxfq}(t)$ is a co-sinusoidal quantity with frequency equal to f. Also, the peak magnitudes of $i_{lxf}(t)$ and $i_{lxfq}(t)$ are equal. The peak magnitude, I_{lxp}, can be determined as in Eq. (3). Lastly, the product of I_{lxp} and the UVT provides the reference value of source current.

$$I_{lxp} = \sqrt{i_{lxf}(t)^2 + i_{lxfq}(t)^2} \tag{3}$$

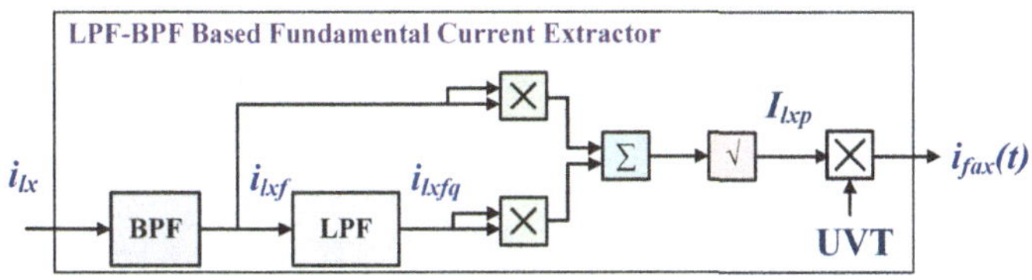

Fig. 2 Schematic diagram representation of LPF-BPF based extractor

3.2 3-Phase LPF-BPF Fundamental Active Current Extractor Without Clarke Transformation

The schematic diagram representation of the reported 3-phase LPF-BPF fundamental active current extractor without Clarke transformation is shown in Fig. 3. i_{la}-i_{lb}-i_{lc} are independently processed by the LPF-BPF fundamental active current extractor to generate the respective fundamental in-phase and quadrature components of the 3-phase currents. The fundamental peak amplitudes of i_{la}-i_{lb}-i_{lc} are determined as I_{lpa}-I_{lpb}-I_{lpc} and the average of I_{lpa}-I_{lpb}-I_{lpc}, $I_{lp\mathrm{AVG}}$ is determined as

$$I_{lp\mathrm{AVG}} = \left(I_{lpa} + I_{lpb} + I_{lpc}\right)/3 \tag{4}$$

The instantaneous reference source currents for *a*-*b*-*c* phases, $i_{sar}(t)$-$i_{sbr}(t)$-$i_{scr}(t)$ is given as

$$i_{sar}(t) = I_{lp\mathrm{AVG}} \times \sin(\theta_a) = I_{lp\mathrm{AVG}} \times \sin(2\pi f t) \tag{5}$$

$$i_{sbr}(t) = I_{lp\mathrm{AVG}} \times \sin(\theta_b) = I_{lp\mathrm{AVG}} \times \sin\left(2\pi f t - 120^{\circ}\right) \tag{6}$$

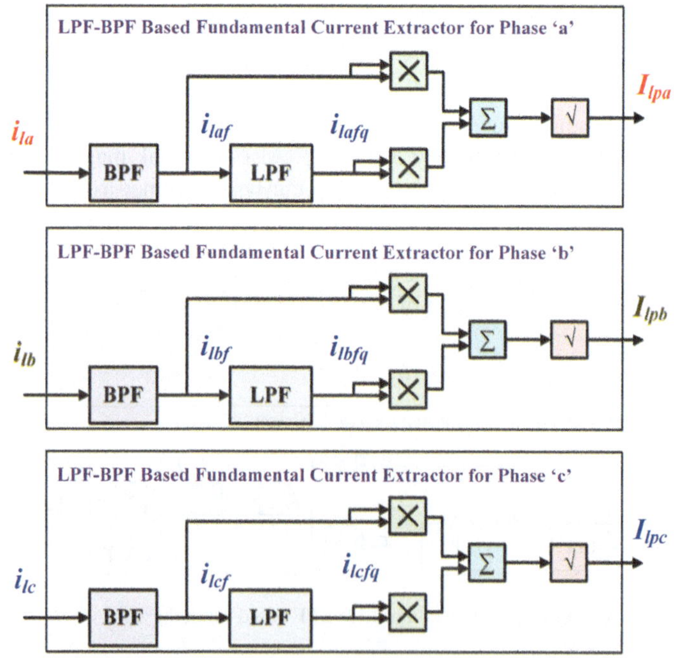

Fig. 3 Schematic diagram representation of LPF-BPF-based extractor for 3-phase SAF without Clarke transformation

$$i_{scr}(t) = I_{lp\mathrm{AVG}} \times \sin(\theta_c) = I_{lp\mathrm{AVG}} \times \sin(2\pi f t - 240^\circ) \tag{7}$$

where, $\sin(\theta_a) = \sin(\omega t)$, $\sin(\theta_b) = \sin(\omega t - 120^\circ)$, and $\sin(\theta_c) = \sin(\omega t - 240^\circ)$ are the UVT for phase *a-b-c* derived from the grid voltages, v_{ga}-v_{gb}-v_{gc}, with the help of PLL. The only demerit is that as a part of fundamental active current extraction, three quantities need to be processed by the LPF-BPF algorithm. This adds to the computational complexity. In order to reduce the computational burden Clarke transformation is utilized. This computationally simpler scheme is explained in the next subsection.

3.3 3-Phase LPF-BPF Fundamental Active Current Extractor with Clarke Transformation

The schematic diagram representation of the proposed 3-phase LPF-BPF fundamental active current extractor with Clarke transformation is shown in Fig. 4. With the incorporation of this transformation, i_{la}-i_{lb}-i_{lc} are resolved into 2-phase quantities $i_{l\alpha}-i_{l\beta}$, as shown in Eq. (8), where K is Clarke transformation matrix. i_{s0r} is the zero sequence component and it is zero for balanced supply and balanced loading. $i_{l\alpha}-i_{l\beta}$ are independently processed by the LPF-BPF fundamental active current extractor.

$$\begin{bmatrix} i_{l\alpha} \\ i_{l\beta} \\ i_{s0r} \end{bmatrix} = [K] \begin{bmatrix} i_{la} \\ i_{lb} \\ i_{lc} \end{bmatrix} \tag{8}$$

Thus, with Clarke transformation instead of three only two quantities, namely $i_{l\alpha}-i_{l\beta}$, are required to be processed by LPF-BPF algorithm for the generation of two sets of in-phase and quadrate fundamental components $i_{l\alpha f}-i_{l\alpha fq}$ and $i_{l\beta f}-i_{l\beta fq}$, respectively. The peak amplitude of fundamental component of $i_{l\alpha}-i_{l\beta}$, $I_{l\alpha p}-I_{l\beta p}$, are determined as

$$I_{l\alpha p} = \sqrt{i_{l\alpha f}(t)^2 + i_{l\alpha fq}(t)^2} \tag{9}$$

$$I_{l\beta p} = \sqrt{i_{l\beta f}(t)^2 + i_{l\beta fq}(t)^2} \tag{10}$$

With the help of the $I_{l\alpha p}-I_{l\beta p}$, and UVT, the instantaneous reference source current in $\alpha-\beta$ reference frame, $i_{s\alpha r}(t)-i_{s\beta r}(t)$, are determined as

$$i_{s\alpha r}(t) = I_{l\alpha p} \times \sin(\theta_a) = I_{l\alpha p} \times \sin(2\pi f t) \tag{11}$$

$$i_{s\beta r}(t) = I_{l\beta p} \times \sin(\theta_b) = I_{l\beta p} \times \sin(2\pi f t + 90^\circ) \tag{12}$$

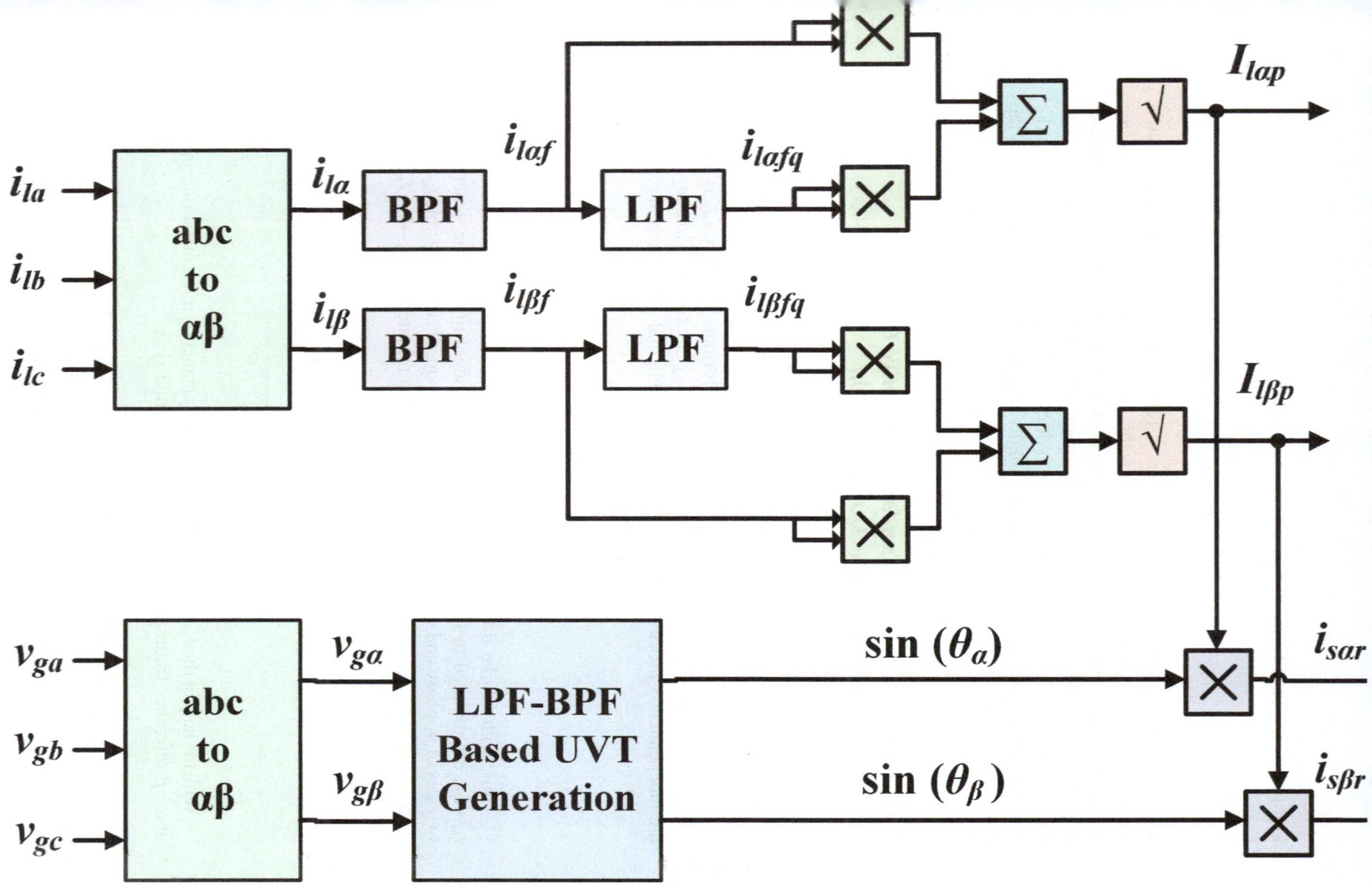

... LPF-BPF based extractor for 3-phase SAF with Clarke transformation

Further, the instantaneous reference source currents for *a-b-c* phases, $i_{sar}(t)$-$i_{sbr}(t)$-$i_{scr}(t)$ is given as in Eq. (13).

$$\begin{bmatrix} i_{sar} \\ i_{sbr} \\ i_{scr} \end{bmatrix} = [K]^{-1} \begin{bmatrix} i_{s\alpha r} \\ i_{s\beta r} \\ i_{s0r} \end{bmatrix} \tag{13}$$

4 Results

The grid tied PV system being controlled for 3-phase shunt active filtering, with the help of LPF-BPF estimator during night times is modeled in MATLAB/SIMULINK software. The performance of grid tied PV system employing 3-phase SAF, controlled using the proposed LPF-BPF based fundamental current estimator with Clarke transform, during the night times is evaluated for steady-state operation under sinusoidal grid voltages and nonlinear load as well as inductive load. Likewise, the dynamic performance evaluation of the PV system interfaced with the grid employing shunt active filtering with proposed algorithm under dark conditions is carried out by increasing the linear load. The 3-phase diode bridge rectifier (DBR) feeding *R–L* load of $(50 + j0.2\omega)\ \Omega$ is connected at a PCC as a nonlinear load. In parallel to DBR, 3-phase *R–L* load having $R = 50\ \Omega$ and $L = 200$ mH is connected. The switch, *sw* connecting grid tied inverter with PV array is kept open for operating PV inverter as a SAF. The values of coupling inductors are 10 mH for all the three phases. The reference value of the *dc*-link voltage, v_{dc}, is 200 V.

4.1 Case—A: Steady-State Performance with Sinusoidal Grid Voltages

Figure 5a reveals the steady-state operation of the grid tied PV inverter employing shunt active filtering using the proposed LPF-BPF based fundamental current extractor using Clarke transform. The balanced and non-distorted source voltage has as RMS value 70.7 V. The grid is feeding 3-phase nonlinear as well as inductive load. The peak amplitude of the load currents, i_{la}-i_{lb}-i_{lc} are 4.297 A, 4.292 A, and 4.287 A, respectively. The total harmonic distortion (THD) of the load currents, i_{la}-i_{lb}-i_{lc} are 11.16%, 11.21%, and 10.82%, respectively.

5th and 7th are the dominant harmonics present in currents drawn by 3-phase load. The peak amplitude of the 5th harmonic in load currents are 0.43 A, 0.42 A, and 0.42 A, respectively. The determined peak amplitude of the 3-phase load currents, $I_{l\alpha p}$ and $I_{l\beta p}$ are 4.364 A and 4.21 A in α-β coordinates. The PV inverter currents, i_{fia}, i_{fib}, and i_{fic} have the peak values 2.76 A, 2.73 A, and 2.71 A. The 3-phase currents

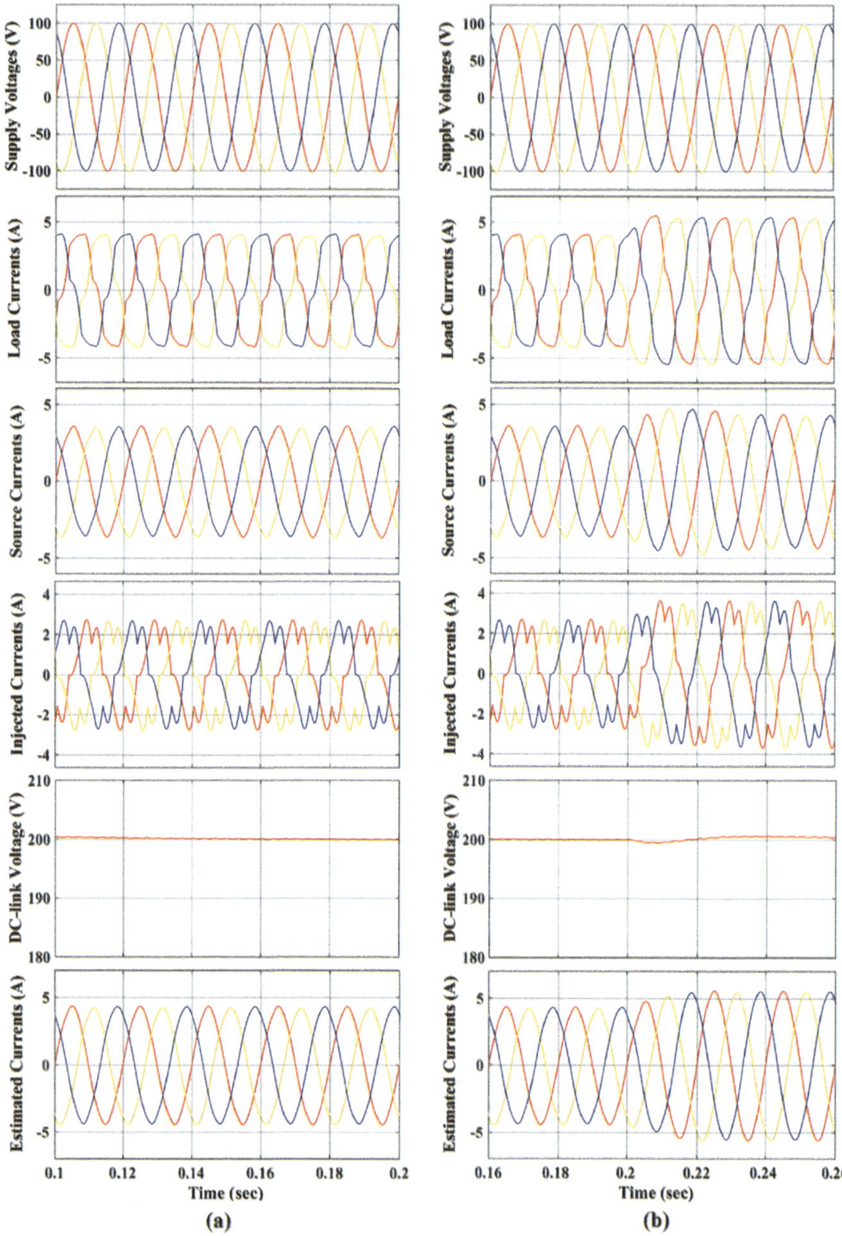

Fig. 5 **a** Steady-state performance of grid tied PV inverter for shunt active filtering. **b** Dynamic performance of grid tied PV inverter for shunt active filtering under load change

drawn from the grid, i_{ga}, i_{gb}, and i_{gc} are sinusoidal having the peak amplitude 3.505 A, 3.52 A, and 3.47 A with the THD values 2.22%, 2.21%, and 2.21%. Moreover, the power factor at the PCC is maintained at unity.

4.2 Case—B: Dynamic Performance Under Increased Loading Condition

Figure 5b depicts the waveforms for the dynamic performance evaluation of the grid integrated PV inverter for the shunt active filtering. The 3-phase inductive load having $R = 50\ \Omega$ and $L = 200$ mH is connected with the grid at the time, t =0.2 s for dynamic performance analysis. The peak amplitude of the 3-phase load currents, i_{la}-i_{lb}-i_{lc} are increased to 5.36 A, 5.39 A, and 5.37 A, respectively. The THD values of the 3-phase load currents, i_{la}-i_{lb}-i_{lc} are 8.71%, 8.68%, and 8.73%, respectively. The determined peak amplitude of the 3-phase load currents, $I_{l\alpha p}$ and $I_{l\beta p}$ are 4.364 A and 5.417 A in α-β coordinates. The PV inverter currents, i_{fia}, i_{fib}, and i_{fic} have the peak values 3.418 A, 3.421 A, and 3.420 A. The peak values of the inverter currents are increased as reactive power demanded by the load is increased. The active current components drawn from the source, i_{ga}, i_{gb}, and i_{gc} are increased to supply active power to the load. The source currents, i_{ga}, i_{gb}, and i_{gc} are sinusoidal having peak amplitudes 4.302 A, 4.292 A, and 4.287 A with the THD values 1.43%, 1.43% and 1.43%. The control of PV inverter for shunt active filtering has resulted into UPF at the supply side.

5 Conclusion

The control algorithm, based on LPF-BPF fundamental current extractor is developed for the grid tied PV system employing shunt active filtering at night times. The proposed algorithm is analyzed using MATLAB/SIMULINK for accurate estimation and fast response. As grid tied PV inverter is being idle at night time, it is used as a SAF for power quality enhancement. Fundamental current extractor is an essential element of the control structure of SAF. The LPF-BPF based fundamental active current extractor for 3-phase using Clarke transformation has resulted into reduced computation efforts. Also, the control algorithm determines the fundamental components of the 3-phase load current accurately and ensuring that the harmonics as well as reactive components of the load are being contributed by the grid tied PV system operating as a SAF. The source currents are sinusoidal regardless of the characteristics of the connected load. The developed algorithm is validated with the results which revels UPF is maintained at point of interconnection and THD of the 3-phase source currents are kept within 5%. The control of PV inverter as a SAF with proposed algorithm ensures IEEE-519 comply operation of the load.

References

1. Sreeraj, S.E., Chatterjee, K., Bandyopadhyay, S.: One-cycle-controlled single-stage single-phase voltage-sensorless grid-connected PV system. IEEE Trans. Ind. Electron. **60**, 1216–1224 (2013). https://doi.org/10.1109/TIE.2012.2191755
2. Sant, A.V., Khadkikar, V., Xiao, W., Zeineldin, H., Al-Hinai, A.: Adaptive control of grid connected photovoltaic inverter for maximum VA utilization. In: IECON 2013 - 39th Annual Conference of the IEEE Industrial Electronics Society, pp. 388–393 (2013)
3. Patel, K.C., Sant, A.V., Gohil, M.H.: Shunt active filtering with NARX feedback neural networks based reference current generation. In: 2017 International Conference on Power and Embedded Drive Control (ICPEDC), pp. 280–285 (2017)
4. Bhalja, H.S., Sant, A.V., Markana, A., Bhalja, B.R.: Microgrid with five-level diode clamped inverter based hybrid generation system. In: 2019 IEEE International Conference on Electrical, Computer and Communication Technologies (ICECCT), pp. 1–7 (2019)
5. Agarwal, R.K., Hussain, I., Singh, B.: Composite observer based control technique for single-phase solar PV-Grid tied system. In: 2016 IEEE 7th Power India International Conference (PIICON), pp. 1–6 (2016)
6. Jain, S., Shadmand, M.B., Balog, R.S.: Decoupled active and reactive power predictive control for PV applications using a grid-tied quasi-Z-source inverter. IEEE J. Emerg. Sel. Top. Power Electron. **6**, 1769–1782 (2018). https://doi.org/10.1109/JESTPE.2018.2823904
7. Raja Mohamed, S., Jeyanthy, P.A., Devaraj, D.: Novel control of grid-tied solar PV for enhancing transient stability limit at zero and high penetration levels. In: 2017 International Conference on Intelligent Computing, Instrumentation and Control Technologies (ICICICT). pp. 132–137 (2017)
8. Li, J., Zhuo, F., Wang, X., Wang, L., Ni, S.: A grid-connected PV system with power quality improvement based on boost + dual-level four-leg inverter. In: 2009 IEEE 6th International Power Electronics and Motion Control Conference, pp. 436–440 (2009)
9. Savitha, P.B., Shashikala, M.S., Putta Buddhi, K.L.: Modeling of photovoltaic array and control of grid connected photovoltaic system to provide quality power to grid. In: 2016 International Conference on Electrical, Electronics, Communication, Computer and Optimization Techniques (ICEECCOT), pp. 97–101 (2016)
10. Sant, A.V., Gohil, M.H.: ANN based fundamental current extraction scheme for single phase shunt active filtering. In: 2019 IEEE International Conference on Electrical, Computer and Communication Technologies (ICECCT), pp. 1–6 (2019)
11. Nie, X., Liu, J.: Current reference control for shunt active power filters under unbalanced and distorted supply voltage conditions. IEEE Access. **7**, 177048–177055 (2019). https://doi.org/10.1109/ACCESS.2019.2957946
12. Dash, A.R., Panda, A.K., Lenka, R.K., Patel, R.: Performance analysis of a multilevel inverter based shunt active filter with RT-EMD control technique under ideal and non-ideal supply voltage conditions. IET Gener. Transm. Distrib. **13**, 4037–4048 (2019). https://doi.org/10.1049/iet-gtd.2018.7060
13. Gohil, M., Sant, A.V.: 5-level cascaded inverter based D-STATCOM with LPF-BPF fundamental active current extractor. In: 2017 Third International Conference on Advances in Electrical, Electronics, Information, Communication and Bio-Informatics (AEEICB), pp. 237–241 (2017)

Modelling and Optimization of Novel Solar Cells for Efficiency Improvement

Chandana Sasidharan and Som Mondal

Abstract Our world requires economic, efficient, and sustainable sources of energy. There is a scope of improvement in the efficiency of energy generated from Solar Photovoltaics, the current prime mover of global energy transformation. The paper is associated with modelling and simulation of Interdigitated Back Contact (IBC) Solar Cells. The paper finds that the efficiency of Interdigitated IBC cells can be improved by optimization of the backside. The key steps are design alternation by eliminating the gap region, and subsequently optimizing for unit cell size and emitter to Back Surface Field (BSF) ratio. When optimized for IBC modelled in Quokka reveals that, at the best unit cell size, maintaining a definite emitter to BSF ratio and contact ratio improves efficiency. The simulation results obtained for an IBC with unit cell dimensions $900 \times 500 \times 90$ μm with an *n*-type bulk region with Emitter to BSF ratio 80:20 and 200 μm contacts is 22.4% efficient.

Keywords Modelling · Optimization · IBC · Quokka · Solar cell

1 Introduction

Humankind is experiencing the harsh aftermath of fossil-fuel-based energy sources. The world requires economic, efficient, and sustainable sources of energy. Solar Photovoltaic based generation is making headlines globally as it is the lynchpin of the clean energy transition [1]. Research on efficient and cost-effective solar cells is the need of the hour as the world is going for an exponential increase in solar deployment.

The history of solar photovoltaics research has effectively been associated with two factors: efficiency and cost. Though the costs have significantly reduced with economies of scale, the efficiency levels had hit glass ceiling without significant improvement [2]. Modelling and optimization of solar cells have been of prime importance to research community for efficiency improvement.

C. Sasidharan (✉) · S. Mondal
TERI School of Advanced Studies, Delhi, New Delhi, India
e-mail: chandana.sasidharan@gmail.com

M. Sahni et al. (eds.), *Mathematical Modeling, Computational Intelligence Techniques and Renewable Energy*, Advances in Intelligent Systems and Computing 1287,
https://doi.org/10.1007/978-981-15-9953-8_32

A solar cell is essentially a *p–n* junction diode which generates power when light shines on it based on photovoltaic effect. Every solar cell has a base region, an emitter region, and metal contacts. There are many design variations possible for arranging base and emitter and contacts, based on choice of materials and geometric structure. The model chosen for the study is IBC where all the contacts are in the back [3].

IBC cells are inherently energy efficient. The front surface losses of IBC cells are lower as the optical losses from front contacts are reduced. Further, reduction in front surface losses is possible with pyramid surface with a two-layer anti-reflection coating of SiN_x and SiO_x. Improved surface passivation leads to lower dark current saturation density (J_0) values as low as 5–15 fA/cm^2. However, there is a scope of further improvement in efficiency of these cells. Procelet al. [4] had already reduced the problem of the efficiency of IBC cells to optimization of the backside. In this paper, modelling of IBC cells is undertaken using Quokka, and optimization of Back Surface is performed. The paper provides information on the basics of solar cell modelling along with the fundamentals of IBC cells in the discussion section to aid future researchers. The results of the optimization are presented subsequently.

2 Literature Review

The schematic representation of IBC cells in Fig. 1 shows that the back surface of typical IBC cell has three regions; the emitter, Back Surface Field, and gap. Within the emitter and BSF region, there are contacted and non-contacted surface regions. Thus, there are five distinct regions in the back. Tracing the best efficient design for an IBC cell through the literature of recent years shows that the focus is on improving the efficiency of cells using standard industrial technologies focusing on optimization of the five distinct regions.

IBC introduced by Lammert and Scharwtz in 1977 is not a new concept. They performed modelling and simulations to identify the effects of parameters, including

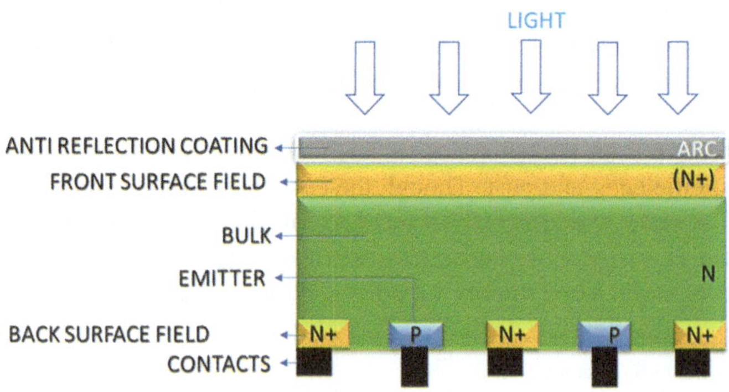

Fig. 1 Schematic representation of IBC solar cell

bulk lifetime, front and back surface recombination velocity, and thickness. The significance of bulk lifetime is iterated as the collection efficiency increased from 72 to 95% as bulk lifetime increased from 10 to 100 μs. It was found that for a value less than 10 cm/s, the front surface velocities have no effect on efficiency. Similarly, back surface velocities less than 100 cm/s have no effect on efficiency [4].

Australian National University (ANU) has focused on back surface design as it has championed IBC cells with an N-type bulk region. Their $2 \times 2\ \text{cm}^2$ cells achieved high efficiency of 24.4% primarily due to the high current density of 41.95 mA/cm^2. Franklin et al. [5] identifies five interfaces in the back between the bulk undiffused region and the passivation film, n^+ doped region and the passivation film, n^+ doped region and metal at contacts, p^+ doped region and passivation film and p^+ doped region and metal at contacts.

Quokka being an inhouse software has been extensively used by researchers at ANU. Fong et al. [6] performed optimization of n^+ region and the diffusion of IBC Cells with a N-type bulk region using Quokka. The paper considered localized and sheet-based diffusion beneath the metal contacts. They find that large area sheet diffusion requires light doping to reduce recombination losses and if doping is higher than a benchmark level, the recombination losses are higher than the gain from reduced resistive losses. For localized diffusion, the effect of diffusion region in recombination becomes less significant. Localized n^+ design has more potential for efficiency improvement if J_0 is less than 20 fA/cm^2 on the metal surface. Having a large contact area with light diffusions is undesirable.

There are many recent works on modelling and optimization of IBC cells performed using Quokka. Mewe et al. [7] also employed Quokka to perform simulations of IBC cells with BSF islands to improve efficiency. They found smaller BSF regions brings down the recombination activity, but sufficient area should be maintained for effective transport and contact alignment. They had arrived at a BSF fraction of 20% per unit cell and had modelled BSF regions of square and circular shapes using Quokka. They concluded that BSF island design boosts the short circuit current (J_{sc}) and open-circuit voltage (V_{oc}) of the cell, if the design has a lightly doped front floating emitter.

One of the ways for simplifying the manufacturing process is by eliminating the gap region. IBC cells without gap region also called Zebra Cells [8] developed using industrial processing techniques has proven efficiencies of average 21%. This design however suffers from Tunnel Junction Shunting between the n^+ and p^+ region. Dong et al. [9] proved that by limiting boron concentration to $10^{19}\ \text{cm}^{-3}$ the tunnel junction shunting could be avoided.

Trina Solar is also invested in the production of high-efficiency IBC Cells using low-cost industrial techniques. Li et al. reported that the 6-inch pilot cells were of 23% efficiency [10]. Li et al. have used Quokka to perform detailed loss analysis on their screen-printed contact IBC cells and found that reduction in J_0 of emitter and passivated area of BSF results in efficiency improvement. Xu et al. reported 23.5% efficiency for 6-inch cells by optimizing contact size for screen printed contacts [11].

3 Materials and Methods

In this paper, simulation performed in Quokka to optimize the efficiency of the IBC structure is presented. IBC cells without gap region with characteristics similar to the screen-printed IBC cells of Trina Solar are simulated. The size of the unit cell and emitter to the BSF ratio are identified as the most important design parameters. The Free Energy Loss Analysis (FELA) function of Quokka is used to analyze the resistive and recombination losses. Co- optimization is possible in a sequence identifying the best unit cell size and identifying emitter to BSF ratio optimization balancing the recombination and resistive losses in the contacted regions.

4 Simulation Results and Discussion

4.1 Modelling of an IBC Solar Cell

There are three basic sets of equations for modelling solar cells. These equations connect the electrical field and current density to charge carrier density and rate of recombination and generation. The equations can be represented in simplified form for one dimension (dx) as:

1. Poisson's equation relating the gradient of Electric field in direction x to the density of charge carriers
2. Current density equations including drift and diffusion phenomena
3. Continuity equations balancing generation and recombination

$$\frac{\mathrm{d}E}{\mathrm{d}x} = \frac{q}{\varepsilon}(p - n - N_a + N_d) \tag{1}$$

$$Jh = q\mu pE + qD\frac{\mathrm{d}p}{\mathrm{d}x} \tag{2}$$

$$Je = q\mu nE + qD\frac{\mathrm{d}n}{\mathrm{d}x} \tag{3}$$

$$\frac{1}{q}\frac{Je}{\mathrm{d}x} = U - G \tag{4}$$

$$\frac{1}{q}\frac{Jh}{\mathrm{d}x} = -(U - G) \tag{5}$$

where,

E—Electric field strength in x direction

ε—Permittivity of the material
p—Hole density
n—Electron density
N_a—Acceptor density
N_d—Donor density
q—Basic charge of electron
Je and *Jh* are current density of electron and hole respectively
μ is mobility
D is diffusion coefficient
U is rate of recombination of charge carriers
G is rate of generation of charge carriers.

The generic equations are generally solved iteratively, posing it as a boundary value problem using software like Quokka. In the IBC structure, the electron-hole pairs formed in the bulk region is collected in the diffused junction at the back. Both the p and n junction being in the back, the contacts are automatically in the backside of the cell. Thus, the optical losses resulting from shadowing and reflection from the metallic grid and bus bar structure in the front is eliminated [3]. Both the contacts in the back also simplifies module assembly steps like tabbing. In IBC cells, the internal resistance caused by high resistivity of the thin front layer is eliminated. There is no lateral current in the diffused layer hence the flow of current is perpendicular to the front surface. Thus, the internal series resistance of IBC cells will be proportional to the excess carrier concentration as the resistance incurred due to the lateral flow of current in the thin front emitter is eliminated.

The complexities on the front surface of conventional cells to simultaneously satisfy multiple demands of emitter recombination, surface recombination, lateral conductivity, and external contact is eliminated in an IBC cell. Superior passivation of the front surface is possible in IBC cells [5]. The carriers generated in the front and bulk regions are collected in the back. IBC cells need high diffusion length which translates to a higher lifetime in the bulk region for minority carriers. Thus, minority carrier lifetime in the bulk region becomes a critical parameter for IBC cells. *P*-type CZ crystals suffer from a low lifetime in bulk region due to recombination centers created in the bulk by Boron-Oxygen (BO) complexes [12].

The Boron-Oxygen defects occur in the material that is both Boron-doped and Oxygen-rich. These defects are quadratically proportional to the concentration of Oxygen and increase linearly with oxygen concentration [13]. BO defects are activated by illumination but research shows that they can be permanently converted to a passive state by annealing under illumination [14].

4.2 Loss Design in An IBC Solar Cell

There are a number of fundamental loss mechanisms which are unavoidable for a solar cell. Firstly, light with energy less than the band gap is not absorbed (transmission losses), which represents a loss of about 27% at the AM 1.5 G spectrum for a band gap of 1.12 eV for a Silicon Solar cell. For light with energy greater than the bandgap the extra energy of the photons is released as heat (thermalization losses) which accounts for about 28%. The transmission and thermalization losses bring down the efficiency of a solar cell to about 45%. The best achievable efficiency limit for a solar cell explained by the Shockley–Queisser limit (29%) is further low, taking into account the loss mechanisms that occur in the solar cell. Swanson [15] proposed that 26% was the practical silicon production efficiency limit. The fundamental losses that occur in solar cells can be broadly classified into Optical, Resistive, and Recombination losses.

The losses in an IBC solar cell is presented in Fig. 2. The optical losses generally arise from shading and reflection of the incident light by the metallic grid which does not occur in IBC cells. The losses that occur due to absorption and reflection of light in the Anti Reflective Coating (ARC) is applicable to IBC cells. The other problem arises from imperfect trapping of light within the cell especially at the rear. Application of a dual SiO_2 and SiN_x has shown to reduce the optical losses while providing passivation at surfaces. Therefore, it is logical to focus on methods to reduce the resistive and recombination losses for an IBC cell to achieve higher efficiencies.

The resistive losses consist of resistive drops occurring in bulk, emitter, contacts, and metal. The resistive losses in IBC cells arise not only from the resistance offered by the semiconductor. The contact resistivity between semiconductor and metal results in losses which can be reduced by high doping near the contacting region. The series resistance arising metal fingers and metallic busbars can be reduced by proper busbar design which is decoupled from the optical performance. In IBC cells, current does not flow laterally through the thin diffused layer which causes series resistance in conventional solar cells.

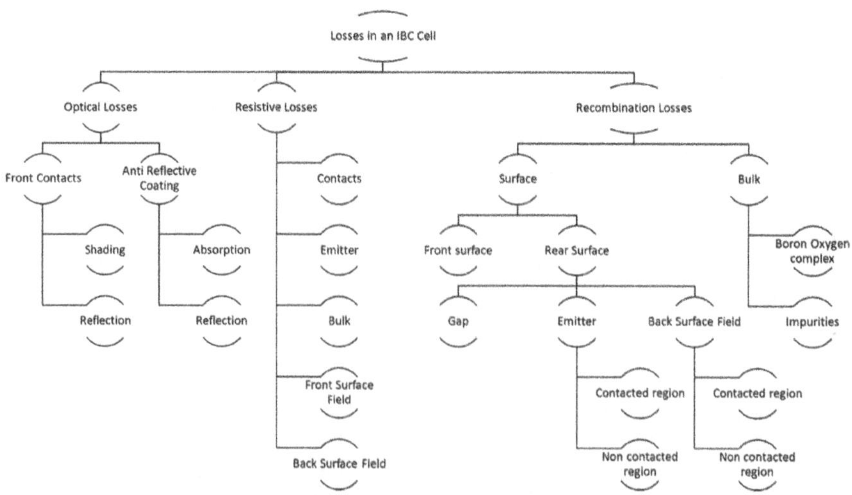

Fig. 2 Classification of losses in IBC cell

The emitter region is significantly small, and the contact resistance can be brought down by excess diffusion. The bulk resistive losses alone account for most of the resistive losses and primarily depend on excess carrier concentration in IBC cells. Bulk resistive losses are reduced via conductivity.

The recombination phenomena in a solar cell arise from intrinsic and extrinsic recombination that occurs in the bulk region as well as the recombination occurring at the front and back surfaces. The intrinsic phenomena, including Auger and Radiative recombination are fundamental processes in a solar cell. The extrinsic recombination losses arise from defect levels are studied as Shockley–Reed–Hall (SRH) recombination arising from impurities. In the case of *p*-type CZ crystals, the extrinsic recombination losses due to BO complex is also significant.

The resistive and recombination losses are often intertwined. The significant contact resistivity between silicon and aluminum can be reduced by creating a Back-Surface Field. But the additional diffusion for creating the field leads to increase in recombination losses of the cell. It is important to obtain a balance between the recombination losses and resistive losses while optimizing the IBC cell for best performance. The complex recombination phenomena deserve much attention than the resistive losses which is more or less from bulk resistivity alone.

4.3 Modelling and Simulation Using Quokka

The tool that was used for simulation is Quokka, which is implementation of simplified solar cell model in MATLAB. Quokka is a freeware that helps in fast simulations of solar cells in 2D/3D using the two basic assumptions of Quasi neutrality for bulk and Conductive boundary approach for diffusion regions. The 3D mesh structure developed in Quokka is given in Fig. 3. One of the limitations for Quokka also arises from the conductive boundary approach which is also the main modelling difference to most other device simulation software.

In comparison with other freeware like PC2D the computational speed is higher for Quokka which also allows flexible meshing. Comparing it with proprietary software like Sentaurus the computational time for Quokka is very less with little loss of accuracy. However, Quokka lacks sophisticated optics models. Hence, it needs a two-step modelling approach separating optical modeling and electrical modelling, unlike Sentaurus [16].

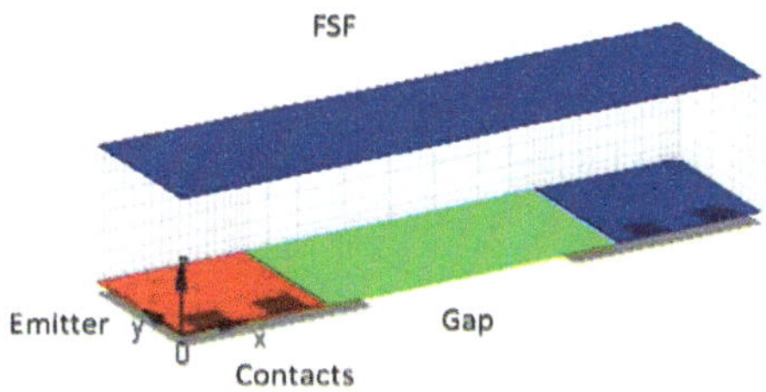

Fig. 3 Three-dimensional mesh structure generated in Quokka

Quasi neutrality of bulk region, one of the primary assumptions of Quokka to simplify the boundary value problem has already been employed by researchers beginning with Lammert and Schwartz [3]. Treating the diffusion at the surface as a conductive boundary is the main highlight of modelling using Quokka. Similarly, a non-conductive boundary in Quokka accounts for undiffused surfaces.

The model generated using Quokka is an electrical model without employing complex optical modeling using ray tracing. The facet angle is fixed at 54.7°, and transmission is fixed at 0.9. It is possible to upload the optical modelling done through ray tracing to Quokka. For the purpose of this study, the focus is more on recombination losses which did not necessitate complicated optical modelling.

The injection dependent conductivities are calculated using Klaassen's [17] mobility model which takes into account all the main contributions to mobility including scattering caused by electron-hole, acceptor and donor. The mobility model considers electron and hole mobility as functions of local variables and is in agreement with experimental data.

Auger modelling was done based on the parameterization done by Ritcher et al. [18]. Optical modelling is done combining optical properties of silicon from Nguyen [19] with the data from Green [20] to cover the full wavelength range.

Quokka essentially solves for the steady-state electrical characteristics of the device to obtain various typical solar cell characteristics such as J_{sc}, V_{oc}, maximum-power-point (MPP) conditions. Quokka can generate light- and dark I–V curves as well as quantum efficiency curve. Power loss analysis is easy with the FELA option of Quokka as it separately resistive and recombination losses in results. [16]

4.4 Modelling Parameters

4.4.1 Bulk Recombination Parameters

Understanding recombination mechanism in the bulk region and surfaces is the key to improve the efficiency of IBC cells. The most relevant parameter for characterizing recombination is carrier lifetime. The efficiency of the IBC cells is dependent on minority carrier lifetime, especially in the bulk region. In the bulk region impurities, like Iron and Chromium have a significant effect on lifetime. Using Quokka SRH defects can be added and its effect on efficiency can be simulated.

For the *p*-type bulk region, the effect BO complex on efficiency is another concern. Since the variation of efficiency is quadratic with oxygen concentration and linearly with boron concentration, the former is selected as a parameter to study the impact of BO complexes. Quokka allows another additional parameter called bulk background lifetime for modelling of the bulk region.

4.4.2 Surface Recombination Parameter

There are two alternatives generally used in modelling as alternatives to minority carrier lifetime τ in describing recombination. One of the parameters used for modeling is using J_0, which is dark saturation current density. The name comes from the assumption that under reverse bias and absence of light, the output current should saturate to a value of J_0. In simple terms, J_0 can be explained as the factor that when multiplied by the normalized *p–n* product gives the total recombination of a region under consideration.

Experimentally determined value of J_0 can be input directly in Quokka for modeling. It is important to note that effective life time τ_{eff} is independent on dimensions of the wafer unlike the dark current density J_0 value [21].

An alternative to J_0 is Surface Recombination Velocity (S), especially for modelling recombination happening at the front and rear surface. The surface recombination phenomena can be well understood using this single parameter, S which is roughly the inverse of carrier lifetime, i.e., low surface recombination velocity is equivalent of high lifetime. Surface recombination velocity was used as the modelling parameter for front and back surfaces by researchers beginning with Lammert and Schwartz [4]. The J_0 for the surface and S are equivalent for ideal recombination and low injection-related by the following equation:

$$J_0 = qS\frac{n_i^2}{N_A} \tag{6}$$

The second limitation with Quokka was that modelling of doping profile or junction depth was not possible at FSF, BSF, or emitter. Using surface recombination velocity for the model, which effectively captures the doping, helps overcome the challenge. More focus is given to recombination happening in non-contacted regions in the rear surface as they will vary with change in surface recombination velocity.

4.4.3 Using S for Emitter Modelling

Emitters are generally challenging to model and analyze because of high dopant concentration. The J_0 parameter is generally used to model emitter assuming Auger recombination as the dominant mechanism. The thickness of the emitter for an IBC cell is quite small compared to the thickness of the IBC cell. The contact area of thickness is designed small to reduce losses, but the emitter is kept sufficiently large to avoid current crowding. The predominant recombination loss at the emitter is assumed to happen at the non-contacted surface of the emitter. For the purpose of the study, the modelling is done using effective surface recombination velocity as focusing more on the recombination happening at the surface.

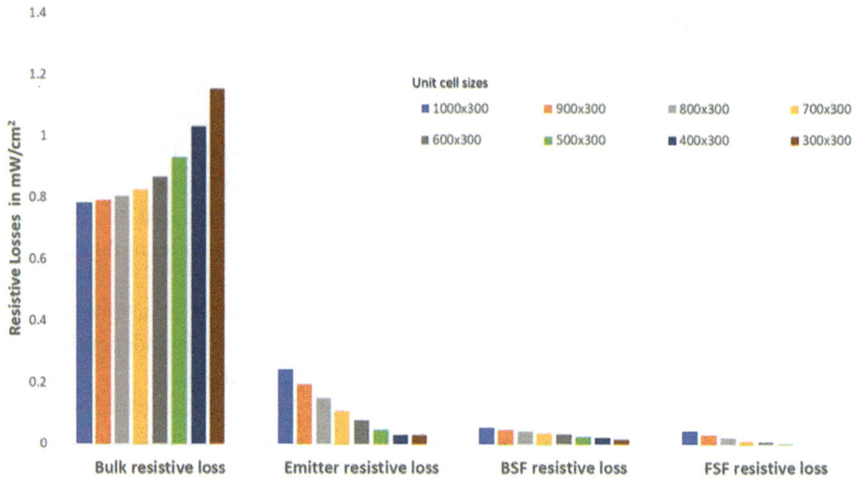

Fig. 4 Resistive losses for different unit cell dimensions

4.5 *Loss Distribution and Efficiency*

Modelling and simulations show that in cells without gap region at fixed emitter to BSF ratio, the efficiency improves at a specific unit size, and vice versa. Figure 4 shows the variation of recombination and resistive losses for the different unit cell dimensions for an IBC cell maintaining Emitter to BSF ratio 80:20. It is evident from Fig. 4 that bulk resistive loss increases and emitter resistive loss decrease with the decrease in unit cell size. For the FSF and BSF regions, the unit cell dimensions have no impact on efficiency. All together, the bulk resistive losses are the most significant resistive losses.

Figure 5 shows the variation of recombination losses for different unit cell dimensions. It is evident from Fig. 5 that recombination losses for bulk, emitter, and BSF region are most significant. In the case of the emitter and the bulk region, the recombination losses typically decrease with a decrease in unit cell size. But in the BSF region, the effect is vice versa, and recombination losses increase with reduction of unit cell size. For the FSF region, the unit cell dimensions does not have much impact on recombination losses.

Results show that further improvement in efficiency requires intervention in reducing the bulk losses in terms of recombination and resistive losses. Losses, when optimized for at the best unit cell size, maintaining a definite emitter to BSF ratio and contact ratio, improves efficiency. The simulation results obtained for an IBC with unit cell dimensions 900 × 500 × 190 μm with the *n*-type bulk region with Emitter to BSF ratio 80:20 and 200 μm contacts improves to 22.4% efficiency. The two possible variables suitable for sensitivity analysis are wafer resistivity and effective life time and the results of their analysis are shown in Fig. 6.

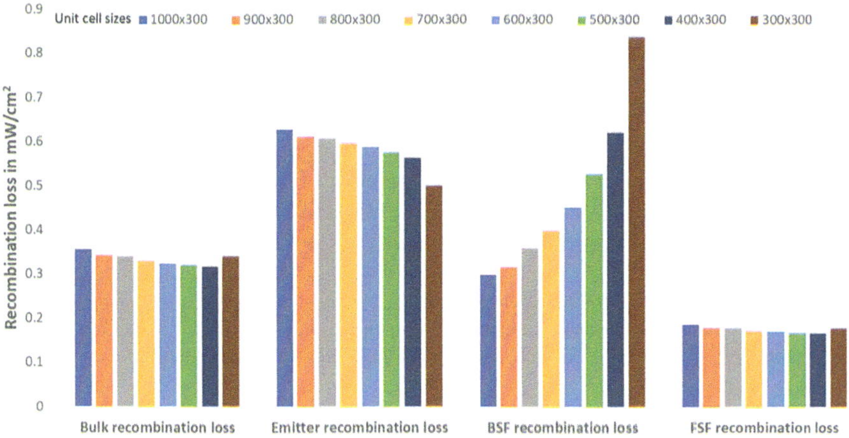

Fig. 5 Variation of recombination losses with different unit cell dimensions

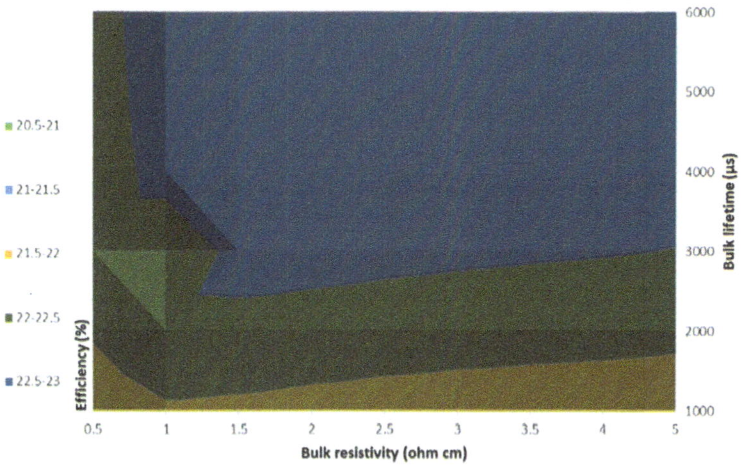

Fig. 6 Sensitivity of efficiency to bulk resistivity and lifetime

5 Conclusion

The study presents a modelling and optimization exercise for IBC solar cells using Quokka. Quokka does not allow modelling of junction depth and doping concentration for IBC cells. Though surface recombination velocity used in modelling depends on doping level and passivation at the surface, the precise effect of doping level on losses could not be considered. The best efficiency obtained for an IBC cell model with unit cell dimensions 900 × 500 × 190 μm with the n-type bulk region with Emitter to BSF ratio 80:20 and 200 μm contacts is 22.4% efficiency. The paper concludes that it is possible to improve the efficiency of the IBC solar cells focusing on optimizing the back region. Research has already shown that gapless back design and proper Emitter BSF ratio enhances efficiencies independently. Combining the above and adapting a suitable unit cell size is the optimal solution for efficient IBC cell design. The design changes should be undertaken in two steps, first by eliminating the gap region in design, and subsequently optimizing the losses. Both unit cell size and emitter to BSF ratio can be optimized. For IBC modelled in Quokka, the paper finds that at the best unit cell size, maintaining a definite emitter to BSF ratio and contact ratio improves the efficiency.

Acknowledgements The authors wish to express thanks to Andreas Fell and the team of Australian National University who developed Quokka. The authors are also thankful to the team behind PVLighthouse.com for their online resources on Quokka and setting file generator.

References

1. International Energy Agency: World Energy Outlook (2019)
2. Green, M.A., Emery, K., Hishikawa, Y., Warta, W., Dunlop, E.D.: Solar cell efficiency tables (Version 45). Prog. Photovoltaics Res. Appl. **23**, 1–9 (2015)
3. Lammert, M.D., Schwartz, R.J.: The interdigitated back contact solar cell: a silicon solar cell for use in concentrated sunlight. IEEE Trans. Electron Devices **24**(4), 337–342 (1977)
4. Procel, P., Ingenito, A., De Rose, R., Pierro, S., Crupi, F., Lanuzza, M., Cocorullo, G., Isabella, O., Zeman, M.: Opto-electrical modelling and optimization study of a novel IBC c-Si solar cell. Prog. Photovoltaics Res. Appl. **25**(6), 452–469 (2017)
5. Franklin, E., Fong, K., McIntosh, K., Fell, A., Blakers, A., Kho, T., Walter, D., Wang, D., Zin, N., Stocks, M., Wang, E.C.: Design, fabrication and characterisation of a 24.4% efficient interdigitated back contact solar cell. Prog. Photovoltaics Res. Appl. **24**(4), 411–427 (2016)
6. Fong, K.C., Teng, K., McIntosh, K.R., Blakers, K.W., Franklin, E., Zin, N., Fell, A.: Optimisation of n+ diffusion and contact size of IBC solar cells. In: 28th European Photovoltaic Solar Energy Conference and Exhibition, Munich (2013)
7. Mewe, A.A., Guillevin, N., Cesar, I., Burgers, A.R.: BSF islands for reduced recombination in IBC cells. In: IEEE 44th Photovoltaic Specialists Conference, Washington (2017)
8. Halm, A., Mihailetchi, V.D., Galbiati, G., Koduvelikulathu, L.J., Roescu, R., Comparotto, C., Kopecek, R., Peter, K., Libal, J.: The zebra cell concept-large area n-type interdigitated back contact solar cells and one-cell modules fabricated using standard industrial processing equipment. In 27nd EU-PVSEC, Frankfurt (2012)
9. Dong, J., Tao, L., Zhu, Y., Yang, Z., Xia, Z., Sidhu, R., Xing, G.: High-efficiency full back contacted cells using industrial processes. IEEE J. Photovoltaics **4**(1), 130–133 (2013)
10. Li, Z., Yang, Y., Zhang, X., Liu, W., Chen, Y., Xu, G., Shu, X., Chen, Y., Altermatt, P.P., Feng, Z., Verlinden, P.J.: Pilot production of 6" IBC solar cells yielding a median efficiency of 23% with a low-cost industrial. In: Proceedings of the 25th European PV Solar Energy Conference, Valencia (2010)
11. Xu, G., Yang, Y., Zhang, X., Chen, S., Liu, W., Chen, Y., Li, Z., Chen, Y., Altermatt, P.P., Verlinden, P.J., Feng, Z.: June. 6 inch IBC cells with efficiency of 23.5% fabricated with low-cost industrial technologies. In: 2016 IEEE 43rd Photovoltaic Specialists Conference (PVSC), pp. 3356–3359. IEEE. Portland (2016)
12. McIntosh, K., Cudzinovic, M., Smith, D., Mulligan, W.: The choice of silicon wafer for the production of low-cost rear-contact solar cells. In: Proceedings of 3rd World Conference on Photovoltaic Energy Conversion, Osaka (2003)

13. Macdonald, D., Rougieux, F., Cuevas, A., Lim, B., Schmidt, J. Di Sabatino, M., Geerligs, L.J.: Light-induced boron-oxygen defect generation in compensated p-type Czochralski silicon. J. Appl. Phys. **105**, 093704 (2009)
14. Herguth, A., Schubert, G., Kaes, M., Hahn, G.: Investigations on the long time behavior of the metastable boron–oxygen complex in crystalline silicon. Prog. Photovoltaics **16**, 135–140 (2008)
15. Swanson, R.M.: Approaching the 29% limit efficiency of silicon solar cells. In: 31st IEEE Photovoltaic Specialists Conference, Florida (2005)
16. Fell, A., McIntosh, K.R., Abbott, M., Walter, D.: Quokka version 2: selective surface doping, luminescence modeling and data fitting. IEEE J. Photovoltaics **4**(4), 1040–1045 (2014)
17. Klaassen, D.B.M.: A unified mobility model for device simulation—I. Model equations and concentration dependence. Solid State Electron **35**(7), 953–959 (1992)
18. Richter, A., Glunz, S.W., Werner, F., Schmidt, J., Cuevas, A.: Improved quantitative description of Auger recombination in crystalline silicon. Phys. Revision **86**, 165–202 (2012)
19. Nguyen,H.T., Rougieux, F.E., Mitchell, B., Macdonald, D.: Temperature dependence of the band-band absorption coefficient in crystalline silicon from photoluminescence. J. Appl. Phys. **115**, 305–309 (2014)
20. Green, M.A.: Self-consistent optical parameters of intrinsic silicon at 300 K including temperature coefficients. Sol. Energy Mater. Sol. Cells **92**, 1305–1310 (2008)
21. Cuevas, A.: The recombination parameter J0. In: 4th International Conference on Silicon Photovoltaics, 25–27 March 2014, Netherlands (2014)

Co-Ordination of a Two-Echelon Supply Chain with Competing Retailers Where Demand Is Sensitive to Price and Quality of the Product

Rubi Das, Pijus Kanti De, and Abhijit Barman

Abstract Supply Chain Management (SCM) is the set of approaches utilized for the appropriate integration and utilization of suppliers, manufacturers, warehouses, and retailers to ensure the production and delivery of products to end-users in the correct amounts and at the perfect time. In real-life situations, designing and utilizing different supply chains help companies to achieve their goals and improve their profitability. In this paper, an inventory mechanism of a single-product two-stage supply chain including a single manufacturer and two retailers has been introduced. The random demand focused by the manufacturer and retailers is sensitive to both price and the product quality. We have designed the manufacturer and retailers optimality system as a profit maximization problem to determine the stochastic order quantity and selling price of the retailers. A centralized case of the supply chain has been analyzed by using the normal distribution. The overall supply chain profit is evaluated by one numerical example and then the effect of some parameters on the optimality of decision variables is analyzed.

Keywords Stochastic demand · Quality · Centralized supply chain · Normal distribution

1 Introduction

To survive in this ambiance, companies and businesses are vigorously struggling and competing. Mainly supply chain profit is affected by various parameters like demand, stock of inventory, and quality of the product, holding cost, market potential, carrying

R. Das (✉) · P. K. De · A. Barman
Department of Mathematics, National Institute of Technology Silchar, Assam 788010, India
e-mail: rubidas1118@gmail.com

P. K. De
e-mail: pijusde@gmail.com

A. Barman
e-mail: abhijitmath93@gmail.com

M. Sahni et al. (eds.), *Mathematical Modeling, Computational Intelligence Techniques and Renewable Energy*, Advances in Intelligent Systems and Computing 1287,
https://doi.org/10.1007/978-981-15-9953-8_33

cost, cost of transportation, and many others. We are trying to employ various parameters to improve supply chain performance and reduce costs in complex and challenging circumstances. Maximizing profit and minimizing cost as the main purpose of a supply chain model. By aiming these facts we have modeled a common supply chain mechanism having a manufacturer and two retailers. Retail price is the key parameter of the competition. Iyer [1] discussed the coordination mechanism for the multi-echelon channel under channel competitions. Tsay and Agrawal [2] designed a sales price-sensitive supply chain distribution system for a single product between one manufacturer and two-retailers. This study accepts a random demand for both the retailers and displays the consequences of uncertain demand on decision variables. Since uncertainty frequently appears shortage and overload may happen that makes it important to incorporate the stock choices for retailers. Whitin [3] first introduced a price-effective newsvendor problem. Whitin [3] modified the newsvendor problem by including probabilistic price-dependent demand. Model formulation of Mills [4] should be mentioned among others.

1.1 Literature Review

Sarkar [5] discussed an EMQ model where demand is affected by both price and advertisement of the item. The market price of materials, manufacturing, and rework cost is considered as reliability variables in this model. Kang et al. [6] introduced a single-step mathematical formulation based on minimizing total average cost considering stock, order quantity, and backorders. Ouyang and Chang [7] mentioned that the production of products always not performed well for real production systems because of imperfect product quality, the defective process of production, etc. Jazinaninejad et al. [8] developed a new method for the first time based on credit option under stochastic price-sensitive demand considering pricing; product and period review inventory Liu et al. [9] discussed the effect of the return policy on retailer's order amount on both centralized and decentralized case. Jana et al. [10] introduced integration and channel competition of a low-quality and high-quality items of manufacturers Xiao and Shi [11] explored channel priority and pricing strategy in supply chain mechanism having dual-channel facing manufacturer shortage originated by uncertain demand. The randomness of demand may observe a notable dispersion but in an application, their exit limited information about the random demand. Moon et al. [12] explored normal distribution in their problem. The effect of non-price parameters such as quality of product pays attention to the modeling of supply chains. When a single product is delivered through two retailers' quality improvement can be recognized. Chen et al. [13] discussed the effect of product quality in their supply chain. Taleizadeh et al. [14] examined a centralized approach for their inventory model and they attained the optimal solutions of the sales price, wholesale prices, and order amounts.

The rest part of the work is categorized as Sect. 2 introduces the formulation of the model. In Sect. 3 the formulation and analysis of the model have been evaluated. In

Fig. 1 Supply chain structure

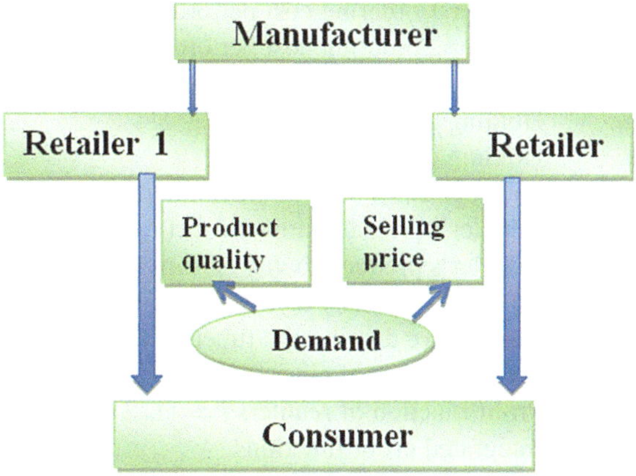

Sect. 4 we have demonstrated the model with one numerical example and the results have been analyzed. A sensitivity analysis has been shown in this section. Section 5 gives some outcomes and future research directions.

2 Problem Description

2.1 Assumptions

- The supply chain mechanism contains one manufacturer and two retailers.
- The model is evaluated for a single product produced by the manufacturer.
- Demand is uncertain which is sensitive to retailers selling price and product quality.
- Shortages at the retailers are permitted.
- Lead time is negligible (Fig. 1).

2.2 Notations

c	Production cost for manufacturer
w	Wholesale price of the manufacturer to the retailer (per unit)
q_i	Order quantity of both the retailers $i = (1, 2)$
d_i	Deterministic demand rate $i = (1, 2)$
x_i	Apart from demand quantity, denotes a stochastic variable considering normal distribution $i = (1, 2)$
k_i	Product quality of both the retailers $i = (1, 2)$
e_k	Cost of improving quality degree of product
A_{m}	Manufacturer ordering cost per order

A_{r_1}	Retailer 1 ordering cost per order
A_{r_2}	Retailer 2 ordering cost per order
h_m	Holding cost of manufacturer
h_{r_1}	Holding cost of retailer 1
h_{r_2}	Holding cost of retailer 2
s_i	Shortage costs of both the retailers $i = (1, 2)$
β_i	Retailer's price sensitivity $i = (1, 2)$
γ_i	Retailer's product quality sensitivity $i = (1, 2)$
$f_i(x_i)$	Probability density distribution function of x_i,$i = (1, 2)$
$F_i(x_i)$	Cumulative distribution function of $x_i i = (1, 2)$
ψ_m	Profit function of manufacturer
ψ_{r_i}	Profit function of retailers $i = (1, 2)$
ψ_{eip}	Integrated profit function

Decision variables

p_i	Retailers selling price $i = (1, 2)$
z_i	Stochastic portion of demand $i = (1, 2)$

The profit of our two-stage manufacturer-two retailers supply chain system is calculated following Ernst [15], $z = q - y(p)$. We stated order quantity as $q_i = d_i + z_i (i = 1, 2)$. According to Taleizadeh et al. [16], both the retailers facing the following demand functions.

$$d_1 = a_1 - \beta_1 p_1 + \gamma_1 k_1 \tag{1}$$

$$d_2 = a_2 - \beta_2 p_2 + \gamma_2 k_2 \tag{2}$$

Also the stochastic portion of demand z_1 is considered as a random variable with range $[a_1, b_1]$ which has mean μ_1 and variance σ_1 and z_2 is considered as a random variable with range $[a_2, b_2]$ which has mean μ_2 and variance σ_2.

The profit function of manufacturer is formulated as follows

$$\psi_m = (w - c)(q_1 + q_2) - h_m \frac{(q_1 + q_2)}{2} - e_k k - A_m \left(\frac{d_1}{q_1} + \frac{d_2}{q_2} \right) \tag{3}$$

where $w(q_1 + q_2)$ is the yearly profit, $c(q_1 + q_2)$ denotes the production investment, $h_m \frac{q_1+q_2}{2}$ is the annual average holding cost, $e_k k$ displays the investment against product quality, $A_m \left(\frac{d_1}{q_1} + \frac{d_2}{q_2} \right)$ is the yearly ordering cost.

The overall profit of both the retailers are described as follows

$$\psi_{r_2} = p_2 d_2 - w q_2 - h_{r_2} \int_{a_2}^{z_2} (z_2 - x_2) f(x_2) \mathrm{d}x_2$$

$$-s_2\int_{z_2}^{b_2}(x_2-z_2f(x_2)\mathrm{d}x_2-A_{r_2}\frac{d_2}{q_2} \tag{4}$$

$$\psi_{r_1}=p_1d_1-wq_1-h_{r_1}\int_{a_1}^{z_1}(z_1-x_1)f(x_1)\mathrm{d}x_1$$

$$-s_1\int_{z_1}^{b_1}(x_1-z_1)f(x_1)\mathrm{d}x_1-A_{r_1}\frac{d_1}{q_1} \tag{5}$$

where p_1d_1 and p_2d_2 are the annual revenue of both the retailers. wq_1 and wq_2 are the annual purchasing cost.$A_{r_1}\frac{d_1}{q_1}$ and $A_{r_2}\frac{d_2}{q_2}$ displays the annual ordering cost for both the retailers. If the selection of $z_i(i=1,2)$ is higher than the recognized value of $x_i(i=1,2)$, then leftovers occur; if the selection of $z_i(i=1,2)$ is smaller than the realized value of $x_i(i=1,2)$, then shortages occur.

3 Centralized Decision-Making Model

In this section, a two-layered supply chain mechanism is established to investigate the effect of the channel on profit maximization criteria. In this case, a single decision maker is controlled by the supply chain's determined result. The individuals of a supply chain maximize their overall profit by consulting among their selves. Combining overall revenue of manufacturer and both the retailers we get overall integrated profit. The overall profit functions of the Supply chain calculated by combining Eqs. (3–5) given as follows

$$\psi_{eip}=-c(q_1+q_2)-h_m\frac{q_1+q_2}{2}-e_kk-(A_m+A_{r_1})\frac{d_1}{q_1}-(A_m+A_{r_2})\frac{d_2}{q_2}$$

$$+p_1d_1+p_2d_2-h_{r_1}\int_{a_1}^{z_1}(z_1-x_1)f(x_1)\mathrm{d}x_1-s_1\int_{z_1}^{b_1}(x_1-z_1)f(x_1)\mathrm{d}x_1$$

$$-h_{r_2}\int_{a_2}^{z_2}(z_2-x_2)f(x_2)\mathrm{d}x_2-s_2\int_{z_2}^{b_2}(x_2-z_2)f(x_2)\mathrm{d}x_2 \tag{6}$$

Now, The necessary conditions of ψ_{eip} are

$$\frac{\partial\psi_{eip}}{\partial z_1}=-c-\frac{h_{\mathrm{m}}}{2}+(A_{\mathrm{m}}+A_{r_1})\frac{d_1}{(d_1+z_1)^2}-h_{r_1}f_1(z_1)+s_1[1-f_1(z_1)] \tag{7}$$

$$\frac{\partial \psi_{eip}}{\partial z_2} = -c - \frac{h_{\mathrm{m}}}{2} + (A_{\mathrm{m}} + A_{r_2})\frac{d_2}{(d_2 + z_2)^2} - h_{r_2} f_2(z_2) + s_2[1 - f_2(z_2)] \quad (8)$$

$$\frac{\partial \psi_{eip}}{\partial p_1} = \beta_1 c + \frac{h_{\mathrm{m}}\beta_1}{2} + \frac{(A_{\mathrm{m}} + A_{r_1})\beta_1 z_1}{(a_1 - \beta_1 p_1 + \gamma_1 k_1 + z_1)^2} + a_1 - 2\beta_1 p_1 + \gamma_1 k_1 \quad (9)$$

$$\frac{\partial \psi_{eip}}{\partial p_2} = \beta_2 c + \frac{h_{\mathrm{m}}\beta_2}{2} + \frac{(A_{\mathrm{m}} + A_{r_2})\beta_2 z_2}{(a_2 - \beta_2 p_2 + \gamma_2 k_2 + z_2)^2} + a_2 - 2\beta_2 p_2 + \gamma_2 k_2 \quad (10)$$

Thus, by equating Eqs (7), (8), (9) and (10) are equal to zero and solving we get the optimal price $\left(p_1^*, p_2^*\right)$ of both the retailers and optimal order quantity $\left(z_1^*, z_2^*\right)$.

Lemma 1 The expected integrated profit of the chain ψ_{eip} is concave in p_1, p_2, z_1, z_2 if $a_1 + \gamma_1 k_1 > \beta_1 p_1$ and $\beta_1^2 p_1 + \beta_1 z_1 > a_1\beta_1 + \beta_1\gamma_1 k_1$.

Proof Hessian matrix for p_1, p_2, z_1, z_2 is established.

$$H = \begin{pmatrix} \frac{\partial^2 \psi_{eip}}{\partial z_1^2} & \frac{\partial^2 \psi_{eip}}{\partial z_1 \partial z_2} & \frac{\partial^2 \psi_{eip}}{\partial z_1 \partial p_1} & \frac{\partial^2 \psi_{eip}}{\partial z_1 \partial p_2} \\ \frac{\partial^2 \psi_{eip}}{\partial z_2 \partial z_1} & \frac{\partial^2 \psi_{eip}}{\partial z_2^2} & \frac{\partial^2 \psi_{eip}}{\partial z_2 \partial p_1} & \frac{\partial^2 \psi_{eip}}{\partial z_2 \partial p_2} \\ \frac{\partial^2 \psi_{eip}}{\partial p_1 \partial z_1} & \frac{\partial^2 \psi_{eip}}{\partial p_1 \partial z_2} & \frac{\partial^2 \psi_{eip}}{\partial p_1^2} & \frac{\partial^2 \psi_{eip}}{\partial p_1 \partial p_2} \\ \frac{\partial^2 \psi_{eip}}{\partial p_2 \partial z_1} & \frac{\partial^2 \psi_{eip}}{\partial z_2 \partial p_2} & \frac{\partial^2 \psi_{eip}}{\partial p_1 \partial p_2} & \frac{\partial^2 \psi_{eip}}{\partial p_2^2} \end{pmatrix}$$

Here,

$$\frac{\partial^2 \psi_{eip}}{\partial z_1^2} = -(A_{\mathrm{m}} + A_{r_1})\frac{2d_1(d_1 + z_1)}{(d_1 + z_1)^4} - h_{r_1} f_1(z_1) - s_1 f_1(z_1) \quad (11)$$

$$\frac{\partial^2 \psi_{eip}}{\partial z_2^2} = -(A_{\mathrm{m}} + A_{r_2})\frac{2d_2(d_2 + z_2)}{(d_2 + z_2)^4} - h_{r_2} f_2(z_2) - s_2 f_2(z_2) \quad (12)$$

$$\frac{\partial \psi_{eip}^2}{\partial p_1^2} = \frac{2\beta_1^2 z_1 (A_{\mathrm{m}} + A_{r_1})(a_1 - \beta_1 p_1 + \gamma_1 k_1 + z_1)}{(a_1 - \beta_1 p_1 + \gamma_1 k_1 + z_1)^4} - 2\beta_1 \quad (13)$$

$$\frac{\partial^2 \psi_{eip}}{\partial p_2^2} = \frac{2(A_{\mathrm{m}} + A_{r_2})\beta_2^2 z_2 (a_2 - \beta_2 p_2 + \gamma_2 k_2)}{(a_2 - \beta_2 p_2 + \gamma_2 k_2 + z_2)^4} - 2\beta_2 \quad (14)$$

If Δ_m(mth order principal minor) satisfies the sign $(-1)^m$ then ψ_{eip} is a concave function i.e. it becomes optimum at $(p_1^*, p_2^*, z_1^*, z_2^*)$.

4 A Case Example and Sensitivity Analysis

A case example is given here in order to establish the use of the model. The magnitude of demand of retailer 1 follows normal distribution N [2, 16] and the magnitude of

Table 1 The equilibrium results of the centralized models

Optimal values	p_1^*	p_2^*	z_1^*	z_2^*	ψ_m^*	ψ_{r1}^*	ψ_{r2}^*	ψ_{eip}^*
Centralized	42.07	39.54	16.29	17.33	2603.35	1197.30	896.42	4697.07

demand of retailer 2 follows normal distribution N [3]. The manufacturer unit cost component of production, wholesale, ordering, and holding are $c = 5$,$w = 20$, $A_m = 6$, $h_m = 0.3$. The retailers unit cost component of ordering, holding, shortage, and product quality are $A_{r_1} = 2$, $A_{r_2} = 2.1$, $h_{r_1} = 0.01$, $h_{r_2} = 0.03$, $s_1 = 15$, $s_2 = 16$, $k_1 = 0.04$, $k_2 = 0.9$. We use degree of product quality $k = 0.6$ and cost used for improving quality degree of product $e_k = 8$. Price sensitivity of both the retailers $\beta_1 = 1.9$, $\beta_2 = 2.1$. Product quality sensitivity of both the retailers $\gamma_1 = 0.2$, $\gamma_2 = 0.25$. The market potential of both the retailers $a_1 = 150$, $a_2 = 155$. Also, the eigenvalues of the hessian matrix are $\Delta_1 = -11.664 < 0$, $\Delta_2 = 6.664 > 0$, $\Delta_3 = -11.24 < 0$, $\Delta_4 = 51.322 > 0$ (Table 1).

Using these value of decision variables with some changes in model parameter are studied in the following section (Table 2).

We have listed the following observation after analysis of some major parameters-

- For the increasing value of a_1,γ_1 expected integrated profit is increasing simultaneously. And for increasing the value of β_1 expected integrated profit is decreasing.
- For the increasing value of β_2, expected integrated profit is increasing firstly and then decreasing. For increasing value of γ_2, the expected integrated profit is increasing simultaneously. And also for increasing value of a_2 expected integrated profit is firstly decreasing and then increasing.
- Expected integrated profit is not very much sensitive to changes of γ_1.
- Expected integrated profit is very much sensitive for changes of a_2 and γ_2.

5 Conclusion

In this competing market situation, the retailers are active to survive in a supply chain model. Retailers are always competing with each other considering various factors like price, quality, advertisement of the product, lead time, etc. The prime objective of our work is to calculate the optimum ordering quantity and optimum selling price for one product that maximizes the overall profit for the centralized channel by minimizing the inventory-related cost, shortage cost, etc. A centralized system provides an optimum solution by giving equal importance to all the supply chain members. In our work, we proposed a supply chain mechanism of two-stage containing one manufacturer and two retailers where random demand functions of both the retailers are sensitive to the selling price and product quality. One portion of the demand of both the retailers is deterministic and another portion is stochastic. By combining both the portions of the demand we get the stochastic order quantity. Since

Table 2 Sensitivity analyses

% change of parameters	p_1	p_2	q_1	q_2	ψ_m	ψ_{r_1}	ψ_{r_2}	ψ_{eip}
−4%	43.72	39.54	16.29	17.33	2606.2	1317.39	896.42	4820.01
−2%	42.88	39.54	16.29	17.33	2604.86	1256.11	896.42	4757.39
$\beta_1 = 1.9$	42.07	39.54	16.29	17.33	2603.35	1197.30	896.42	4697.07
+2%	41.29	39.54	16.29	17.33	2602.04	1140.19	896.42	4638.65
+4%	40.55	39.54	16.29	17.33	2600.55	1086.49	896.42	4583.46
−4%	42.07	41.08	16.29	17.33	2610.35	1197.30	1138.18	4945.83
−2%	42.07	40.29	16.29	17.33	2608.87	1197.30	1078.95	4885.12
$\beta_2 = 2.1$	42.07	39.54	16.29	17.33	2603.35	1197.30	896.42	4697.07
+2%	42.07	38.81	16.29	17.33	2605.6	1197.30	967.79	4770.69
+4%	42.07	38.11	16.29	17.33	2604.12	1197.30	915.52	4716.94
−4%	40.49	39.54	16.29	17.33	2558.96	1024.92	896.42	4480.3
−2%	41.28	39.54	16.29	17.33	2581.24	1109.86	896.42	4587.52
$a_1 = 150$	42.07	39.54	16.29	17.33	2603.35	1197.30	896.42	4697.07
+2%	42.86	39.54	16.29	17.33	2625.75	1286.04	896.42	4808.21
+4%	43.65	39.54	16.29	17.33	2647.99	1378.89	896.42	4923.3
−4%	42.07	38.06	16.29	17.33	2561.22	2603.35	859.72	6024.29
−2%	42.07	38.79	16.29	17.33	2583.9	2603.35	939.88	6127.13
$a_2 = 155$	42.07	39.54	16.29	17.33	2603.35	1197.30	896.42	4697.07
+2%	42.07	40.27	16.29	17.33	2630.39	2603.35	1106.84	6340.58
+4%	42.07	41.01	16.29	17.33	2653.36	2603.35	1193.90	6450.61
−40%	42.06	39.54	16.29	17.33	2603.30	1196.18	896.42	4695.90
−20%	42.07	39.54	16.29	17.33	2603.35	1197.30	896.42	4697.07
$\gamma_1 = 0.2$	42.07	39.54	16.29	17.33	2603.35	1197.30	896.42	4697.07
+20%	42.08	39.54	16.29	17.33	2603.45	1197.80	896.42	4697.67
+40%	42.08	39.54	16.29	17.33	2603.45	1197.80	896.42	4697.67
−40%	42.07	39.54	16.29	17.33	2605.07	1197.30	1020.63	4823.00
−20%	42.07	39.53	16.29	17.33	2606.78	1197.30	1021.20	4825.28
$\gamma_2 = 0.25$	42.07	39.54	16.29	17.33	2603.35	1197.30	896.42	4697.07
+20%	42.07	39.55	16.29	17.33	2607.49	1197.30	1023.58	4828.37
+40%	42.07	39.56	16.29	17.33	2607.84	1197.30	1024.78	4829.92

demands are uncertain in nature, retailers have to pay holding costs and shortage costs for unsold products. We have included the quality of the product in demand function and also we have included ordering costs in our model. Our article has the following novelty.

Firstly, a two-layer supply chain model is designed for uncertain demand with a manufacturer and two competing retailers considering various costs like holding cost, ordering cost, shortage cost, etc. Lastly, the normal distribution is applied for a centralized structure which has a severe impact on today's business.

Future scope The present model can be extended for a decentralized case as well as for different types of demand patterns etc. Any other concept like buyback policy, joint return policy, advertisement effort may be included in further research.

Acknowledgements First author is thankful to MHRD to giving financial support for carrying out research at NIT Silchar and present this paper in MMCITRE-2020 at Gujarat.

References

1. Iyer, G.: Coordinating channels under price and nonprice competition. Mark. Sci. **17**(4), 338–355 (1998)
2. Tsay, A.A., Agrawal, N.: Channel dynamics under price and service competition. Manuf. Serv. Oper. Manage. **2**(4), 372–391 (2000)
3. Whitin, T.M.: Inventory control and price theory. Manage. Sci. **2**(1), 61–68 (1955)
4. Mills, E.S.: Uncertainty and price theory. Q. J. Econ. **73**(1), 116–130 (1959)
5. Sarkar, B., Chaudhuri, K., Moon, I.: Manufacturing setup cost reduction and quality improvement for the distribution free continuous-review inventory model with a service level constraint. J. Manuf. Syst. **34**, 74–82 (2015)
6. Kang, C.W., Ullah, M., Sarkar, B.: Optimum ordering policy for an imperfect single-stage manufacturing system with safety stock and planned backorder. Int. J. Adv. Manuf. Technol. **95**(1–4), 109–120 (2018)
7. Ouyang, L.Y., Chuang, B.R.: (Q, R, L) inventory model involving quantity discounts and a stochastic backorder rate. Prod. Plann. Control **10**(5), 426–433 (1999)
8. Jazinaninejad, M., Seyedhosseini, S.M., Hosseini-Motlagh, S.M., Nematollahi, M.: Coordinated decision-making on manufacturers EPQ-based and buyer's period review inventory policies with stochastic price-sensitive demand: a credit option approach. RAIRO Oper. Res. **53**(4), 1129–1154 (2019)
9. Liu, J., Xiao, T., Tian, C., Wang, H.: Ordering and returns handling decisions and coordination in a supply chain with demand uncertainty. Int. Trans. Oper. Res. **27**(2), 1033–1057 (2020)
10. Jena, S.K., Sarmah, S.P., Sarin, S.C.: Price competition between high and low brand products considering coordination strategy. Comput. Ind. Eng. **130**, 500–511 (2019)
11. Xiao, T., Shi, J.J.: Pricing and supply priority in a dual-channel supply chain. Eur. J. Oper. Res. **254**(3), 813–823 (2016)
12. Moon, I., Yoo, D.K., Saha, S.: The distribution-free newsboy problem with multiple discounts and upgrades. Math. Probl. Eng. (2016)
13. Chen, J., Liang, L., Yao, D.Q., Sun, S.: Price and quality decisions in dual-channel supply chains. Eur. J. Oper. Res. **259**(3), 935–948 (2017)
14. Taleizadeh, A.A., Noori-daryan, M., Cárdenas-Barrón, L.E.: Joint optimization of price, replenishment frequency, replenishment cycle and production rate in vendor managed inventory system with deteriorating items. Int. J. Prod. Econ. **159**, 285–295 (2015)
15. Ernst, R.: A linear inventory model with a monopoly firm. Doctoral dissertation, Ph.D. thesis, University of California, Berkeley (1970)
16. Taleizadeh, A.A., Rabiei, N., Noori-Daryan, M.: Coordination of a two-echelon supply chain in presence of market segmentation, credit payment, and quantity discount policies. Int. Trans. Oper. Res. **26**(4), 1576–1605 (2019)

Determining the Best Strategy for the Network Administrator in Dynamic Environment Through Game Theory

Jatna Bavishi, Mohammed Saad Shaikh, and Samir Patel

Abstract Communication between individuals, organization, etc., takes place over a network through various protocols so it is necessary to protect the network from various kinds of security attacks from the attacker. Hence, risk assessment must be done by the security team of the organization to look for the security threats on the network and also protect it during the time of attack. Thus, during the attack, network administrator must apply the best strategies against the attacker. Hence, in this paper, a model is proposed wherein firstly, the risk assessment is done with risk rank algorithm and Monte Carlo simulation and based on this risk, a decision is taken in a view to mitigate this risk using conditional random fields (CRF). Further, during the attack, using game theoretic approaches, the network administrator determines the best strategy to protect the network. Then, to adapt to dynamically changing environment, advantage learning is used to learn the new risks, attacks and ways to tackle the same.

Keywords Game theory · Reinforcement learning · Network security · Monte carlo simulation · Risk assessment · Conditional random fields

1 Introduction

Security is a very critical issue in an organization and every type of organization, whether it is an IT company, a government body, etc. suffers from security issues. Organizations possess some amount of risk in its each business unit and also the

J. Bavishi (✉) · M. S. Shaikh · S. Patel
Department of Computer Science and Engineering, Pandit Deendayal Petroleum University, Raisan, Gandhinagar, India
e-mail: jatnabavishi@gmail.com

M. S. Shaikh
e-mail: shaikhsaad133@gmail.com

S. Patel
e-mail: Samir.Patel@sot.pdpu.ac.in

M. Sahni et al. (eds.), *Mathematical Modeling, Computational Intelligence Techniques and Renewable Energy*, Advances in Intelligent Systems and Computing 1287,
https://doi.org/10.1007/978-981-15-9953-8_34

business units are interdependent on each other, so the risk is transferred from one business unit to another. There have been several cases of data leakage, security breach, loss of confidentiality and integrity of an organization which is a serious threat to the organization because the obtained data can be misused by the black hat hackers. This happens because the organization is vulnerable to some security threats.

So, a proper risk assessment must be done in order to identify the risk of the organization and to reach to a decision to mitigate that risk. Further, if there is any attack on the network of organization, the network administrator must follow the proper strategy so that the effect of the attack is minimum and the cost of protecting the security of the organization by network administrator is also minimum. On the other hand, the loss incurred to the attacker is also maximum and we must make sure that all the security attacks of the attacker must be in vain. So, there must be proper approaches to determine such a strategy which might be considered best for the network administrator and worst for the attacker.

Secondly, there are many types of environment to which the network of the organization might be exposed. The solutions and strategies developed by network administrator for tackling the attacker must be modelled in such a way that it adapts to the environment. So, a game like situation between the attacker and the network administrator is created which can be solved using game theory [1]. According to the environment, the non-cooperative game (there is a competition between attacker and network administrator) [1] can be classified as in [1]:

- The game of complete information: Here, all the strategies of all the players are known [1].
- The game of incomplete information: All the strategies of at least one player is not known by at least one player [1].

Both the types of games are further classified into following two categories as mentioned in [1]:

- The game of perfect information: All the previous actions of all the players which are part of the game are known by every player [1].
- The game of imperfect information: At least one player is unaware about the previous attack of at least one opponent in a multiplayer game and in two players game, all the previous actions of the competitor is not known [1].

Further classification as shown in [1] is:

- Static game: In this type of game, all the players take their actions at the same time and so it is a one-shot game [1].
- Dynamic game: Here, the game can be considered of finite or infinite moves that is, it is a multistage game where the environment keeps on changing according to the strategy played by each player at each stage of the game [1].

Therefore, the type of environment and game has to be considered to select the particular strategy.

Hence, the novel contribution of our paper are:

- We rebuild the risk rank model [2] for risk assessment and rank them in order of their risks and inculcate Monte Carlo simulation [3] in it to consider all the possibilities of risks and determine minimum and maximum risks. We also consider the dependencies between different parts of an organization and include that risk as well.
- An approach to study risk assessment, transform the stages which includes discrete information to the stages involving categorical information and take the decision based on many previous stages which can mitigate those risks by considering various factors using conditional random fields (CRF) [4].
- We built a game theoretic approach in which the network administrator which takes care of the network of the organization and the attacker which tries to breach the security of the organization acts as the players of the game and the network administrator selects the best possible strategy to mitigate the effect of loss caused by network attacker.
- Finally, considering all these factors and taking the dynamic environment into consideration, wherein multiple stages are involved and attacker and network administrator plays their moves, we include advantage learning [6] which is the type of reinforcement learning [5], where the network administrator gets reward for mitigating the effect an attack and gets penalized if there is a successful attack on the network and security is breached. After getting penalties and rewards, the network administrator learns and inculcates the same in the payoff matrix of game theory [1] and takes care of the same another time the same attack happens. So, the network administrator learns from the previous actions in dynamic environment from advantage learning.

So, considering this model, the organization of the paper is done as follows: Sect. 2 of the paper describes the risk assessment and Sect. 3 describes the game theoretic approach when the attack on the network takes place. Section 4 describes the advantage learning in the dynamic environment where the network administrator evolves with new attacks, and finally, Sects. 5 and 6 describes the conclusion and future scope, respectively.

2 Risk Assessment

2.1 Risk Assessment Framework

We model the risk assessment framework by making minor tweaks in risk rank algorithm. So, firstly we calculate the risk that different units an organization possesses with the help of risk rank algorithm and see which business unit has more probability of risk in future. Dependencies between the business units of an organization is also considered in risk rank algorithm. Modelling of risk in risk rank algorithm takes place as follows as in [5]. An organization can be modelled as a graph in which the business units are taken as nodes and edges between the nodes indicate the dependencies between the nodes that is, the business units of an organization. Now, each business unit possess some amount of risk of its own and some risk is also carried from one business unit to another due to the dependencies between them. Hence, some correlation needs to be created between the business units so that risk assessment can be done. The risk possessed by each business unit of an organization at time *t* can be represented as given in [2]:

$$vX(t) = [vX1(t), \ldots, vXi, \ldots, vX \quad MX + 1(t)] \tag{1}$$

Also, depending on the present state of the business unit, the future state of the business unit can be determined by examining what effect is produced due to correlation between the business units and previous risk value of business unit which can be given as below as in [2]:

$$vX(t+1) = HvX(t) \tag{2}$$

Here, the risk sink is taken into account because it might also happen that the risk might decrease with time, so that particular amount of risk is to be removed and has to be put into the sink. And, the total risk in an organization is the summation of proportion of risk carried by each business unit multiplied by risk on that node [2]. Now, risk rank algorithm similar to page rank algorithm [2] is applied to find out which business unit of an organization carries the maximum risk so that we can mitigate it. So, the risk vector of the future state can be given as below which considers the balance between current and future risk:

$$vX(t+1) = \alpha HvX(t) + \beta vX(0) \tag{3}$$

Thus, we got the maximum risk probability of the specific business unit of an organization. Further, many more simulations are done with the help of Monte Carlo simulation. Monte Carlo simulation can be used to understand the effect of risk under uncertainties. By performing Monte Carlo simulation within the node, various risk values of the business unit can be estimated. Similarly, different risk values for all the business units of an organization can be simulated. Now, by applying the simulations on the effects and dependencies on business units, different results of risks can be estimated. Hereby, the effects of both intra-dependencies and inter-dependencies can be known. So, with the help of Monte Carlo simulation, we can get the risk of each business units under various uncertainties so that all the possibilities are considered and the minimum, maximum and average value of risk of an organization can be estimated and appropriate actions must be taken.

Hence, we modelled the risk of the organization by using risk rank algorithm and introducing Monte Carlo simulation within it so that we can get the probabilistic range of the risk within the organization. Now, we need to control the risk by examining that which action must be taken to reduce the cost of the risk. This leads us to markov decision process (MDP) [7]. But we know that MDP considers only the previous state of the system to compute the risk of the present state, hence we use a different methodology to calculate the risk.

2.2 Controlling Risk Dynamics and Taking Decisions

Time is to be discretized in order to represent the states of the system. Also, we have to estimate that the risk is high, low or average, and hence, we cannot use the risk vectors which are obtained from probability risk model directly. Firstly, they are to be converted to a different state which accumulates the nodes which are to be classified as low as one state, high risk as another state and average risk as third state and so on. Thus, we get a new set of states $s1$, $s2$,…, si.

Now, as we know that MDP considers the probability of the risk of previous state and predicts the risk of the next state. But, we are not always sure that the risk of the previous state only contributes to the risk of the next state. This is the dynamics of risk wherein all the previous states contributes differently to the next state. If we consider only one state, the effect of the previous states goes on reducing or exceeding on reaching the next state which might not always be the case. It may also happen that some of the previous states may contribute less and some may contribute more to the next state. So, a series of previous states must be considered. This can be fulfilled by using conditional random fields (CRF) instead of MDP. CRF can be used for pattern recognition which can be used to model dependencies between the states of the system [8]. CRF can be used in various ways as described in [8]:

- For learning of model: If the risk of some states are given, the model can be trained in such a way, that some previous states are considered and the next states can be predicted.
- For predicting the maximum likelihood of risk: If the risk of the previous states are known, the maximum probability of risk of previous state can be computed.

The probability of new state can be determined as follows:

$$P[s(t+1)\,|P[s(t)],\,P[s(t-1)],\;P[s(t-i)]]$$

CRF are discriminative models which provides the confidence measure of the outcome along with the outcome itself [4] so that we can be sure that this amount of risk is present in the state and the state consists of different business units. So, if the probability of risk of state is classified as high, all the business units of the states try to reduce the risks by looking through previous states and reducing the risks of the previous states. HMM is a generative model and it does not provide the confidence measure [4]. Another advantage of using CRF is that the previous states has the flexibility of being chosen over time and it has relaxation of independence assumption, wherein it is assumed that all the states are independent [4]. So, CRF would lead to better results than MDP because of its conditional nature to determine the risk of the state.

Now, after minimizing the risk, there is some attacker attacking the system. Based on probabilistic risk framework, we can know the probability of the risk on each business unit and its dependencies. Similarly, as the attacker goes on attacking and the defender goes on reducing the risk on the network of organization and thereby increasing the security by trying to cause the minimal effect, we compute the dynamic risk through CRF and finally, on the basis of the results of this risk, with the help of game theory, an appropriate action of the defender is selected to minimize the risk of security causing maximum loss to the attacker.

3 Solving Zero-Sum Game

In this section, we will formulate the problem of attacker and defender as a game using game theoretic approach [9]. Here, the attacker is any outsider or insider who wants to breach security premises of any organization while defender is network administrator. We define our zero-sum two player game similar to [9] as a set consisting $\{S, P, Ai,\}$, where

$$S = \{s1, s2, \ldots sk\} \tag{4}$$

is a non-empty set of finite state space of the game.

$$P = \{Pi\} \;\; i = 1, 2 \tag{5}$$

is a set of finite players playing in the game (here, attacker and defender).

$$Ai = \{a1, a2, \ldots an\} \tag{6}$$

is a set of finite actions defined for each player $pi \in P$.

In zero-sum two player game, the two players either want to maximize the minimum gain or minimize the maximum loss. So here, the network administrator will try to minimize the maximum loss, i.e., the network administrator will try to minimize the impact of loss that attacker is trying on organization. While attacker will try to maximize the gain, i.e., attacker will try to breach security to maximum possible extent. We consider two players in a game as $t1$ and $t2$ where $t1$ is a network administrator and $t2$ is attacker. It can be easily inferred that set of actions of two players will be set of possible attacks and defence moves. And in this game, we consider each move (attack or defence) as an action. Hence, action sate space for $t1$ and $t2$ can be defined as:

$$t1 => A1 = \{a1, a2, \ldots an\} \tag{7}$$

$$t2 => A2 = \{a1, a2, \ldots an\} \tag{8}$$

Consider a stochastic game *H*, consisting of a finite number of states

$$S = \{s1, s2, \ldots sk\} \tag{9}$$

With each state *Sm*, game matrix *Hm* is associated. State transition from *Sm* to *Sn* is contingent on outcome of *Hm* and probability of transition from state *Sm* to *Sn*, *P*(*Sn*)*m*.

Where *Pm* is a probability distribution over state space and hence it holds $0 \leq Pm \leq 1$.

Here, we consider defender as the row player and attacker as the column player in the game *Hm* which is defined in 2-D matrix form as:

$$H(m) = \left(a_{i,j}^{m} + \sum_{i}^{N} P_i^{(l)} H(l) \right) \text{ for } m = 1, 2 \ldots N \tag{10}$$

We can interpret that when players choose row *i* and column *j* of matrix, it will lead attacker to gain amount *ai*, *j* causing defender to lose the same amount. The game may transit to other state or itself or may even terminate. To find saddle point, we use approach as illustrated in [9], where it is considered that each game *Hm* has value associated with it *V*(*k*). Furthermore, it is defined as below:

$$V(k) = \left(a_{i,j}^{m} + \sum_{i}^{N} P_i^{(l)} V(l) \right) \tag{11}$$

In stochastic games, the games may itself be dependent on other games and outcome of specific choice of pure strategies of player may be dependent on probability of playing other game. As stated earlier, every game has a value associated with it, which can be found out using min-max theorem. "If $V = 0$ the game is said to be fair, if $V > 0$ the game is said to favour player 1 and if $V < 0$ the game is said to favour player 2 [10]."

In order to find the value of the game, first we need to find saddle point from the matrix, if we can find saddle point than that point is the value. If we cannot find, then we may convert the matrix into linear programming (LP) problem and then solve with respect to defender's game as it is converted into min programming problem and it is standard formulation of LP problem. After that we apply simplex method in order to solve LP problem and finally attain the value of the game. According to [10], the LP formulation to find average gain *V* with respect to defender is given by:

$Z =$ maximum *v*,

Choose *p*1, …, *pm* and *v*

Subject to the constraints,

$$v \leq \sum_{j=1}^{n} pj\, a1j, \ldots, v \leq \sum_{j=1}^{n} pj\, amj.$$

$$p1 + \cdots + pm = 1,\ pj \geq 0 \quad \text{for j} = 1,2,\ldots \text{n.}$$

Similarly, by converting this problem to dual [10], we can find LP with respect to attacker. After solving both LP problems for attacker and defender, we will end up with optimal strategies for both.

4 Using Advantage Learning

Reinforcement learning is a notion applied for making machines intelligent by combining two concepts, dynamic programming and supervised learning. It solves the problems which cannot be solved using either of the two mentioned concepts. Dynamic programming is a mathematical method for solving problems of optimization and control. But it suffers from limitation of size and complexity of problems that it can successfully elucidate. Supervised learning is a type of methodology which maps inputs to outputs with respect to training samples on which model is trained. Furthermore, this can be done using neural networks which makes approximate function which takes input, processes it and maps into output [11]. Here, we may not exactly know strategies of attacker and defender, i.e., we cannot use supervised learning alone to predict action with respect to attacker or defender.

Hence, we use reinforcement learning as used in [5] in order to solve problem of dynamically changing environment, i.e., dynamically changing strategies of attacker and defender, and solve the game accordingly. In reinforcement learning, the model is given a desired goal state and this model tries to achieve goal by exploring and exploiting states through time.

As stated earlier, the states of both attacker and defender are evolving and changing with respect to time, in other words transition probabilities, i.e., transition from current state to the next state of attacker or defender, may not be known. The environment is non-deterministic; hence, we need to make model accurate where both agents will understand and learn situation better over time. There are numerous approaches to this type of learning, in [5] they used Q-learning to solve this problem. In Q-learning, the model updates Q-function iteratively through time in order to converge to optimal policies of attacker and defender. In traditional value-iteration process, model will find mapping from state to state values while Q-learning maps from state to Q-values using Q-function. After many iterations, convergence of Q-value of specific action will attain to optimal Q-value.

As Q-learning is advantageous over value-iteration process in non-deterministic environment, it suffers from one major limitation and as stated in future work of [5]. When the state space is larger or when there is smaller increment in time between actions, convergence to optimal Q-value is attained after performing significantly large number of iterations, and hence it scales poorly.

Advantage learning can be used to overcome this problem. In advantage learning, it calculates advantage (utility value) rather than Q-value. It finds function similar to Q-learning that is based on state/action pairs which is called advantage function. As stated in [6], advantage learning stores information of two types. Firstly, for every state x, it stores $V(x)$ (value of that state) which is defined by value corresponding to maximum advantage at that state. Secondly, for every action u and state x, it stores $A(x, u)$ which represents summation of value of that state and advantage of executing an action u over the action which is currently accepted best. Thus, we can infer that if u is the optimal action itself than $A(x, u)$ will be zero, conferring there is no advantage of an action when compared with itself. And $A(x, u)$ will be having negative value for any suboptimal state u as suboptimal action will have advantage which will be negative with respect to action considered to be best. In [12], the updates in advantage learning have been proved to learn faster compared to Q-learning, meaning it will perform smaller number of iterations comparative to Q-learning.

The advantage function $A^*(x, u)$ for state x and corresponding action u can be defined by [6]:

$$A*(x,u) = \max_u A*(x,u) + \frac{\left[R + \gamma \Delta t A*\left(x', u\right)\right] - A*(x,u)}{\Delta t K} \tag{12}$$

where $\gamma \Delta t$ is the discount factor per time step,

R is the reinforcement received on transition from state x to state x' by executing action u.

We can infer that for optimal action $A^*(x, u)$ is 0 which indicates that action's value is similar to value of the state while for actions which are suboptimal $A^*(x, u)$ < 0 implying that impact of that action is negative relative to optimal action.

In order to converge the algorithm, we can use bellman residual algorithm for advantage learning [6]. This is called residual gradient algorithm as it learns advantage function approximation by effectuating gradient descent. The bellman equation for advantage learning as given in [6] is defined as:

$$A*(x,u) = \left[\mathrm{R} + \gamma \Delta t \mathrm{max} u A*\left(x', u\right)\right]\frac{1}{\Delta t K} + \left(1 - \frac{1}{\Delta t K}\right)\max_u A*(x,u) \tag{13}$$

The mean squared bellman residual, if $A(x, u)$ is approximation of $A^*(x, u)$ is given by:

$$E = \left[\left(\left[R + \gamma \Delta t \mathrm{max} u A*\left(x', u\right)\right]\frac{1}{\Delta t K} + \left(1 - \frac{1}{\Delta t K}\right)\max_u A(x,u) - A(x,u)\right)^2\right. \tag{14}$$

In the above equation, first part represents total expected value for performing particular action u and the overall expression represents expected value across every possible actions and states.

5 Conclusion

We tried to build a mathematical approach, wherein risk assessment is done with risk rank algorithms and considering various cases for risk using Monte Carlo simulation which will consider the broad state space for risky situations. Later, for mitigating those risks, CRF is used which might prove an advantage over MDP for taking decisions for the risk due to its nature of considering many previous states and relaxation from independence assumption. Further, during the attack on the network of organization, the strategy of the network administrator is selected using game theoretic approach and the learning from dynamic environment takes place using advantage learning that is a form of reinforcement learning which proves to be better than Q-learning or any other form of reinforcement learning because of its special features. This helps learn the situations which are not faced earlier by the network administrator whose knowledge can be used when the similar kinds of situations are encountered.

6 Future Scope

Current work describes the model where only, the strategies can be determined when there is a single attacker at a time. But if there are multiple attackers, the approach needs to be changed and this can be taken as future work. Also, reducing the complexity of risk assessment can be done.

References

1. Liang, X, Xiao, Y.: Game theory for network security. IEEE Commun. Surv. Tutorials **15**(1), First Quarter (2013)
2. Alpcan, T., Bambos, N.: Modeling dependencies in security risk management. In: 2009 Fourth International Conference on Risks and Security of Internet and Systems (CRiSIS 2009)
3. Vrugt, J.A.: Markov chain Monte Carlo simulation using the DREAM software package: theory, concepts, and MATLAB implementation. Environ. Model Softw. **75**, 273–316 (2016)
4. Zhang, X., Aberdeen, D., Vishwanathan, S.V.N.: Conditional random fields for multi- agent reinforcement learning. In: Proceedings of the 24th International Conference on Machine Learning, Corvallis, OR (2007)
5. Bommannavar, P., Alpcan, T., Bambos, N.: Security risk management via dynamic games with learning. In: 2011 IEEE International Conference on Communication (ICC), pp. 1–6 (2011)
6. Harmon, M.E., Baird, L.C.: Multi-Player Residual Advantage Learning With General Function Approximation. Technical Report, Wright Laboratory, WL/AACF, Wright- Patterson Air Force Base, OH, 45433-7308 (1996)
7. Wei, Z., Xu, J., Lan, Y., Guo, J., Cheng, X.: Reinforcement learning to rank with Markov decision process. In Proceedings of the 40th International ACM SIGIR Conference on Research and Development in Information Retrieval, pp. 945–948 (2017)
8. Sutton, C., McCallum, A.: An introduction to conditional random fields. Found. Trends Mach. Learn. **4**(4), 267–373 (2012)

9. Ibidunmoye, E.O., Alese, B.K., Ogundele, O.S.: Modeling Attacker-Defender Interaction as a Zero-Sum Stochastic Game. Journal of Computer Sciences and Applications **1**(2), 27–32 (2013)
10. Ferguson S. T.: Game Theory II – Two-Person Zero-Sum Games (2007)
11. Harmon, M.E., Harmon S.S.: Reinforcement Learning: A Tutorial. Technical Report, Wright Lab Wright-Patterson AFB OH (1997)
12. Watkins, C.J.C.H.: Learning from delayed rewards. Doctoral thesis, Cambridge University, Cambridge, England (1989)

Thermal Modeling of Laser Powder-Based Additive Manufacturing Process

Harsh Soni, Meet Gor, Gautam Singh Rajput, and Pankaj Sahlot

Abstract Additive manufacturing (AM) is getting wide acceptance for various applications in the industries to fabricate 3D complex components with least material wastage. However, it is difficult to build the defect-free components with nominal residual stresses and high mechanical strength due to the high thermal gradient during the AM process. Hence, it is important to study temperature distribution on a layer by layer deposition to get defect-free components. In this paper, 3-D transient heat transfer numerical model is developed for laser powder based additive manufacturing process to compute the thermal behavior. The conservation of energy is solved to compute the temperature distribution on each layer during AM processes. Thermal cycles show variation in the peak temperature as the subsequent layer gets deposited over previous layer. The peak temperature is highest in the beginning and decreases as the more layers get deposited during the AM process. The heating and cooling gradient is also observed more in the beginning and successively reduces with the time. The developed model will be useful to reduce residual stresses and improve geometrical stability. This study will also be helpful to understand the AM process and to produce defect-free components.

Keywords Additive manufacturing · Thermal modeling · Thermal cycles · Peak temperature

1 Introduction

Additive Manufacturing (AM) has revolutionized the manufacturing industry in the recent past. It is widely adopted in different industries such as aviation, aerospace, automobile, and biomedical [1, 2]. Powder based fusion technique is one type of AM processes which is defined by the ASTM [3]. According to ASTM standard, the seven

H. Soni · M. Gor · G. S. Rajput · P. Sahlot (✉)
Mechanical Engineering, School of Technology, Pandit Deendayal Petroleum University, Gandhinagar, Gujarat 382007, India
e-mail: pankaj.sahlot@sot.pdpu.ac.in

M. Sahni et al. (eds.), *Mathematical Modeling, Computational Intelligence Techniques and Renewable Energy*, Advances in Intelligent Systems and Computing 1287,
https://doi.org/10.1007/978-981-15-9953-8_35

processes of AM is powder bed fusion, binder jet, sheet laminating process, directed energy deposition, material jet, vat photopolymerization, and material extrusion [1, 3]. In powder bed fusion, high energy density source is required like EBM and laser to melt or sinter the metallic powder [2, 4, 5]. The laser-based AM is getting more popular nowadays due to its ability to produce complex shapes. The main benefit of this process is to provide freedom in design for complex shapes with minimum material wastage and less expense over conventional manufacturing process [4, 6]. The metal powder melts with the help of laser beam and deposits it layer by layer to get the desired shape [6–8]. The powder metal, in this case, undergoes rapid heating and cooling (melting and solidification) so a large amount of thermal gradient is developed in the part, which leads to the formation of residual stresses and distortion [5, 7, 8]. As additive manufacturing involves many process variables, it is challenging to measure the temperature variation experimentally and physical behavior throughout the process. Numerical modeling is very useful to provide insight for a complex process by performing only a few expensive experiments [9–11]. There are various manufacturing processes like welding, forging, casting where the computation of the temperature distribution helped to understand the processes in a better way [9, 12, 13].

The modeling of AM process is the key to analyze the entire process closely. As the temperature distribution plays a major role to predict the build component properties, many researchers proposed modeling of additive manufacturing and mainly focused on temperature distributions during the process. Ayalew and Cao investigated powder deposition of various I/P and O/P model of a laser. They observed the distribution of temperature and layer thickness during the process [14]. Toyserkani et al. proposed a three-dimensional to resolve the issue with heat, by using a combine physics module. They employed thermal gradient analysis for predicting the molten pool shape [15]. Batut et al. proposed an analytical model to study temperature behavior for direct deposition method. They predicted temperature distribution by the transient heat source [16]. Foteinopoulos et al. introduced a model that is capable of accurately simulate and store full temperature history of a 2 D component. They used an adaptive meshing strategy to develop [17]. As the temperature distribution and peak temperature play a critical role to decide the properties of build structure, and limited research has been done on variation in peak temperature with time on a particular layer, which helps to analyze thermo-mechanical behavior of additive manufacturing. In this paper, three-dimensional model by moving heat source is developed for laser powder based AM process to compute the thermal behavior. The developed model will be helpful to improve geometrical stability.

2 Numerical Modeling of AM Process

In this research, a 3-Dimensional heat transfer model is used to study the temperature distribution during laser power based AM process. Finite element method based

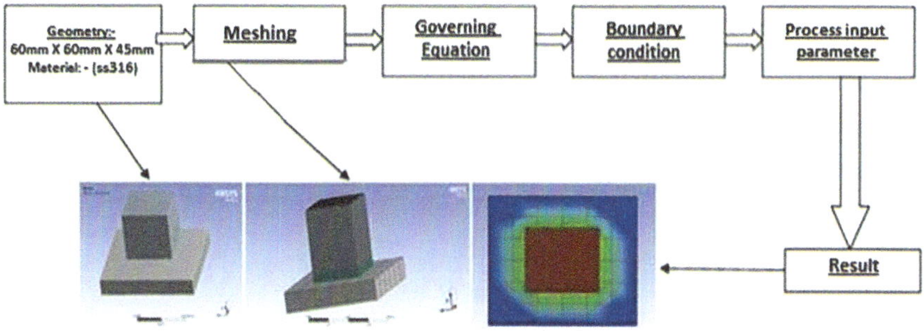

Fig. 1 Modeling flow

model is evolved using the ANSYS WORKBENCH 19.2. The modeling flow with various followed steps to develop the model for AM processes is shown in Fig. 1.

2.1 *Geometry and Meshing*

The aim of this study is to develop a heat transfer model for building a cuboid geometry of size 60 mm × 60 mm × 45 mm over a base material as shown in Fig. 2 with meshing performed over the domain. The different mesh size has been applied for a base and build material. Currently, this model is developed by considering the coarse mesh to reduce the computational time. Stainless steel (SS) 316 is used as a build and base material. The physical properties of SS316 used in this model are Density (8 g/cm^3), Melting point (1400 °C), modulus of elasticity (193 GPa), Thermal conductivity (16.30 W/m K), and Thermal expansion (15.9 × 10^{-6} k^{-1}).

Fig. 2 Geometry with meshing

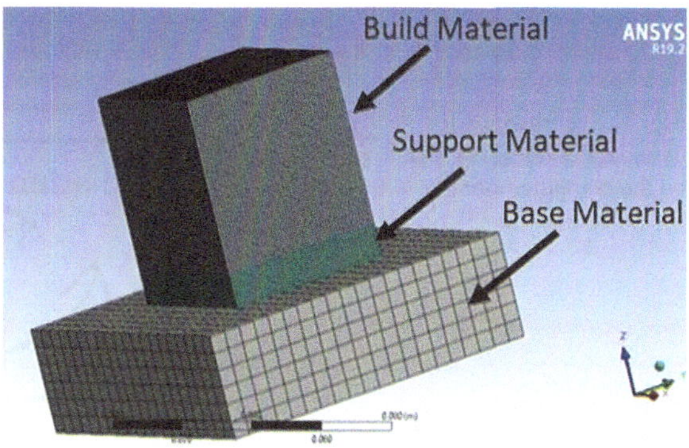

2.2 Governing Equation

These equations are used to predict the general physical behavior of the process. In this study governing equation is solved for transient thermal analysis where the material is assumed to be incompressible. The conservation energy equation which used to govern the physical behavior in whole process is described below in Eq. 1.

$$\rho + \mathrm{Cp}\frac{DT}{Dt} = Q'' + \nabla(k\nabla T) \tag{1}$$

where, T, k and ρ are temperature, thermal conductivity, and density, respectively.

2.3 Boundary Equation

There are mainly three types of boundary conditions are applied, convection, radiation, and ambient temperature over the computed domain. Figure 3 shows all the boundary conditions.

(i) Heat dissipation by convection: According to Newton's law of Cooling, some amount of heat is dissipated to the surrounding by convection as mentioned in Eq. 2. Where h represent coefficient of transfer, and it is considered as 10 w/m^2.

$$Q = h(T1-T2) \tag{2}$$

(ii) Heat dissipation by radiation: The fraction of heat is also transferred by radiation which follows Stefan–Boltzmann's law. Where ε is emissivity and value is considered 0.7 and σ *is* Stefan–Boltzmann constant.

$$Q = \sigma\varepsilon(T4-T04) \tag{3}$$

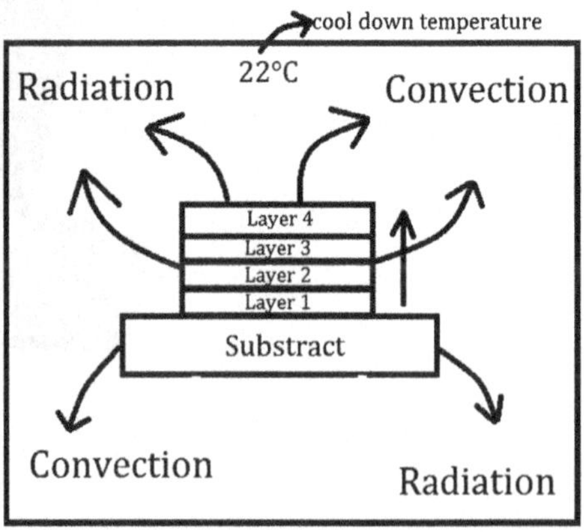

Fig. 3 Section view for boundary condition applied on the computed domain

Table 1 Laser process based input parameters

Parameters	Value
Preheat temperature(°C)	100
Deposition thickness(m)	4.e-005
Laser scanning speed(m/s)	1.2
Hatch spacing(m)	1.3e-004
Dwell time(sec)	10

(iii) Ambient temperature: Surrounding temperature of 22 °C is applied at the outer domain of geometry.

2.4 *Laser Process Based Input Parameters*

Additive manufacturing (AM) involves various process parameters which directly affects mechanical properties [14]. Hence, it is important to select appropriate parameters such as preheating temperature, deposition layer thickness, laser speed, hatch spacing, and dwell times for AM process. The selected parameters in this research are given in Table 1.

3 Results and Discussion

The developed 3D transient heat transfer model is used to determine the temperature variation during laser powder based AM process. Figure 4 shows isometric and section view of temperature distribution in the AM component. The isometric view shows 3-D temperature distribution during 1st, 5th, and 10th layer of deposition. More clear temperature distribution can be observed in section view during deposition for all the three different layers. The sequence of these layer depositions creates the fluctuation in the temperature during the deposition. As initially powder is considered at room temperature and heating is provided with the help of laser heat source to melt the powder. Thereafter, molten material starts getting solidifies during the dwell time. During melting, temperature will be highest at the deposited layer and least at the base material as shown in Fig. 4. As heat starts diffusing due to conduction toward base material. The amount of heat transfer to the base plate will be more during deposition of first layer. However, the amount of heat transfer reduces as the subsequent deposition layer takes place. The deposition of one layer over the other takes place with some interval of dwell time to get cool down the previous layer.

Figure 5 shows the variation in the thermal cycles during deposition for 1st, 5th, and 10th layer. Each thermal cycle depict heating and cooling nature due to melting and solidification of the deposited layer. The beginning of 1st layer start melting as the heat energy is given to the metallic powder and temperature of the deposited

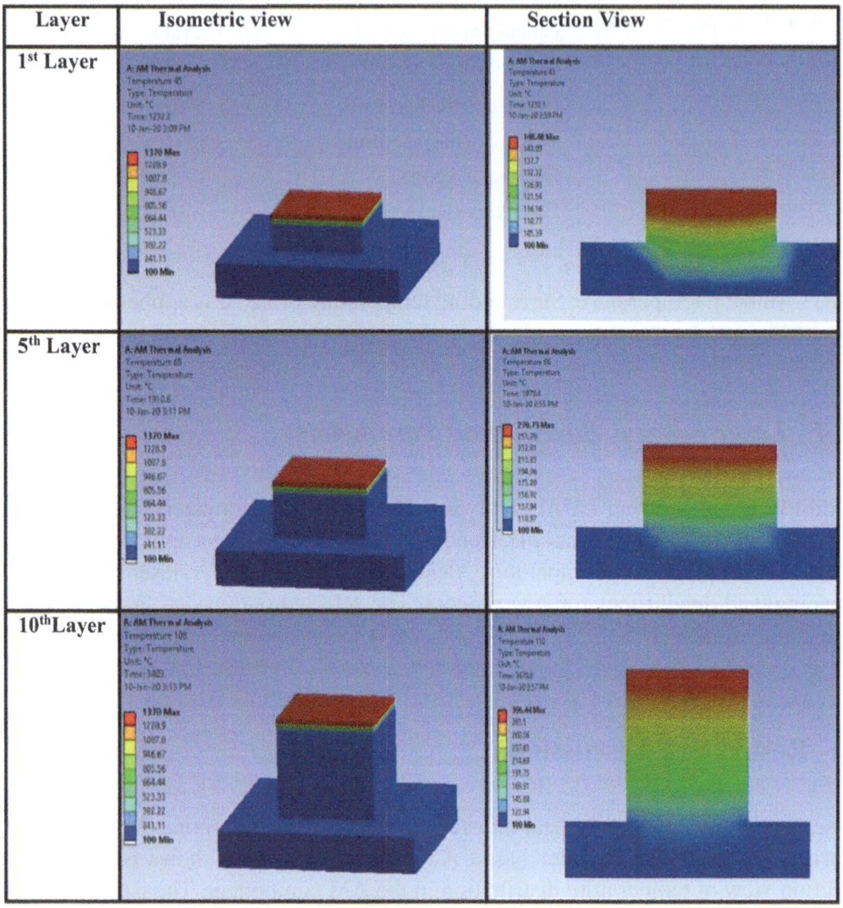

Fig. 4 Temperature distribution for various layers during deposition of SS 316 material over a base plate during AM process

layer increases up to melting point and then temperature decreases as material start solidifying as time is given for solidification. Then again, new layer of metallic powder is spread to the previous layer and similar heating and cooling cycles take place for deposited layer. The nature of temperature distribution leads to fluctuation in the temperature during each layer deposition. The variation of temperature at each layer will be more in the beginning and reduces as more layer get added. Similar results also observed by other researchers [18, 19].

Figure 6 shows the variation in thermal cycle at the base plate surface during deposition of material layer by layer. Thermal cycle depicts the variation in the temperature with the time as the material gets deposited over the base plate. Initially, the temperature reaches around melting point of the deposited material as the first layer gets deposited. The temperature reduces as the solidification of materials take

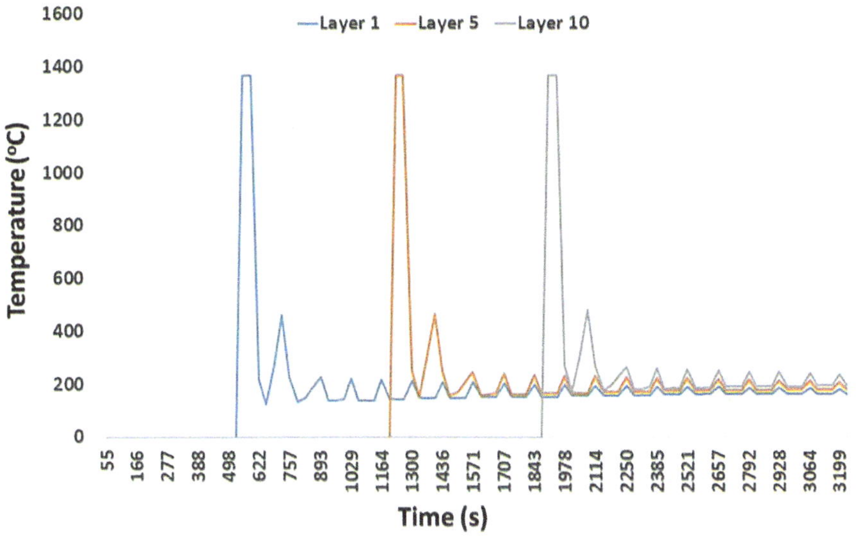

Fig. 5 Temperature variation for 1st, 5th, 10th layers for laser powder based AM process

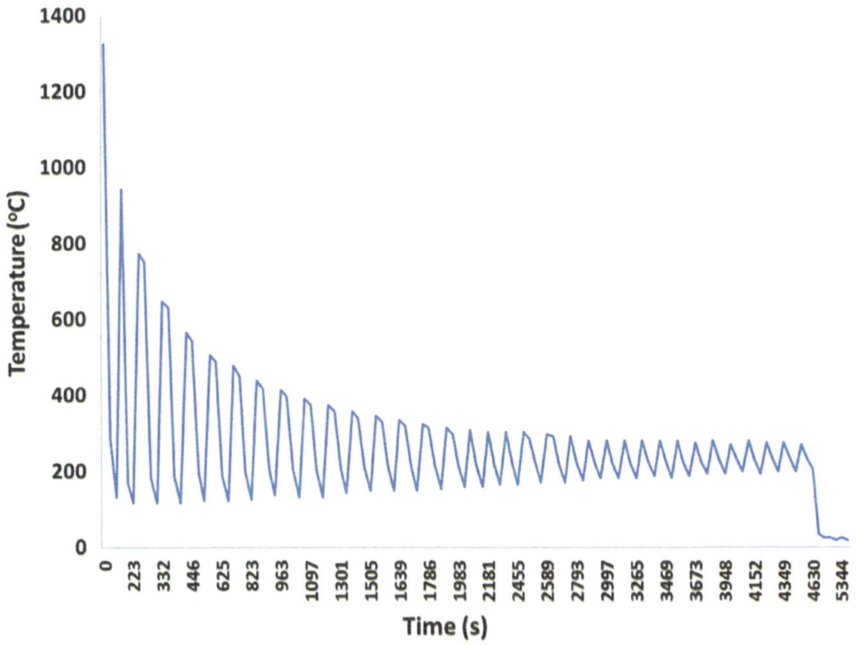

Fig. 6 Thermal cycle at the top surface of base plate during deposition of material

place due to cooling. This cycle repeats every time as the new layer get deposited over the previous layer. The peak temperature is more in the beginning and reduces as the deposition takes place. The difference in the peak and least temperature is also more in the beginning as compared to the later stage. This is due to the transfer of heat to the base plate is more at the beginning of the deposition processes and decreases as the more layer get deposited. The similar thermal cycle will be observed by all the previously deposited layers. This variation in thermal cycle leads to induce the residual stresses inside the material and may result in delamination of the deposited layer.

4 Conclusions

A numerical model is developed for predicting the temperature distribution on the build part in laser powder based additive manufacturing process. The proposed model depicts multilayer deposition for AM process. Following are the main points of this research. The developed model shows successful deposition of each layer for laser powder based additive manufacturing processes. The heating and cooling cycles during process also predicted for melting and solidification of the deposited material for various layers. The heating and cooling gradient is more in the beginning and decreases with the time. Thermal cycle is computed at the base plate during deposition of various layers. The thermal cycle shows fluctuation in peak temperature as layer gets deposited one over the other. The peak temperature decreases as more and more layer get deposited. The difference in the maximum and minimum temperature is more in the beginning of the deposition and decreases as the subsequent layer gets deposited. In this preliminary study, model simulates the physical behavior of laser-based additive manufacturing process. However, a more reliable model will be developed in future by improving some factors and then validation will be done by performing the experiments.

References

1. Uriondo, A., Esperon-Miguez, M., Perinpanayagam, S.: The present and future of additive manufacturing in the aerospace sector: a review of important aspects. Proc. Inst. Mech. Eng. Part G J. Aerosp. Eng. **229**, 2132–2147 (2015)
2. Yadroitsev, I., Krakhmalev, P., Yadroitsava, I.: Selective laser melting of Ti6Al4V alloy for biomedical applications: temperature monitoring and microstructural evolution. J. Alloy Compd. **583**, 404–409 (2014)
3. ASTM International: F2792-12a-standard terminology for additive manufacturing technologies. Rapid Manuf. Assoc. 10–12 (2013)
4. Xia, M., Gu, D., Yu, G., Dai, D., Chen, H., Shi, Q.: Influence of hatch spacing on heat and mass transfer, thermodynamics and laser processability during additive manufacturing of Inconel 718 alloy. Int. J. Mach. Tools Manuf **109**, 147–157 (2016)

5. Mukherjee, T., Zuback, J.S., Zhang, W., DebRoy, T.: Residual stresses and distortion in additively manufactured compositionally graded and dissimilar joints. Comput. Mater. Sci. **143**, 325–337 (2018)
6. Raghavan, A., Wei, H.L., Palmer, T.A., DebRoy, T.: Heat transfer and fluid flow in additive manufacturing. J. Laser Appl. **25**, 052006 (2013)
7. Greer, C., Nycz, A., Noakes, M., Richardson, B., Post, B., Kurfess, T., Love, L.: Introduction to the design rules for metal big area additive manufacturing. Addit. Manufact. **27**, 159–166 (2019)
8. Gusarov, A.V., Pavlov, M., Smurov, I.: Residual stresses at laser surface remelting and additive manufacturing. Phys. Procedia. **12**, 248–254 (2011)
9. Sahlot, P., Arora, A.: Numerical model for prediction of tool wear and worn-out pin profile during friction stir welding. Wear **408–409**, 96–109 (2018)
10. Arora, A., De, A., DebRoy, T.: Toward optimum friction stir welding tool shoulder diameter. Scripta Mater. **64**, 9–12 (2011)
11. Das, H., Jana, S.S., Pal, T.K.: Numerical and experimental investigation on friction stir lap welding of aluminium to steel **19**, 69–75 (2014)
12. Singh, A.K., Sahlot, P., Paliwal, M., Arora, A.: Heat transfer modeling of dissimilar FSW of Al 6061/AZ31 using experimentally measured thermo-physical properties. Int. J. Adv. Manuf. Technol. **105**, 771–783 (2019)
13. Soleymani Yazdi, M.R., Seyed Bagheri, G., Tahmasebi, M.: Finite volume analysis and neural network modeling of wear during hot forging of a steel splined hub. Arab. J. Sci. Eng. **37**, 821–829 (2012)
14. Cao, X., Ayalew, B.: Control-oriented MIMO modeling of laser-aided powder deposition processes. In: Proceedings of the American Control Conference (ACC), pp. 3637–3642 (2015)
15. Toyserkani, E., Khajepour, A., Corbin, S.: 3-D finite element modeling of laser cladding by powder injection: effects of laser pulse shaping on the process. Opt. Lasers Eng. **41**, 849–867 (2004)
16. de La Batut, B., Fergani, O., Brotan, V., Bambach, M., El Mansouri, M.: Analytical and numerical temperature prediction in direct metal deposition of Ti6Al4V. J. Manuf. Mater. Process. **1**, 3 (2017)
17. Foteinopoulos, P., Papacharalampopoulos, A., Stavropoulos, P.: On thermal modeling of additive manufacturing processes. CIRP J. Manuf. Sci. Technol. **20**, 66–83 (2018)
18. Kirka, M.M., Nandwana, P., Lee, Y., Dehoff, R.R.: Solidification and solid-state transformation sciences in metals additive manufacturing. Scripta Mater. **135**, 130–134 (2017)
19. Debroy, T., Wei, H.L., Zuback, J.S., Mukherjee, T., Elmer, J.W., Milewski, J.O., Beese, A.M., Wilson-heid, A., De, A., Zhang, W.: Additive manufacturing of metallic components—process, structure and properties. Prog. Mater Sci. **92**, 112–224 (2018)

Modelling and Simulation of Helical Coil Embedded Heat Storage Unit Using Beeswax/Expanded Graphite Composite as Phase Change Material

Abhay Dinker, Madhu Agarwal, and G. D. Agarwal

Abstract Latent heat storage systems are the sustainable methods to store energy both from conventional as well as renewable sources of energy. The geometry of thermal storage system as well as the phase change materials are important parameters to design an efficient thermal energy storage unit. In this paper modelling and Simulation of a shell and tube types thermal storage unit was performed using COMSOL Multiphysics using beeswax and its composite with expanded graphite as phase change materials. The beeswax was filled in the shell part of the thermal storage unit (TSU) while the heat transfer fluid is passing through the helical tube embedded in the phase change material. The model of the prepared experimental setup was simulated on the COMSOL Multiphysics and it was observed that the simulation data are in good agreement with the experimental data with only 6% and 7.5% difference for beeswax and composite material respectively. The enthalpy method was used to simulate the melting of beeswax in the thermal storage unit. It was also observed that the pattern of heat flow as well as solid-liquid interface moves in similar pattern in case of both experimental and modeled parts.

Keywords Phase change material · Multi-physics · Shell and tube · Bees wax · Helical tube

---

A. Dinker (✉)
Department of Engineering and Physics Sciences, IAR, Gandhinagar, Gujarat 382426, India
e-mail: abhay.dinker@iar.ac.in

M. Agarwal
Department of Chemical Engineering, MNIT, Jaipur, Rajasthan 302017, India

G. D. Agarwal
Department of Mechanical Engineering, MNIT, Jaipur, Rajasthan 302017, India

M. Sahni et al. (eds.), *Mathematical Modeling, Computational Intelligence Techniques and Renewable Energy*, Advances in Intelligent Systems and Computing 1287,
https://doi.org/10.1007/978-981-15-9953-8_36

1 Introduction

In recent years, thermal energy storage has been the most exploring field due to increasing energy demand and depletion of conventional energy sources. Thermal energy storage systems have shown significant utility in various fields of applications such as solar air heater, solar water heart, concentrated solar plants, etc. [1–3]. A number of experimental studies have been done to study the thermal performance and efficiency of heat storage units with different geometrical configurations like shell and tube type [4, 5], flat plat type [6], concentric tube type [7] and triplex tube type [8] configurations. Various works have also been reported on the analytical solutions of latent heat storage unit based on the geometry and the type of materials used in the unit. Analytical study on shell and tube type latent heat storage unit considering the effect of various parameters like flow rate, outer tube diameter and tube length on the charging time of the paraffin wax and the solution is obtained by exponential integral function and variable separation method [9]. Results of the study showed that as compared to the numerical solutions, analytical solutions showed reduced charging time of PCM and geometrical parameters such as tube diameter and tube length played important role in optimization of thermal storage performance. In another numerical and analytical study in which the phase change material was filled inside a vertical cylindrical enclosure integrated with concentric heater and it was observed that the performance of thermal storage unit was dependent on geometrical parameters like spiral pitch, diameter of the spiral coil, and number of spiral turns [10].

A study was carried out by Pirasaci and Goswami [11] to observe the effect of heat storage unit design on its performance and they used a novel model in which both latent heat as well as sensible heat of phase change material was considered and it was found that the length of the storage unit, diameter of the tubes and flow rate of heat transfer fluid had a significant effect on the effectiveness of the tube while distance between the tubes did not have much effect on the performance of thermal storage unit. A mathematical model was prepared to study the performance characteristics of a shell and tube type latent heat storage system using effective heat capacity method in which latent heat of material and Boussinesq approximation was used to add the buoyancy effect in the molten layer of PCM [12]. The governing equations involved in this model were solved by finite element based software COMSOL Multiphysics 4.3a and results showed that the charging of PCM was a convection dominant process and discharging was a conduction dominant process.

Aadmi et al. [13] studied the melting characteristics of PCM based on epoxy resin paraffin wax both experimentally and numerically. COMSOL Multiphysics 4.3 was used to simulate the thermal storage system and it was found that natural convection, contact melting, and melting temperature of PCM had evident effect on melting process. Simulation results showed good agreement with the numerical values. In another study Manfrid et al. [14] conducted a simulation study of solar power plant associated with a latent heat storage and a ORC (Organic Rankine Cycle) unit using TRANSYS and found that the present system was able to provide power for 78.5%

of time with weekly average efficiency of 13.4% for ORC unit and 3.4% for whole plant.

Mousa and Gujarathi [15] developed a mathematical model for desalination of water using PCM and effect of various parameters like melting point, flow rate, and solar intensity on the productivity of unit was expressed in terms of fresh water produced per day. From this study, it was observed that with decrease in flow rates and increase in solar intensity, the productivity of system increased. Other many similar studies were carried out on latent heat storage systems using various simulators like COMSOL, TRANSYS, FLUENT, CFD, MATLAB, etc. and in most of the cases simulated results were in good agreement with the experimental results. However, very little work has been found on the COMSOL simulation of phase change materials and thermal energy storage units.

2 Materials and Methods

Modelling and simulation of small and large thermal storage unit was carried out for predicting thermal performance using COMSOL Multiphysics® software version 5.2a and simulation results were validated with experimental results. The COMSOL software is a simulator and solver package based on finite element method that is used effectively to solve large number of engineering problems.

Input geometrical parameters used for creating model geometry in COMSOL software for TSU were taken from the Table 1. In actual experimental set-up, PCM was filled in the shell part of thermal storage unit and a helical tube was passed through the center of the unit. In modelling of TSU the dimension of PCM block was taken for the study rather than the dimension of TSU. Thermo-physical properties of beeswax and composite materials were defined in the blank material section of COMSOL software. 'Heat study in Fluids' physics was selected as study module for PCM materials and 'Heat transfer in solid' physics was selected for natural graphite and during the simulation following assumptions were made such as tube surface is maintained at uniform temperature throughout the tube length, Pressure drop across tube length is negligible, The velocity profile of the fluid is fully developed and

Table 1 Geometrical property for thermal storage units

Geometrical property	Large TSU
L XB X H	500 mm × 340 mm × 250 mm (excluding insulation)
Inner diameter of tube	8 mm
Outer diameter of tube	9 mm
Number of tubes	1
Geometry of tube	Helical
Mass of PCM	18 kg

remains same for complete tube length, Fluid is incompressible and flow in the tube is laminar.

All the simulation studies were carried out at the constant flow rate of 0.5 LPM and fixed inlet fluid temperature of 80 °C and simulation results were validated against the 3 experimental data sets for beeswax and composite materials. Initial temperature of PCM domain and packed bed of natural graphite was taken equal to ambient temperature of 25 °C. Initial boundary condition for the tube was taken as 0.5 LPM flow rate and 80 °C as fluid temperature. Free triangular mesh was selected for all the geometric model consists of 73,042 domain elements and 9878 boundary elements. The computation time for PCM simulation and natural graphite simulation was 7 h and 3.5 h respectively for an DELL workstation with CORE i5 processor.

3 Modelling Equations

To simulate the heat transfer fluid flowing through the copper tube, inbuilt continuity Eq. (1) and momentum Eq. (2) in COMSOL software were used:

$$\frac{\partial \rho}{\partial t} + \nabla \cdot (\rho u) = 0 \tag{1}$$

$$\frac{\partial \rho u}{\partial t} + \rho(u \cdot \nabla)u - \nabla \cdot [\mu(\nabla u) + (\nabla u)^T] + \nabla p = 0 \tag{2}$$

Heat transfer from HTF to the inner wall of the tube was through forced convection and energy balance equation for this is given by Eq. 3.

$$\rho C_p \frac{\partial T}{\partial t} + \rho C_p u \cdot \nabla T = \nabla \cdot (k \nabla T) \tag{3}$$

After the convective heat transfer from HTF to tube wall, the heat transfer further takes place from tube wall to PCM through conduction as represented by Eq. 4

$$\rho C_p \frac{\partial T}{\partial t} = \nabla \cdot (k \nabla T) \tag{4}$$

After heat conduction to PCM, the melting of phase change material occurred, (if φ is the fraction of liquid phase of PCM) then the heat enthalpy equation for the phase change material is expressed by Eq. 5.

$$H = (1 - \varphi)H_{P1} + \varphi H_{P2} \tag{5}$$

where H_{P1} = Enthalpy of phase 1 (kJ kg^{-1}), H_{P2} = Enthalpy of phase 2 (kJ kg^{-1}).

Differentiation of Eq. 5 with respect to temperature will provide the equation for specific heat as:

$$C_{\mathrm{eff}} = (1-\varphi)\frac{\mathrm{d}H_{P1}}{\mathrm{d}T} + \varphi\frac{\mathrm{d}H_{P2}}{\mathrm{d}T} + \left(H_{P2} - H_{p1}\right)\frac{\mathrm{d}\varphi}{\mathrm{d}T} \tag{6}$$

Equation 6 can be represented as:

$$C_{\mathrm{eff}} = (1-\varphi)C_{p1} + \varphi C_{p2} + \left(H_{P2} - H_{p1}\right)\frac{\mathrm{d}\varphi}{\mathrm{d}T} \tag{7}$$

$$C_{\mathrm{eff}} = (1-\varphi)C_{p1} + \varphi C_{p2} + \left(H_{P2} - H_{p1}\right)\frac{\mathrm{d}\alpha}{\mathrm{d}T} \tag{8}$$

The effective thermal conductivity of PCM during phase change process is given by Eq. 9.

$$k = (1-\varphi)k_1 + \varphi k_2 \tag{9}$$

where, k_1 = Thermal conductivity of phase 1 (W m^{-1} K^{-1}), k_1 = Thermal conductivity of phase 2 (W m^{-1} K^{-1}).

Effective density of PCM during the melting process is given by Eq. 10.

$$\rho = \frac{(1-\alpha)\rho_1 C_{P1} + \alpha\rho_2 C_{P2}}{(1-\alpha)C_{P1} + \alpha C_{P2}}, \tag{10}$$

where, ρ_1 = density of PCM in solid phase (kg m^{-3}), ρ_2 = density of solid in liquid phase (kg m^{-3}), C_{p1} = specific heat of solid phase (J kg^{-1} K^{-1}), C_{p2} = specific heat of liquid phase (J kg^{-1} K^{-1}).

Using these modeling equations from 1 to 10, a mathematical model for the TSU having PCM with embedded tube was developed for both TSU and simulated. The simulation results obtained were validated with experimental results.

4 Results and Discussion

The geometrical model of large TSU developed using COMSOL Multiphysics software is represented in Fig. 1, in which PCM was filled in the outer rectangular shell while the HTF was passed through the helical coil of the thermal storage unit. Parameters used in the simulation study are taken from the actual experimental setup and simulation study was carried out with fluid flow rate of 0.5 LPM and 80 °C inlet fluid temperature.

Temperature variation in the volume of beeswax at four different times is presented in Fig. 2 and it is observed that the matrix temperature of beeswax increased from inner side of the tube and then through conduction the heat is moving towards the outer side of the TSU. The conduction is followed by the convection of hot fluid as the temperature of the matrix reaches the melting point of the beeswax. This due to the fact that as the hot inlet fluid starts flowing in the helical tube the heat starts

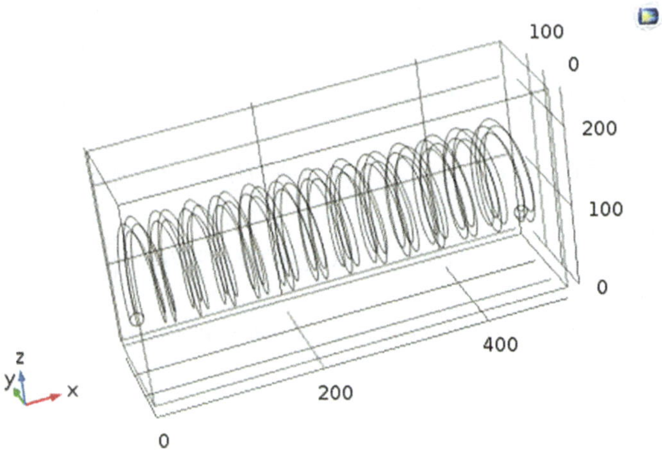

Fig. 1 Geometric model of large TSU

conducted from the hot fluid to the PCM due to large temperature difference. This movement of heat begins the melting of beeswax near to coil and then moves towards the outer side of thermal storage unit.

The charging time of beeswax in large TSU from simulation study was found to be 1515 min which is 6% smaller as compared to experimental value (105 min). This may be due to the presence of heat losses such as conduction losses, convection losses, etc., as well as this may be due to the presence of impurities in the phase change material which affects its property in the experimental system. Figure 3 representing the melting pattern of beeswax in large TSU at different times and it is observed that during charging the solid-liquid interface moves in upward direction due to the dominance of convection heat transfer and the complete melting of beeswax occurred in 660 min. The comparison between experimental results and simulated results for temperature variation at position T3 with respect to time is shown in Fig. 4 and it is found that simulation results are in line with experimental results. Lower values of temperature in experimental study are due to the heat losses taking place.

Simulation study on beeswax/expanded graphite composite material was also performed on the developed 3D geometric model for large TSU. Charging time of composite material is 1040 min which is found to be lower as compared to pure beeswax i.e. 1515 min and which is also lower as compared to experimental value of 1125 min as shown in Fig. 5. The melting profile of composite material is presented in Fig. 6 and complete melting of composite material took place in 700 min. Comparison of temperature variation with time at position T3 of TSU from experimental study and simulation analysis is shown in Fig. 7. Since both the temperature profiles are almost similar prove that the simulation results are in agreement with experimental results. Lower values (7.5%) of temperature in experimental study are due to heat losses taking place from TSU and helical tubes.

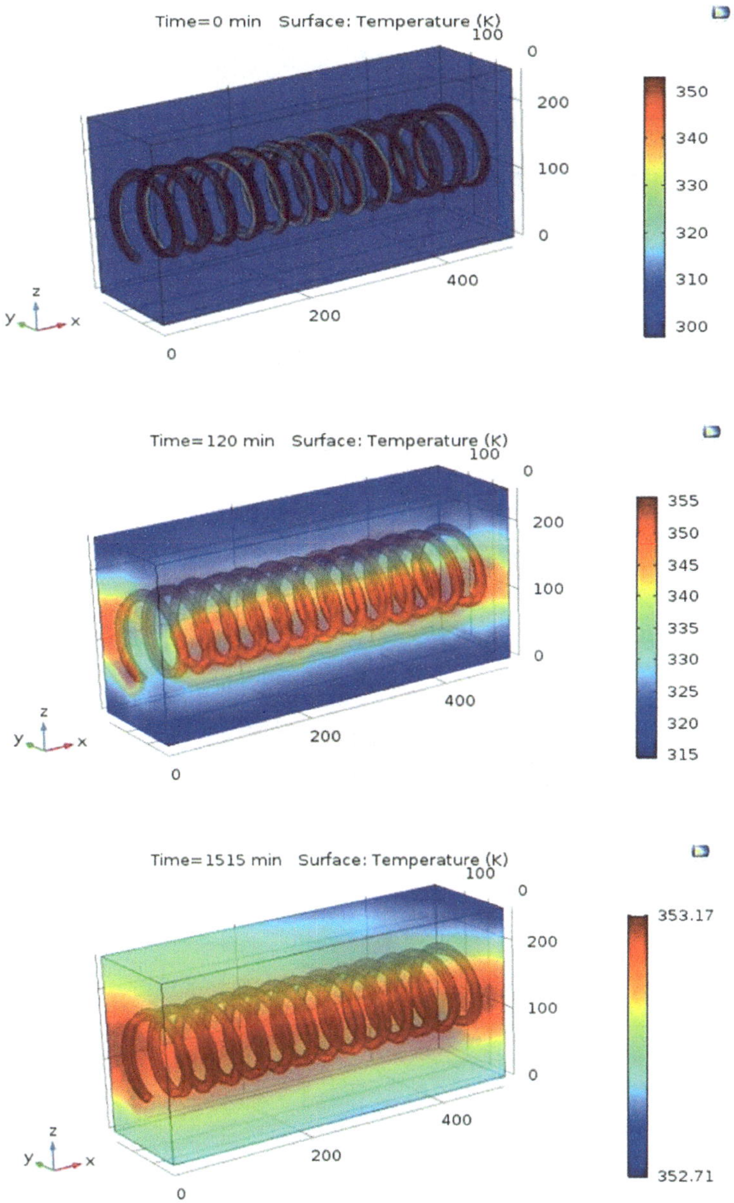

Fig. 2 Simulated temperature variation during charging of beeswax in large TSU

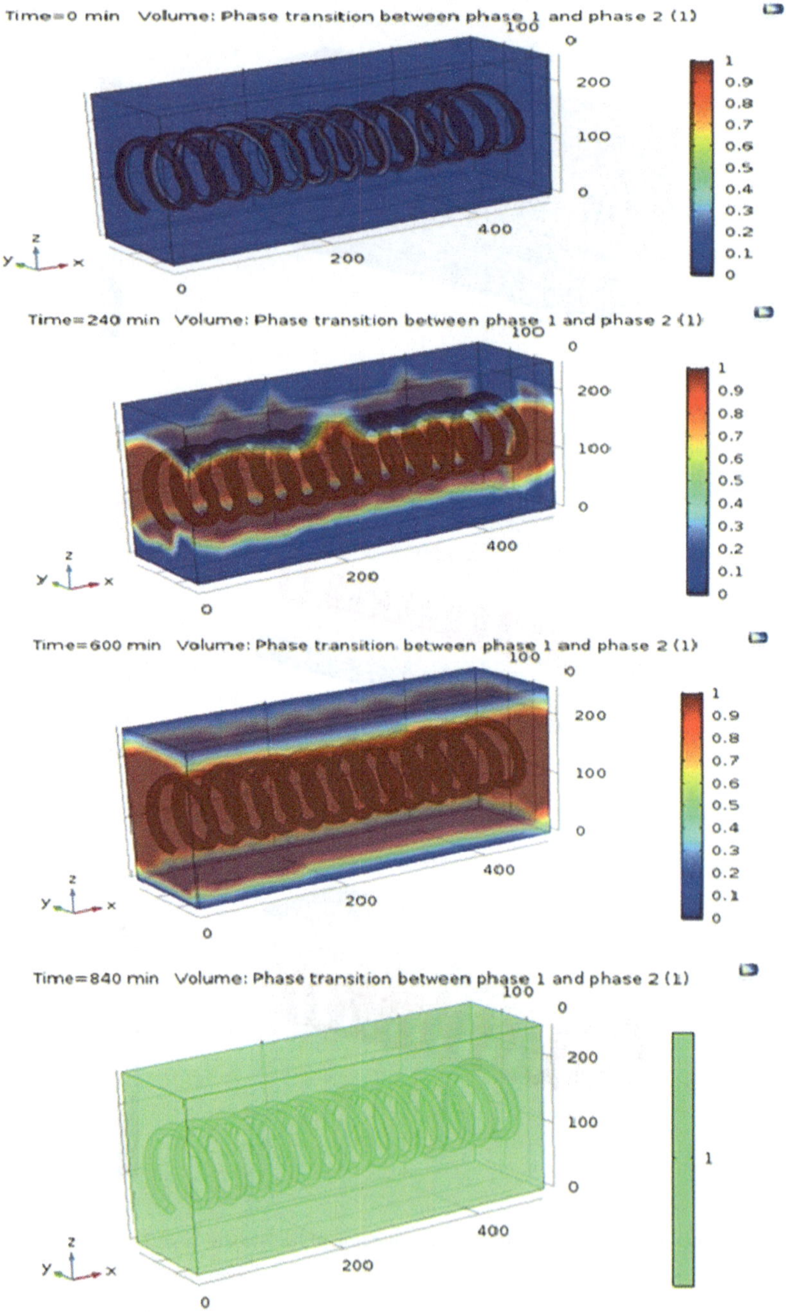

Fig. 3 Melting profile of beeswax at different time during charging

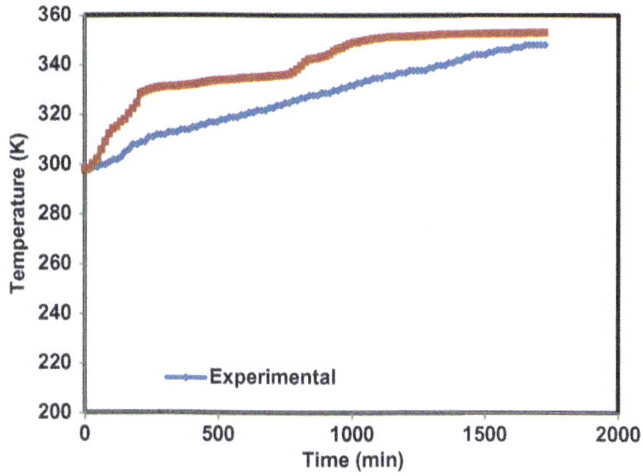

Fig. 4 Comparison of experimental and simulated temperature variation of beeswax in large TSU

5 Conclusion

Modeling and Simulation of beeswax and its composite with expanded graphite in a shell and tube type thermal storage unit was performed using COMSOL Multiphysics. The enthalpy method was used to simulate the melting behavior of both the phase change materials in thermal storage unit. The phase change material was filled in the shell part of the thermal storage unit (TSU) while the heat transfer fluid is passing through the helical tube embedded in the phase change material. The model of the prepared experimental setup was simulated on the COMSOL Multiphysics and it was observed that the simulation data are in good agreement with the experimental data with only 6% difference for beeswax and 7.5% difference for composite material. The charging time of beeswax and composite material through simulation was found to be 1515 min and 1040 min respectively. It was also observed that the pattern of heat flow as well as solid-liquid interface moves in similar pattern in case of both experimental and modeled part of unit with beeswax and composite material as phase change material.

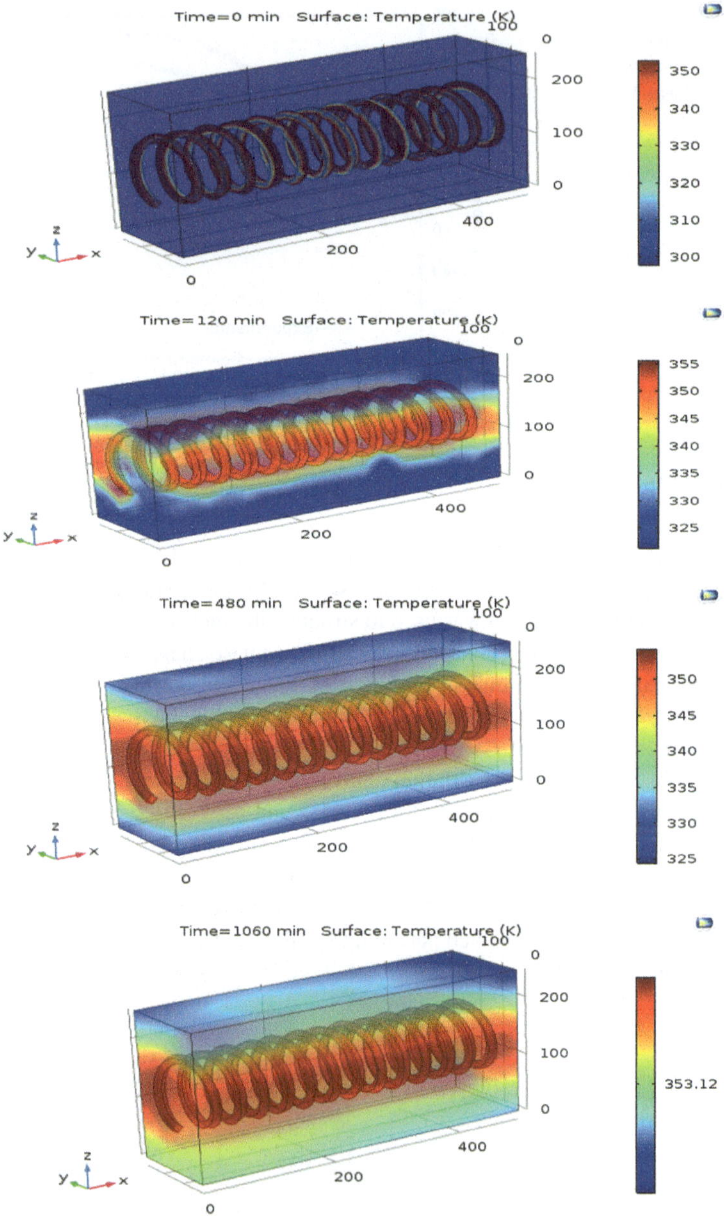

Fig. 5 Simulated temperature variation during charging of composite in large TSU

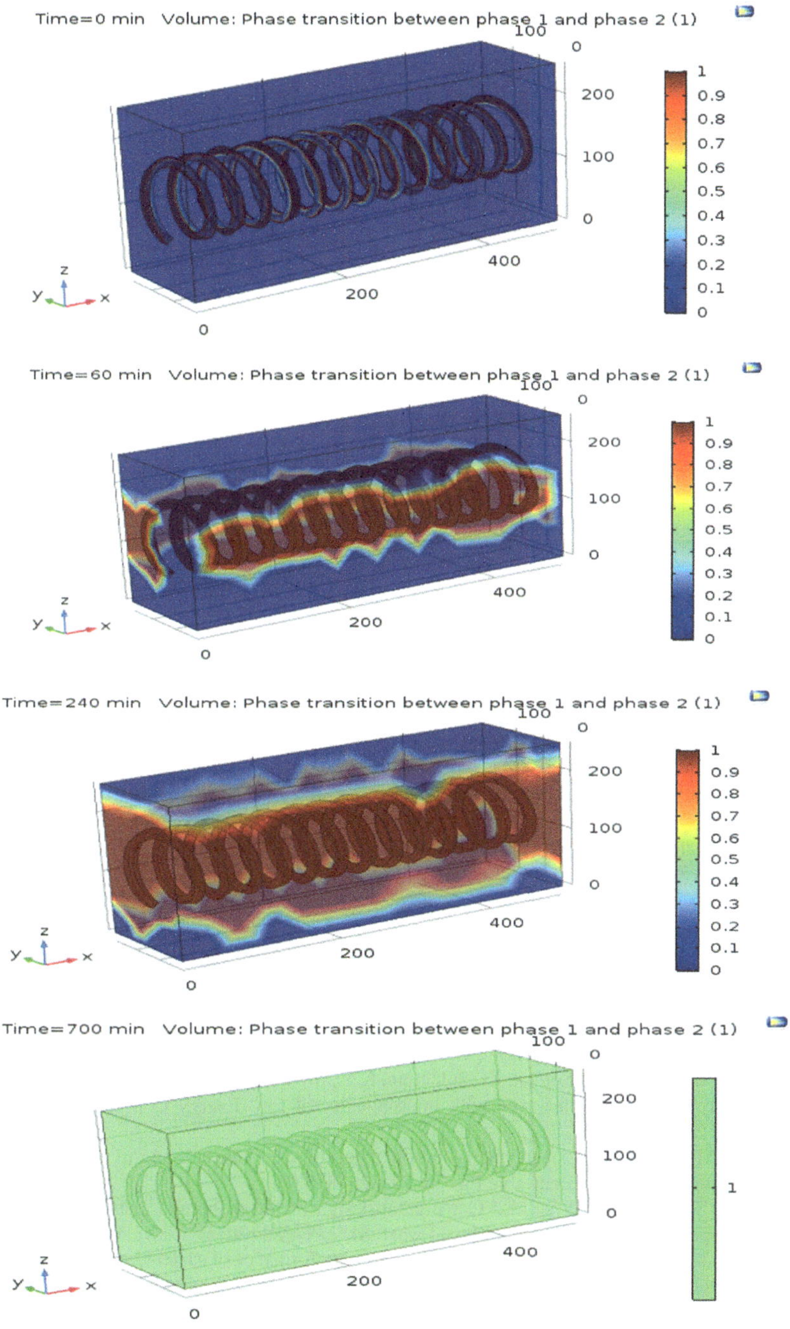

Fig. 6 Melting profile of composite material from simulation analysis

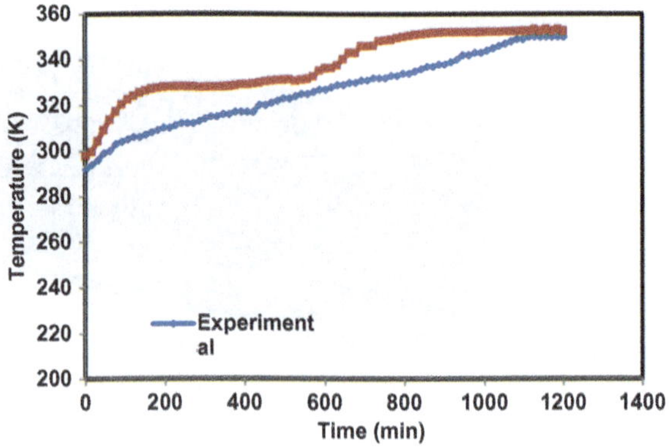

Fig. 7 Comparison of experimental and simulated temperature variation of composite in large TSU

Acknowledgements This work was supported by MNIT Jaipur, India for COMSOL software support.

References

1. Dinker, A., Agarwal, M., Agarwal, G.D.: Heat storage materials, geometry and applications: a review. J. Energy Inst. **90**, 1–11 (2017)
2. Sharma, S.D., Sagara, K.: Latent heat storage materials and systems: a review. Int. J. Green Energy **2**, 1–56 (2007)
3. Zalba, B., Marín, J.M., Cabeza, L.F., Mehling, H.: Review on thermal energy storage with phase change materials. Heat transfer analysis and applications. 00192-8 (2003)
4. Akgün, M., Aydin, O., Kaygusuz, K.: Thermal energy storage performance of paraffin in a novel tube-in-shell system. Appl. Therm. Eng. **28**, 405–413 (2008)
5. Dinker, A., Agarwal, M., Agarwal, G.D.: Experimental assessment on thermal storage performance of beeswax in a helical tube embedded storage unit. Appl. Therm. Eng. **111**, 358–368 (2017)
6. Peng, D., Chen, Z.: Numerical simulation of phase change heat transfer of a solar flat-plate collector with energy storage. Build. Simul. **2**, 273–280 (2009)
7. Medrano, M., Yilmaz, M.O., Nogués, M., Martorell, L., Roca, J., Cabeza, L.F.: Experimental evaluation of commercial heat exchangers for use as PCM thermal storage systems. Appl. Energy **86**, 2047–2055 (2009)
8. Mat, S., Al-Abidi, A.A., Sopian, K., Sulaiman, K.Y., Mohammad, A.T.: Enhance heat transfer for PCM melting in triplex tube with internal-external fins. Energy Convers. Manage. **74**, 223–236 (2014)
9. Bechiri, M., Mansouri, K.: Analytical solution of heat transfer in a shell-and-tube latent thermal energy storage system. Renewable Energy **74**, 825–838 (2015)
10. Lorente, S., Bejan, A., Niu, J.L.: Constructal design of latent thermal energy storage with vertical spiral heaters. Int. J. Heat Mass Transf. **81**, 283–288 (2015)
11. Pirasaci, T., Goswami, D.Y.: Influence of design on performance of a latent heat storage system for a direct steam generation power plant. Appl. Energy **162**, 644–652 (2016)
12. Niyas, H., Prasad, S., Muthukumar, P.: Performance investigation of a lab–scale latent heat storage prototype—numerical results. Energy Convers Mange **135**, 188–199 (2017)
13. Aadmi, M., Karkri, M., El Hammouti, M.: Heat transfer characteristics of thermal energy storage of a composite phase change materials: numerical and experimental investigations. Energy **72**, 381–392 (2014)

14. Manfarida, G., Secchi, R., Stanczyk, K.: Modelling and simulation of phase change materials latent heat storage applied to a solar powerd organic Rankine cycle. Applied Energy **170**, 378–388 (2016)
15. Mousa, H., Gujarathi, A.M.: Modeling and analysis the productivity of solar desalination units with phase change materials. Renewable Energy **95**, 225–232 (2016)

Graph Theory and Fuzzy Theory Applications

Graphical Representation of a DNA Sequence and Its Applications to Similarities Calculation: A Mathematical Model

Majid Bashir and Rinku Mathur

Abstract In this study, DNA sequence composed of four nucleotides is interpreted into directed weighted graph based on the presence of nucleotides positions in the sequence and hence its adjacency distance matrix and fuzzy membership distance matrix has been calculated. The distance matrices obtained for each sequence are then used to calculate similarities among different DNA sequences by using Euclidean distance metric. The model has been implemented on the published dataset of sequences and results obtained were found to be in agreement with the published work of phylogenetic relationships.

Keywords DNA sequence · Directed weighted graph · Distances matrices · Similarities · Euclidean distance metric

1 Introduction

Evolutionary history of number of species or organisms can be predicted by evaluating the similarities or dissimilarities among DNA or protein sequences of species. The DNA sequence is a sequence of the four types of nucleotide bases adenine (A), thymine (T), guanine (G), and cytosine (C). But it is very difficult to get the information about the organisms directly from the sequences. In recent years, several authors [1–4] presented various methods to predict the similarity between the DNA sequences. Graphical technique is one of them which comes out or emerged as a very powerful approach for analyzing and visualizing the long DNA sequence. Moreover, graphical representation of DNA sequences provides fruitful perception about the local and global properties, occurrences, differences, and repetition of nucleotides along a sequence, which are not obtained easily by other methods. On the basis of

M. Bashir · R. Mathur (✉)
Department of Mathematics, Lovely Professional University, Phagwara, Jalandhar, India
e-mail: rinkumathur56@gmail.com

M. Bashir
e-mail: majidbashir331@gmail.com

M. Sahni et al. (eds.), *Mathematical Modeling, Computational Intelligence Techniques and Renewable Energy*, Advances in Intelligent Systems and Computing 1287,
https://doi.org/10.1007/978-981-15-9953-8_37

graphical approach, Nandy [5] at first proposed the graphical representation of DNA sequences. The different graphical representation of the DNA sequences has been suggested by Lia et al., Randic et al., Liao and Wang, Wang and Song [6–9] who contributed a lot to this graphical methodology. Many authors have presented their views to incorporate mathematical concepts and description over the DNA sequences, to find out the comparison and similarities among the sequences concerned [9–11].

In this paper, an alignment-free mathematical model has been presented in which the DNA sequence is represented graphically and interpreted into directed weighted graph, and then authors calculated its adjacency distance matrix and fuzzy membership distance matrix to calculate the similarities among the sequences. At last, Euclidean distance metric has been used to obtain the similarities among the different DNA sequences taken into consideration.

2 Methodology

As we know that the DNA sequence consists of four nucleotide bases namely adenine, cytosine, guanine, and thymine. In order to elaborate the procedure of obtaining dissimilarity or distance matrix of DNA sequence, authors translated the directed weighted graph of a DNA sequence into its adjacency matrix. Let us consider a sequence (GATCC), we observe that the position of G at 1, A is at 2, T is at 3 and C is at 4 and 5 also. So, the DWG for the sequence GATCC is given below [12] (Fig. 1).

The above DWG graph is then converted into distance matrix. As there are two edges in the same direction among the same vertices and hence all the weights of the edges directing in the same direction can be added and the distance matrix obtained is as:

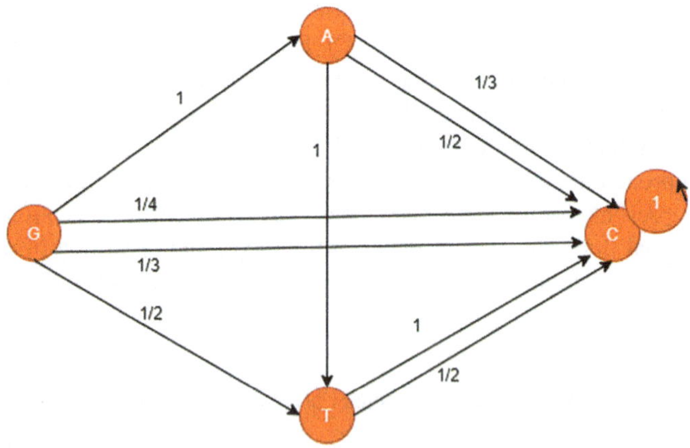

Fig. 1 Directed weighted graph w.r.t nucleotides position in the sequence GATCC

$$\begin{array}{c} \\ A \\ G \\ C \\ T \end{array} \begin{array}{c} \begin{array}{cccc} A & G & C & T \end{array} \\ \begin{bmatrix} 0 & 0 & 0.83 & 1 \\ 1 & 0 & 0.58 & 0.5 \\ 0 & 0 & 1 & 0 \\ 0 & 0 & 1.5 & 0 \end{bmatrix} \end{array}$$

Now, a fuzzy relational membership matrix is generated based on the distance matrix by defining the function [13].

$\mu_f : X \times X \to [0, 1]$ and the function has been defined as:

$$\mu_f(i, j) = \frac{1}{1 + d(i, j)} \times d(i, j) \tag{1}$$

The fuzzy distance matrix obtained for the sequence GATCC by using the function (1) is an under:

$$\begin{array}{c} \\ A \\ G \\ C \\ T \end{array} \begin{array}{c} \begin{array}{cccc} A & G & C & T \end{array} \\ \begin{bmatrix} 0 & 0 & 0.45 & 0.5 \\ 0.5 & 0 & 0.36 & 0.33 \\ 0 & 0 & 0.5 & 0 \\ 0 & 0 & 0.6 & 0 \end{bmatrix} \end{array}$$

Lot of methods are available to calculate the similarities among different DNA sequences like hamming distance, correlation angle, Euclidean distance, and so on [13, 14]. But in this work, Euclidean distance metric has been used to compute the similarities among sequences. The Euclidean distance between two fuzzy matrices has been computed by the following formula [12].

$$d(A, B) = \sqrt{\sum_{i=1}^{4} \sum_{j=1}^{4} [W_{ij}(A) - W_{ij}(B)]^2} \tag{2}$$

where $W_{ij}(A) - W_{ij}(B)$ represent the fuzzy differences between the same entries of the matrix A and B respectively.

3 Validation of Proposed Model

3.1 Data Used

In this work, authors considered the set of nine hypothetical sequences that were already published by Mathur and Adlakha [15] and the information of the sequences is provided in the Table 1.

Table 1 Input data of DNA sequences with their labels [15]

Label	DNA sequence
Seq I	ACAAG
Seq II	ACCAG
Seq III	GACAA
Seq IV	AGCAA
Seq V	AGACA
Seq VI	AGATA
Seq VII	CGATA
Seq VIII	AGGTA
Seq IX	CGGTA

3.2 *Data Analysis*

The adjacency distance matrix of the order 4 × 4 is therefore obtained for each DNA sequence. The directed weighted graph (DWG) for the Seq I, i.e., ACAAC has been presented in Fig. 2.

The adjacency distance matrix corresponding to the above directed weighted graph has been obtained as:

$$\begin{array}{c} \\ A \\ G \\ C \\ T \end{array} \begin{array}{c} \begin{array}{cccc} A & G & C & T \end{array} \\ \begin{bmatrix} 1.83 & 1.75 & 1 & 0 \\ 0 & 0 & 0 & 0 \\ 1.5 & 0.33 & 0 & 0 \\ 0 & 0 & 0 & 0 \end{bmatrix} \end{array}$$

In the similar fashion, authors calculated the adjacency distance matrices for other sequences, which are shown below:

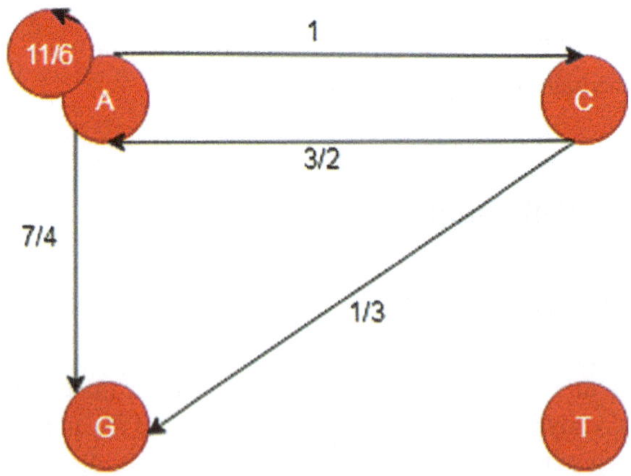

Fig. 2 Directed weighted graph for the sequence ACAAG

$$\begin{array}{c} \\ A \\ G \\ C \\ T \end{array}\begin{array}{c} \begin{array}{cccc} A & G & C & T \end{array} \\ \begin{bmatrix} 0.33 & 1.25 & 1.5 & 0 \\ 0 & 0 & 0 & 0 \\ 1.5 & 0.83 & 1 & 0 \\ 0 & 0 & 0 & 0 \end{bmatrix} \end{array} \quad \begin{array}{c} \\ A \\ G \\ C \\ T \end{array}\begin{array}{c} \begin{array}{cccc} A & G & C & T \end{array} \\ \begin{bmatrix} 1.83 & 0 & 1 & 0 \\ 1.58 & 0 & 0.5 & 0 \\ 1.5 & 0 & 0 & 0 \\ 0 & 0 & 0 & 0 \end{bmatrix} \end{array}$$

$$\begin{array}{c} \\ A \\ G \\ C \\ T \end{array}\begin{array}{c} \begin{array}{cccc} A & G & C & T \end{array} \\ \begin{bmatrix} 1.58 & 1 & 0.5 & 0 \\ 0.83 & 0 & 1 & 0 \\ 1.5 & 0 & 0 & 0 \\ 0 & 0 & 0 & 0 \end{bmatrix} \end{array} \quad \begin{array}{c} \\ A \\ G \\ C \\ T \end{array}\begin{array}{c} \begin{array}{cccc} A & G & C & T \end{array} \\ \begin{bmatrix} 1.25 & 0 & 1.33 & 0 \\ 1.33 & 0 & 0.5 & 0 \\ 0.5 & 0 & 0 & 0 \\ 0 & 0 & 0 & 0 \end{bmatrix} \end{array}$$

$$\begin{array}{c} \\ A \\ G \\ C \\ T \end{array}\begin{array}{c} \begin{array}{cccc} A & G & C & T \end{array} \\ \begin{bmatrix} 1.25 & 1 & 0 & 1.33 \\ 1.33 & 0 & 0 & 0.5 \\ 0 & 0 & 0 & 0 \\ 1 & 0 & 0 & 0 \end{bmatrix} \end{array} \quad \begin{array}{c} \\ A \\ G \\ C \\ T \end{array}\begin{array}{c} \begin{array}{cccc} A & G & C & T \end{array} \\ \begin{bmatrix} 0.5 & 0 & 0 & 1 \\ 1.33 & 0 & 0 & 0.5 \\ 0.75 & 1 & 0 & 0.33 \\ 1 & 0 & 0 & 0 \end{bmatrix} \end{array}$$

$$\begin{array}{c} \\ A \\ G \\ C \\ T \end{array}\begin{array}{c} \begin{array}{cccc} A & G & C & T \end{array} \\ \begin{bmatrix} 0.25 & 1.5 & 0 & 0.33 \\ 0.83 & 1 & 0 & 1.5 \\ 0 & 0 & 0 & 0 \\ 1 & 0 & 0 & 0 \end{bmatrix} \end{array} \quad \begin{array}{c} \\ A \\ G \\ C \\ T \end{array}\begin{array}{c} \begin{array}{cccc} A & G & C & T \end{array} \\ \begin{bmatrix} 0 & 0 & 0 & 0 \\ 0.83 & 1 & 0 & 1.5 \\ 0.25 & 1.5 & 0 & 0.33 \\ 1 & 0 & 0 & 0 \end{bmatrix} \end{array}$$

The above adjacency distance matrices for all the sequences are then converted into fuzzy distance matrices by using Eq. (1).

$$\begin{array}{c} \\ A \\ G \\ C \\ T \end{array}\begin{array}{c} \begin{array}{cccc} A & G & C & T \end{array} \\ \begin{bmatrix} 0.64 & 0.63 & 0.5 & 0 \\ 0 & 0 & 0 & 0 \\ 0.6 & 0.23 & 0 & 0 \\ 0 & 0 & 0 & 0 \end{bmatrix} \end{array} \quad \begin{array}{c} \\ A \\ G \\ C \\ T \end{array}\begin{array}{c} \begin{array}{cccc} A & G & C & T \end{array} \\ \begin{bmatrix} 0.24 & 0.55 & 0.6 & 0 \\ 0 & 0 & 0 & 0 \\ 0.6 & 0.45 & 0.5 & 0 \\ 0 & 0 & 0 & 0 \end{bmatrix} \end{array}$$

$$\begin{array}{c} \\ A \\ G \\ C \\ T \end{array}\begin{array}{c} \begin{array}{cccc} A & G & C & T \end{array} \\ \begin{bmatrix} 0.64 & 0 & 0.5 & 0 \\ 0.61 & 0 & 0.33 & 0 \\ 0.60 & 0 & 0 & 0 \\ 0 & 0 & 0 & 0 \end{bmatrix} \end{array} \quad \begin{array}{c} \\ A \\ G \\ C \\ T \end{array}\begin{array}{c} \begin{array}{cccc} A & G & C & T \end{array} \\ \begin{bmatrix} 0.61 & 0.5 & 0.33 & 0 \\ 0.45 & 0 & 0.5 & 0 \\ 0.60 & 0 & 0 & 0 \\ 0 & 0 & 0 & 0 \end{bmatrix} \end{array}$$

$$\begin{array}{c} \\ A \\ G \\ C \\ T \end{array}\begin{array}{c} \begin{array}{cccc} A & G & C & T \end{array} \\ \begin{bmatrix} 0.55 & 0.5 & 0.57 & 0 \\ 0.57 & 0 & 0.33 & 0 \\ 0.33 & 0 & 0 & 0 \\ 0 & 0 & 0 & 0 \end{bmatrix} \end{array} \quad \begin{array}{c} \\ A \\ G \\ C \\ T \end{array}\begin{array}{c} \begin{array}{cccc} A & G & C & T \end{array} \\ \begin{bmatrix} 0.55 & 0.5 & 0 & 0.57 \\ 0.57 & 0 & 0 & 0.33 \\ 0 & 0 & 0 & 0 \\ 0.5 & 0 & 0 & 0 \end{bmatrix} \end{array}$$

$$\begin{array}{c} \\ A \\ G \\ C \\ T \end{array}\begin{array}{c} \begin{array}{cccc} A & G & C & T \end{array} \\ \begin{bmatrix} 0.33 & 0 & 0 & 0.5 \\ 0.57 & 0 & 0 & 0.33 \\ 0.27 & 0.5 & 0 & 0.24 \\ 0.5 & 0 & 0 & 0 \end{bmatrix} \end{array} \quad \begin{array}{c} \\ A \\ G \\ C \\ T \end{array}\begin{array}{c} \begin{array}{cccc} A & G & C & T \end{array} \\ \begin{bmatrix} 0.2 & 0.6 & 0 & 0.24 \\ 0.45 & 0.5 & 0 & 0.6 \\ 0 & 0 & 0 & 0 \\ 0.5 & 0 & 0 & 0 \end{bmatrix} \end{array}$$

$$\begin{array}{c} \\ A \\ G \\ C \\ T \end{array}\begin{array}{c} \begin{array}{cccc} A & G & C & T \end{array} \\ \begin{bmatrix} 0 & 0 & 0 & 0 \\ 0.45 & 0.5 & 0 & 0.6 \\ 0.2 & 0.6 & 0 & 0.24 \\ 0.5 & 0 & 0 & 0 \end{bmatrix} \end{array}$$

3.3 Similarity Calculation

In order to calculate the similarities among these sequences, Euclidean distance between fuzzy matrices obtained corresponding to the DNA sequences has been computed by using the formula (2). The similarities obtained among the nine sequences considered is shown in Table 2.

The results obtained by [15] for the same set of sequences by using the approach of hamming distance has been presented in the Table 3.

On observing the results of Table 2, it has been found that the more similar pair of sequences are (Seq IV, Seq V), (Seq III, Seq IV), (Seq I, Seq II), (Seq I, Seq IV) and the less similar pair of sequences are (Seq II, Seq VIII), (Seq IV, Seq V).

In the same way, results of Table 3 suggests that (Seq I, Seq II), (Seq V, Seq VI), (Seq VI, Seq VII), (Seq VI, Seq VIII) are more similar and the distance among all these pairs equals to unity. In addition to it less similar pair of sequences in Table 3 are (Seq II, Seq IX), (Seq II, Seq VII) and (Seq II, Seq VIII). On comparing the results of both the approaches, the less and more similar pairs of DNA sequences are almost same such as (Seq II, Seq VIII) and (Seq I, Seq II). So, an alignment-free model is implemented to find out the similarities among the sequences of different lengths which suggests that purposed model is better than the existing models. The model gives the same kind of results obtained by the other existing methods in very less time as multiple alignment of sequences is not required in this approach.

Table 2 Similarities obtained among the fuzzy matrices of nine sequences

Sequence	ACAAG	ACCAG	GACAA	AGCAA	AGACA	AGATA	CGATA	ACCTA	CGGTA
ACAAG	0	0.6859	0.9092	0.7461	0.7708	1.3043	1.3823	1.4080	1.5704
ACCAG		0	1.1857	1.0570	1.0157	1.5099	1.4224	1.6778	1.5075
GACAA			0	0.7514	0.5809	1.2893	1.2173	1.4885	1.4981
AGCAA				0	0.4211	1.1919	1.3114	1.3470	1.5194
AGACA					0	1.1074	1.2854	1.2677	1.5149
AGATA						0	0.8269	0.7607	1.2931
CGATA							0	1.0864	0.8432
ACCTA								0	0.9566
CGGTA									0

Table 3 Hamming distance matrix among nine sequences

Sequence	ACAAG	ACCAG	GACAA	AGCAA	AGACA	AGATA	CGATA	ACCTA	CGGTA
ACAAG	0	1	4	4	4	5	6	6	7
ACCAG		0	3	3	5	6	7	7	8
GACAA			0	2	4	5	6	6	7
AGCAA				0	2	3	4	4	5
AGACA					0	1	2	2	3
AGATA						0	1	1	2
CGATA							0	2	1
ACCTA								0	1
CGGTA									0

4 Conclusion

This study introduced a graphical representation of DNA sequence by considering directed weighted graph and correspondingly distance matrices and fuzzy distance matrices has been calculated for every sequence. So, an alignment-free mathematical model for DNA sequences of different lengths is proposed which also focuses on the individual information of every DNA sequence in the form of distances. The similarities among the nine DNA sequences have been calculated by using the Euclidean distance measure. The results obtained by the proposed model are promising and in agreement with the already published results on the same set of nine DNA sequences. In the end, it is also expected that this model will be beneficial to the biological community to find the similarities/dissimilarities among different organisms without going directly to the wet lab.

References

1. Jafarzadeh, N., Iranmanesh, A.: C-curve: a novel 3D graphical representation of DNA sequence based on codons. Math. Biosci. **241**(2), 217–224 (2013)
2. Jafarzadeh, N., Iranmanesh, A.: A novel graphical and numerical representation for analyzing DNA sequences based on codons. Match Commun. Math. Comput. Chem. **68**(2), 611–620 (2012)
3. Liao, B., Ding, K.Q.: A 3D graphical representation of DNA sequences and its application. Theor. Comput. Sci. **358**(1), 56–64 (2006)
4. He, P.H., Li, D., Zhang, Y.P., Wang, X., Yao, Y.: A 3D graphical representation of protein sequences based on the Gray code. J. Theor. Biol. **304**, 81–87 (2012)
5. Nandy, A.: A new graphical representation and analysis of DNA sequence structure: methodology and application to globin genes. Curr. Sci. 309–314 (1994)
6. Lia, B., Xilin, Q., Cai, L., Cao, Z.: A new graphical coding of DNA sequence and its similarity calculation. Phys. A **392**(19), 4663–4667 (2013)
7. Randic, M., Vracko, M., Lers, N., Plavis, D.: Novel 2-D graphical representation of DNA sequences and their characterization. Chem. Phys. Lett. **368**(14), 1–6 (2003)
8. Liao, B., Wang, T.M.: New 2D graphical representation of DNA sequences. J. Comput. Chem. **25**(11), 1364–1368 (2004)
9. Liao, B., Wang, T.M.: A 3D graphical representation of RNA secondary structures. J. Biomol. Struct. Dyn. **21**(6), 827–832 (2004)
10. Mathur, R., Adlakha, N.: A new measure of dissimilarity and fuzzy linear programming model to construct phylogenetic network among DNA sequences. Netw. Biol **9**(4), 78–95 (2019)
11. Angel, A.: A graph theoretical approach for node covering in tree-based architectures and its applications to bioinformatics. Netw. Model. Anal. Health Inform. Bioinform. **8**(12), 65–78 (2019)
12. Mathur, R., Adlakha, N.: A graphical theoretic model for predication of reticulation events and phylogenetic networks for DNA sequences. Egypt. J. Basic Appl. Sci. **3**(3), 263–271 (2016)
13. Mathur, R., Adlakha, N.: A fuzzy weighted least squares approach to construct phylogenetic networks among subfamilies of grass species. J. Appl. Math. Bioinform. **3**(2), 137–158 (2013)
14. Xu, L., Shao, X.: Methods of Chemometrics in Science. Beijing (1995)
15. Mathur, R., Adlakha, N.: Binary sequences-based approach for construction of evolutionary network. Int. J. Biomath. **2**(7), 1–14 (2014)

Applications of Petri Net Modeling in Diverse Areas

Gajendra Pratap Singh, Madhuri Jha, and Mamtesh Singh

Abstract Mathematical modeling is playing a very important role in clarifying, analyzing, and drawing of qualitative and quantitative results. In almost every area of sciences, the complex system is being modeled using some mathematical modeling using a graphical approach, differential equation approach, statistical approach, and many more. In this chapter, we are focusing on the graphical approach using Petri Nets (PN). Petri net came in focus in the early '90s after Carl Adam introduced this method in his Ph.D. thesis. By then, it is being widely used to model the complex systems in engineering, computational, biological, and many more fields. Initially, it was accounted as a modeling tool in Computer Science (CS), but later, PN method started to model systems in Automatic Control (AC) too for automation after that it was being used in the background of Operations Research (OR). Therefore, this method recognized as the Discrete Event Dynamic Systems (DEDS) theory, at the intersection of CS, AC, and OR. Moreover, different kinds of hybrid PNs like Time Petri Net (TPN), Stochastic Petri Net (SPN) are being extensively studied today. PN is also being proved to be a useful tool to model the biological system like cell cycle, pathways in diseases to treat, and many more. It will help to find several structural and behavioral properties of the system.

Keywords Graph theory · Petri net · Reachability tree · Marking vector · Tuberculosis · Cognitive science

1 Introduction

Rapid development in mathematical modeling in a diverse area can be noticed since the nineteenth century. Mathematical modeling provides a way to design any system

---

G. P. Singh (✉) · M. Jha
School of Computational and Integrative Sciences, Jawaharlal Nehru University, New Delhi 110067, India
e-mail: gajendra@mail.jnu.ac.in

M. Singh
Department of Zoology, Gargi College(University of Delhi), Delhi, India

M. Sahni et al. (eds.), *Mathematical Modeling, Computational Intelligence Techniques and Renewable Energy*, Advances in Intelligent Systems and Computing 1287,
https://doi.org/10.1007/978-981-15-9953-8_38

in more controllable with mathematical formulation and concepts on the way to draw numerical and behavioral results. Several complex systems in a diverse area can be better explained and verified with the use of suitable mathematical modeling. In this chapter, we are discussing a model with a Discrete Event System (DES), Petri Net (PN). Petri net is a mathematical model providing a graphical approach to intimate any complex system with logical and conceptual thinking. PN is first introduced by Carl Adam in his Ph.D. thesis in the early '90s [1]. PN is very much efficient in dealing with the system with complexity like concurrency and synchronization. Earlier, it was considered that PN is only the technique to deal with the computational system but slowly it has been used to model other processes like metabolic pathways in biology, networks in communications, and many more. PN can also be used in binary vectors generations using star graph modeling [2–4]. PN is also used as a recommender system to several complex systems that behave as a filtering system which gives preference to the words given by the user [5, 6]. From the last 30 years, PN is being highly used in modeling complex biological processes explaining the metabolic pathways and the molecular mechanism behind these processes. This process covers plant science, medical science, neural networks, cardiovascular disease network, tuberculosis drug pathways, and many more [7–11]. In this chapter, some of the basic application of Petri net in the diverse area is being discussed, and the properties of Petri net is connected and explained on the way of the original process. Some diverse area has been considered as the basic life cycle of butterfly, networks in communication, and the blood donation and receiving rule. Generally, places are taken as the stages coming in the system, and the conditions responsible for those stages are modeled as transitions. The resources on any place are also a fact to be considered to analyze any system, and PN modeling provides it comfortably as tokens on any place. Petri net has much more flexibility in its places and transitions, where time or any condition is to be considered to know the behavior of the system in a different situation.

2 Method of Modeling

2.1 Petri Net

In mathematical modeling, the Petri Net (PN) plays an important role to connect the different streams of knowledge. It is a kind of directed bipartite graph containing two sets of bipartitions that are by definition disjointed from each other. Formally, it is a 5-tuple set (see Fig. 1 for graphical representation) defined as:

$$\mathrm{PN} = \left(\mathrm{PL}, \mathrm{TN}, I^{+}, I^{-}, \mu^{0}\right)$$

where

PL non-empty set of places (represented as circles)

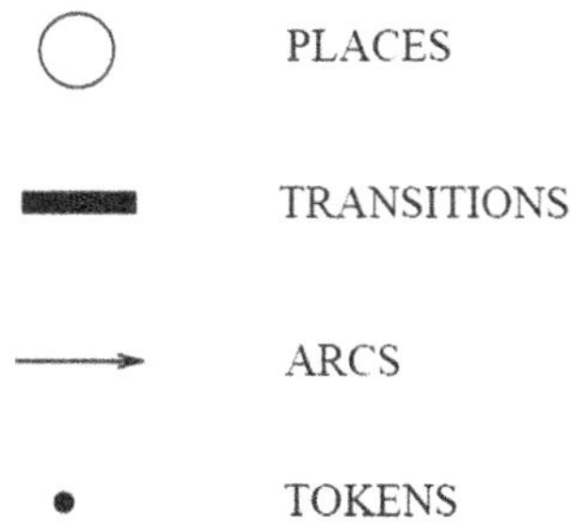

Fig. 1 Elements of Petri nets

TN non-empty set of transitions (represented as a rectangle)
I^+ set of arcs from places to transitions (represented as arrows)
I^- set of arcs from transitions to places (represented as arrows)
μ^0 initial marking on the places (represented as black dots on the place).

The Petri net structure consists of only the 4-tuple excluding initial marking. Sometimes, it is beneficial to study the structure modeling with initial marking, as marking varies according to the need of the user. With some marking on the places, it shows the availability of the resource in that place. The two nodes have distinct features depending on the process to be modeled. Generally, places denote the events while the transitions denote the condition by the event occurs. The token on the place can be transferred to other places with a suitable condition of the firing of transitions. Any transition fires to move a token from any place once it satisfies the enabling condition, and that the number of incoming arcs on the transition should be less than or equal to the number of tokens in that place.

Any transition t will fire if,

$$I^-(p, t) \leq \mu^0(p), \quad \forall\, p \in \text{PL}$$

After firing of t the token moves by the following equation,

$$\mu^1(p) = \mu^0(p) - I^-(p, t) + I^+(p, t)$$

This determines the next marking μ^1 from μ^0, and read as μ^1 is reachable from μ^0. Tracing all the marking with the predecessor and successor marking depicts a directed graph, where corners (vertices) represent the marking, and arcs represent the connection. This graph is called reachability graph (RG(PN, μ^0)) of the corresponding net and with the corresponding marking [12–15].

Petri net theory has many more approaches to allocate tokens on the places according to the demand of the system which is named as Timed Petri Net (TPN), Hybrid Petri Net (HPN), Stochastic Petri Net (SPN), Fuzzy Petri Net (FPN), Colored Petri Net (CPN), etc., [16, 17].

Several open software is available to formulate the Petri net model and validate it according to the actual process. It is also used to simulate the results with real data for finding some interesting results without performing the real process. In this chapter, PIPE v4.3.0 and Woped software are being used to draw the net and for finding qualitative results [18, 19].

2.2 Structural and Behavioral Properties of Petri Net

Petri net proves as a powerful tool to model any system which helps in drawing useful future properties that the system can behave. The properties are structural and behavioral both and can be applied as per requirement. These properties are also called power of Petri nets as it behaves as a power in classifying any process to the next level. The main properties are:

1. **Reachability tree/graph of a Petri net**: This is the most important property depicted by the modeling with the Petri net. This is the graph that shows all the markings as vertices, and the dependencies are shown by edges. If PN is the Petri net, then reachability graph is represented by $R(\text{PN}, \mu^0)$, with the initial marking μ^0.
   Mathematically, $R(\text{PN}, \mu^0) = (M, K)$, where M is the set of all reachable markings from μ^0, and K is the set of directed arcs from one marking to another [20].
2. **Bounded Petri net**: A Petri net is said to be bounded PN, if the count of tokens on any place never exceeds to a finite amount. In Fig. 2, the number of tokens on any place never exceeds the value two, so it can be identified that the PN is 2-bounded.

Example:

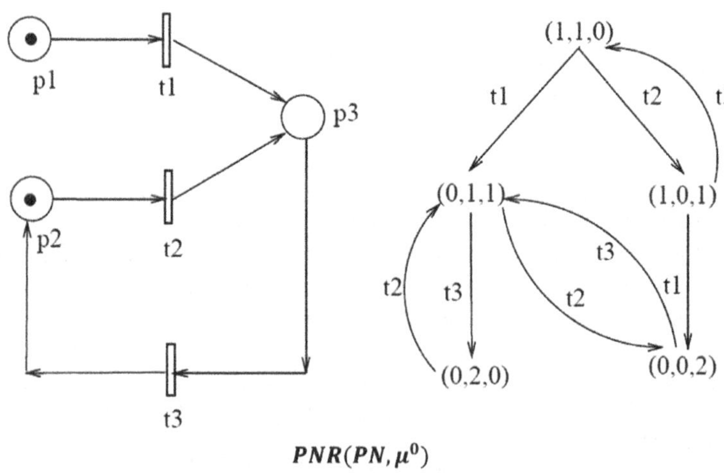

Fig. 2 A simple model example of a Petri net and its reachability graph

3. **Safe Petri net**: A Petri net is safe if the number of tokens on any place never exceeds one. In other words, this net is classified as 1-bounded. In Fig. 2, the net is not safe.
4. **Live Petri net**: A Petri net is live if at any stage there is at least one transition that is enabled and can fire to get the next marking. In other words, the net should not attain any deadlock stage. In Fig. 2, the PN is live as at any marking stage, there is some transition enable to get the next marking stage.
   Example:
5. **Invariant properties**: Invariant property in any Petri net is applied for both places and transitions. For places, it is p-invariant which means the set of places with which the sum of tokens is constant at every stage. On the other hand, t-invariant is the set of transitions by the firing of which the initial marking is obtained again. p-invariant is a vector $x \in N^p$ where $p = |P|$ and satisfies the equation $M'.x = 0$. M is the incidence matrix, and M' is the transpose of the incidence matrix [21].
6. **Marked graph and state machine**: A Petri net is a marked graph if each place is input for exactly one transition and also output for exactly one transition; on the other hand, the Petri net in which each transition is strictly having one input, and one output place is the state machine.

3 Application of Petri Net in the Diverse Area Modeling

3.1 *Life Cycle of a Butterfly*

Here, we are exploring the life cycle of a butterfly, which includes all the stages of its life with the conversion states. The butterfly life cycle is a basic and simple one to comprehend and is extraordinary for understudy ventures. Mainly, there are four life stages and that is egg, larva, pupa, and adult butterfly. Each state has its significance and responsible for the next state. Figure 3 shows all the stages in a butterfly life from egg to adult [22].

Modeling this cycle with Petri net, the first is to select the places and transitions. Here, the four main stages are taken as places while the intermediate stages as transitions. The token on the places represents the present stage of butterfly life. The flow of token represents the conversion of one stage to another. Petri net modeling of the life cycle is shown in Fig. 4.

In this model, set of places is $|\text{PL}| = (p_1, p_2, p_3, p_4)$, and set of transition is $|\text{TN}| = (t_1, t_2, t_3, t_4)$. Here, $\mu^0(p_1) = 1$, and for rest places, μ^0 is zero.

Initially, t_1 is enabled as $I^-(p, t_1) \leq \mu^0(p), \forall\ p \in \text{PL}$. The transition t_1 will fire to shift the token on p_1 to the next place according to the rule.

$$\begin{aligned}\mu^1(p_2) &= \mu^0(p_2) - I^-(p_2, t_1) + I^+(p_2, t_1)\\ &= 0 - 0 + 1 = 1\end{aligned}$$

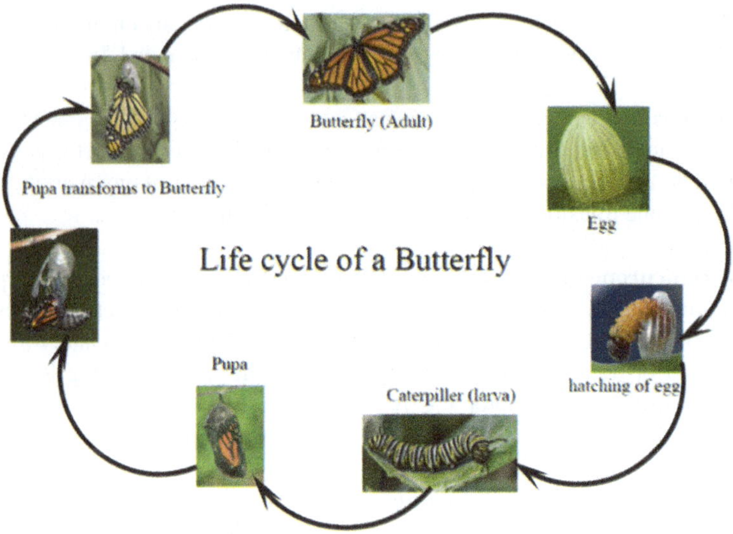

Fig. 3 Stages in the life cycle of a butterfly

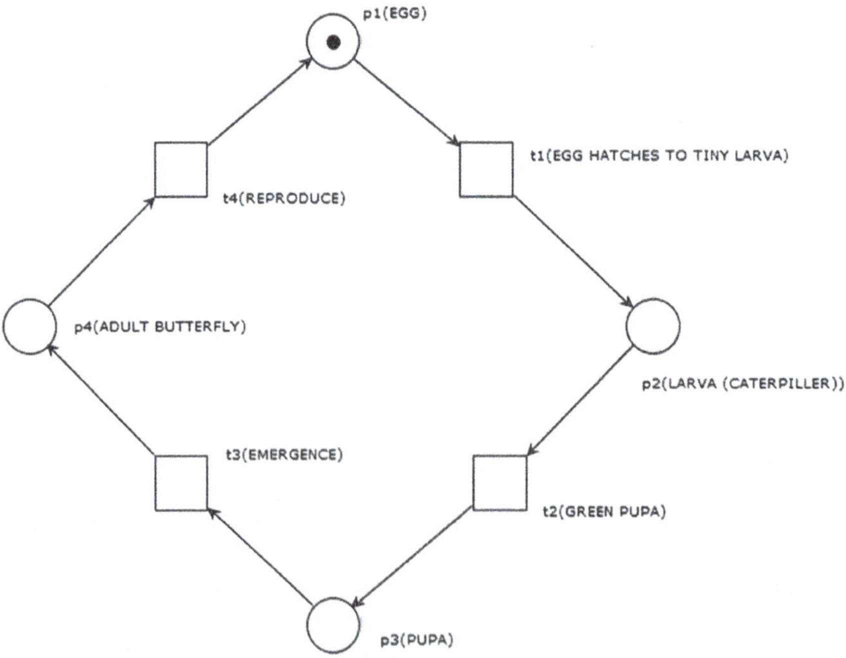

Fig. 4 Petri net model of the butterfly life cycle

Similarly, the firing of t_2, t_3, t_4 follows to change the markings. It explains the full life cycle of a butterfly with a mathematical approach. One can extend this model in more number of stages according to the need in the research area to study the intermediate stages in butterfly life.

3.2 Wheel Network in Communication

The communication network is a large hub of networks, where connection flows from one end to another with some pattern of direction. There are mainly five types of networks followed in communication, i.e., chain network, wheel network, circle network, Y network, and star network. The pattern of network is decided according to the need of the organization. All these networks can be modeled using Petri net to get the qualitative and quantitative results which can be used further for improving the network. In this chapter, wheel network is being modeled with Petri net. In a wheel network, statistics/data/information flows to and fro by a single person in the middle. Employees in the team can talk mainly with that middle one through which they can interact with each other. Such a conversation community is a speedy capacity of getting data/information to employees, considering the individuals in the middle of the wheel can do so immediately and efficiently. The wheel network depends on the leader or the middle individual to act as the central channel for the whole group's interaction or communication [23].

In Fig. 5, a wheel network has shown with five members, and a leader who is communicating with each of the members and is the only responsible person for any group communication.

This network is being modeled with Petri net to get a star Petri net graph. The modeled net is shown in Fig. 6, where the members are taken as transition, and the

Fig. 5 Wheel network with six members

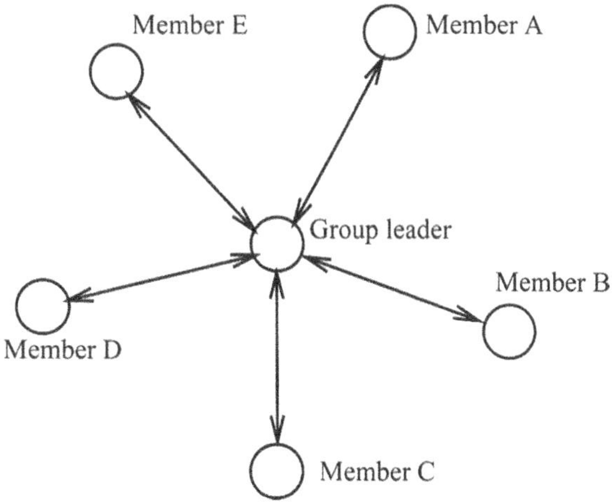

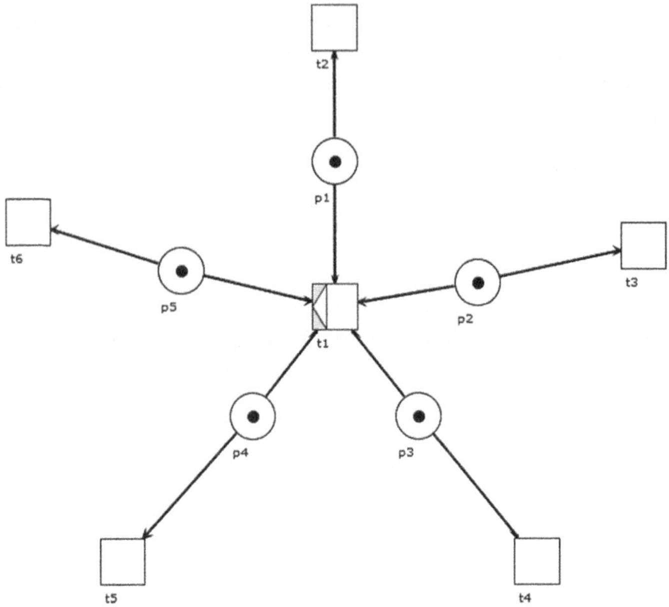

Fig. 6 Petri net modeling of the wheel network with five members and a group leader

exchange of information between them is taken as places.

The PN in Fig. 6 has $|\mathrm{PL}| = (p_1, p_2, p_3, p_4, p_5)$ and $|\mathrm{TN}| = (t_1, t_2, t_3, t_4, t_5, t_6)$, where t_1 is OR–JOIN transition which means it can fire if any of the connected places have token and the rest of the transition follows normal firing rule. Also, initially, one token is placed on each place which is showing the availability of information between the group leader and the corresponding member.

Therefore, $\mu^0(p) = 1, \forall\ p \in \mathrm{PL}$. Initially, all the transitions are enabled as satisfying the enabling condition. It can be observed that $I^-\left(p_i, t_j\right) \leq \mu^0(p_i), \forall\ p_i\ \in \mathrm{PL}$ and $\forall\ t_j\ \in \mathrm{TN}$, so transitions can fire either simultaneously or one by one. When t_1 will fire, it will change the marking of that place via which it is firing if it is firing toward p_1, then it will change the marking of p_1 only, as

$$\begin{aligned}\mu^1(p_1, t_1) &= \mu^0(p_1) - I^-(p_1, t_1) + I^+(p_1, t_1)\\ &= 1 - 1 + 0 = 0\end{aligned}$$

So the next marking will be $\mu^1(p_1, t_1) = (0, 1, 1, 1, 1)$ and similarly with others places.

Next, when t_2 will fire the marking will change only for p_1 as,

$$\begin{aligned}\mu^1(p_1, t_2) &= \mu^0(p_1) - I^-(p_1, t_2) + I^+(p_1, t_2)\\ &= 1 - 1 + 0 = 0\end{aligned}$$

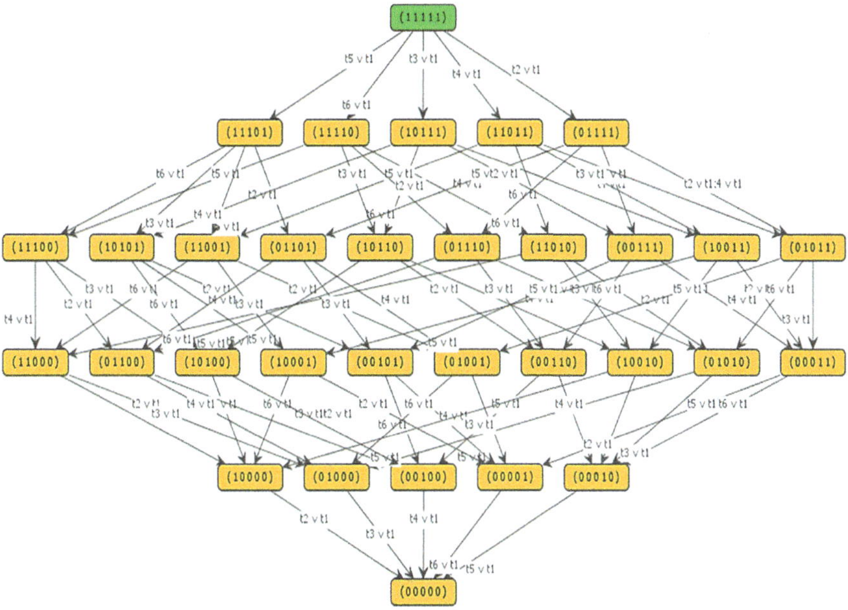

Fig. 7 Reachability graph of the wheel network Petri net

Hence, the next marking will be $\mu^1(t_2) = (0, 1, 1, 1, 1)$.

Similarly, after firing of t_3, next marking will be $\mu^1(t_3) = (1, 0, 1, 1, 1)$; after firing of t_4, the next marking will be $\mu^1(t_4) = (1, 1, 0, 1, 1)$; after firing of t_5, the next marking will change as $\mu^1(t_5) = (1, 1, 1, 0, 1)$, and finally, by firing t_6, the next marking will change as $\mu^1(t_6) = (1, 1, 1, 1, 0)$. One can observe that these five markings are shown in the first stage of the reachability graph of the net in Fig. 7; each can be obtained by two ways of firing.

After this, with each marking t_1 is enabled but t_2, t_3, t_4, t_5, t_6 are not enabled at $\mu^1(t_2)$, $\mu^1(t_3)$, $\mu^1(t_4)$, $\mu^1(t_5)$, and $\mu^1(t_6)$, respectively.

To the next second level firing, one can observe ten marking stages which are obtained as μ^2; similarly, ten marking stage in third level firing is obtained as μ^3; on the fourth level marking, five marking stages are obtained as μ^4, and finally, only one stage is obtained as $\mu^5 = (0, 0, 0, 0, 0)$ as the dead stage, where no transitions are enabled. By observing the reachability graph, one can conclude that all the possible binary vector sequences are generated in this Petri net. There are 5-tuples and $2^5 = 32$, binary vectors are obtained.

Table 1 Rule of donation and acceptance of different blood group

S. No.	Blood group	Donor (D)	Receiver (R)
1	A^+	A^+, AB^+	A^+, A^-, O^+, O^-
2	A^-	A^+, A^-, AB^+, AB^-	A^-, O^-
3	B^+	B^+, AB^+	B^+, B^-, O^+, O^-
4	B^-	$B^+, B^- AB^+, AB^-$	B^-, O^-
5	AB^+	AB^+	$A^+, A^-, B^+, B^-, AB^+, AB^-, O^+ O^-$
6	AB^-	AB^+, B^+, O^+, AB^-	A^-, B^-, AB^-, O^-
7	O^+	A^+, AB^+	O^+, O^-
8	O^-	$A^+, A^-, B^+, B^-, AB^+, AB^-, O^+, O^-$	O^-

3.3 Blood Type Chart

Detecting blood type is the principal blood test that will decide whether your blood is perfect with the potential contributor's blood. If the donor's blood matches with the receiver's blood, then only the donation of blood is allowed.

There is eight type of blood groups in human in which group O is the most common blood group throughout the globe. In Table 1, all the blood groups are mentioned with the donor and acceptor blood groups.

While modeling this process the different blood, the group is taken as places with a donor (D) and receiver (R) type while the donation and acceptor phase is taken as the transition. The tokens on any place show the availability of the blood of that particular blood group and firing of transition show the donation and receiving ability of the blood group. We can check the maximum incoming arcs are on AB^+ that infers that this can accept all type of blood group. On the other hand, the maximum outgoing arcs from the transition t_6, which is attached to O^- blood group, which is the universal donor group [24]. Figure 8 shows the Petri net modeling of this blood donation network.

Initially, one token is being placed on each place showing the availability of all types of blood in a blood bank. With this, all the transitions are enabled here satisfying the enabling condition. Let us consider firing of one transition say t_6, as it is connected with all the receiver places so it will transfer the token to all the places.

$$I^-\left(O^-(D), t_6\right) \leq \mu^0\left(O^-(D)\right)$$

$$\begin{aligned}\mu^1(R) &= \mu^0(R) - I^-(R, t_6) + I^+(R, t_6) \\ &= 0 - 0 + 1 = 1\end{aligned}$$

This modeled net can also be analyzed by certain structural properties shown in Fig. 9 and behavioral property shown in Fig. 10.

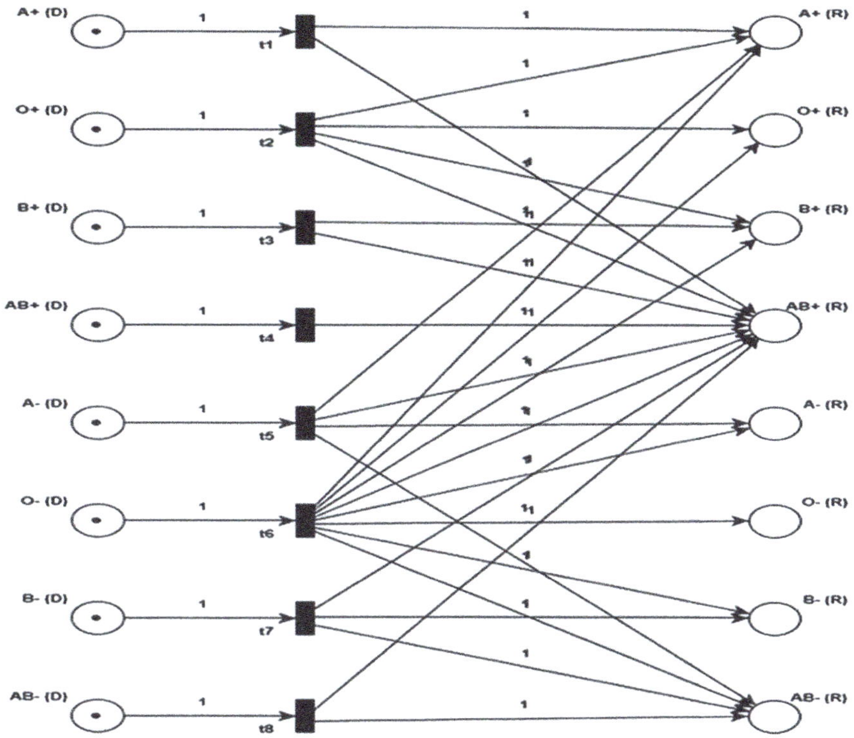

Fig. 8 Petri net modeling of the different acceptor and donor capability of blood type

Fig. 9 Classification of net shown in Fig. 8

Petri net classification results

State Machine	false
Marked Graph	false
Free Choice Net	true
Extended Free Choice Net	true
Simple Net	true
Extended Simple Net	true

4 Conclusion and Scope

Petri net is an efficient tool for the system having conflict and synchronized property and is used to draw useful structural and behavioral properties about the modeled system. In all most every field of sciences, it is widely used to model complex systems. It is efficient because of its elements, i.e., places and transitions which can

P-Invariants

A+ (D)	A+ (R)	AB+ (D)	AB+ (R)	AB- (D)	AB- (R)	A- (D)	A- (R)	B+ (D)	B+ (R)	B- (D)	B- (R)	O+ (D)	O+ (R)	O- (D)	O- (R)
1	1	0	0	0	0	1	0	0	0	0	0	1	0	1	0
1	0	1	1	1	0	1	0	1	0	1	0	1	0	1	0
0	0	0	0	1	1	1	0	0	0	1	0	0	0	1	0
0	0	0	0	0	0	1	1	0	0	0	0	0	0	1	0
0	0	0	0	0	0	0	0	1	1	0	0	1	0	1	0
0	0	0	0	0	0	0	0	0	0	1	1	0	0	1	0
0	0	0	0	0	0	0	0	0	0	0	0	1	1	1	0
0	0	0	0	0	0	0	0	0	0	0	0	0	0	1	1

The net is covered by positive P-Invariants, therefore it is bounded.

P-Invariant equations

M(A+ (D)) + M(A+ (R)) + M(A- (D)) + M(O+ (D)) + M(O- (D)) = 4
M(A+ (D)) + M(AB+ (D)) + M(AB+ (R)) + M(AB- (D)) + M(A- (D)) + M(B+ (D)) + M(B- (D)) + M(O+ (D)) + M(O- (D)) = 8
M(AB- (D)) + M(AB- (R)) + M(A- (D)) + M(B- (D)) + M(O- (D)) = 4
M(A- (D)) + M(A- (R)) + M(O- (D)) = 2
M(B+ (D)) + M(B+ (R)) + M(O+ (D)) + M(O- (D)) = 3
M(B- (D)) + M(B- (R)) + M(O- (D)) = 2
M(O+ (D)) + M(O+ (R)) + M(O- (D)) = 2
M(O- (D)) + M(O- (R)) = 1

Fig. 10 Invariant property of net in Fig. 8

be reformed according to the requirement. In this chapter, some of the application of Petri net modeling has shown and also the criteria in the selection of places and transition in the Petri net model. The software PIPE is being used to get some properties used to analyze the system more clearly which is shown here with some examples. In the present research field, each stream of sciences is intermixing with each other to have a better understanding of the subject. Petri net modeling plays an important role in mathematical modeling because of its easy engulfing nature with any type of problem. In the future, one can model any problem or system accordingly and can get useful results.

Acknowledgements The first author the manuscript is thanks to the funding agencies Science and Engineering Research Board, Govt. of India, India (file no. ECR/2017/003480/PMS), Department of Science and Technology Purse grant, and University Grant Commission (Project id. 257 under University of Potential Excellence Scheme-II) for providing the financial help and instrumentation support.

References

1. Brauer, W., Reisig, W.: Carl adam Petri and "Petri nets". Fundam. Concepts Comput. Sci. **3**(5), 129–139 (2009)
2. Kansal, S., Acharya, M., Singh, G.P.: Boolean Petri nets. In: Petri Nets-Manufacturing and Computer Science. IntechOpen (2012)
3. Kansal, S., Singh, G.P., Acharya, M.: On Petri nets generating all the binary n-vectors. Sci. Math. Japonicae **71**(2), 209–216 (2010)
4. Singh, G.P., Kansal, S., Acharya, M.: Construction of a crisp Boolean Petri net from a 1-safe Petri net. Int. J. Comput. Appl. **73**(17) (2013)

5. Gupta, S., Singh, G.P., Kumawat, S.: Petri net recommender system to model metabolic pathway of polyhydroxyalkanoates. Int. J. Knowl. Syst. Sci. **10**(2), 42–59 (2019)
6. Singh, G.P., Singh, S.K.: Petri net recommender system for generating of perfect binary tree. Int. J. Knowl. Syst. Sci. **10**(2), 1–12 (2019)
7. Singh, G.P., Gupta, A.: A Petri net analysis to study the effects of diabetes on cardiovascular diseases. IEEE Xplore. ISBN: 978-93-80544-36-6 (2019)
8. Singh, G.P., Jha, M., Singh, M., Naina.: Modeling the mechanism pathways of the first-line drug in Tuberculosis using Petri nets. Int. J. Syst. Assur. Eng. Manag. 1–12 (2020)
9. Singh, G.P., Jha, M., Singh, M.: Petri net modeling of clinical diagnosis path in tuberculosis. In: International Conference on Recent Trends in Engineering, Technology and Business Management (ICRTETBM). Springer, (Amity) Noida (2019)
10. Olszak, J., Radom, M., Formanowicz, P.: Some aspects of modeling and analysis of complex biological systems using time Petri nets. Bull. Pol. Acad. Sci. Tech. Sci. **66**(1) (2018)
11. Gupta, S., Kumawat, S., Singh, G.P.: Fuzzy Petri net representation of fuzzy production propositions of a rule-based system. In: International Conference on Advances in Computing and Data Sciences, pp. 197–210. Springer, Singapore (2019)
12. Singh, G.P., Kansal, S.: Basic results on crisp boolean Petri nets. In: Modern Mathematical Methods and High-Performance Computing in Science and Technology, pp. 83–88. Springer, Singapore (2016)
13. Singh, G.P.: Applications of Petri nets in electrical, electronics and optimizations. In: International Conference on Electrical, Electronics, and Optimization Techniques (ICEEOT) IEEE, pp. 2180–2184 (2016)
14. Singh, G.P., Kansal, S., Acharya, M.: Embedding an arbitrary 1-safe Petri net into a boolean Petri net. Int. J. Comput. Appl. **70**(6), 7–9 (2013)
15. Herdal, S., Mouline, S.: Modelling and simulation of biochemical processes using Petri nets. Processes **6**(8), 97 (2018)
16. Rovetto, C., Cano, E., Ojo, K., Tuon, M., Montes, H.: Coloured petri net model for remote monitoring of cardiovascular dysfunction. In: Memorias de Congresos UTP, pp. 405–411 (2018)
17. Finkel, A.: The minimal coverability graph for Petri nets. In: International Conference on Application and Theory of Petri Nets, pp. 210–243. Springer, Berlin, Heidelberg (1991)
18. Dingle, N.J., Knottenbelt, W.J., Suto, T.: PIPE2: a tool for the performance evaluation of generalised stochastic Petri Nets. ACM SIGMETRICS Perform. Eval. Rev. **36**(4), 34–39 (2009)
19. Eckleder, A., Freytag, T.: WoPeD 2.0 goes BPEL 2.0. AWPN **380**, 75–80 (2008)
20. Murata, T.: Petri nets: properties, analysis, and applications. Proc. IEEE **77**(4), 541–580 (1989)
21. Peterson, J.L.: Petri nets. ACM Comp. Surv. (CSUR) **9**(3), 223–252 (1977)
22. Cinici, A.: From caterpillar to butterfly: a window for looking into students' ideas about life cycle and life forms of insects. J. Biol. Educ. **47**(2), 84–95 (2013)
23. Franceschetti, M., Meester, R.: Random Networks for Communication: From Statistical Physics to Information Systems, p. 24 (2008)
24. Istratoaie, M.L., Istratoaie, O.: Organ donation, brief medical and legal perspective. In: International Conference Education and Creativity for a Knowledge-Based Society, p. 181 (2015)

Edgeless Graph: A New Graph-Based Information Visualization Technique

Mahipal Jadeja and Rahul Muthu

Abstract Information visualization is the study of visual representations of abstract data in order to strengthen the human understanding of the data. Graph visualization is one of the subfields of information visualization. It is used for the visualization of structured data, i.e. for inherently related data elements. In the traditional graph visualization techniques, nodes are used to represent data elements, whereas edges are used to represent relations. In this paper, our focus is on the representation of structured data. According to us, key challenges for any graph-based visualization technique are related to edges. Some of them are planar representation, minimization of edge crossing, minimizing the number of bends, distinguish between the vertices and the edges, etc. In this paper, we propose two methods that use the same underlying idea: Assignment of unique label to each vertex of the graph and remove all the edges. Nodes are adjacent if and only if their corresponding labels are disjoint. Our proposed representations do not have edges so we do not need to consider the challenges related to edges that is the biggest advantage. The algorithm for obtaining valid labelling as well as procedures related to dynamic changes (addition/removal of edges and/or vertices) is explained in detail. The space complexities of the proposed methods are O(n^2) and O(n^3) where n denotes the number of nodes. Application of our proposed methods in the analysis of a social network site is also demonstrated. Characteristics of these methods are highlighted along with possible future modifications.

Keywords Graph visualization · Information visualization · Social network analysis · Graph labelling

---

M. Jadeja (✉)
Malaviya National Institute of Technology, Jaipur, India
e-mail: mahipaljadeja.cse@mnit.ac.in

R. Muthu
Dhirubhai Ambani Institute of Information and Communication Technology, Gandhinagar, India
e-mail: rahul_muthu@daiict.ac.in

M. Sahni et al. (eds.), *Mathematical Modeling, Computational Intelligence Techniques and Renewable Energy*, Advances in Intelligent Systems and Computing 1287,
https://doi.org/10.1007/978-981-15-9953-8_39

1 Introduction

Graph-based information visualization techniques are very well studied in the literature [1–5].

Applications of graph visualization are in many areas. Some of them are: Representation of hierarchical structures using trees website maps History of Internet browsing data Biology and chemistry (for the representation of phylogenetic trees, genetic maps, molecular maps, etc.) and other applications include data flow diagrams, E-R diagrams, logic programming, etc.

Size of a graph is one of the major challenges in the field of graph visualization because it is challenging to display a large graph on the available display platform. Viewability issue: for the large graph, sometimes it becomes almost impossible to distinguish between the vertices and the edges of the graph. The visualizations of large graphs are difficult to comprehend. Test for planarity and if the graph is planar, then planar embedding of the graph is another challenge.

Recent work related to graph visualization is discussed in [6–10]. Existing algorithms are too complicated for the practical applications. Other challenges are listed below:

1. It is difficult to interact with the dense graph representation.
2. Predictability: Representations of two similar (isomorphic) graphs must be same not radically different.
3. Proper prioritizing of various aesthetics. According to Purchase [11], reducing the crossings is by far the most important aesthetic, while minimizing the number of bends and maximizing symmetry have a lesser effect.
4. Time complexity: Real-time interaction is necessary for any visualization technique. Therefore, it is desirable to have proper estimation of time complexity of the underlying technique because this factor is crucial for the implementation of large systems. Optimal and polynomial time solutions for the minimization of edge crossing are not possible even for the restricted case where only consecutive layers are considered. In general, the problem is NP-hard and the decision version is NP-complete.

We try to address the above-mentioned challenges using a special type of graph labelling.

1.1 Discussion on the Labelling Scheme

Proposed Labelling Scheme

A set labelling of a graph $G(V, E)$ is a function $f\colon V \rightarrow \mathcal{P}(\{1, 2, \ldots, k\}) - \{\phi\}$ where $k \in N$ with following properties:

1. Two distinct vertices must not get the same set label (one-one function f).
2. $\forall x, y \in V, (x, y) \in E \Leftrightarrow f(x) \cap f(y) = \phi$.

Universe Size Number (USN) of G: Least $k \in N$ for which a set labelling of G can be obtained.

The closely related concept is Knesser graphs [12] $KG_{n,k}$ where vertices correspond to all possible k cardinality subsets of the n cardinality set and adjacency of vertices coincides with disjointness of these subsets. It is possible to solve many graph problems using this set theoretic definition of Knesser graphs and that too, very efficiently. The natural generation is to label any arbitrary graph in this manner, i.e. for a given graph, identify an underlying universal set as well as unique subsets which can associate with vertices in order to preserve the "Knesser" property. Due to the potential irregular structure of the given arbitrary graph, selection of the same cardinality subsets as labels will not work.

(USN) is the smallest possible cardinality of universal set S such that distinct subsets of S can be assigned as labels of vertices of the graph and adjacency of vertices coincides with subsets disjointness. **Universe size number** (USN) and its applications are very well studied in [13–15].

2 Results on USN

Theorem 1 USN(G') =USN$(G) + 1$, *if* $degree(v) = |V(G| = n$, *where* $G' = G + v$.

Proof Here, the degree of the vertex v is n (in G'), and hence, it is adjacent to all vertices of the graph. Therefore, according to the definition of set labelling, it is not allowed to use any of the element of existing labels in the label of v, i.e. USN(G') > USN(G). Here, the objective is to do set labelling of the graph G' using as few elements as possible, and since empty set is not allowed, at least an element is required as a valid set label for any vertex. There is no need for adding anything extra in the set label of v.

Theorem 2 USN$(K_n) = n$.

Proof Corollary of Theorem 1.

Using the following algorithm, we can compute a valid labelling of any arbitrary graph.

Algorithm 1 To obtain a valid (not necessarily optimal) set labelling for any given arbitrary graph

1. Consider the complete graph on n vertices corresponding to any given arbitrary graph on n vertices. Using Theorem 2, obtain the optimal labelling of K_n with USN $= n$. The total number of distinct elements used after this step is exactly n.
2. Covert the K_n into the given arbitrary graph by deleting the necessary edges. Perform the deletion in one by one manner. In order to preserve dis-adjacency, corresponding to the each deleted edge, add an extra (new) label element to the

labels of both edge endpoints. Since K_n has $\binom{n}{2}$ edges, in the worst case (where the given graph is empty), $\binom{n}{2}$ additional elements are required corresponding to this edge(s) deletion step.

Hence, Algorithm 1 will generate a valid labelling of any given arbitrary graph with at most $n + \binom{n}{2} = \sum n$ elements.

Theorem 3 $G' = G + v$ *where* $|V(G)| = n$.
$\text{USN}(G) \leq \text{USN}(G') \leq \text{USN}(G) + (n - 1)$.

Proof Consider a graph $G(n, m)$ with USN k. After adding a new vertex v and m' edges between v and some m' distinct vertices of G, consider $G'(n + 1, m + m')$. Observation: $\text{USN}(G')$ is at least k.

This can be rephrased as a lower bound on increment in USN, upon adding a vertex is 0.

Proof by contradiction:

Suppose $\text{USN}(G') \leq k - 1$.

Now after removing vertex v (and all edges incident on it), the remaining graph $= G$ has a valid labelling with $\text{USN}(G') = k - 1$ which is not possible because optimal labelling has $\text{USN}(G) = k$. (Initial assumption)

Upper bound on increment: $n - d(v)$.

Idea:

A label of v should have some elements from the labels of all the vertices which are non-adjacent to it. Number of non-neighbours are $n - d(v)$ for v, and hence, in the worst case, $n - d(v)$ new elements are required.

The tightness of the upper bound result is shown using the following example. Consider the worst case where the graph G has a clique of size $n - 1$ and one vertex with degree 0, say v_n. Naturally, $|l(v_n)| =$USNG. The newly added degree 1 vertex is v which has v_n as the only neighbour. Since there are no connections between $n - 1$ size clique and v_n, to preserve non-adjacency, $n - 1$ extra elements are required (see Fig. 1).

The next theorem is essentially a rewording of previous result as deletion of a vertex is a reversal of addition of a vertex.

Theorem 4 $G' = G - v$ *where* $|V(G)| = n$.
$\text{USN}(G) - (n - 1) \leq \text{USN}(G') \leq \text{USN}(G)$.

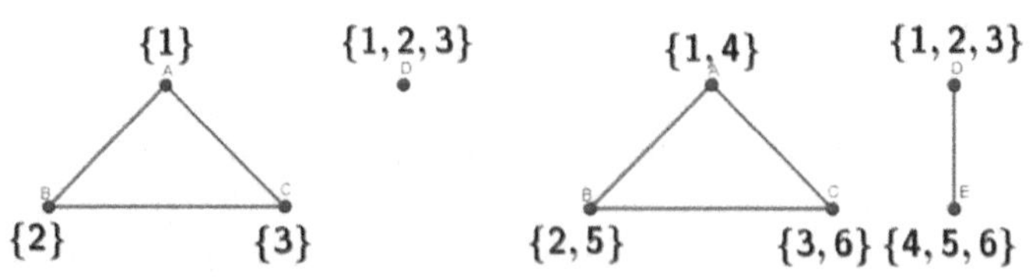

Fig. 1 Increase (decrease) in USN after adding (removing) a vertex

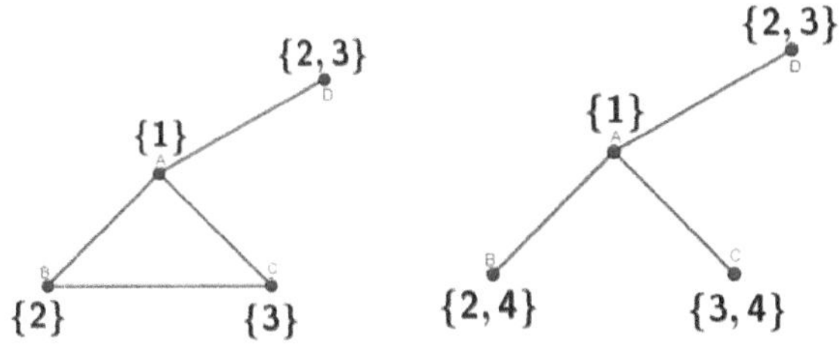

Fig. 2 Lower bound on increase (decrease) in USN after removing (adding) an edge

Theorem 5 *$G' = G \setminus e$,*
$\text{USN}(G) - \min(x, y) \le \text{USN}(G') \le \text{USN}(G) + 1$, *where* $e = (v_a, v_b) \in E(G)$ *and* $x = |l(v_a)|$ *and* $y = |l(v_b)|$.

Theorem 6 *$G' = G + e$,*
$\text{USN}(G) - 1 \le \text{USN}(G') \le \text{USN}(G) + \min(x, y)$, *where* $e = (v_a, v_b) \notin E(G)$ *and* $x = |l(v_a)|$ *and* $y = |l(v_b)|$ *(Fig. 2).*

Proof Theorem 6 is effectively just a rewording of Theorem 5 because the operations they deal with are addition and deletion of an edge respectively which are reversal of each other. The first inequality of Theorem 5 is the same as the second inequality of Theorem 6. Similarly, the second inequality of Theorem 5 is the same as the first inequality of Theorem 6. Hence, we provide a combined prove of both theorems. It is easier to prove the first inequality of Theorem 5 in the framework of edge addition while the second inequality of Theorem 5 in the framework of edge deletion. Thus, we follow this scheme.

Assume k elements are common between labels of v_a and v_b where $1 \le k \le min(|l(v_a)|, |l(v_b|)$.

Procedure 1: Obtaining a valid labelling using k additional labels:

1. Compare the cardinalities of the labels of the two endpoints of the newly added edge. The label with higher cardinality is retained as it is.
2. Replace $C = \{1, 2, \ldots, k\}$, the common elements with $C' = \{1', 2', \ldots, k'\}$ in the label of the vertex which has lower cardinality.
3. For all other labels, if they contain any non-empty subset of C, then add the corresponding non-empty subset from C' to their labels for preservation of non-adjacency with the lower cardinality edge endpoint.

After deletion (removal) of an edge, USN may increase by at most one. Since in the worst case, addition of only one new element is sufficient in the labels of the endpoints (to preserve non-adjacency) (Fig. 3).

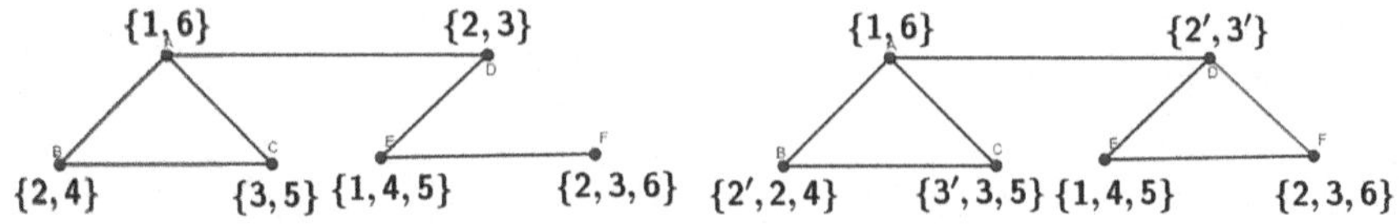

Fig. 3 Upper bound on increase (decrease) in USN after adding (removing) an edge

3 Brief Descriptions of Our Proposed Methods

Both of our proposed methods use the valid labelling described in the previous section.

Method-1: This method requires an n by n grid in which all labels are partitioned based on their cardinalities (see Fig. 6). Value of x-axis represents the cardinality value, i.e. all nodes which are present on the line $x = i$ will have labels of size i. Value of y-axis represents the number of elements which are present in each same cardinality group. Largest possible cardinality for any individual label is n (see Algorithm 1). Therefore, the upper limit on the value of x-axis is n. At most n elements can be present in one group. Therefore, the upper limit on the value of y-axis is also n.

Method-2: This method requires an n by $O(n^2)$ grid in which all labels are partitioned based on their individual label structures (see Fig. 9). Labels which are present in the xth column will have xth element common in them. All the elements which are less than x will be missing in these labels, i.e. $\{1, 2, \ldots, x-1\}$. Value of y-axis represents the number of elements which are present in that group. Largest possible cardinality of USN is $O(n^2)$ (see Algorithm 1). Therefore, the upper limit on the value of x-axis is $O(n^2)$. At most n elements can be present in one group. Therefore, the upper limit on the value of y-axis is n.

4 Application: Social Network Analysis

In this section, we describe how our proposed technique can be used for the analysis of a social network.

One sample graph which represents the friendship relationship between 6 people is shown in Fig. 4, i.e. if edge (u, v) is present, then person u and v are friends on social network sites like Facebook/Twitter.

Few iterations of the Algorithm 1, for the construction of the valid labelling of the given graph are shown in Figs. 4 and 5. Our proposed visualization method is shown in Fig. 6.

Key characteristics of our proposed method are listed below:

1. Space complexity: n by n grid is required for the visualization technique. This is because each vertex can be non adjacent to at most remaining $n-1$ vertices, and

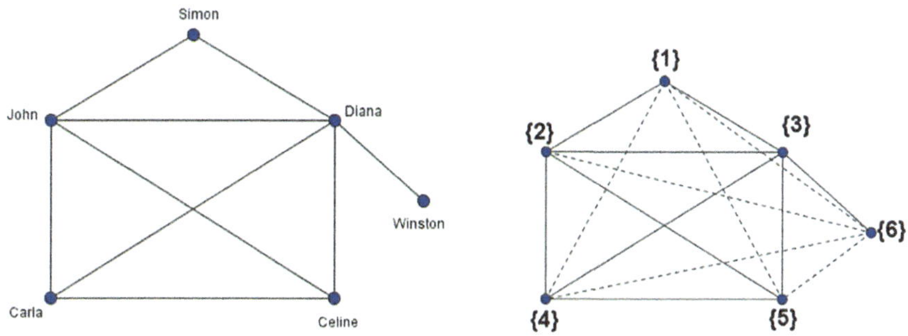

Fig. 4 **a** Social network graph and **b** its corresponding complete graph

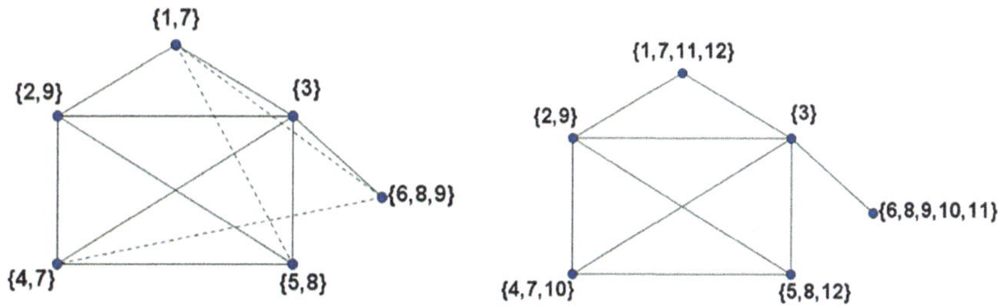

Fig. 5 Steps for obtaining valid labelling of the social network graph

Fig. 6 Edgeless graph: our proposed method

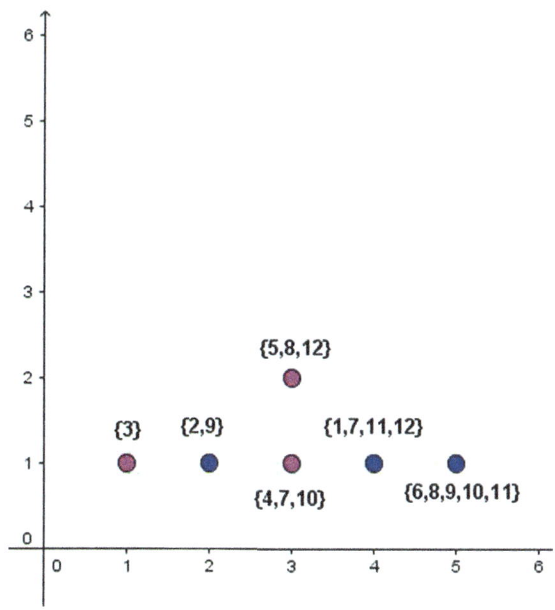

therefore, size of individual label can not exceed n (see Algorithm 1). Total number of vertices and hence total number of distinct labels are exactly n. Therefore, total number of labels with the same cardinality can be at most n.

2. x-axis represents the cardinality of the individual label, i.e. classification is done based upon the size of the label. y-axis represents the number count corresponding to the each group.
3. It is easy to identify whether two nodes are related or not by comparing their corresponding labels. Nodes are related if and only if their corresponding labels are disjoint.
4. In the visualization, number of friends of a node tends to be inversely proportional to its individual label size. Here, label 3 has the minimum cardinality, and therefore, it is very likely that the corresponding person (Diana) has the most number of friends in the network which is actually true (see Fig. 4). Diana is the friend of all others.
5. Here, value of $|l| - 1$ for any node denotes minimum number of non-neighbours for that particular node. For example, for $l = \{2, 9\}$ value of $|l| - 1$ is $2 - 1 = 1$ from which we can infer that John has at least one non-neighbour in the graph which is true. See Fig. 2, John and Winson are not related.
6. Vertices which are present in the same vertical line have at least k non neighbours where $k = x - 1$ where x denotes the value of X coordinate.
7. People who are present in first column tend to be the most popular, whereas those who are present in the last column tend to be the least popular.
8. Identification of the clusters with same popularity in the underlying social network graph is possible using this visualization.
9. Further details can be obtained by assigning colours to the node. Here in our example, for female gender nodes, pink colour is assigned whereas for males blue colour is assigned. This will help to analyse gender specific patterns in the underlying social network graph.

Using our proposed visualization method (edgeless graph), we will be able to analyse following crucial questions related to the given social network.

1. Identification of the users who are the most active as well as who are most inactive. This is necessary for promotional activities.
2. It can help the social networking site to improve their "People you may know" feature.
3. With the use of labels, it is easier to verify whether any given two persons are connected or not.
4. It is easy to identify all people with specific number of friends.

4.1 Study of Dynamic Changes

Social networks are highly dynamic in the nature. Using procedure 1 and 2 described in the previous section, we can efficiently and effectively add (remove) new vertices and/or edges in our existing visualization.

In Fig. 7, modified graph is shown after adding 2 new vertices and 4 edges. The modified visualization is shown in Fig. 8 which has following interesting features:

1. No additional space is required (in this particular case). Number of nodes are $n + 2$ but still the n by n grid is sufficient because no new elements were added during the procedure.

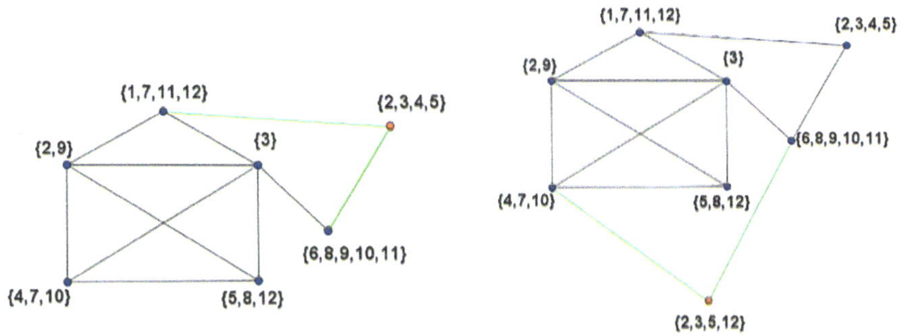

Fig. 7 Valid labelling of the graph after addition of 2 new vertices and 4 edges

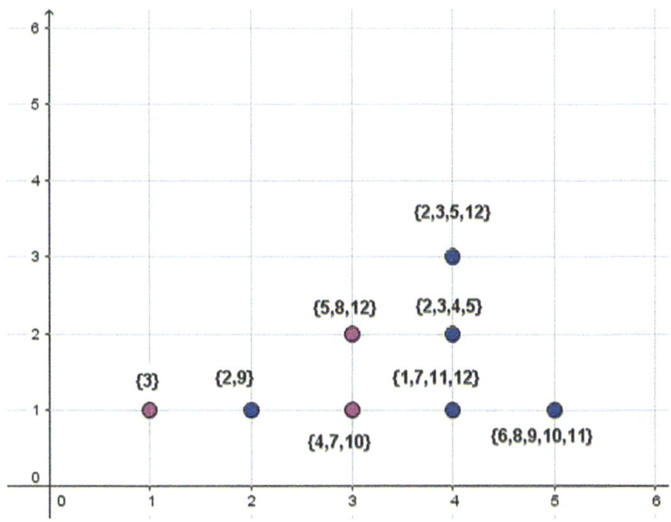

Fig. 8 Representation of the graph after addition of 2 new vertices and 4 edges

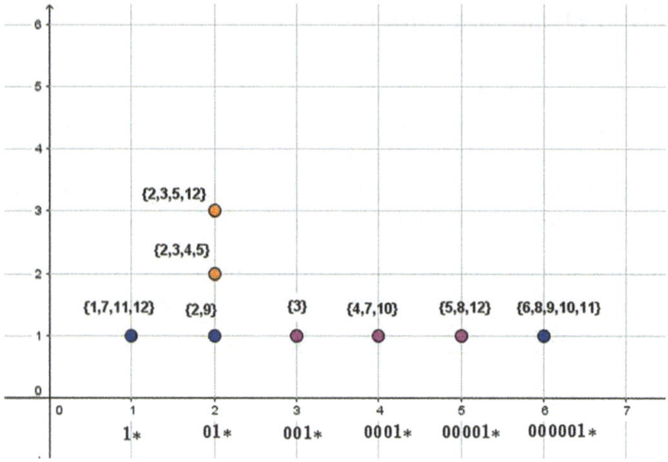

Fig. 9 Proposed method-2 for identify collection of nodes who are mutually non-adjacent

2. It is easy to identify the correct location for the newly added vertex based upon its label size. Here, both of the newly added vertices have individual label size 4, and therefore, they are placed in the 4th column.
3. One can also highlight the newly added vertices with some different colour in order to compare their initial popularity in the underlying social network.

The visualization of the same graph (shown in Fig. 5) using our proposed method-2 is shown in Fig. 9. The properties of this method are listed below.

- Elements which are vertically aligned have at least one element in common, and therefore, they are mutually non-adjacent.
- In order to identify, collection of mutually adjacent vertices (cliques), one can plot valid labelling of the complement graph using this method.
- Space complexity: n by $O(n^2)$ grid is required because size of individual label is at most n and the size of underlying labelling set is $O(n^2)$ (see Algorithm 1).
- Here, newly added vertices are represented using orange colour and they are vertically aligned which says that the newly added vertices are non-adjacent to each other.

5 Conclusions and Future Work

Our proposed data visualization techniques have almost all the necessary characteristics. It shows all the data without any information loss. Using our technique, it is possible to display large data sets coherently. The method encourages the human eye to compare different objects because each object is represented by unique label.

In future, we plan to design a more efficient interactive version of the proposed techniques in which we would incorporate the following features.

- In the interactive version, we want to highlight edges which are present within certain vertical regions of the representation whenever the user clicks on it. Using this feature, users can understand whether those nodes which belong to the same cardinality group are related to each other or not.
- Within one vertical region of the representation, we will put disjoint labels, i.e. adjacent neighbours within the alpha-neighbourhood of them. This will help users to identify connections rapidly.
- Whenever the user clicks on a particular node, we will highlight all the neighbours of the node by doing real-time computation on labels and we will also highlight the corresponding edges so that users can find out the neighbours.

References

1. Herman, I., Melançon, G., Marshall, M.S.: Graph visualization and navigation in information visualization: a survey. IEEE Trans. Visual Comput. Graphics **6**(1), 24–43 (2000)
2. Battista, G.D., Eades, P., Tamassia, R., Tollis, I.G.: Graph Drawing: Algorithms for the Visualization of Graphs. Prentice Hall PTR, Upper Saddle River, NJ (1998)
3. Novak, O.: Visualization of Large Graphs. Doctoral dissertation, PhD thesis, Master's thesis, Czech Technical University in Prague (2002)
4. Lee, B.: Interactive Visualizations for Trees and Graphs. Doctoral dissertation (2006)
5. Erdos, P., Goodman, A.W., Pósa, L.: The representation of a graph by set intersections. Can. J. Math. **18**(106–112), 86 (1966)
6. Debusscher, B., Landuyt, L., Van Coillie, F.: A visualization tool for flood dynamics monitoring using a graph-based approach. Remote Sens. **12**(13), 2118 (2020)
7. Imre, M., Tao, J., Wang, Y., Zhao, Z., Feng, Z., Wang, C.: Spectrum-preserving sparsification for visualization of big graphs. Comput. Graphics **87**, 89–102 (2020)
8. Telea, A.: Image-based graph visualization: advances and challenges. In: International Symposium on Graph Drawing and Network Visualization, pp. 3–19. Springer (2018)
9. Lu, J., Si, Y.W.: Clustering-based force-directed algorithms for 3d graph visualization. J. Supercomput. **76**, 9654–9715 (2020)
10. Walsh, K., Voineagu, M.A., Vafaee, F., Voineagu, I.: Tdaview: an online visualization tool for topological data analysis. Bioinformatics (2020)
11. Purchase, H.: Which aesthetic has the greatest effect on human understanding? In: International Symposium on Graph Drawing, pp. 248–261. Springer (1997)
12. Van Dam, E.R., Haemers, W.H.: An odd characterization of the generalized odd graphs. J. Comb. Theory Ser. B **101**(6), 486–489 (2011)
13. Jadeja, M., Muthu, R.: Labeled object treemap: a new graph-labeling based technique for visualizing multiple hierarchies. Ann. Pure Appl. Math. **13**, 49–62 (2017)
14. Jadeja, M., Muthu, R., Sunitha, V.: Set labelling vertices to ensure adjacency coincides with disjointness. Electron. Notes Discrete Math. **63**, 237–244 (2017)
15. Jadeja, M., Muthu, R.: Uniform set labeling vertices to ensure adjacency coincides with disjointness. J. Math. Comput. Sci **7**(3), 537–553 (2017)

Floyd's Algorithm for All-Pairs Interval-Valued Neutrosophic Shortest Path Problems

Nayankumar Patel, Ritu Sahni, and Manoj Sahni

Abstract Neutrosophic set theory becomes an important tool in almost every real-world problems and therefore every field of mathematics taken the advantage of this theory. Present work is based on one of the classical problem of a branch of mathematics, known as graph theory. This subject contains the methods for finding minimal path of any network. Although, many algorithms are available for finding shortest path, but to increase its applicability in real-life problems, it is important to generalize it in those environments, which considers most of the real-life situations. In this work, we have generalized one of the popular algorithms, called Floyd's algorithm using interval-valued neutrosophic set. The graph of five nodes is considered with the weights between them as interval-valued neutrosophic numbers, and the shortest path matrix is formed using the score function. The final score and transition path matrix provides the minimum value and the path between any pair of nodes.

Keywords Interval-valued neutrosophic set · Floyd's algorithm · All-pair shortest path problem · Minimum spanning tree

1 Introduction

Fuzzy set theory is prevalent in making our daily decisions and dealing with day-to-day incomplete and vague informations. This theory is credited to Zadeh [1], a famous mathematician, computer scientist, electrical engineer, a great researcher from the University of California, Berkeley. This theory is based on graded system that means it provides membership values to each of the information within the

N. Patel
Adani Institute of Infrastructure Engineering, Ahmedabad, Gujarat, India

R. Sahni
Institute of Advanced Research, Gandhinagar, Gujarat, India

M. Sahni (✉)
Pandit Deendayal Petroleum University, Gandhinagar, Gujarat, India
e-mail: manojsahani117@gmail.com

M. Sahni et al. (eds.), *Mathematical Modeling, Computational Intelligence Techniques and Renewable Energy*, Advances in Intelligent Systems and Computing 1287,
https://doi.org/10.1007/978-981-15-9953-8_40

interval [0, 1], in spite of classical set theory, which tells that received information is either true, denoted by 1 or false denoted by 0 and nothing in between. In real-life situation, it is not possible always to describe the information precisely, so the data obtained is not in the form of crisp values and in that case conventional logic fails to model the problem in a precise manner. Fuzzy logic allows vague informations with membership values, which provides degree of belongingness, and thus able to model the situation. In all the areas where informations are collected through human judgment, fuzzy theory serves as an efficient tool to cope with insufficient, imprecise data [2–5]. However, fuzzy theory deals with only belongingness and cannot express non-belongingness of the data. During analysis of the problem, most of the time we need non-membership values also. In view of this lacking in fuzzy set theory, Atanassov [6] described intuitionistic fuzzy set theory, which incorporate non-membership values also and in such a way that after adding membership and non-membership values, we get the resultant value within the interval [0, 1]. As the extension of these fuzzy sets, Gorzalczany [7] introduced interval-valued fuzzy set and Atanassov and Gargov [8] gave the concept of interval-valued intuitionistic fuzzy set theory, so that they allow the interval of membership values and non-membership values. In this way, this theory generalizes and extended in many ways by different researchers in order to cover and solve most of the real-life problems. In this sequence, a new emerging tool namely neutrosophic set theory for dealing with uncertain data is came into existence. This is the generalization of all previous fuzzy theories, as it deals with neutral, indeterminate, and unknown informations. Smarandache [9] coined neutrosophic logic and neutrosophic set theory. Neutrosophic set allows truth values, indeterminacy, false values within the non-standard unit interval $]0^{-}, 1^{+}[$, which is obvious generalization of standard unit interval [0, 1]. But this non-standard interval create problems while dealing with real applications. Hence, single-valued neutrosophic set [10] were introduced to overcome this problem. Again, in real situations, the degree of truthness, indeterminacy, and falsity sometimes cannot be given in the form of single value, but it is given in the form of several possible interval values. In this way, neutrosophic set generalizes to interval-valued neutrosophic sets by Wang et al. [11], just like interval-valued fuzzy set and intuitionistic fuzzy set. This interval-valued neutrosophic set theory applied to many real-life problems after its inception, like in decision-making problems, computer science problems, pattern recognition problems, artificial intelligence problems, operation research problems, and many more.

Network flow problem is one of the important problem used in many day-to-day situations such as in communication, transportation, and economical problems. Graph theory plays an important role in dealing with these kinds of problems, as it contains many algorithms in which using graph models many real-world systems are described. Finding shortest paths between two junctions denoted by nodes or vertices has been very important topic, as it is easily reformulated and frequently used in calculating minimum distance, least cost, minimum time, etc., from the source vertex to destination vertex. These values are generally imprecise in nature as it fluctuates from situation to situation. In that case, fuzzy theory plays an important role to describe those fluctuated information in a precise manner. Dealing with shortest path problem by applying fuzzy set theory was first initiated by Dubois and Prade [12] using Floyd's and Ford's algorithm. Fuzzy linear programming approach was considered by Lin and Chen [13] to solve shortest path problems. Fuzzy shortest path problem for multiple constraints network was discussed by Dou et al. [14]. Deng et al. [15] presented modified version of Dijkstra algorithm for solving fuzzy shortest path problems.

In this direction, we have developed an algorithm for finding shortest path of any network by considering famous Floyd's algorithm [16] and generalize it using interval-valued neutrosophic set theory. We have presented the working of the algorithm and then used an example for finding the shortest route between any two nodes, where the interval-valued neutrosophic numbers are used to denote the weights on the graph or network.

2 Some Basic Definitions

Smarandache [17] defined the neutrosophic set in the following way:

Definition 2.1 (*Neutrosophic set*) [17]. Let $N = \{< x : T_N(a), I_N(a), F_N(a) >, a \in X\}$, be any set defined over universe of discourse X, and let $T, I, F : X \rightarrow]^-0, 1^+[$ are the functions which describes the degree of membership, the degree of indeterminacy, and the degree of non-membership, respectively of the element $a \in X$. Further, these functions satisfy the following condition:

$^-0 \leq T_N(a) + I_N(a) + F_N(a) \leq 3^+$. Then, A is known as neutrosophic set.

This kind of neutrosophic set is problematic when we try to apply them on practical problems, because here all the values are assumed within the non-standard interval $[^-0, 1^+]$. Therefore, for calculation purpose, Wang et al. [11] replaced the interval $]^-0, 1^+[$ with standard interval [0, 1] and defined single-valued neutrosophic set as follows:

Definition 2.2 (*Single-valued neutrosophic set*) [10]. Let $N = \{< x : T_N(a), I_N(a), F_N(a) >, a \in X\}$ be any set defined over universe of discourse X, and let $T, I, F : X \rightarrow [0, 1]$ are the functions which describes the degree of membership, the degree of indeterminacy, and the degree of non-membership, respectively of the element $a \in X$. Further, these functions satisfy the following condition: $0 \leq T_N(a) + I_N(a) + F_N(a) \leq 3$, then N is called a single-valued neutrosophic set.

Definition 2.3 (*Interval-valued neutrosophic set*) [11]. Let set N is defined over universe of discourse X in the following way:

$$N = \left\{ < a : [T_N^L(a), T_N^U(a)], [I_N^L(a), I_N^U(a)], [F_N^L(a), F_N^U(a)] >, a \in X \right\},$$

where $T_N(a)$, $I_N(a)$ and $F_N(a)$ describe degree of truth-membership, degree of indeterminacy, and degree of falsity, respectively, with $T_N(a), I_N(a), F_N(a) : X \to [0, 1]$ satisfying

$$0 \leq T_N(a) + I_N(a) + F_N(a) \leq 3.$$

Then, N is called an interval-valued neutrosophic set.

Score function is used to compare various single-valued neutrosophic numbers. It is defined in the following way:

Definition 2.4 *Score function for IVNSs*: Let set N is defined over universe of discourse X, the formula for calculating score function for interval-valued neutrosophic set is given as follows:

$$M(N) = \frac{2 + \left(T_N^L + T_N^U\right) - \left(I_N^L + I_N^U\right) - \left(F_N^L + F_N^U\right)}{3},$$

where $M(N)$, represents the score values.

3 Word Done

There are many algorithms available to determine the shortest path of many network problems, in which one has to find the shortest route between any two vertices. Floyd's algorithm is one of them, which deals with both directed and undirected routes. The present work considers Floyd's algorithm with weighted graph or network, where weights or distances are represented in the form of interval-valued neutrosophic numbers, and the algorithm works in the following way:

Suppose that network contains 'n' vertices and represented in the form of matrix containing n rows and n columns. The distance between two vertices a and b is denoted by M_{ab}, that is pair (a, b) exists in the matrix (i) if nodes a and b are directly related to each other, i.e., there is a finite distance between two nodes, (ii) otherwise, the distance is infinite. Now we consider an example in which a graph contains three vertices a, b, and c, and the problem is to find the shortest distance between the vertices a and c. Suppose a is directly connected to c with the distance M_{ac}, but as the weight is also assigned to it, so while calculating the distance in terms of weight, we found that the distance from a to c via b is shorter than directly a to b and $M_{ab} + M_{bc} < M_{ac}$. So, it is optimal to replace direct route $a \to c$ with indirect route $a \to b \to c$. This is called transitive relation for finding optimal route.

Floyd's algorithm is a dynamic programming problem. This is a method for finding the shortest path between two locations using following steps:

The Floyd's algorithm finds the shortest path between every two pair of matrices.

Step 1: Let the graph G has n number of nodes, and we find the shortest route between any two vertices.

It is known that the square matrix of $n \times n$ will be constructed in which diagonal element will be left blank because no loops are considered. The other off-diagonal cells of the matrix are filled with score function values using definition 2.4. Initially, setting the node $k = 1$, we get the score matrix and initial path matrix depicted in Tables 1 and 2, respectively.

Similarly by taking $k = 2$ to n, and using the transitive operation $M_{ab}+M_{bc} < M_{ac}$, and a and c varying from 1 to n, we are able to find the shortest path from one node to another using the intermediate nodes. In this if the transitive operation is true, then the path changes using the node k; otherwise, it will remain same. Hence, in this way, we are able to construct the shortest path matrix and score values matrix N_k and M_k, respectively, at the last matrix for the last node, i.e., at $k = n$.

Step 1: Given any problem whose weights are defined in terms of interval-valued neutrosophic numbers, the problem is to calculate the shortest distance between every two vertices.

Table 1 Score matrix between every pair of nodes

M_0	1	2	3	…	k	…	n
1	–	M_{12}	M_{13}	…	M_{1k}	…	M_{1n}
2	M_{21}	–	M_{23}	…	M_{2k}	…	M_{2n}
3	M_{31}	M_{32}	–	…	M_{3k}	…	M_{3n}
…	…	…	…	…	…	…	…
k	M_{k1}	M_{k2}	M_{k3}	…	–	…	M_{kn}
…	…	…	…	…	…	…	…
n	M_{n1}	M_{n2}	M_{n3}	…	M_{nk}	…	–

Table 2 Matrix showing transition of path from the initial node

N_0	1	2	3	…	k	…	n
1	–	2	3	…	k	…	n
2	1	–	3	…	k	…	n
3	1	2	–	…	k	…	n
…	…	…	…	…	…	…	…
k	1	2	3	…	–	…	n
…	…	…	…	…	…	…	…
n	1	2	3	…	k	…	–

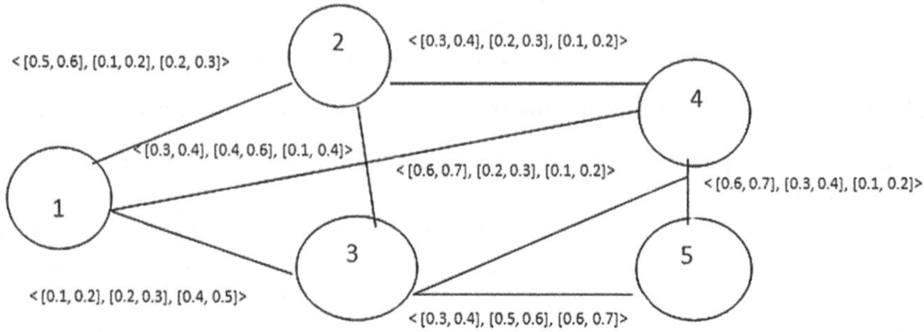

Fig. 1 Undirected graph between five nodes with weight in the form of IVN numbers

Let us consider undirected graph of five nodes having the weights defined in terms of interval-valued neutrosophic numbers between the nodes shown in Fig. 1.

Solution:

The interval-valued neutrosophic numbers between two nodes are defined as:

$$< 1, 2 >=< [0.5, 0.6], [0.1, 0.2], [0.2, 0.3] >$$
$$< 1, 3 >=< [0.1, 0.2], [0.2, 0.3], [0.4, 0.5] >$$
$$< 1, 4 >=< [0.3, 0.4], [0.4, 0.6], [0.1, 0.4] >$$
$$< 2, 3 >=< [0.2, 0.3], [0.4, 0.5], [0.5, 0.6] >$$
$$< 2, 4 >=< [0.3, 0.4], [0.2, 0.3], [0.1, 0.2] >$$
$$< 3, 4 >=< [0.6, 0.7], [0.2, 0.3], [0.1, 0.2] >$$
$$< 3, 5 >=< [0.3, 0.4], [0.5, 0.6], [0.6, 0.7] >$$
$$< 4, 5 >=< [0.6, 0.7], [0.3, 0.4], [0.1, 0.2] >$$

Step 1: First we convert the interval-valued neutrosophic numbers into a single digit using the score function defined as:

$$M(N) = \frac{2 + \left(T_N^L + T_N^U\right) - \left(I_N^L + I_N^U\right) - \left(F_N^L + F_N^U\right)}{3}$$

Construct matrices M_0 and N_0, and using the corresponding score function, the initial path matrix is formed. As cells (1, 5), (5, 1), (2, 5), and (5, 2) have no line segment, hence, the initial values is assigned as infinity.

The calculation of values of score matrix (M_0) are as follows:

For node (1, 2), $M_{12} = \frac{2+(0.5+0.6)-(0.1+0.2)-(0.2+0.3)}{3} = 0.77$

For node (1, 3), $M_{13} = \frac{2+(0.1+0.2)-(0.2+0.3)-(0.4+0.5)}{3} = 0.30$

For node (1, 4), $M_{14} = \frac{2+(0.3+0.4)-(0.4+0.6)-(0.1+0.4)}{3} = 0.40$

For node (2, 3), $M_{23} = \frac{2+(0.2+0.3)-(0.4+0.5)-(0.5+0.6)}{3} = 0.17$

For node (2, 4), $M_{24} = \frac{2+(0.3+0.4)-(0.2+0.3)-(0.1+0.2)}{3} = 0.63$

For node (3, 4), $M_{34} = \frac{2+(0.6+0.7)-(0.2+0.3)-(0.1+0.2)}{3} = 0.83$,
For node (3, 5), $M_{35} = \frac{2+(0.3+0.4)-(0.5+0.6)-(0.6+0.7)}{3} = 0.10$
For node (4, 5), $M_{45} = \frac{2+(0.6+0.7)-(0.3+0.4)-(0.1+0.2)}{3} = 0.77$
As there is no direct connectivity between the nodes (1, 5) and (2, 5), so the score value is being assigned as infinity (Inf).

Step 2: Setting $k = 1$ in Table 3, where the pivot row and pivot column are first row and first column, respectively. The working for the next matrix is (Table 4).

As $C_{21} + C_{13} < C_{23}$, $C_{21} + C_{14} < C_{24}$, $C_{31} + C_{12} < C_{32}$, $C_{41} + C_{12} < C_{42}$ are not true and $C_{41} + C_{13} < C_{43}$ and $C_{31} + C_{14} < C_{34}$ is true as $0.40 + 0.30 < 0.83$, then setting $S_{43} = 1$, and $S_{34} = 1$, a new table of score values and transition path matrix is formed and is given in Tables 5 and 6, respectively.

Step 3: Setting $k = 2$, then the second row and second column become the pivot row and pivot columnm and no other changes are observed, i.e., Tables 5 and 6

Table 3 Initial score matrix between every pair of nodes

M_0	1	2	3	4	5
1	–	0.77	0.30	0.40	Inf
2	0.77	–	0.17	0.63	Inf
3	0.30	0.17	–	0.83	0.10
4	0.40	0.63	0.83	–	0.77
5	Inf	Inf	0.10	0.77	–

Table 4 Initial path matrix between every pair of nodes

N_0	1	2	3	4	5
1	–	2	3	4	5
2	1	–	3	4	5
3	1	2	–	4	5
4	1	2	3	–	5
5	1	2	3	4	–

Table 5 Score matrix between every pair of nodes for $k = 1$

M_1	1	2	3	4	5
1	–	0.77	0.30	0.40	Inf
2	0.77	–	0.17	0.63	Inf
3	0.30	0.17	–	0.70	0.10
4	0.40	0.63	0.70	–	0.77
5	Inf	Inf	0.10	0.77	–

Table 6 Transition path matrix from the node $k = 1$

N_1	1	2	3	4	5
1	–	2	3	4	5
2	1	–	3	4	5
3	1	2	–	1	5
4	1	2	1	–	5
5	1	2	3	4	–

Table 7 Score matrix between every pair of nodes for $k = 2$

C_2	1	2	3	4	5
1	–	0.77	0.30	0.40	Inf
2	0.77	–	0.17	0.63	Inf
3	0.30	0.17	–	0.70	0.10
4	0.40	0.63	0.70	–	0.77
5	Inf	Inf	0.10	0.77	–

Table 8 Transition path matrix from the node $k = 2$

C_2	1	2	3	4	5
1	–	2	3	4	5
2	1	–	3	4	5
3	1	2	–	1	5
4	1	2	1	–	5
5	1	2	3	4	–

remain same. Tables 7 and 8 represent the score values and transition path matrix, respectively, by using nodes 1 and 2.

The calculation is shown below:

$$C_{12} + C_{23} < C_{13},\quad C_{12} + C_{24} < C_{14},\quad C_{12} + C_{25} < C_{15},\quad C_{32} + C_{21} < C_{31},$$
$$C_{32} + C_{24} < C_{34},\quad C_{32} + C_{25} < C_{35},\quad C_{42} + C_{21} < C_{41},\quad C_{42} + C_{23} < C_{43},$$
$$C_{42} + C_{25} < C_{45}.$$

As all of them are not true, hence Tables 5 and 6 does not change.

Step 4: Now, setting $k = 3$, the third row and third column become the pivot row and pivot column, and the matrix cells (1, 2), (2, 1), (1, 5), (5, 1), (2, 5), and (5, 2) score values get changes with the inclusion of node 3. The changes are given in Tables 9 and 10, respectively, which shows the score matrix and transition path matrix, respectively. The calculation of the changes is shown below:

Table 9 Score matrix between every pair of nodes for $k = 3$

C_3	1	2	3	4	5
1	–	0.47	0.30	0.40	0.40
2	0.47	–	0.17	0.63	0.27
3	0.30	0.17	–	0.70	0.10
4	0.40	0.63	0.70	–	0.77
5	0.40	0.27	0.10	0.77	–

Table 10 Transition path matrix from the node $k = 3$

S_3	1	2	3	4	5
1	–	3	3	4	3
2	3	–	3	4	3
3	1	2	–	1	5
4	1	2	1	–	5
5	3	3	3	4	–

$$C_{13} + C_{32} < C_{12}, \quad C_{23} + C_{31} < C_{21}, \quad C_{13} + C_{35} < C_{15}$$
$$C_{53} + C_{31} < C_{51}, \quad C_{53} + C_{32} < C_{52}, \quad C_{13} + C_{35} < C_{15}.$$

As all of them are true, hence, the matrix cells listed above is true.

Step 5: Now, setting $k = 4$, the fourth row and fourth column become the pivot row and pivot column, and there is no changes in other matrix cells. So, Tables 11 and 12 are same as that of 9 and 10, respectively.

Step 6: Now, setting $k = 5$, the fourth row and fourth column become the pivot row and pivot column, and there is no changes in other matrix cells. So, Tables 13 and 14 are same as that of 11 and 12, respectively.

Table 13 represents the final score matrix which have the minimum values between any two pair of matrices. Table 14 represents the transition path matrix, i.e., the route to reach the initial node to the final destination node.

Table 11 Score matrix between every pair of nodes for $k = 4$

C_4	1	2	3	4	5
1	–	0.47	0.30	0.40	0.40
2	0.47	–	0.17	0.63	0.27
3	0.30	0.17	–	0.70	0.10
4	0.40	0.63	0.70	–	0.77
5	0.40	0.27	0.10	0.77	–

Table 12 Transition path matrix from the node $k = 4$

S_4	1	2	3	4	5
1	–	3	3	4	3
2	3	–	3	4	3
3	1	2	–	1	5
4	1	2	1	–	5
5	3	3	3	4	–

Table 13 Score matrix between every pair of nodes for $k = 5$

C_5	1	2	3	4	5
1	–	0.47	0.30	0.40	0.40
2	0.47	–	0.17	0.63	0.27
3	0.30	0.17	–	0.70	0.10
4	0.40	0.63	0.70	–	0.77
5	0.40	0.27	0.10	0.77	–

Table 14 Transition path matrix from the node $k = 5$

S_5	1	2	3	4	5
1	–	3	3	4	3
2	3	–	3	4	3
3	1	2	–	1	5
4	1	2	1	–	5
5	3	3	3	4	–

4 Conclusions

In the present work, finding a minimum route from initial node to the final node between every pair of matrix is discussed using Floyd's algorithm by taking weights in the form of interval-valued neutrosophic numbers. An example is used to describe the working of this generalized algorithm. The fuzzified Floyd's algorithm is simple and efficient in handling real-world network problems. This Floyd's algorithm can also be extended to bipolar neutrosophic sets and can be used in various network problems, such as in transportation problem, traffic management problem, and electric circuit problem as a future scope.

References

1. Zadeh, L.A.: Fuzzy sets. Inf. Control **8**, 338–353 (1965)
2. Sahni, M., Mandaliya, A., Sahni, R.: Evaluation of teachers' performance based on students' feedback using aggregator operator. WSEAS Trans. Math. **18**, 85–90 (2019)
3. Sahni, M., Sahni, R., Verma, R., Mandaliya, A., Shah, D.: Generalized trapezoidal intuitionistic fuzzy number for finding radial displacement of a solid disk. WSEAS Trans. Math. **18**, 105–111 (2019)
4. Verma, R., Merigó, J.M., Sahni, M.: On generalized fuzzy jensen-exponential divergence and its application to pattern recognition. In: IEEE Symposium Series on Computational Intelligence (SSCI), pp. 1515–1519 (2018)
5. Verma, R., Merigó, J.M., Sahni, M.: Pythagorean fuzzy graphs: some results. arXiv preprint arXiv:1806.06721
6. Atanassov, K.T.: Intuitionistic fuzzy sets. Fuzzy Sets Syst. **20**(1), 87–96 (1986)
7. Gorzalczany, M.B.: A method of inference in approximate reasoning based on interval-valued fuzzy sets. Fuzzy Sets Syst. **21**, 1–17 (1987)
8. Atanassov, K., Gargov, G.: Interval-valued intuitionistic fuzzy sets. Fuzzy Sets Syst. **31**, 343–349 (1989)
9. Smarandache, F.: A unifying field in logics: neutrosophic logic. Multiple-Valued Logic **8**, 385–438 (2002)
10. Wang, H., Smarandache, F., Zhang, Y., Sunderraman, R.: Single Valued Neutrosophic Sets. Multispace Multistructure **4**, 410–413 (2010)
11. Wang, H., Smarandache, F., Zhang, Y., Sunderraman, Q.: Interval neutrosophic sets and logic: theory and applications in computing. Hexis, Phoenix, AZ (2005)
12. Dubois, D., Prade, H.: Fuzzy sets and systems: theory and applications. Academic Press, New York (1980)
13. Lin, K., Chen, M.: The fuzzy shortest path problem and its most vital arcs. Fuzzy Sets Syst. **58**(3), 343–353 (1994)
14. Dou, Y., Zhu, L., Wang, H.S.: Solving the fuzzy shortest path problem using multi-criteria decision method based on vague similarity measure. Appl. Soft Comput. **12**(6), 1621–1631 (2012)
15. Deng, Y., Chen, Y., Zhang, Y., Sankaran, M.: Fuzzy Dijkstra algorithm for shortest path problem under uncertain environment. Appl. Soft Comput. **12**(3), 1231–1237 (2012)
16. Jeyanthi, V., Radhika, V.S.: Applying Floyd's algorithm for solving neutrosophic shortest path problems. Int. J. Math. Trends Technol. **61**(1), 58–63 (2018)
17. Smarandache, F.: Neutrosophic set, a generalization of intuitionistic fuzzy sets. Int. J. Pure Appl. Math. **24**, 287–297 (2005)

Optimization of an Economic Production Quantity Model with Three Levels of Production and Demand as a Time Declining Market in Crisp and Fuzzy Environment

Renu Sharma, Ritu Arora, Anand Chauhan, and Anubhav Pratap Singh

Abstract A model of economic production quantity plays a crucial role in the field of manufacturing and service products. In the modern era, it is very important to study the management and safeguarding of deteriorating items in production inventory. In the proposed study, we introduce an EPQ model with three different production rates, and there is demand as a decreasing function of time. For the fuzzy model, the cost of production, the cost of ordering, the cost of carrying, and the cost of shortage are taken as in the form of a trapezoidal fuzzy number. Total cost is defuzzify by using signed distance technique. The paper's aim is to reduce the overall the overall cost to the manufacturer. A comparative study of the model in a crisp and fuzzy environment has done through a numerical example and sensitivity analysis demonstrates the effects of the model's parameters on the optimum result of the crisp and fuzzy model.

Keywords Economic production quantity · Trapezoidal fuzzy number · Signed distance method · Demand declining market · Three levels of production

---

R. Sharma (✉) · R. Arora
Department of Mathematics and Statistics, Gurukul Kangri (Deemed to be University), Haridwar, India
e-mail: km.renuvishwkarma@gmail.com

R. Arora
e-mail: ritu.arora29@gmail.com

A. Chauhan
Department of Mathematics, Graphic Era University, Dehradun, India
e-mail: dranandchauhan83@gmail.com

A. P. Singh
Department of Mathematics, S.G.R.R. (PG) College, Dehradun, India
e-mail: drapsingh78@gmail.com

M. Sahni et al. (eds.), *Mathematical Modeling, Computational Intelligence Techniques and Renewable Energy*, Advances in Intelligent Systems and Computing 1287,
https://doi.org/10.1007/978-981-15-9953-8_41

1 Introduction

Due to the advancement of the multinational markets of developing countries, a proper strategy for holding the inventory economically needs to be established. Different levels of production are one of the most realistic conditions in the inventory system. When the demand for the products increases, the manufacturer increases the rate of production, and when the demand for the products decreases, the manufacturer again decreases the rate of production. Thus, in this article, three levels of production are used so that the manufacturer gets consumer's satisfaction with minimizing his total cost and earns a potential profit.

In most of the EPQ models, there is a constant demand, but in actual situations, demand is changeable. To make the inventory models more realistic, several researchers have employed other demand types. Time-dependent demand is more appropriate in different types of inventory systems. Therefore, demand is taken as a decreasing function of time in this article. The parameters are definite in the crisp inventory models, such kinds of the model provide only general understating of the behavior of inventory system which is not accomplished the practical condition of the inventory system. Using the definite values of the parameters in the inventory system leads to incorrect decisions. Therefore, in the case of uncertainty, the flexibility in parameters can make the decision more practical. Many authors describe their inventory models using fuzzy approach. In recent year, Biswas and Islam [1] estimate the overall cost and level of inventory of production model for different environment crisp and fuzzy. They take as constant holding cost and shortages are not permitted. The cost parameters are assumed as triangular number of fuzzy and to defuzzify the total cost, and the method of signed distance is taken by them in their paper.

Islam and Biswas [2] gave a deterministic model of EOQ with deterioration as Weibull distribution, demand as exponential function, and holding cost is time dependent in different environments of crisp and fuzzy. Each cycle permitted shortages for partially backlogging. Different costs are taken as triangular fuzzy numbers. For defuzzifying the model's optimum valves, graded mean integration method is applied. This model helps in reducing the total cost of inventory by exploring the optimum time period in the environment of crisp and fuzzy. Karthikeyan and Viji [3] introduced an EPQ model with three different production rates which leads to balancing the stock of manufacturing items in order to reducing the holding cost. The difference in the rate of production gives an approach to increase the consumer satisfaction and gaining profit by minimizing the total cost. Three levels of production inventory model for deteriorative objects are assumed by Krishnamoorthin and Sivashankari [4]. Two levels of production are taken, when shortages are not allowed and next level of production is considered with shortages in their model.

In the study of Mahata and Goswmi [5], a realistic inventory model with vague parameters within economic manufacturing quantity under trade credit policy is described. In realistic condition, production rate never remains same throughout the whole production period. Further, the study on policy of trade credit is taken by Majumder et al. [6] in their research work. They employed demand as a fuzzy triangular number with considering upper α—*cut* and lower α—*cut* of the fuzzy number. Then, the total cost is split in two parts, i.e., α—*cut* and lower α—*cut*. To exchange the multiple objective function to single objective function they used weighted sum method. After that, Pal et al. [7] described an EPQ with ramp type demand rate. Production is for a single object with finite rate of production, which is directly proportional to the market pace two-parameter Weibull's distribution is assumed as the deterioration rate. Under finite time horizon, the inflation effect is taken when the model is without shortage. The optimum solution is formed in both the environment of crisp and fuzzy. Rajput et al. [8] investigated the model in three different cases demand about pharmaceutical inventory for comparison by using fuzzy and crisp parameters. Moreover, in the research study of Rajput et al. [9], they compare the optimum defuzzified result of graded mean integration methodology and singed distance technique. The fuzzifying variables used in their study are the cost of ordering, cost of shortage, cost of decaying, and cost of holding.

Sahni et al. [10] evaluated the teacher's presentation using the concept of metric through a questionnaire of fifteen assigned a weight depending on the importance. Their study is helpful to the faculties to know the weakness and strength in each category so that it can be corrected. Sahni et al. [11] used the fuzzy number of generalized trapezoidal intuitionistic to find a solid disk's radial displacement. Different rates of production are used due to transform of the pattern of demand with respect to time and market fluctuations. Therefore, Sarkar and Chakrabarti [12] studied a single-item EPQ model with different rates of production in which the deterioration rate as Weibull distribution with two parameters. Demand is taken as a fuzzy triangle number for uncertain market demand. The global criteria methodology is used to solve multi-objective equation through the generalized reduced gradient approach.

In the research study of Sayal et al. [13], an EOQ system is developed using unpreserved items with demand rate of the ramp type. In this case, the deterioration of commodities also depends on time. They considered that the production rate is proportional to the demand rate in their study and the rate of deterioration is also proportional to time. In crisp and fuzzy inference, Sayal et al. [14] studied EOQ system without shortage to reduce the system's overall inventory cost. They used triangular fuzzy number in cost parameters. Shah et al. [15] provided a new fuzzified iterative method to solve the fuzzy nonlinear and transcendental equations by using triangular fuzzy number, which is the modification of Newton–Raphson method.

Sivasankari and Panayappan [16] introduced an inventory model with two different rates of productions. It is considered that in starting the production rate is low and after some time production rate will be increase in their EPQ model. Sujatha and Parvathi [17] built a fuzzy inventory model with demand levels as Weibull distribution of two parameters which depends on time. Each rotation has shortages with partially backlogged in current scenario. A time varying holding cost is used in their research work with trade credit policy. The shortage cost and opportunity cost are used and decaying cost is fuzzy trapezoidal number. The signed distance technique is used to defuzzify the framework. In EPQ model of Viji and Karthikeyan [18], three rates of production inventory are taken and the Weibull distribution of two-parameter deterioration rate is considered.

The manuscript structure is as follows: The associated definitions are mentioned in preliminaries Sect. 2. The notations and assumptions used in the article are set out in Sect. 3. In Sect. 4, the formulation of the mathematical model in the different environment has been derived. The computational algorithm is described in Sect. 5. Section 6 provides a numerical example to illustrate this model. At the end of the manuscript, sensitivity analysis is described in Sect. 7 and the conclusion in Sect. 8.

2 Preliminaries

Definition 2.1. Fuzzy Number [1] Consider the fuzzy set $\tilde{F}$ on R of real number. The membership grade of these functions has the form $\mu_{\tilde{F}} : R \to [0, 1]$. Then, $\tilde{F}$ acts as a fuzzy number if: $\tilde{F}$ is normal fuzzy set, i.e., $\mu_{\tilde{F}}(\lambda) = 1$ for at least one $\lambda \in \tilde{F}$ and the support of $\tilde{F}$ must be bounded, i.e., a fuzzy number must be 1. α is closed for every $\alpha \in [0, 1]$, i.e., $\alpha \neq 0$ with $\tilde{F}$ is convex fuzzy set.

Definition 2.2. Trapezoidal Fuzzy Number [9] Let $\tilde{F}(l, m, n, o), l < m < n < o$ be a fuzzy set on $R = (-\infty, \infty)$. It is a trapezoidal fuzzy number if the membership functions of $\tilde{F}$ is defined as;

$$\mu_{\tilde{F}}(\lambda) = \begin{cases} \frac{\lambda - l}{m - l}, & l \leq \lambda \leq m \\ 1, & m \leq \lambda \leq n \\ \frac{o - \lambda}{o - n}, & n \leq \lambda \leq o \\ 0, & \text{otherwise} \end{cases}$$

Definition 2.3. Signed Distance Defuzzification Method [9] For $\tilde{F} \in F$, the distance signed from $\tilde{F}$ to 0 is described as: $d(\tilde{F}, 0) = \frac{1}{2}\left[\int_0^1 F_L(\alpha) + F_R(\alpha)\right]d\alpha$, where $\alpha \in [0, 1]$; $F_L(\alpha)$ = left distance value from $\tilde{F}$ to 0; $F_R(\alpha)$ = right distance value from $\tilde{F}$ to 0. For trapezoidal fuzzy number, $F_L(\alpha) = l + \alpha(m - l)$; $F_R(\alpha) = o - \alpha(o - n)$

Then,

$$d(\tilde{F}, 0) = \frac{1}{4}(l + m + n + o).$$

Definition 2.4. Fuzzy Arithmetical Operations [9] Let $\tilde{\tau} = (\tau_1, \tau_2, \tau_3, \tau_4)$ and $\tilde{\omega} = (\omega_1, \omega_2, \omega_3, \omega_4)$ are two trapezoidal fuzzy numbers where $\tau_1, \tau_2, \tau_3, \tau_4, \omega_1, \omega_2, \omega_3, \omega_4$ are positive real numbers, then fuzzy operations of arithmetic are;

(i) $\tilde{\tau} \oplus \tilde{\omega} = (\tau_1 + \omega_1, \tau_2 + \omega_2, \tau_3 + \omega_3, \tau_4 + \omega_4)$

(ii) $\tilde{\tau} \Theta \tilde{\omega} = (\tau_1 - \omega_4, \tau_2 - \omega_3, \tau_3 - \omega_2, \tau_4 - \omega_1)$

(iii) $\tilde{\tau} \otimes \tilde{\omega} = (\tau_1\omega_1, \tau_2\omega_2, \tau_3\omega_3, \tau_4\omega_4)$

(iv) $\tilde{\tau} \oslash \tilde{\omega} = \left(\frac{\tau_1}{\omega_4}, \frac{\tau_2}{\omega_3}, \frac{\tau_3}{\omega_2}, \frac{\tau_4}{\omega_1}\right)$

(v) $\alpha\tilde{\tau} = \begin{cases} (\alpha\tau_1, \alpha\tau_2, \alpha\tau_3, \alpha\tau_4), \text{ for } \alpha \geq 0 \\ (\alpha\tau_4, \alpha\tau_3, \alpha\tau_2, \alpha\tau_1), \text{ for } \alpha < 0 \end{cases}$ where α is any real number.

3 Notation and Assumption

3.1 Assumption

1. Constant rate of production with three levels is considered. The production rate is bigger than the rate of demand rate.
2. The rate of demand is diminishing time function, i.e., $D = (a - b\,t)$ and the rate of deterioration is constant.
3. A finite time horizon of this inventory system of production is considered with shortage which is completely backlogged.

3.2 Notations

- $S(t)$: Inventory point at time T.
- Δ: Production rate.
- K: Maximum shortage.
- O: Optimum production quantity.
- ψ: Deterioration rate of the items.
- L, M, and N are maximum inventory height at time T_1, T_2, and T_3, respectively.
- α: Production cost factor in crisp and $\tilde{\alpha}$ is a production cost factor in fuzzy.
- β: Ordering cost factor in crisp and $\tilde{\beta}$ is ordering cost factor in fuzzy.
- γ: Carrying cost factor in crisp and $\tilde{\gamma}$ is holding cost factor in fuzzy.
- ϕ: Shortage cost factor and $\tilde{\phi}$ is shortage cost factor in fuzzy.
- TC: Total cost in crisp and $\widetilde{TC}$ is total cost in fuzzy.
- T: Cycle time and T_i denotes the unit time in period ($i = 1, 2, 3, \ldots$).

4 Model Formulation

Figure 1 depicts the manufacturing model with three production rates, allowable

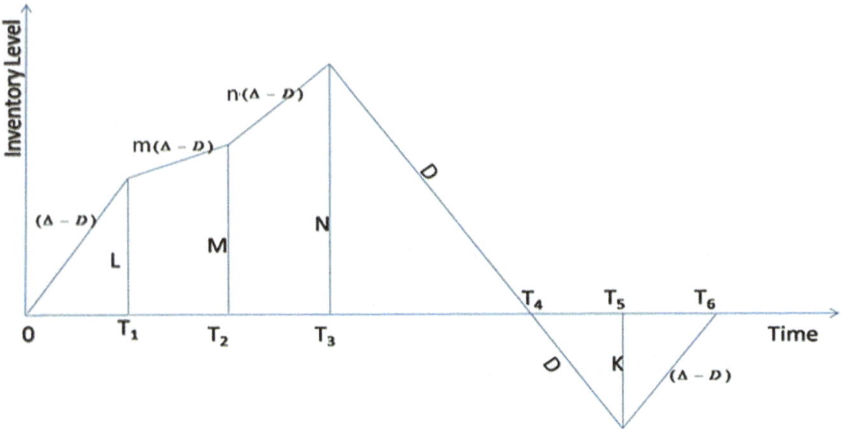

Fig. 1 Inventory level of defective items with respect to time

shortages. Firstly, find the inventory at any time 't' with the formulation of the inventory system. In $[0, T_1]$, the produced items increase the inventory with rate Δ and decrease due to customer's demand and deteriorative items. Hence, the inventory height in this period is

$$\frac{dS(t)}{dt} + \psi S(t) = \Delta - (a - bt); \quad 0 \le t \le T_1 \tag{1}$$

The next level of production increases the inventory with 'm' times of production rate Δ in $[T_2, T_3]$. Then, the reduction of the inventory is due to deterioration rate and 'm' times of demand rate. Thus, the inventory point in this period can be formulated by the subsequent differential Eq. (2),

$$\frac{dS(t)}{dt} + \psi S(t) = m\{\Delta - (a - bt)\}; \quad T_1 \le t \le T_2 \tag{2}$$

Further, in interval $[T_3, T_4]$, the next level of production increases the inventory level with 'n' times of production rate Δ and the reduction of the inventory is due to 'n' times of demand rate and decaying rate. Thus, in this period, the inventory can be governed by the Eq. (3),

$$\frac{dS(t)}{dt} + \psi S(t) = n\{\Delta - (a - bt)\}; \quad T_2 \le t \le T_3 \tag{3}$$

The inventory reduces in period $[T_3, T_4]$ because of demand and decaying rate. Therefore, the inventory is

$$\frac{dS(t)}{dt} + \psi S(t) = -(a - bt); \quad T_3 \le t \le T_4 \tag{4}$$

In period $[T_4, T_5]$, the producer would have no inventory on hand. Thus, the shortage occurs. The differential equation in this period is

$$\frac{dS(t)}{dt} = -(a - bt); \quad T_4 \le t \le T_5 \tag{5}$$

In $[T_5, T]$, the production starts with rate Δ and decrease due to customer's demand. Hence, in this period, the inventory level is denoted by Eq. (6),

$$\frac{dS(t)}{dt} = \Delta - (a - bt); \quad T_5 \le t \le T \tag{6}$$

The initial conditions are

$$S(0) = 0;\ S(T_1) = L;\ S(T_2) = M;\ S(T_3) = N;\ S(T_4) = 0;\ S(T_5) = K;\ S(T) = 0. \tag{7}$$

Solving Eq. (1) with initial condition $S(0) = 0$, we get

$$S(t) = -\frac{b}{\psi^2} + \frac{be^{-t\psi}}{\psi^2} - \frac{a}{\psi} + \frac{ae^{-t\psi}}{\psi} + \frac{bt}{\psi} + \frac{\Delta}{\psi} - \frac{e^{-t\psi}\Delta}{\psi};$$

By expansion of $e^{-\psi t}$ up to two terms for small value of ψ, we get

$$S(t) = t(\Delta - a); \tag{8}$$

similarly, by solving the Eqs. (2), (3) with initial condition $S(0) = 0$

$$S(t) = mt(\Delta - a); \tag{9}$$

$$S(t) = nt(\Delta - a); \tag{10}$$

By solving the Eq. (4) with $S(T_4) = 0$

$$S(t) = (a - bT_4)(T_4 - t); \tag{11}$$

By solving the Eq. (5) using $S(T_4) = 0$

$$S(t) = a(T_4 - t) - \frac{b}{2}(T_4^2 - t^2); \tag{12}$$

By solving the Eq. (6) using $S(T) = 0$

$$S(t) = -(\Delta - a)(T - t) - \frac{b}{2}(T^2 - t^2); \tag{13}$$

Using the boundary condition $S(T_1) = L$, the maximum inventory level

$$L = T_1(\Delta - a); \tag{14}$$

Using the boundary condition $S(T_2) = M$, the maximum inventory level

$$M = mT_2(\Delta - a); \tag{15}$$

Using the boundary condition $S(T_3) = N$, the maximum inventory level

$$N = nT_3(\Delta - a); \tag{16}$$

With the help of the above-mentioned function of the inventory level, the subsequent relations can be accomplished from Fig. 1.

(i) The producer's economic production quantity is

$$O = \Delta T \tag{17}$$

(ii) The inventory level in interval $[T_4, T_5]$ is equal to the inventory level in interval $[T_5, T]$ at T_5. Therefore, from Eq. (7), (12), and (13), we get the relationship between T_5, T_4, and T such as

$$S(T_5) = K = a(T_4 - T_5) - \frac{b}{2}(T_4^2 - T_5^2) \text{ and}$$

$$S(T_5) = K = -(\Delta - a)(T - T_5) - \frac{b}{2}(T^2 - T_5^2)$$

Therefore,

$$T_5 = T + \frac{(T_4 - T)(2a - bT - bT_4)}{2\Delta} \tag{18}$$

The producer's production model consists of the following different costs which are formulated below:

(1) Production cost/unit time $= \alpha[a - bT]$ (19)

(2) Ordering cost/unit time $= \frac{\beta}{T}$ (20)

(3) Carrying cost/unit time $= \frac{\gamma}{T}\left[\int_0^{T_1} S(t)\mathrm{d}t + \int_{T_1}^{T_2} S(t)\mathrm{d}t\right.$

$$+\int_{T_2}^{T_3} S(t)\mathrm{d}t + \int_{T_3}^{T_4} S(t)\mathrm{d}t\Bigg]$$

$$= \frac{\gamma}{T}\left[(\Delta - a)\left\{\frac{T_1^2}{2} + \frac{m}{2}(T_2^2 - T_1^2) + \frac{n}{2}(T_3^2 - T_2^2)\right\}\right.$$

$$\left. + (a - bT_4)\left(\frac{T_4^2}{2} + \frac{T_3^2}{2} - T_4T_3\right)\right]$$

(4) Deteriorating cost $= \dfrac{\alpha}{T}\left[\displaystyle\int_0^{T_1} \psi S(t)\mathrm{d}t + \int_{T_1}^{T_2} \psi S(t)\mathrm{d}t\right.$

$$\left. + \int_{T_2}^{T_3} \psi S(t)\mathrm{d}t + \int_{T_3}^{T_4} \psi S(t)\mathrm{d}t\right]$$

$$= \frac{\alpha\psi}{T}\left[\{(\Delta - a)\left\{\frac{T_1^2}{2} + \frac{m}{2}(T_2^2 - T_1^2)\right.\right.$$

$$\left.\left. + \frac{n}{2}(T_3^2 - T_2^2)\right\} + (a + bT_4)\left(\frac{T_4^2}{2} + \frac{T_3^2}{2} - T_4T_3\right)\right]$$

(5) Shortage cost $= \dfrac{\phi}{T}\left[\displaystyle\int_{T_4}^{T_5} S(t)\mathrm{d}t + \int_{T_5}^{T} S(t)\mathrm{d}t\right]$

$$= \frac{\phi}{T}\left[\frac{a(T_4 - T)(2T_5 - T_4 - T)}{2}\right.$$

$$- \frac{b(T_4 - T)T_5(T_4 + T)}{2}$$

$$+ \frac{b}{3}(T_4 - T)(T_4^2 + T^2 + T_4T)\}$$

$$\left. - \Delta\left(\frac{T_5^2}{2} + \frac{T^2}{2} - T_5T\right)\right]$$

Using Eq. (18)

$$= \frac{\phi(T_4 - T)}{T}\left[\frac{a}{2}\left\{T - T_4 + \frac{(T_4 - T)(2a - bT - bT_4)}{\Delta}\right\}\right.$$

$$- \frac{b((T_4 + T)}{2}\left\{T + \frac{(T_4 - T)(2a - bT - bT_4)}{2\Delta}\right\} + \frac{b}{3}(T_4^2 + T^2 + T_4T)$$

$$\left. + \frac{(T_4 - T)(2a - bT - bT_4)^2}{8\Delta}\right] \tag{23}$$

Since, total inventory cost is sum of the cost of holding, cost of shortage, cost of production, cost of ordering, and cost of deterioration. Let us consider $T_1 = xT_4$, $T_2 = yT_4$, $T_3 = zT_4$. Thus, total cost TC (T_4, T)

Therefore,

$$\begin{aligned}\text{Total cost TC}(T_4,\ T) = \alpha[a - bT] + \frac{\beta}{T} + \frac{\gamma}{T}\bigg[\frac{(\Delta - a)T_4^2}{2}\{x^2 + m(y^2 - x^2)\\ + n(z^2 - y^2)\} + \frac{(a - bT_4)T_4^2}{2}(1 + z^2 - z)\bigg]\\ + \frac{\alpha\psi}{T}\bigg[\frac{(\Delta - a)T_4^2}{2}\{x^2 + m(y^2 - x^2)\\ + n(z^2 - y^2)\} + (a - bT_4)\frac{T_4^2}{2}(1 + z^2 - z)\bigg]\\ + \frac{\phi(T_4 - T)}{T}\bigg[\frac{a}{2}\bigg\{T - T_4 + \frac{(T_4 - T)(2a - bT - bT_4)}{\Delta}\bigg\}\\ - \frac{b((T_4 + T)}{2}\bigg\{T + \frac{(T_4 - T)(2a - bT - bT_4)}{2\Delta}\bigg\}\\ + \frac{b}{3}(T_4^2 + T^2 + T_4T)\\ + \frac{(T_4 - T)(2a - bT - bT_4)^2}{8\Delta}\bigg]\end{aligned} \tag{24}$$

In fuzzy environment, the total costs are formulated as:

$$\begin{aligned}\widetilde{TC}(T_4, T) = \tilde{\alpha} \otimes [a - bT] \oplus \frac{\tilde{\beta}}{T} \oplus \frac{\tilde{\gamma}}{T} \otimes \bigg[\frac{(\Delta - a)T_4^2}{2}\{x^2 + m(y^2 - x^2)\\ + n(z^2 - y^2)\} + \frac{(a - bT_4)T_4^2}{2}(1 + z^2 - z)\bigg]\\ \oplus \frac{\tilde{\alpha} \otimes \psi}{T}\bigg[\frac{(\Delta - a)T_4^2}{2}\{x^2 + m(y^2 - x^2) + n(z^2 - y^2)\}\\ + (a - bT_4)\frac{T_4^2}{2}(1 + z^2 - z)\bigg] \oplus \frac{\tilde{\phi} \otimes (T_4 - T)}{T}\bigg[\frac{a}{2}\{T - T_4\\ + \frac{(T_4 - T)(2a - bT - bT_4)}{\Delta}\bigg\}\\ - \frac{b((T_4 + T)}{2}\bigg\{T + \frac{(T_4 - T)(2a - bT - bT_4)}{2\Delta}\bigg\}\\ + \frac{b}{3}(T_4^2 + T^2 + T_4T) + \frac{(T_4 - T)(2a - bT - bT_4)^2}{8\Delta}\bigg]\end{aligned} \tag{25}$$

Using signed distance method for trapezoidal fuzzy number in Eq. (29), we get

$$\widetilde{TC}(T_4, T) = \Bigg\{ \frac{(\alpha_1 + \alpha_2 + \alpha_3 + \alpha_4)(a - bT)}{4T} + \frac{(\beta_1 + \beta_2 + \beta_3 + \beta_4)}{4T}$$
$$+ \frac{(\gamma_1 + \gamma_2 + \gamma_3 + \gamma_4)}{4T}((\Delta - a)\frac{T_4^2}{2}(x^2 + m(y^2 - x^2) + n(z^2 - y^2))$$
$$+ (a - bT_4)\frac{T_4^2}{2}(1 - z)) + \frac{(\alpha_1 + \alpha_2 + \alpha_3 + \alpha_4)\psi}{4T}((\Delta - a)\frac{T_4^2}{2}(x^2$$
$$+ m(y^2 - x^2) + n(z^2 - y^2)) + (a - bT_4)\frac{T_4^2}{2}(1 - z))$$
$$+ \frac{(\phi_1 + \phi_2 + \phi_3 + \phi_4)(T_4 - T)}{4T}\left(\frac{a}{2}(T - T_4)\right.$$
$$\left.+ \frac{(T_4 - T)(2a - bT - bT_4)}{\Delta}\right) + \frac{b(T_4^2 + T^2 + T_4 T)}{3}$$
$$- \frac{b(T_4 + T)}{2}\left(T + \frac{(T_4 - T)(2a - bT - bT_4)}{2\Delta}\right)$$
$$+ \left(\frac{(T_4 - T)(2a - bT - bT_4)^2}{8\Delta}\right) \tag{26}$$

5 Computational Algorithm

We need to decide the optimum value of T_4 and T to maximize the overall cost, so that TC (T_4, T) is minimum. The computational algorithm is as follows to reduce the total cost of TC (T_4, T) in crisp and fuzzy model:

Step 1: Start with differentiating TC (T_4, T) with respect to T_4 then equate it to zero, i.e., $\frac{\partial TC(T_4,T)}{\partial T_4} = 0$.
Step 2: Differentiate TC (T_4, T) from T and compare it with zero, i.e., $\frac{\partial TC(T_4,T)}{\partial T} = 0$.
Step 3: Then, by solving these equations, we get the values of T_4 and T. These points are critical points.
Step 4: Take the second TC (T_4, T) with respect to T_4 and T, respectively, i.e., $\frac{\partial^2 TC(T_4,T)}{\partial T_4^2}$ and $\frac{\partial^2 TC(T_4,T)}{\partial T^2}$ and also find $\left(\frac{\partial^2 TC(T_4,T)}{\partial T_4 \partial T}\right)^2$.
Step 5: If condition $\left(\frac{\partial^2 TC(T_4,T)}{\partial T_4^2}\right)\left(\frac{\partial^2 TC(T_4,T)}{\partial T^2}\right) - \left(\frac{\partial^2 TC(T_4,T)}{\partial T_4 \partial T}\right)^2 > 0$ is satisfied at critical points T_4 and T, then total cost is minimum.
Step 6: By putting the value of T_4 and T in Eqs. (24) and (26), we obtain the minimum total cost for crisp and fuzzy model.

6 Numerical Example

To exemplify the result of the planned model, the following data in crisp environment

$$\Delta = 1000\,\text{units/year}, a = 900\,\text{units/year}, b = 0.1,$$
$$\alpha = \text{Rs}\,50/\text{unit}, \beta = \text{Rs}\,100/\text{order}, \gamma = \text{Rs}\,0.6/\text{unit},$$
$$\phi = \text{Rs}\,10/\text{unit}, \psi = 0.05, m = 2, n = 1, x = 0.2, y = 0.4, z = 0.6.$$

By using these values in Eq. (24) and apply the computational algorithm step by step, we will get the optimum T_4 value and T as 0.38381 and 0.41556, respectively, and optimum total cost is 45481.2848 (Table 1).

In fuzzy environment, the input data is as follows:

$$\Delta = 1000\,\text{units/year}, a = 900\,\text{units/year}, b = 0.1, \alpha_1 = 44, \alpha_2 = 45,$$
$$\alpha_3 = 55, \alpha_4 = 56, \beta_1 = 70, \beta_2 = 80, \beta_3 = 120, \beta_4 = 130, \psi = 0.05,$$
$$\gamma_1 = 0.3, \gamma_2 = 0.4, \gamma_3 = 0.8, \gamma_4 = 0.9, \phi_1 = 7, \phi_2 = 8, \phi_3 = 12,$$
$$\phi_4 = 13, x = 0.2, y = 0.4, z = 0.6, m = 2, n = 1.$$

Using these values in Eq. (26) and apply the computational algorithm in similar manner, then the optimum value of T_4 and T are 8.12280 and 8.79509, respectively and optimum total cost is 10,255.4539 (Tables 2, 3 and Fig. 2).

Table 1 Optimum solution in crisp environment

T	O	T_1	T_2	T_3	T_4	T_5	L	M	N	TC
0.415	373.989	0.076	0.153	0.230	0.383	0.386	7.676	30.704	69.085	45481.3

Table 2 Optimum solution in fuzzy environment

T	O	T_1	T_2	T_3	T_4	T_5	L	M	N	TC
8.795	7915.216	1.625	3.249	4.873	8.123	8.190	879.509	1759.018	879.509	10255.4

Table 3 Comparison of optimal solutions in crisp and fuzzy environment

	T_4	T	$\text{TC}(T_4, T)$	O
Crisp	0.38381	0.41556	45481.3	373.989
Fuzzy	8.12280	8.79509	10255.4	7915.216

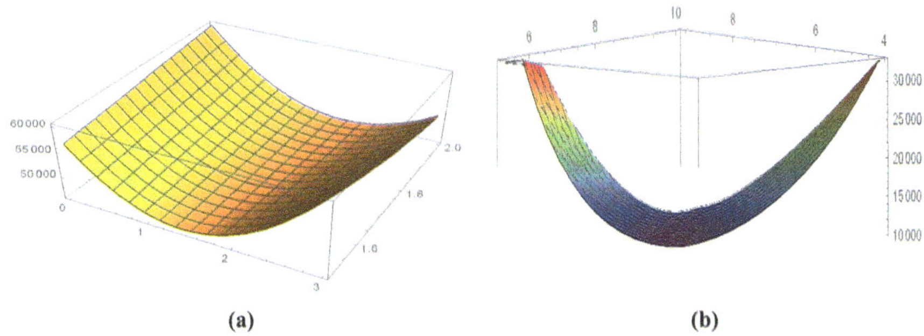

Fig. 2 Convex behavior of the total cost with respect to T_4 and T for fuzzy and crisp

7 Sensitivity Analysis

7.1 The effects of making changes in crisp model's parameters are shown in Table 4 and its description is as follows;

- As we increase the production cost factor (α), holding cost factor (γ), production rate (Δ), and deterioration rate (Ψ), then we observe that the cycle time (T), time of production (T_1, T_2, T_3), maximum inventories (L, M, N), and optimum quantity (O) are decreased, but the total cost TC (T_4, T) is increased.
- With the increase in the ordering cost factor (β), we observe that the cycle time (T), time of production (T_1, T_2, T_3), maximum inventories (L, M, N), optimum quantity (O), and total cost TC (T_4, T) are also increases.
- As we increase the demand's fixed factor (a) and the shortage cost factor ($\emptyset$), then we notice that production time, maximum inventories (L, M, N), and total cost TC (T_4, T) are increase, but optimum quantity (O) and cycle time (T) are decreased.
- As we increase in the demand's time-dependent factor (b), we find that cycle time (T), production time (T_1, T_2, T_3), maximum inventories (L, M, N), and optimum quantity (O) are also increased, but total cost TC (T_4, T) is decreased.

7.2 The effects of making changes in the fuzzy model's parameters are shown in Table 5 and its description is as follows:

- As we increase the demand's fixed factor (a), then cycle time (T), time of production (T_1, T_2, T_3), maximum inventories (L, M, N), optimum quantity (O), and overall the total cost TC (T_4, T) is increased.
- As increasing in the demand's time-dependent factor (b), then time of production (T_1, T_2, T_3), cycle time (T), optimum inventories (L, M, N), and optimum quantity (O) are increased, but the total cost TC (T_4, T) is decreased.
- With an increase in the production rate (Δ) and the deterioration rate (Ψ), we observed that the time of production (T_1, T_2, T_3), time of cycle (T), optimum

Table 4 Effect of different parameters on the optimum value in crisp environment

Parameter	% change in parameter	Optimum Values										
		T	O	T_1	T_2	T_3	T_4	T_5	L	M	N	TC
α	10%	0.4010	360.9076	0.0736	0.1472	0.2208	0.3681	0.3714	7.3627	29.4510	66.2648	49998.7
	5%	0.4080	367.2541	0.0751	0.1503	0.2254	0.3757	0.3789	7.5152	30.0608	67.6369	47740.1
	−5%	0.4235	381.1566	0.0784	0.1569	0.2354	0.3924	0.3955	7.8479	31.3916	70.6311	43222.2
	−10%	0.4320	388.8009	0.0803	0.1606	0.2409	0.4015	0.4045	8.0300	32.1203	72.2707	40963.0
β	10%	0.4358	392.2450	0.0805	0.1610	0.2415	0.4025	0.4059	8.0513	32.2053	72.4620	45504.8
	5%	0.4258	383.2256	0.0786	0.1573	0.2359	0.3933	0.3965	7.8662	31.4648	70.7959	45493.2
	−5%	0.4050	364.5200	0.0748	0.1496	0.2244	0.3741	0.3772	7.4822	29.9289	67.3401	45469.1
	−10%	0.3942	354.7969	0.0728	0.1456	0.2184	0.3641	0.3671	7.2826	29.1307	65.5441	45456.6
γ	10%	0.4118	370.6676	0.0759	0.1519	0.2279	0.3798	0.3830	7.5972	30.3888	68.3749	45485.6
	5%	0.4137	372.3163	0.0763	0.1527	0.2291	0.3818	0.3850	7.6366	30.5465	68.7297	45483.4
	−5%	0.4174	375.6885	0.0771	0.1543	0.2315	0.3858	0.3890	7.7172	30.8688	69.4548	45479.1
	−10%	0.4193	377.4128	0.0775	0.1551	0.2327	0.3879	0.3910	7.7583	31.0334	69.8252	45476.9
ϕ	10%	0.4141	372.6781	0.0770	0.1540	0.2311	0.3851	0.3880	7.7031	30.8127	69.3286	45483.0
	5%	0.4148	373.3036	0.0769	0.1538	0.2307	0.3845	0.3875	7.6905	30.7620	69.2145	45482.2
	−5%	0.4164	374.7471	0.0766	0.1532	0.2298	0.3830	0.3864	7.6613	30.6455	68.9524	45480.3
	−10%	0.4173	375.5868	0.0764	0.1528	0.2293	0.3822	0.3857	7.6445	30.5782	68.8010	45479.2
a	10%	0.4155	374.0155	0.0781	0.1562	0.2344	0.3907	0.3931	7.8140	31.2560	70.3260	49981.3
	5%	0.4154	373.8454	0.0774	0.1549	0.2324	0.3874	0.3902	7.7479	30.9916	69.7311	47731.5
	−5%	0.4161	374.5320	0.0759	0.1519	0.2279	0.3799	0.3835	7.5986	30.3947	68.3881	43230.6
	−10%	0.4173	375.6183	0.0752	0.1504	0.2256	0.3760	0.3801	7.5199	30.0797	67.6794	40929.2

(continued)

Table 4 (continued)

Parameter	% change in parameter	Optimum Values										
		T	O	T_1	T_2	T_3	T_4	T_5	L	M	N	TC
b	10%	0.4155	373.9894	0.0767	0.1535	0.2303	0.3838	0.3870	7.6766	30.7064	69.0895	45481.3
	5%	0.4156	374.0713	0.0767	0.1535	0.2303	0.3839	0.3870	7.6783	30.7132	69.1047	45481.2
	−5%	0.4154	373.9084	0.0767	0.1535	0.2302	0.3837	0.3869	7.6749	30.6998	69.0746	45481.4
	−10%	0.4153	373.8265	0.0767	0.1534	0.2302	0.3836	0.3868	7.6732	30.6931	69.0595	45481.5
Δ	10%	0.3975	357.7397	0.0717	0.1435	0.2152	0.3587	0.3626	7.1752	28.7011	64.5775	45503.1
	5%	0.4067	366.0454	0.0743	0.1487	0.2231	0.3718	0.3753	7.4378	29.7514	66.9407	45491.7
	−5%	0.4263	383.6477	0.0795	0.1590	0.2386	0.3977	0.4005	7.9547	31.8188	71.5923	45469.1
	−10%	0.4383	394.5174	0.0825	0.1651	0.2476	0.4127	0.4153	8.2555	33.0223	74.3002	45456.2
Ψ	10%	0.4008	360.7627	0.0736	0.1472	0.2207	0.3679	0.3712	7.3598	29.4392	66.2382	45498.9
	5%	0.4079	367.1776	0.0751	0.1502	0.2254	0.3756	0.3789	7.5136	30.0546	67.6229	45490.2
	−5%	0.4236	381.2421	0.0785	0.1569	0.2354	0.3924	0.3956	7.8496	31.3986	70.6469	45472.1
	−10%	0.4322	388.9818	0.0803	0.1606	0.2410	0.4016	0.4047	8.0338	32.1353	72.3045	45462.7

Table 5 Effect of different parameters on optimum values of the inventory system in fuzzy environment

Parameter	% change in parameter	Optimum values										
		T	O	T_1	T_2	T_3	T_4	T_5	L	M	N	TC
a	10%	9.2243	8301.5137	1.7343	3.4686	5.2028	8.6714	8.7272	922.433	1844.866	922.433	10754.3
	5%	9.0085	8107.2576	1.6801	3.3602	5.0403	8.4005	8.4618	900.848	1801.696	900.848	10512.2
	−5%	8.5852	7726.2872	1.5674	3.1347	4.7021	7.8369	7.9123	858.516	1717.032	858.516	9981.9
	−10%	8.3801	7541.7418	1.5082	3.0164	4.5245	7.5409	7.6255	838.010	1676.020	838.010	9689.1
b	10%	8.7959	7915.9535	1.6247	3.2494	4.8741	8.1235	8.1913	879.591	1759.182	879.591	10254.5
	5%	8.7955	7915.5845	1.6246	3.2493	4.8739	8.1231	8.1909	879.550	1759.100	879.550	10255.0
	−5%	8.7947	7914.8555	1.6245	3.2490	4.8735	8.1225	8.1903	879.469	1758.938	879.469	10255.9
	−10%	8.7943	7914.4865	1.6244	3.2489	4.8733	8.1221	8.1899	879.428	1758.856	879.428	10256.4
Δ	10%	8.4153	7573.4203	1.5188	3.0377	4.5565	7.5942	7.6769	841.530	1683.060	841.530	10718.1
	5%	8.5942	7734.4319	1.5700	3.1400	4.7100	7.8499	7.9250	859.421	1718.842	859.421	10495.0
	−5%	9.0207	8118.2821	1.6832	3.3663	5.0495	8.4158	8.4768	902.073	1804.146	902.073	9999.1
	−10%	9.2746	8346.7366	1.7465	3.4930	5.2396	8.7326	8.7873	927.458	1854.916	927.458	9725.7
Ψ	10%	8.4863	7637.3353	1.5579	3.1159	4.6738	7.7896	7.8599	848.632	1697.264	848.632	10628.6
	5%	8.6361	7772.1311	1.5903	3.1806	4.7709	7.9515	8.0205	863.610	1727.220	863.610	10444.3
	−5%	8.9643	8067.4705	1.6609	3.3219	4.9828	8.3046	8.3712	896.427	1792.854	896.427	10061.9
	−10%	8.1448	7329.9365	1.6996	3.3992	5.0988	8.4981	8.4625	814.475	1628.950	814.475	9863.3

inventory (L, M, N), and optimum quantity (O) decreases, but the total cost TC (T_4, T) increases.

8 Conclusion

This article is about a manufacturing model of imperfect products with three production stages. Demand rate decreases time function and the constant rate of deterioration is assumed. Demand rate is decreasing function of time and the constant deterioration rate is taken. There is a desirable situation that production starts at a low rate, which leads to a reduction in holding costs. When the demand for the products increases, the manufacturer increases the rate of production, and when the demand of the products decreases, the manufacturer again decreases the rate of production. Thus, the manufacturer gets consumer satisfaction by minimizing his total cost and earns a potential profit. In this proposed article, we established a mathematical model for crisp and fuzzy inference. To validate this production model, a numerical example has given and analyzed the optimum solution in both environments. After that, a comparative study between the optimum solutions of the crisp model and the fuzzy model has done. This comparison shows that the minimum cost has obtained in a fuzzy environment. To see the effect of different parameters on total cost, the sensitivity analysis is represented through Tables 4 and 5 for the crisp and fuzzy model, respectively. This model can be extended in many ways for further analysis by changing the demand factor (linear, quadratic, cubic, stock dependent, etc.), by taking three parameters of Weibull deterioration rate, rework of defective products, etc.

References

1. Biswas, A.K., Islam, S.: A fuzzy EPQ model for non-instantaneous deteriorating items where production depends on demand which is proportional to population, selling price as well as advertisement. Independent J. Manag. Prod. **10**(5), 1679–1703 (2019). https://doi.org/10.14807/ijmp.v10i5.897
2. Islam, S., Biswas, A.K.: Fuzzy inventory model having exponential demand with Weibull distribution for non-Instantaneous deterioration, shortages under partially backlogging and time dependent holding cost. Int. J. Adv. Res. Comput. Sci. Softw. Eng. **7**(6), 434–443 (2017). https://doi.org/10.23956/ijarcsse/V7I6/0184
3. Karthikeyan, K., Viji, G.: Economic production quantity inventory model for three levels of production with deteriorative item. Int. J. Appl. Eng. Res. **10**(55), 3717–3722 (2015)
4. Krishnamoorthi, C.C., Sivashankari, C.K.: Production inventory models for deteriorative items with three levels of production and shortages. Yugoslav J. Oper. Res. **27**(4), 499–519 (2017). https://doi.org/10.2298/YJOR150630014K
5. Mahata, G.C., Goswami, A.: The optimal cycle for EPQ inventory model of deteriorating items trade credit financing in fuzzy sense. Int. J. Oper. Res. **7**(1), 26–40 (2010)
6. Majumder, P., Bera, U.K., Maiti, M.: An EPQ model of deteriorating items under partial trade credit financing and demand declining market in crisp and fuzzy environment. In: International Conference on Advanced Computing Technologies and Applications, vol. 45, pp. 780–789 (2015). https://doi.org/10.1016/j.procs.2015.03.154

7. Pal, S., Mahapatra, G.S., Samanta, G.P.: An EPQ model of ramp type demand with Weibull deterioration under inflation and finite horizon in crisp and fuzzy environment. Int. J. Prod. Econ. **156**, 159–16 (2014). https://doi.org/10.1016/j.ijpe.2014.05.007
8. Rajput, N., Pandey, R.K., Singh, A.P., Chauhan, A.: An Optimization of fuzzy EOQ model in healthcare industries with three different demand pattern using signed distance technique. MESA **10**(2), 205–218 (2019)
9. Rajput, N., Singh, A.P., Pandey, R.K.: Optimize the cost of a fuzzy inventory model with shortage using signed distance method. Int. J. Res. Advent Technol. **7**(5), 198–202 (2019)
10. Sahni, M., Mandaliya, A., Sahni, R.: Evaluation of teachers' performance based on students' feedback using aggregator operator. WSEAS Trans. Math. **18**, 85–90 (2019)
11. Sahni, M., Sahni, R., Verma, R., Mandaliya, A., Shah, D.: Second order cauchy euler equation and its application for finding radial displacement of a solid disk using generalized trapezoidal intuitionistic fuzzy number. WSEAS Trans. Math. **18**, 37–45 (2019)
12. Sarkar, S., Chakrabarti, T.: An EPQ model with two-component demand under fuzzy environment and Weibull distribution deterioration with shortages. Adv. Oper. Res. pp. 1–22 (2012)
13. Sayal, A., Singh, A.P., Aggarwal, D.: Crisp and fuzzy EOQ model for perishable items with ramp type demand under shortages. Int. J. Agric. Stat. Sci. **14**(1), 441–452 (2018)
14. Sayal, A., Singh, A.P., Aggarwal, D.: Inventory model in fuzzy environment without shortage using triangular fuzzy number with sensitivity analysis. Int. J. Agric. Stat. Sci. **14**(1), 391–396 (2018)
15. Shah, D., Sahni, M., Sahni, R.: Solution of algebraic and transcendental equations using fuzzified he's iteration formula in terms of triangular fuzzy numbers. WSEAS Trans. Math. **18**, 91–96 (2019)
16. Sivasankari, C.K., Panayappan, S.: Production inventory model for two level productions with deteriorative items and shortage. Int. J. Adv. Manuf. Technol. **76**, 2003–2014 (2015). https://doi.org/10.1007/s00170-014-6259-8
17. Sujatha, J., Parvathi, P.: Fuzzy inventory model for deteriorating items with Weibull demand and time varying holding cost under trade credit. Int. J. Innov. Res. Comput. Commun. **3**(11), 11110–11123 (2015)
18. Viji, G., Karthikeyan, K.: An economic production quantity model for three levels of production with Weibull distribution deterioration and shortage. Ain Shams Eng. J. **9**, 1481–1487 (2018). https://doi.org/10.1016/j.asej.2016.10.006

Population Dynamic Model of Two Species Solved by Fuzzy Adomian Decomposition Method

Purnima Pandit, Prani Mistry, and Payal Singh

Abstract Adomian decomposition method (ADM) is powerful method to solve nonlinear functional equations. This method also solves nonlinear differential equations and provides the solution in series form, which effectively and accurately converges very fast to the exact solution, if it exists. In this work, we propose Fuzzy Adomian Decomposition Method (FADM) in parametric form to compute the solution for nonlinear dynamical system. We propose theorem for existence of FADM in parametric form for such system. Population dynamics model of two species; i.e., prey-predator model, involving fuzzy parameters and fuzzy initial condition is solved using proposed method, and results are compared at core.

Keywords Fuzzy Set · Fuzzy Adomian Decomposition Method (FADM) · Non-linear differential equation · Population Dynamics · Semi-Analytic solution

1 Introduction

Prey-predator dynamics is based on Lotka-Volterra model. This model is of lot practical importance and it fits into in many areas like biological model, financial problem, environmental problem, etc. We propose to consider Prey-Predator dynamics in fuzzy setup because there may be manual or machine error in estimating parameters or/and initial condition. So, modelling in fuzzy setup gives more realistic results.

P. Pandit · P. Mistry (✉)
Department of Applied Mathematics, The Maharaja Sayajirao University of Baroda, Vadodara, Gujarat, India
e-mail: praniph15@gmail.com

P. Pandit
e-mail: pkpandit@yahoo.com

P. Singh
Department of Applied Science and Humanities, Parul Institute of Engineering and Technology, Parul University, Vadodara, Gujarat, India
e-mail: singhpayalmath@gmail.com

M. Sahni et al. (eds.), *Mathematical Modeling, Computational Intelligence Techniques and Renewable Energy*, Advances in Intelligent Systems and Computing 1287,
https://doi.org/10.1007/978-981-15-9953-8_42

Many authors have done work in direction of fuzzy Prey-Predator model. Pandit and Singh [1] have solved prey-predator model involving fuzzy initial condition with the help of eigen values and eigen vectors. Da Silva Peixoto et al. [2] used fuzzy rule-based system to demonstrate a predator–prey type of model to study the interaction between aphids-preys and ladybugs-predators in citriculture. Hussin et al. [3] give numerical simulation to solve this model and they have taken initial condition as fuzzy triangular number. M.Z. Ahmad and De Baets [4] have solved the model with initial condition fuzzy by Runge–Kutta 4th order method. Purnima and Payal [5] have presented semi-analytical technique to solve fully fuzzy prey-predator model taking initial condition as well as parameters fuzzy, under new fuzzy derivative named Modified Hukuhara derivative.

In this article, we consider fully fuzzy semi-linear dynamical system,

$$\begin{aligned}\dot{\tilde{X}}(t) &= \tilde{A} \otimes \tilde{X} \oplus \tilde{f}\left(t, \tilde{X}\right)\\ \tilde{X}(0) &= \tilde{X}_0\end{aligned} \tag{1}$$

Prey-predator model is of the form given above, as it is represented as

$$\left.\begin{aligned}\dot{\tilde{x}} &= (\tilde{a} \otimes \tilde{x}) \ominus \left(\tilde{b} \otimes \tilde{x} \otimes \tilde{y}\right)\\ \dot{\tilde{y}} &= \left(\tilde{d} \otimes \tilde{x} \otimes \tilde{y}\right) \ominus (\tilde{c} \otimes \tilde{y})\end{aligned}\right\} \tag{2}$$

involving fuzzy parameters and with fuzzy initial condition $\tilde{x}_0$ and $\tilde{y}_0$.

In system (1), variables $\tilde{x}$ and $\tilde{y}$ represents number of prey and predator populations and their derivatives show change in populations with respect to time. The fuzzy parameters $\tilde{a}, \tilde{b}, \tilde{c}, \tilde{d}$ represent as the following,

$\tilde{a}$ = Possible range of growth rate of prey with enough food, in absence of predator.
$\tilde{b}$ = Possible range of loss rate of prey when they interact with predator.
$\tilde{c}$ = Possible range of loss rate of predator in absence of prey.
$\tilde{d}$ = Possible range of growth rate of predator when they interact with prey.

These parameters basically describe death and birth rate and which may be affected by many reasons like natural calamities, e.g., heavy rain, draught, earthquake, etc. or prey population departed, etc. Simple crisp form may not explain precise changes in death and birth number as time passes. So, considering them fuzzy is more practical.

We propose here to solve system (1) by using Adomian Decomposition Method (ADM), [6] in fuzzy parametric form. This method is semi-analytical technique to solve nonlinear dynamical system. In this technique, we decompose nonlinear term into the fuzzy Adomian polynomials and give solution in series form. The results required, to obtain the solution of system (1) by the proposed method, are also stated and proved in the paper. The following section contains required preliminaries and basic concepts, Sect. 3 contains supportive results in form of lemma and theorem, Sect. 4 contains illustrative application followed by conclusion.

2 Preliminaries

Let $E^n = \{\tilde{u}: R^n \rightarrow [0, 1]$ such that $\tilde{u}$ satisfies following properties$\}$.

- $\tilde{u}$ is normal.
- $\tilde{u}$ is a fuzzy convex.
- Support is bounded.

So, E^n is the collection of fuzzy number.

2.1 *Hausdorff Distance*

The Hausdorff distance defined between fuzzy numbers is given by,

$$d' : E^n \times E^n \rightarrow R_+ \cup \{0\}$$
$$d'(\tilde{u}, \tilde{v}) = \sup_{\alpha \in [0,1]} \max\{|\underline{u} - \underline{v}, \overline{u} - \overline{v}|\}$$

where, $\tilde{u} = (\underline{u}, \overline{u})$, $\tilde{v} = (\underline{v}, \overline{v})$ and d is metric on E^n.

2.2 *Fuzzy Number in Parametric Form*

Parametric form of fuzzy number is an ordered pair of the form ${}^{\alpha}u = (\underline{u}, \overline{u}), 0 \leq \alpha \leq 1$, satisfies the following conditions:

- $\underline{u}$ is a bounded left continuous increasing function in the interval [0, 1].
- $\overline{u}$ is a bounded right continuous decreasing function in the interval [0, 1].
- $\underline{u} \leq \overline{u}$.

For each α, $\underline{u} = \overline{u}$ then,u is crisp number.

For triangular fuzzy number, u can be represented as (l, c, r) with $l \leq c \leq r$ and each $l, c, r \in R$. Then, the parametric form of triangular fuzzy number u becomes,
$\underline{u} = (l + (c - l)\alpha)$ and $\overline{u} = (r - (r - c)\alpha)$.

2.3 *Fuzzy Multiplication, Addition and Difference*

For $\tilde{u}, \tilde{v} \in E^n$ and $k \in R$, the sum $\tilde{u} \oplus \tilde{v}$, the scalar product $k \otimes \tilde{u}$ and the product $\tilde{u} \otimes \tilde{v}$ are defined as

$$\tilde{u} \oplus \tilde{v} = [\underline{u}, \overline{u}] \oplus [\underline{v}, \overline{v}] = [\underline{u} + \underline{v}, \overline{u} + \overline{v}]$$

$$k \otimes \tilde{u} = k \cdot [\underline{u}, \overline{u}] = [k \cdot \overline{u}, k \cdot \underline{u}] \quad \text{for all } \alpha \in [0, 1]$$
$$\tilde{u} \otimes \tilde{v} = [\underline{u}, \overline{u}] \cdot [\underline{v}, \overline{v}]$$
$$= [\min(\underline{u}\overline{v}, \underline{u}\underline{v}, \overline{u}\underline{v}, \overline{u}\overline{v}), \max(\underline{u}\overline{v}, \underline{u}\underline{v}, \overline{u}\underline{v}, \overline{u}\overline{v})]$$

Difference of fuzzy number can be also given by,

$$\tilde{u} \ominus \tilde{v} = [\underline{u}, \overline{u}] \ominus [\underline{v}, \overline{v}] = [\underline{u}, \overline{u}] \oplus [-\overline{v}, -\underline{v}] = [\underline{u} - \overline{v}, \overline{u} - \underline{v}]$$

2.4 Fuzzy Continuity

We extend the definition as in [7], for system of equations as if $\tilde{f} : (0, t) \times E^n \to E^n$ then $\tilde{f}$ is fuzzy continuous at point $(t_0, \tilde{z}_0)$ for any fixed number $\alpha \in [0, 1]$ and any $\varepsilon > 0, \exists\, \delta(\varepsilon, \alpha)$ such that $d'\left(\tilde{f}(t, \tilde{z}), \tilde{f}(t_0, \tilde{z}_0)\right) < \varepsilon$ whenever $|t - t_0| < \delta(\varepsilon, \alpha)$ and $d'([\tilde{z}], [\tilde{z}_0]) < \delta(\varepsilon, \alpha)\, \forall\, t \in (0, t)$ and $\tilde{z} \in E^n$.

2.5 Equicontinuity

As given in [8], equi-continuity of fuzzy function $\tilde{f}$ is given by $\left|\tilde{f}_{\pm}(t, x, y) - \tilde{f}_{\pm}(t, x_1, y_1)\right| < \varepsilon, \forall\, \alpha \in [0, 1]$ whenever $\|(t, x, y) - (t_1, x_1, y_1)\| < \delta$ and $\tilde{f}$ is uniformly bounded on any bounded set.

2.6 Hukuhara Derivative

As given in [9], a fuzzy mapping $\tilde{f} : (0, t) \to E^n$ is said to be Hukuhara differentiable at $t_0 \in (0, t) \exists$ an element $\dot{\tilde{f}}(t_0) \in E^n$ such that for all $h > 0 \,\exists\, \tilde{f}(t_0 + h) \ominus \tilde{f}(t_0), \tilde{f}(t_0) \ominus \tilde{f}(t_0 - h)$ exist and the limits,

$$\lim_{h \to 0+} \frac{\tilde{f}(t_0 + h) \ominus \tilde{f}(t_0)}{h} = \lim_{h \to 0-} \frac{\tilde{f}(t_0) \ominus \tilde{f}(t_0 - h)}{h} = \dot{\tilde{f}}(t_0)$$

Let $\tilde{x}, \tilde{y} \in E^n$. If there exists $\tilde{z} \in E^n$ such that $\tilde{x} = \tilde{y} \oplus \tilde{z}$, then $\tilde{z}$ is called the H-difference of $\tilde{x}$ and $\tilde{y}$ and is denoted by $\tilde{x} \ominus \tilde{y}$ such that $\tilde{x} \ominus \tilde{y} \neq \tilde{x} \oplus (-1)\tilde{y}$. In this paper, "$\ominus$" stands for always H-difference.

2.7 Modified Hukuhara Derivative

To give unique and bounded solution, in [5], new derivative is defined as below,

A function $\tilde{f}\colon (0, t) \rightarrow E^n$ is said to be modified Hukuhara differentiable at $t_0 \in (0, t)$, there exist an element $\dot{\tilde{f}}(t_0) \in E^n$ such that for all $h > 0$ sufficiently small $\exists\ \tilde{f}(t_0 + h) \ominus \tilde{f}(t_0)$, $\tilde{f}(t_0) \ominus \tilde{f}(t_0 - h)$ should exists and the limits,

$$\lim_{h \to 0+} \frac{\tilde{f}(t_0 + h) \ominus \tilde{f}(t_0)}{h} = \lim_{h \to 0-} \frac{\tilde{f}(t_0) \ominus \tilde{f}(t_0 - h)}{h} = \dot{\tilde{f}}(t_0)$$

where parametric representation of

$$\lim_{h \to 0+} \frac{\tilde{f}(t_0+h) \ominus \tilde{f}(t_0)}{h} =$$

$$\left[\min\left\{ \lim_{h \to 0} \frac{(\bar{f}(t_0+h) - \bar{f}(t_0))}{h}, \lim_{h \to 0} \frac{\left(\underline{f}(t_0+h) - \bar{f}(t_0)\right)}{h}, \lim_{h \to 0} \frac{(\bar{f}(t_0+h) - \bar{f}(t_0))}{h}, \lim_{h \to 0} \frac{\left(\bar{f}(t_0+h) - \underline{f}(t_0)\right)}{h} \right\}, \right.$$
$$\left. \max\left\{ \lim_{h \to 0} \frac{\left(\underline{f}(t_0+h) - \underline{f}(t_0)\right)}{h}, \lim_{h \to 0} \frac{\left(\underline{f}(t_0+h) - \bar{f}(t_0)\right)}{h}, \lim_{h \to 0} \frac{(\bar{f}(t_0+h) - \bar{f}(t_0))}{h}, \lim_{h \to 0} \frac{\left(\bar{f}(t_0+h) - \underline{f}(t_0)\right)}{h} \right\} \right]$$

$$\lim_{h \to 0-} \frac{\tilde{f}(t_0) \ominus \tilde{f}(t_0-h)}{h}$$

$$\left[\min\left\{ \lim_{h \to 0} \frac{\left(\underline{f}(t_0) - \underline{f}(t_0-h)\right)}{h}, \lim_{h \to 0} \frac{\left(\underline{f}(t_0) - \bar{f}(t_0-h)\right)}{h}, \lim_{h \to 0} \frac{(\bar{f}(t_0) - \bar{f}(t_0-h))}{h}, \lim_{h \to 0} \frac{\left(\bar{f}(t_0) - \underline{f}(t_0-h)\right)}{h} \right\}, \right.$$
$$\left. \max\left\{ \lim_{h \to 0} \frac{\left(\underline{f}(t_0) - \underline{f}(t_0-h)\right)}{h}, \lim_{h \to 0} \frac{\left(\underline{f}(t_0) - \bar{f}(t_0-h)\right)}{h}, \lim_{h \to 0} \frac{(\bar{f}(t_0) - \bar{f}(t_0-h))}{h}, \lim_{h \to 0} \frac{\left(\bar{f}(t_0) - \underline{f}(t_0-h)\right)}{h} \right\} \right]$$

3 Adomian Decomposition Method for Fuzzy Function in Parametric Form

Consider nonlinear fuzzy differential equation given as,

$$L\tilde{u} \oplus N\tilde{u} \oplus R\tilde{u} = \tilde{g}$$

where L is linear (modified) Hukuhara differentiable operator, N is nonlinear operator, R is the operator of less order than that of L, and $\tilde{g}$ is source term. Then applying,L^{-1}-fuzzy integration operator on both sides we get

$$\tilde{u} = L^{-1}\tilde{g} \ominus L^{-1}(N\tilde{u}) \ominus L^{-1}(R\tilde{u})$$

The parametric form of above equation is given by

$$[\underline{u}, \overline{u}] = L^{-1}\left[\min\left(\underline{g}, \overline{g}\right), \max\left(\underline{g}, \overline{g}\right)\right]$$

$$\ominus L^{-1}\left(\left[\min\left(N\underline{u}, N\overline{u}\right), \max\left(N\underline{u}, N\overline{u}\right)\right]\right)$$
$$\ominus L^{-1}\left(\left[\min\left(R\underline{u}, R\overline{u}\right), \max\left(R\underline{u}, R\overline{u}\right)\right]\right) \tag{3}$$

By ADM as in [6], nonlinear term $\left[\min\left(N\underline{u}, N\overline{u}\right), \max\left(N\underline{u}, N\overline{u}\right)\right]$ in parametric form is decomposed into the series of Adomian polynomials in parametric form and it is given as $[\underline{D}_n, \overline{D}_n]$.

where,

$$\left.\begin{array}{l} \underline{D}_n = \min\left(\frac{1}{n!}\frac{d^n}{d\lambda^n}\left(N\sum\limits_{k=0}^{\infty}\underline{X}_k\lambda^k\right), \frac{1}{n!}\frac{d^n}{d\lambda^n}\left(N\sum\limits_{k=0}^{\infty}\overline{X}_k\lambda^k\right)\right) \\ \overline{D}_n = \max\left(\frac{1}{n!}\frac{d^n}{d\lambda^n}\left(N\sum\limits_{k=0}^{\infty}\underline{X}_k\lambda^k\right), \frac{1}{n!}\frac{d^n}{d\lambda^n}\left(N\sum\limits_{k=0}^{\infty}\overline{X}_k\lambda^k\right)\right) \end{array}\right\} \tag{4}$$

So, solution of Eq. (3) in parametric form is given as,

$$\left[\underline{u}, \overline{u}\right] = \left[\underline{u}_1, \overline{u}_1\right] \oplus \left[\underline{u}_2, \overline{u}_2\right] \oplus \left[\underline{u}_3, \overline{u}_3\right] \oplus \cdots$$

The following result proves the existence of ADM in parametric form for fuzzy semi-linear dynamical system.

3.1 Main Result

Let $\tilde{f}: (0, t) \times E^n \to E^n$ be modified Hukuhara differentiable and using the notation of parametric form of fuzzy number, we can write, $\tilde{f} = \left[\underline{f}, \overline{f}\right]$. Where, $\underline{f} = \min f\left(t, \underline{X}, \overline{X}\right), \overline{f} = \max f\left(t, \underline{X}, \overline{X}\right)$ and the boundary function $\underline{f}, \overline{f}$ are modified Hukuhara differentiable. Let the first derivatives of $\tilde{f}$ and fuzzy variables $\tilde{X}$ be given as.

$$\dot{\tilde{f}} = \left[\underline{\dot{f}}, \dot{\overline{f}}\right], \dot{\tilde{X}} = \left[\underline{\dot{X}}, \dot{\overline{X}}\right] \text{ and } \tilde{X}(t) = \left[\underline{X}, \overline{X}\right].$$

where, $\underline{\dot{f}} = \min \dot{f}\left(t, \underline{X}, \overline{X}\right), \dot{\overline{f}} = \max \dot{f}\left(t, \underline{X}, \overline{X}\right)$.

Consider the fuzzy initial value problem (FIVP) as given by system (1)

$$\dot{X}(t) = \tilde{A} \otimes \tilde{X} \oplus \tilde{f}\left(t, \tilde{X}\right); \ \tilde{X}_0 = \tilde{X}(0)$$

Using above notation, we can convert FIVP as in (1) into system of ODEs.

$$\left[\underline{\dot{X}}, \dot{\overline{X}}\right] = \left[\min\left(\underline{AX}, \underline{A\overline{X}}, \overline{A\underline{X}}, \overline{AX}\right), \max\left(\underline{AX}, \underline{A\overline{X}}, \overline{A\underline{X}}, \overline{AX}\right)\right]$$
$$\oplus \left[\underline{f}, \overline{f}\right]$$

$$[\underline{X}_0, \overline{X}_0] = [\min(\underline{X}(0), \overline{X}(0)), \max(\underline{X}(0), \overline{X}(0))] \tag{5}$$

Comparing components both the sides, we get

$$\begin{aligned} \dot{\underline{X}} &= \min(\underline{AX}, \underline{A}\overline{X}, \overline{A}\underline{X}, \overline{AX}) + \underline{f} \quad \text{where,} \\ \underline{X}_0 &= \min(\underline{X}(0), \overline{X}(0)) \end{aligned} \tag{6}$$

$$\begin{aligned} \dot{\overline{X}} &= \max(\underline{AX}, \underline{A}\overline{X}, \overline{A}\underline{X}, \overline{AX}) + \overline{f} \quad \text{where,} \\ \overline{X}_0 &= \max(\underline{X}(0), \overline{X}(0)) \end{aligned} \tag{7}$$

Lemma *FIVP* (1) *is equivalent to system* (5) *iff* $\tilde{f}$ *is equicontinuous, uniformly bounded and Lipschitz.*

Proof Since, $\tilde{f}$ is equi-continuous and Lipschitz, then

$$\left|\tilde{f}\left(t, \tilde{X}\right) \ominus \tilde{f}\left(t, \tilde{Y}\right)\right| < K\left|\tilde{X} \ominus \tilde{Y}\right|$$

Using the Lipschitz condition in parametric form,

$$\left|\min f\left(t, \underline{X}, \bar{X}\right) - \min f\left(t, \underline{Y}, \bar{Y}\right)\right| < K\left|\left[\underline{X}, \bar{X}\right] - \left[\underline{Y}, \bar{Y}\right]\right|$$

i.e.,

$$\begin{aligned} &\left|\min f(t, \underline{X}, \overline{X}) - \min f(t, \underline{Y}, \overline{Y})\right| \\ &\quad < K\left|\min((\underline{X} - \underline{Y}), (\underline{X} - \overline{Y}), (\overline{X} - \underline{Y}), (\overline{X} - \overline{Y}))\right| \end{aligned}$$

where K is Lipchitz constant.

Thus, Eq. (6) has unique solution.

Similarly,

$$\begin{aligned} &\left|\max f(t, \underline{X}, \overline{X}) - \max f(t, \underline{Y}, \overline{Y})\right| \\ &\quad < K'\left|\max((\underline{X} - \underline{Y}), (\underline{X} - \overline{Y}), (\overline{X} - \underline{Y}), (\overline{X} - \overline{Y}))\right| \end{aligned}$$

where K' is Lipschitz constant.

Hence, Eq. (7) has unique solution.

Thus, if FIVP (1) has unique solution and $\tilde{X}(t)$ is modified Hukuhara differentiable, then $\underline{X}(t)$ and $\overline{X}(t)$ are modified Hukuhara differentiable in sense of alpha-cut and are the solution of system (6) and (7), respectively.

Conversely, let $[\underline{X}(t), \overline{X}(t)]$ be the solution of (5) then $\underline{X}(t)$ and $\overline{X}(t)$ are differentiable and the solution of system (6) and (7), also $\tilde{f}$ is Lipschitz. So, it guarantees the unique solution of (1) given by decomposition theorem as in Klir [10]. We can construct solution $\tilde{X}(t)$ for $\alpha \in [0, 1]$, i.e.,

$$\tilde{X}(t) = \bigcup_{\alpha\in[0,1]} \left[\underline{X}(t), \overline{X}(t)\right]$$

where, $\cup$ denotes fuzzy union.

So, system (1) is equivalent to system (5).

Theorem *Let* $\left[\underline{f}, \overline{f}\right]$ *be analytic function in system* (5), *then system* (5) *has series solution by Adomian decomposition method, i.e.,*

$$\left[\underline{X}(t), \overline{X}(t)\right] = \left[\underline{X}_1(t), \overline{X}_1(t)\right] \oplus \left[\underline{X}_2(t), \overline{X}_2(t)\right] \oplus \left[\underline{X}_3(t), \overline{X}_3(t)\right] + \cdots$$

where,

$$\left[\underline{X}_1(t), \overline{X}_1(t)\right] = \left[\int_0^t \underline{PX}_0 \mathrm{d}t, \int_0^t \overline{PX}_0 \mathrm{d}t\right] \oplus L^{-1}\left[\min\left(\underline{D}_0, \overline{D}_0\right), \max\left(\underline{D}_0, \overline{D}_0\right)\right]$$

$$\left[\underline{X}_2(t), \overline{X}_2(t)\right] = \left[\int_0^t \underline{PX}_1 \mathrm{d}t, \int_0^t \overline{PX}_1 \mathrm{d}t\right] \oplus L^{-1}\left[\min\left(\underline{D}_1, \overline{D}_1\right), \max\left(\underline{D}_1, \overline{D}_1\right)\right] \text{ and so on.}$$

where, $\underline{PX} = \left[\min\left(\underline{AX}, \underline{A}\overline{X}, \overline{A}\underline{X}, \overline{AX}\right)\right]$ and $\overline{PX} = \max\left(\underline{AX}, \underline{A}\overline{X}, \overline{A}\underline{X}, \overline{AX}\right)$ *and L is differential operator.*

Proof System (5) is given as,

$$\left[\underline{\dot{X}}, \dot{\overline{X}}\right] = \left[\min\left(\underline{AX}, \underline{A}\overline{X}, \overline{A}\underline{X}, \overline{AX}\right), \max\left(\underline{AX}, \underline{A}\overline{X}, \overline{A}\underline{X}, \overline{AX}\right)\right] \oplus \left[\underline{f}, \overline{f}\right]$$

$$\left[\underline{X}_0, \overline{X}_0\right] = \left[\min\left(\underline{X}(0), \overline{X}(0)\right), \max\left(\underline{X}(0), \overline{X}(0)\right)\right]$$

Let $\left[\underline{f}, \overline{f}\right]$ be analytic function in system (5) so ADM is applicable. In this technique, nonlinear term can be written in form of Adomian polynomials $\left[\underline{D}_n, \overline{D}_n\right]$.

System (5) can be written as

$$\left[L\underline{X}, L\overline{X}\right] = \left[\min\left(\underline{AX}, \underline{A}\overline{X}, \overline{A}\underline{X}, \overline{AX}\right), \max\left(\underline{AX}, \underline{A}\overline{X}, \overline{A}\underline{X}, \overline{AX}\right)\right]$$

$$\oplus\left[\underline{f}, \overline{f}\right]$$

Now, taking L^{-1} on both sides, where $L^{-1} = \int_0^t (\cdot)\mathrm{d}t$

$$L^{-1}\left[L\underline{X}, L\overline{X}\right] = L^{-1}\left[\min\left(\underline{AX}, \underline{A}\overline{X}, \overline{A}\underline{X}, \overline{AX}\right), \max\left(\underline{AX}, \underline{A}\overline{X}, \overline{A}\underline{X}, \overline{AX}\right)\right] \oplus L^{-1}\left[\underline{f}, \overline{f}\right] \tag{8}$$

In (8), putting, $\underline{PX} = \left[\min\left(\underline{AX}, \underline{A}\overline{X}, \overline{A}\underline{X}, \overline{AX}\right)\right]$ and $\overline{PX} = \max\left(\underline{AX}, \underline{A}\overline{X}, \overline{A}\underline{X}, \overline{AX}\right)$, we get

$$\left[\underline{X}(t), \overline{X}(t)\right] = \left[\underline{X_0}, \overline{X_0}\right] \oplus \left[\int_0^t \underline{PX}(t)\mathrm{d}t, \int_0^t \overline{PX}(t)\mathrm{d}t\right] \oplus L^{-1}\left[\underline{f}, \overline{f}\right] \tag{9}$$

Now, let $\underline{X}(t) = \sum_{n=0}^{\infty} \underline{X}_n(t), \overline{X}(t) = \sum_{n=0}^{\infty} \overline{X}_n(t)$ be the series solution of system (5). Then (9) becomes

$$\left[\sum_{n=0}^{\infty} \underline{X}_n(t), \sum_{n=0}^{\infty} \overline{X}_n(t)\right] = \left[\underline{X_0}, \overline{X_0}\right] \oplus \left[\int_0^t \sum_{n=0}^{\infty} \underline{PX}_n(t)\mathrm{d}t, \int_0^t \sum_{n=0}^{\infty} \overline{PX}_n(t)\mathrm{d}t\right] \oplus L^{-1}\left[\min\left(\sum_{n=0}^{\infty} \underline{D}_n, \sum_{n=0}^{\infty} \overline{D}_n\right), \max\left(\sum_{n=0}^{\infty} \underline{D}_n, \sum_{n=0}^{\infty} \overline{D}_n\right)\right]$$

By Recurrence relation we get,

$$\left[\underline{X}_n(t), \overline{X}_n(t)\right] = \left[\int_0^t \underline{PX}_{n-1}\mathrm{d}t, \int_0^t \overline{PX}_{n-1}\mathrm{d}t\right] \oplus L^{-1}\left[\min\left(\underline{D}_{n-1}, \overline{D}_{n-1}\right), \max\left(\underline{D}_{n-1}, \overline{D}_{n-1}\right)\right]$$

for $n = 1, 2, 3, \ldots$.
Therefore, for different values of n we get,

$$\left[\underline{X}_1(t), \overline{X}_1(t)\right] = \left[\int_0^t \underline{PX}_0\mathrm{d}t, \int_0^t \overline{PX}_0\mathrm{d}t\right] \oplus L^{-1}\left[\min\left(\underline{D}_0, \overline{D}_0\right), \max\left(\underline{D}_0, \overline{D}_0\right)\right]$$

$$[\underline{X}_2(t), \overline{X}_2(t)] = \left[\int_0^t \underline{PX}_1 \mathrm{d}t, \int_0^t \overline{PX}_1 \mathrm{d}t\right] \oplus L^{-1}\left[\min\left(\underline{D}_1, \overline{D}_1\right), \max\left(\underline{D}_1, \overline{D}_1\right)\right]$$

$$[\underline{X}_3(t), \overline{X}_3(t)] = \left[\int_0^t \underline{PX}_2 \mathrm{d}t, \int_0^t \overline{PX}_2 \mathrm{d}t\right] \oplus L^{-1}\left[\min\left(\underline{D}_2, \overline{D}_2\right), \max\left(\underline{D}_2, \overline{D}_2\right)\right]$$

and so on.

The Adomian polynomials, $\left[\underline{D}_n, \overline{D}_n\right]$ can be obtained by the formula as given in (4)

$$\underline{D}_n = \min\left(\frac{1}{n!}\frac{\mathrm{d}^n}{\mathrm{d}\lambda^n}\left(N\sum_{k=0}^{\infty}\underline{X}_k\lambda^k\right), \frac{1}{n!}\frac{\mathrm{d}^n}{\mathrm{d}\lambda^n}\left(N\sum_{k=0}^{\infty}\overline{X}_k\lambda^k\right)\right)$$
$$\overline{D}_n = \max\left(\frac{1}{n!}\frac{\mathrm{d}^n}{\mathrm{d}\lambda^n}\left(N\sum_{k=0}^{\infty}\underline{X}_k\lambda^k\right), \frac{1}{n!}\frac{\mathrm{d}^n}{\mathrm{d}\lambda^n}\left(N\sum_{k=0}^{\infty}\overline{X}_k\lambda^k\right)\right)$$

where, λ is decomposition factor.

So, the solution of system (5) is given by,

$$[\underline{X}(t), \overline{X}(t)] = [\underline{X}_1(t), \overline{X}_1(t)] \oplus [\underline{X}_2(t), \overline{X}_2(t)] \oplus [\underline{X}_3(t), \overline{X}_3(t)] \oplus \cdots$$

Convergence of such a series solution has been discussed in [11].

4 Application

Using the fuzzy ADM technique, we solve the following fully fuzzy prey-predator model involving the fuzzy parameters and fuzzy initial condition given as

$$\left.\begin{aligned}\dot{\tilde{x}} &= (\widetilde{0.1} \otimes \tilde{x}) \ominus \left(\widetilde{0.005} \otimes \tilde{x} \otimes \tilde{y}\right)\\ \dot{\tilde{y}} &= \left(\widetilde{0.008} \otimes \tilde{x} \otimes \tilde{y}\right) \ominus (\widetilde{0.4} \otimes \tilde{y})\end{aligned}\right\} \tag{10}$$

with $\tilde{x}_0 = \widetilde{130} = [120, 150]$ and $\tilde{y}_0 = \widetilde{40} = [20, 50]$.

where,

$$\widetilde{0.1} = [0.05, 0.15], \widetilde{0.005} = [0.004, 0.006],$$
$$\widetilde{0.4} = [0.3, 0.5], \widetilde{0.008} = [0.007, 0.009].$$

Applying the proposed FADM technique, we have

$$\begin{aligned} L[\underline{x}, \overline{x}] &= ([\underline{0.1}, \overline{0.1}] \cdot [\underline{x}, \overline{x}]) \\ &\ominus \left([\underline{0.005}, \overline{0.005}] \cdot [\underline{x}, \overline{x}] \cdot \left[\underline{y}, \overline{y}\right]\right) \\ L\left[\underline{y}, \overline{y}\right] &= \left([\underline{0.008}, \overline{0.008}] \cdot [\underline{x}, \overline{x}] \cdot \left[\underline{y}, \overline{y}\right]\right) \\ &\ominus \left([\underline{0.4}, \overline{0.4}] \cdot \left[\underline{y}, \overline{y}\right]\right) \end{aligned}$$

Applying inverse operator both the sides, we have

$$\begin{aligned} L^{-1}L[\underline{x}, \overline{x}] &= L^{-1}([0.05, 0.15] \cdot [\underline{x}, \overline{x}]) \\ &\ominus L^{-1}\left([0.004, 0.006] \cdot [\underline{x}, \overline{x}] \cdot \left[\underline{y}, \overline{y}\right]\right) \\ L^{-1}L\left[\underline{y}, \overline{y}\right] &= L^{-1}\left(-[0.3, 0.5] \cdot \left[\underline{y}, \overline{y}\right]\right) \\ &\oplus L^{-1}\left([0.007, 0.009] \cdot [\underline{x}, \overline{x}] \cdot \left[\underline{y}, \overline{y}\right]\right) \end{aligned}$$

Thus, by solving above equation, we get

$$\left.\begin{aligned} [\underline{x}(t), \overline{x}(t)] &= [\underline{x}(0), \overline{x}(0)] \oplus L^{-1}([0.05, 0.15] \cdot [\underline{x}, \overline{x}]) \ominus L^{-1}\left([0.004, 0.006] \cdot [\underline{x}, \overline{x}] \cdot \left[\underline{y}, \overline{y}\right]\right) \\ \left[\underline{y}(t), \overline{y}(t)\right] &= \left[\underline{y}(0), \overline{y}(0)\right] \oplus L^{-1}\left([-0.5, -0.3] \cdot \left[\underline{y}, \overline{y}\right]\right) \oplus L^{-1}\left([0.007, 0.009] \cdot [\underline{x}, \overline{x}] \cdot \left[\underline{y}, \overline{y}\right]\right) \end{aligned}\right\}$$

As given in our main result, let $[\underline{x}(t), \overline{x}(t)] = \sum_{n=0}^{\infty} [\underline{x}_n(t), \overline{x}_n(t)]$ and $\left[\underline{y}(t), \overline{y}(t)\right] = \sum_{n=0}^{\infty} \left[\underline{y}_n(t), \overline{y}_n(t)\right]$ be the series solution of system (10) and the nonlinear term be expressed in the form of Adomian polynomial as $\sum_{n=0}^{\infty} [\underline{D}_n(t), \overline{D}_n(t)] = [\underline{x}, \overline{x}] \cdot \left[\underline{y}, \overline{y}\right]$.

So the system of Eq. (10) can be written as

$$\begin{aligned} \sum_{n=0}^{\infty} [\underline{x}_n(t), \overline{x}_n(t)] &= [\underline{x}(0), \overline{x}(0)] \\ &\oplus L^{-1}\left([0.05, 0.15] \cdot \sum_{n=0}^{\infty} [\underline{x}_n(t), \overline{x}_n(t)]\right) \\ &\ominus L^{-1}\left([0.004, 0.006] \cdot \sum_{n=0}^{\infty} [\underline{D}_n, \overline{D}_n]\right) \\ \sum_{n=0}^{\infty} \left[\underline{y}_n(t), \overline{y}_n(t)\right] &= \left[\underline{y}(0), \overline{y}(0)\right] \end{aligned}$$

$$\oplus L^{-1}\left([-0.5, -0.3] \cdot \sum_{n=0}^{\infty}\left[\underline{y}_n(t), \overline{y}_n(t)\right]\right)$$
$$\oplus L^{-1}\left([0.007, 0.009] \cdot \sum_{n=0}^{\infty}\left[\underline{D}_n, \overline{D}_n\right]\right)$$

Comparing both the sides, we get

$$\begin{aligned}
\left[\underline{x}_0(t), \overline{x}_0(t)\right] &= \left[\underline{x}(0), \overline{x}(0)\right] \\
\left[\underline{x}_1(t), \overline{x}_1(t)\right] &= L^{-1}\left([0.05, 0.15] \cdot \left[\underline{x}_0(t), \overline{x}_0(t)\right]\right) \ominus L^{-1}\left([0.004, 0.006] \cdot \left[\underline{D}_0, \overline{D}_0\right]\right) \\
\left[\underline{x}_2(t), \overline{x}_2(t)\right] &= L^{-1}\left([0.05, 0.15] \cdot \left[\underline{x}_1(t), \overline{x}_1(t)\right]\right) \ominus L^{-1}\left([0.004, 0.006] \cdot \left[\underline{D}_1, \overline{D}_1\right]\right) \\
&\vdots \\
\left[\underline{x}_n(t), \overline{x}_n(t)\right] &= L^{-1}\left([0.05, 0.15] \cdot \left[\underline{x}_{n-1}(t), \overline{x}_{n-1}(t)\right]\right) \ominus L^{-1}\left([0.004, 0.006] \cdot \left[\underline{D}_{n-1}, \overline{D}_{n-1}\right]\right)
\end{aligned}$$

Similarly,

$$\begin{aligned}
\left[\underline{y}_0(t), \overline{y}_0(t)\right] &= \left[\underline{y}(0), \overline{y}(0)\right] \\
\left[\underline{y}_1(t), \overline{y}_1(t)\right] &= L^{-1}\left([-0.5, -0.3] \cdot \left[\underline{y}_0(t), \overline{y}_0(t)\right]\right) \oplus L^{-1}\left([0.007, 0.009] \cdot \left[\underline{D}_0, \overline{D}_0\right]\right) \\
\left[\underline{y}_2(t), \overline{y}_2(t)\right] &= L^{-1}\left([-0.5, -0.3] \cdot \left[\underline{y}_1(t), \overline{y}_1(t)\right]\right) \oplus L^{-1}\left([0.007, 0.009] \cdot \left[\underline{D}_1, \overline{D}_1\right]\right) \\
&\vdots \\
\left[\underline{y}_n(t), \overline{y}_n(t)\right] &= L^{-1}\left([-0.5, -0.3] \cdot \left[\underline{y}_{n-1}, \overline{y}_{n-1}\right]\right) \oplus \left([0.007, 0.009] \cdot \left[\underline{D}_{n-1}, \overline{D}_{n-1}\right]\right)
\end{aligned}$$

Adomian polynomial can be defined as

$$\begin{aligned}
\left[\underline{D}_0, \overline{D}_0\right] &= \left[\underline{x}_0(t), \overline{x}_0(t)\right] \cdot \left[\underline{y}_0(t), \overline{y}_0(t)\right] \\
\left[\underline{D}_1, \overline{D}_1\right] &= \left(\left[\underline{x}_0(t), \overline{x}_0(t)\right] \cdot \left[\underline{y}_1(t), \overline{y}_1(t)\right]\right) \\
&\quad \oplus \left(\left[\underline{x}_1(t), \overline{x}_1(t)\right] \cdot \left[\underline{y}_0(t), \overline{y}_0(t)\right]\right) \\
\left[\underline{D}_2, \overline{D}_2\right] &= \left(\left[\underline{x}_0(t), \overline{x}_0(t)\right] \cdot \left[\underline{y}_2(t), \overline{y}_2(t)\right]\right) \\
&\quad \oplus \left(\left[\underline{x}_1(t), \overline{x}_1(t)\right] \cdot \left[\underline{y}_1(t), \overline{y}_1(t)\right]\right) \\
&\quad \oplus \left(\left[\underline{x}_2(t), \overline{x}_2(t)\right] \cdot \left[\underline{y}_0(t), \overline{y}_0(t)\right]\right)
\end{aligned}$$

and so on.

By substituting values, we have

$$\tilde{x}_0 = \widetilde{130} = [120, 150] \quad \text{and} \quad \tilde{y}_0 = \widetilde{40} = [20, 50]$$

The Adomian polynomial

$$\begin{aligned}
\tilde{D}_0 = \tilde{x}_0 \otimes \tilde{y}_0 &= [120, 150] \cdot [20, 50] \\
&= [\min(2400, 6000, 3000, 7500), \max(2400, 6000, 3000, 7500)] \\
&= [2400, 7500] \\
\left[\underline{x}_1, \overline{x}_1\right] &= L^{-1}([0.05, 0.15] \cdot [120, 150]) \\
&\quad \ominus L^{-1}([0.004, 0.006] \cdot [2400, 7500]) \\
&= L^{-1}[\min(6, 7.5, 18, 22.5), \max(6, 7.5, 18, 22.5)] \\
&\quad \ominus L^{-1}[\min(9.6, 30, 14.4, 45), \max(9.6, 30, 14.4, 45)] \\
&= L^{-1}[6, 22.5] \ominus L^{-1}[9.6, 45] \\
&= [-39t, 12.9t] \\
\left[\underline{y}_1, \overline{y}_1\right] &= L^{-1}(-[0.4, 0.3] \cdot [20, 50]) \\
&\quad \oplus L^{-1}([0.007, 0.009] \cdot [2400, 7500]) \\
&= L^{-1}[\min(-10, -25, -6, -15), \max(-10, -25, -6, -15)] \\
&\quad \oplus L^{-1}[\min(16.8, 52.5, 21.6, 67.5), \max(16.8, 52.5, 21.6, 67.5)] \\
&= L^{-1}[-25, -6] \oplus L^{-1}[16.8, 67.5] \\
&= [-8.2t, 61.5t]
\end{aligned}$$

Similarly, for $n = 2, 3, \ldots$, we can compute the solution of system.

$$\begin{aligned}
\tilde{x}(t) = \tilde{x}_0 \oplus \tilde{x}_1 \oplus \tilde{x}_2 \oplus \cdots &= [120, 150] \\
&\oplus [-39t, 12.9t] \oplus \left[-32.535t^2, 10.508t^2\right] \\
&\oplus \left[-18.2037t^3, 17.48138t^3\right] \\
&\oplus \left[-11.4075t^4, 15.2055t^4\right] \oplus \cdots \\
\tilde{y}(t) = \tilde{y}_0 \oplus \tilde{y}_1 \oplus \tilde{y}_2 \oplus \cdots &= [20, 50] \\
&\oplus [-8.2t, 61.5t] \oplus \left[-29.685t^2, 46.465t^2\right] \\
&\oplus \left[-33.1782t^3, 29.8129t^3\right] \\
&\oplus \left[-25.5515t^4, 20.2346t^4\right] \oplus \cdots
\end{aligned}$$

The evolution of prey-predator population at different time is as shown in Table 1. Figures 1 and 2 show the evolution if predator and prey population for time $t = 0$ to 0.5.

5 Conclusion

In this article, we have solved fully fuzzy Prey-Predator that is involving fuzzy initial condition as well as fuzzy parameters by Adomian decomposition method for

Table 1 Evolution on predator and prey data with time

t	Prey-data			Predator-data		
0	$l-120$	$c-130$	$r-150$	$l-20$	$c-40$	$r-50$
0.056	117.7297	129.2541	150.7522	19.4470	41.4411	53.5654
0.11	115.2383	128.4609	151.5894	18.6730	42.9199	57.4510
0.17	112.5032	127.6203	152.5345	17.6354	44.4364	61.6943
0.22	109.4991	126.7323	153.6145	16.2855	45.9907	66.3377
0.28	106.1982	125.7971	154.8593	14.5685	47.5827	71.4281
0.33	102.57	124.8144	156.3027	12.4241	49.2124	77.0168
0.38	98.5814	123.7845	157.9817	9.78599	50.8799	83.1600
0.44	94.1968	122.7072	159.9369	6.5821	52.5851	89.9185
0.5	89.3778	121.5825	162.2124	2.7345	54.328	97.3575

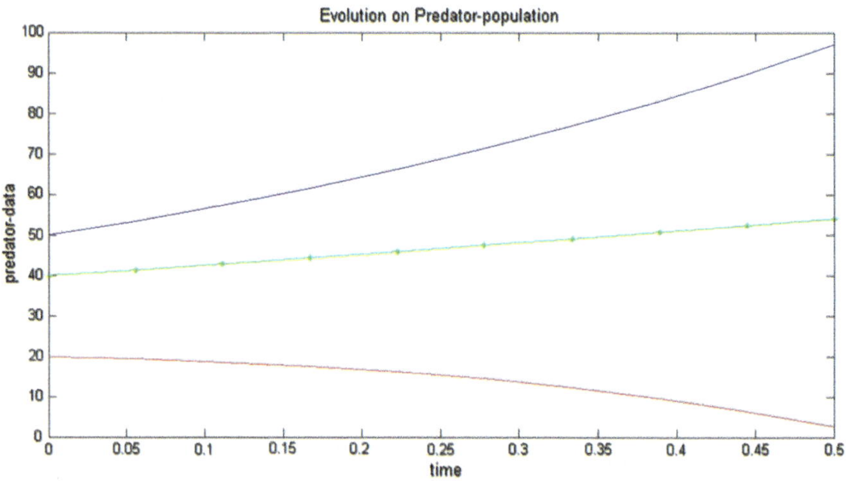

Fig. 1 Evolution of predator population for $t = 0$ to $t = 0.5$

parametric form of fuzzy numbers. It is shown in application how does such model evolves with time. When we take parameters fuzzy as well as initial condition, change in population of preys and predators are obtained with more accuracy. Future work will be in direction of fuzzy Adomian decomposition method applied to the system directly in the fuzzy form.

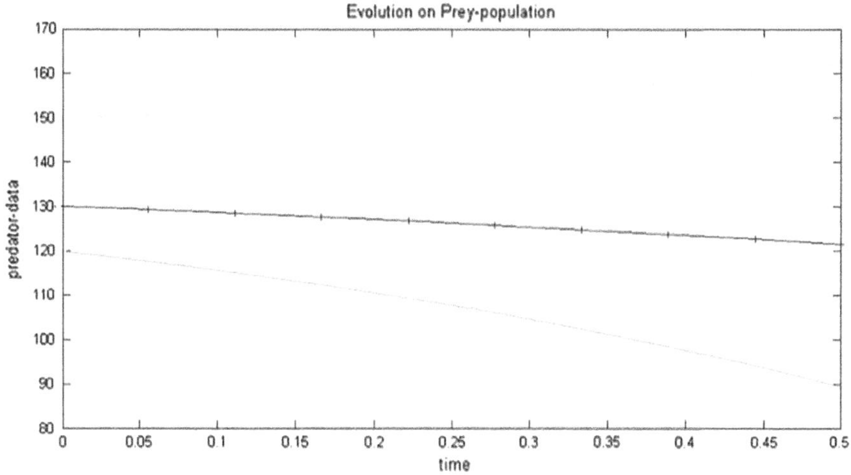

Fig. 2 Evolution of prey population for $t = 0$ to $t = 0.5$

References

1. Purnima, P., Payal, S.: Prey-predator model and fuzzy initial condition. Int. J. Eng. Innov. Technol. (IJEIT) **3**(12), 65–68 (2014)
2. Da Silva Peixoto, M. et al.: Predator-prey fuzzy model. Ecol. Modell. **214**(1), 39–44 (2008)
3. Akin, O., Oruc, O.: A prey predator model with fuzzy initial values, Hacettepe. J. Math. Stat. **41**(3), 387–395 (2012)
4. De Baets, M.Z.Ahmad.: A predator-prey model with fuzzy initial populations, IFSA-EUSFLAT (2009)
5. Purnima, P., Payal, S.: Fully fuzzy Semi-Linear dynamical system solved by fuzzy laplace transform under modified Hukuhara derivative, soft computing on problem solving, vol. 1, AISC, Springer, pp. 155–179 (2019)
6. Adomian, G.: Solving Frontier Problems of Physics: The Decomposition Method. Kluwer Academic Publishers (1994)
7. Song, S., Wu, C.: Existence and uniqueness of solutions to cauchy problem of fuzzy differential equations. Fuzzy Sets Syst. **110**, 55–67 (2000)
8. Bede, B.: Note on numerical solutions of fuzzy differential equations by predictor corrector method. Inf. Sci. **178**, 1917–1922 (2008)
9. Puri, M.L., Ralescu: Differential of fuzzy functions. J. Math. Anal. Appl. **91**, 321–325 (1983)
10. Klir, G.J., Yuan, B.: Fuzzy Sets and Fuzzy Logic: Theory and Applications. Prentice hall, Englewood Cliffs NJ (1995)
11. Abbaoui, K., Cherruault, Y.: New ideas for proving convergence of decomposition methods. Comput. Math. Appl. **29**, 103–108 (1995)

A Fuzzy Logic Approach for Optimization of Solar PV Site in India

Pavan Fuke, Anil Kumar Yadav, and Ipuri Anil

Abstract The addition of renewable energy sources in the electricity grid is increasing day by day because of environmental concern. Due to its easy availability and high potential, solar energy is considered to be one of the most important renewable energy sources in India. Solar photovoltaic (PV) technology is one of the foremost solutions to mitigate energy shortage and rapidly increasing energy demand problems in the country. For the development and economic growth of solar PV, it needs to use solar energy carefully and planned solar power plants at an optimum location. The optimization of site selection is a critical issue for utility-scale power plants which require a large investment. To operate the solar PV system at its highest efficiency, it needs to select the most optimal sites with favorable environmental conditions. Among a wide range of environmental parameters most obvious factors that affect the PV performance are solar irradiance and air temperature. In this study, a fuzzy logic approach is used to propose a decision support system for selecting the most appropriate solar PV sites in India. Because of the huge regional span, to analyze every region, we divide India into six zones based upon climatic and geographical features that are north zone, south zone, east zone, west zone, central zone, and north-east zone. Firstly, we analyze the effect of solar irradiance and temperature on PV performance. Secondly, the variation of these factors for each zone is investigated. Then, a fuzzy logic approach is used to optimize zone wise highest priority sites with high solar PV potential.

Keywords Solar photovoltaic · Environmental factors · Site selection · Fuzzy logic

P. Fuke (✉) · I. Anil
Centre for Energy and Environmental Engineering, NIT Hamirpur, Hamirpur 177005, India
e-mail: fukepavan44@gmail.com

A. K. Yadav
Department of Electrical Engineering, NIT Hamirpur, Hamirpur 177005, India

M. Sahni et al. (eds.), *Mathematical Modeling, Computational Intelligence Techniques and Renewable Energy*, Advances in Intelligent Systems and Computing 1287,
https://doi.org/10.1007/978-981-15-9953-8_43

1 Introduction

Renewable energy is continuously gaining interest and support in all countries. PV has emerged as the most popular and common available source and can convert solar energy directly to electricity. An increasing energy cost and climate change awareness lead to rapid growth in the PV industry in recent years [1]. The PV modules are manufactured at STC but install at various locations with different atmospheric conditions. PV installations are normally anticipated to be designed for maximum yield. The factors that determine optimum yield are PV module characteristics, environmental factor, and PV installation design which are shown in Fig. 1.

As seen in the above figure, there are environmental factors also called meteorological parameters such as solar insolation, ambient temperature, precipitation, wind velocity influence the PV performance in addition to the module characteristics and design parameters [2]. These factors can affect the PV cell performance. The recently researchers in the energy field are focusing on it, many of which are focusing on the impact of meteorological parameters on the performance and productivity of PV cells [3]. After studying the effect of all these meteorological parameters effect on PV modules, it found that solar irradiance and air temperature these two factors are affecting the most. It revealed that an increase in solar insolation intensity productivity of PV modules is increased. By studying the effect of ambient temperature, it found that the rise in ambient temperature leads to a reduction in output power [4].

Due to the rapidly increasing population and fast-growing economy, India's energy needs increase. As a result of this, now, it is necessary to expand and add renewable energy recourses in all areas. As solar energy is one of the cheap and easily available renewable sources of energy in India. So in the current scenario, to attain India's energy needs, solar-based power generation can play an important role in it. In India, where there are about 250–300 clean sunny days/year in most regions of the country with yearly global radiation ranges from 1600 to 2200 kWh/m^2. India has the target to install 100 GW of solar power projects by the end of 2022 under a greatly ambitious JNNSM program [5]. One important aspect of achieving such highly ambient plans to identify the optimum sites for solar PV installation by analyzing daily data of solar

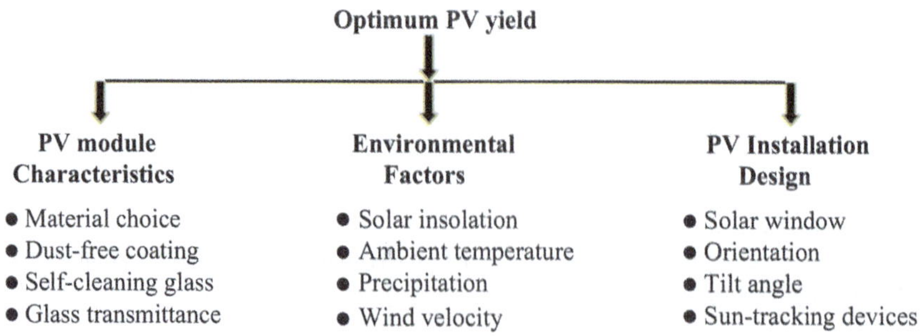

Fig. 1 Factors determining the PV system yield

irradiance and ambient temperature [6]. The purpose of this present study is to use a fuzzy logic approach for optimization of solar PV site selection in India; for that, the zone-wise data of solar irradiance and temperature is analyzed in India.

2 Solar Irradiance

The global radiation, i.e., total radiation reaching to earth surface is a sum of three radiations. Firstly, direct radiation received by earth surface directly, it also called beam radiation. Then, indirect radiation that reached earth's surface after scattering is also called diffuse radiation and albedo radiation that is reflected sunlight from the ground. When total solar radiations passes into the earth's surrounding, it passes through several effects till come into the earth's ground level. The major effects for PV are—

- The total solar radiation power reaches to earth is reduced due to absorption, reflection, and scattering effect caused by atmospheric gases, aerosols, and dust. Specific gases like CO_2, O_3, and H_2O vapor have a very large incorporation of photons [7].
- A significant incorporation of some wavelengths result to variation in the photonic content of the solar radiation. Local variation in the aerosphere factors which have supplementary effects on the incoming energy range.
- Development of a diffuse part in the solar irradiance. On a clear day, around 10% of total irradiance is diffuse irradiance [8].

Approximately 30% of the total energy coming from the sun is either reflected or absorbed by clouds, ocean, and landmasses. The solar energy that strikes the solar cell subjected to a loss in energy absorption and reflection. These losses are approximate of about 15–20% of energy in addition to the mentioned 30% [9]. The resultant current and resultant voltage calculation of the solar module are as below.

$$I = I_L - I_o\left[\exp\left(\frac{qV}{nkT}\right) - 1\right];\ I_L \cong I_{sc} \tag{1}$$

$$V_{oc} = \frac{nkT}{q}\ln\left(\frac{I_{sc}}{I_o} + 1\right) \tag{2}$$

From (1), (2) output current I directly proportional on the current resulted from photons which itself depends on the arriving photons arriving on a solar module. The resultant voltage directly proportional to I_L. As the power yield of solar module is a consequence of V_{oc} and I_{sc}, i.e., it is related to arriving solar radiations on a solar module [10]. The PV curve of solar module with respect to solar irradiance is shown (Fig. 2).

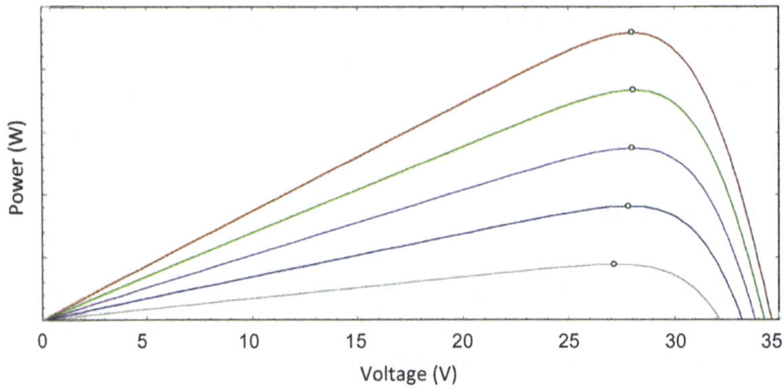

Fig. 2 Power output variation of PV with respect to solar irradiance

3 Temperature

As a solar cell is a semiconductor device so as like other semiconductor devices, it is sensitive to temperature. An increment in temperature leads to a reduction in the semiconductor band gap. Due to this less energy is needed for electron to jump into the conduction band. So the quantity of charge carrier increases with temperature, i.e., short circuit current of PV cell increases with temperature. The open-circuit voltage is most affected parameter due to temperature [11]. The open-circuit voltage reduced with temperature increase. The PV curve of temperature with respect to solar irradiance is shown below in Fig. 3. The pace of variation of open-circuit voltage V_{oc} with respect to temperature gives as—

$$\frac{\mathrm{d}V_{\text{oc}}}{\mathrm{d}t} = -\frac{V_{\text{Go}} - V_{\text{oc}} + \gamma\frac{\text{kT}}{q}}{T} \tag{3}$$

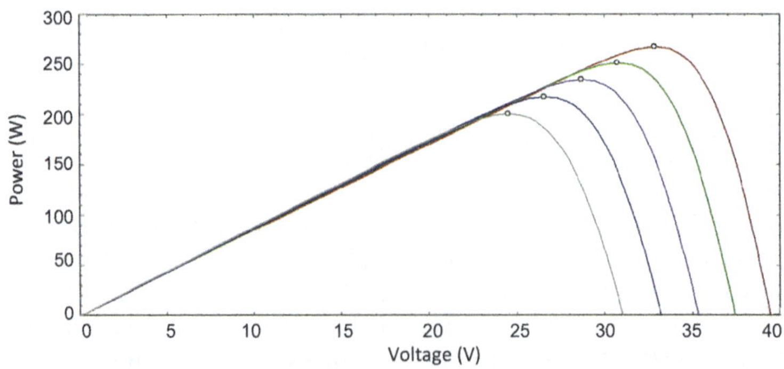

Fig. 3 Power output variation of PV with respect to temperature

As seen in (3), it is clear that the rate of change of open-circuit voltage lessen with an increment in temperature. As power yield of PV is a consequence of current and voltage so due to the increase in temperature, power gets reduced for the same solar insolation [12]. Figure 3 shows the variation of PV output power with respect to temperature.

4 Zone-Wise Solar Irradiance and Temperature Data Analysis of India

India has unnoticeable potential for solar energy applications because of its geographical position within the sunbelt and mainly due to a large number of cloudless days in most of the country. However, the optimization of site selection is a critical issue for large ratting PV power units which require a large investment. To operate a solar PV system at its highest efficiency, it needs to select the most optimal sites with favorable environmental conditions. Among all environmental factors, solar irradiance and temperature affect most to module power output and energy production. So an analysis of all these meteorological factors is required to optimize the highest propriety sites for PV sector enhancement in India [13].

India has a huge regional span, so to give each region and feature due regard India is split into six zones established upon climatic and geological features that are north zone, south zone, east zone, west zone, central zone, and northeast zone. The solar irradiance and temperature data of three major cities in each zone are studied. For analysis, data is taken from NASA POWER site, and the annual mean from years 2015 to 2018 is taken. The monthly mean solar irradiance (S. Irrad.) and temperature (Temp.) of each site of all six zones are illustrated from Tables 1, 2, 3, 4, 5, and 6 and are expressed in watt per meter square and degree Celsius, respectively.

5 Results and Discussion

There are several techniques used for the optimization of site selection for renewable energy sources. Many of these techniques require accurate models and knowledge of the system to get satisfactory results. The principal advantage of the fuzzy logic system is that it not requires knowledge of the system mathematical model [14]. The different fuzzy models are used in renewable energy for performance prediction, microgrid system analysis, hybrid system optimization, etc., [15]. In this study, we used a fuzzy logic approach for the optimization of site selection for solar PV system installation. The fundamental configuration of a fuzzy controller consists of a fuzzy interface engine, membership functions, rule base, defuzzification as arranging below (Fig. 4).

Table 1 Monthly mean data of solar irradiance and temperature for cities in North India

Month	Srinagar		Chandigarh		Lucknow	
	S. Irrad.	Temp.	S. Irrad.	Temp.	S. Irrad.	Temp.
January	0.71	-1.29	3.36	12.27	3.54	15.11
February	4.58	-0.47	4.39	16.20	4.33	21.07
March	4.90	4.41	5.67	21.79	5.88	27.44
April	5.12	8.43	6.27	27.25	6.37	32.12
May	5.24	11.67	6.71	31.50	6.58	36.04
June	5.92	16.66	5.95	32.01	6.07	36.71
July	4.98	17.37	4.84	28.32	3.95	31.28
August	5.59	17.27	4.71	26.96	3.72	28.31
September	5.09	13.88	4.52	24.75	4.50	26.96
October	3.90	7.45	4.90	21.14	5.00	24.45
November	2.30	1.86	3.63	17.02	3.94	20.08
December	2.95	-2.00	3.29	11.89	3.55	14.94

Table 2 Monthly mean data of solar irradiance and temperature for cities in South India

Month	Bangalore		Hyderabad		Madurai	
	S. Irrad.	Temp.	S. Irrad.	Temp.	S. Irrad.	Temp.
January	5.55	20.51	3.36	12.27	5.25	21.62
February	6.43	22.45	4.39	16.20	5.65	24.85
March	6.63	26.32	5.67	21.79	6.31	29.08
April	6.74	29.19	6.27	27.25	6.60	31.77
May	6.04	26.76	6.71	31.50	6.54	32.97
June	4.70	24.62	5.95	32.01	5.20	28.00
July	3.89	24.48	4.84	28.32	3.83	26.34
August	4.07	23.62	4.71	26.96	3.91	25.56
September	5.83	24.98	4.52	24.75	5.02	25.77
October	5.45	23.62	4.90	21.14	5.43	25.41
November	4.71	23.10	3.63	17.02	5.03	24.54
December	4.80	22.95	3.29	11.89	4.19	21.90

The key environmental factors affecting the optimization of solar PV site selection are solar irradiance and air temperature. The fuzzy logic controller uses two inputs, i.e., solar irradiance (S) and temperature (T); it also uses a single output variable probability/priority factor (P). Both of these input variable data for considered sites in all six zones are feed to fuzzy logic block. By evaluating this data, the fuzzy logic block generates an output that can be used to prioritize the solar PV sites and help to choose the highest priority site [16]. In this fuzzy logic controller, the span of interest

Table 3 Monthly mean data of solar irradiance and temperature for cities in Eastern India

Month	Patna		Kolkata		Bhubaneswar	
	S. Irrad.	Temp.	S. Irrad.	Temp.	S. Irrad.	Temp.
January	3.23	14.61	4.18	15.37	4.70	18.36
February	4.54	21.56	4.71	21.46	5.29	23.01
March	5.81	27.67	5.52	27.36	5.99	28.35
April	6.16	31.54	5.75	29.47	6.37	30.15
May	6.21	33.86	5.31	30.81	6.17	30.35
June	5.53	35.65	4.22	29.79	4.41	30.07
July	4.02	32.54	3.66	28.97	3.26	28.45
August	4.23	29.57	4.20	28.45	3.63	28.05
September	4.86	29.05	4.60	27.93	4.51	27.61
October	4.88	26.97	4.47	25.81	4.83	26.22
November	4.07	22.90	4.19	22.70	4.57	23.65
December	3.50	16.90	3.59	17.44	3.78	19.64

Table 4 Monthly mean data of solar irradiance and temperature for cities in Western India

Month	Jodhpur		Gandhinagar		Aurangabad	
	S. Irrad.	Temp.	S. Irrad.	Temp.	S. Irrad.	Temp.
January	4.39	17.46	4.61	21.31	4.97	21.44
February	4.94	21.51	5.26	24.89	5.39	24.98
March	6.17	26.55	6.39	28.89	6.13	28.98
April	6.83	31.86	7.04	32.80	6.74	31.97
May	7.00	35.34	7.28	35.40	6.84	33.64
June	6.35	34.87	6.01	35.15	4.80	29.18
July	5.05	30.83	3.74	29.65	3.00	25.88
August	4.82	29.49	3.61	28.06	3.29	25.44
September	5.15	28.60	4.64	27.50	5.02	25.51
October	5.19	27.79	5.53	28.88	5.56	26.78
November	4.27	22.64	4.62	25.49	4.87	24.54
December	3.98	16.30	4.21	19.87	4.60	20.45

of input and output variables is divided into five linguistic levels as VL, L, M, H, VH which stand for very low, low, medium, high, and very high, respectively.

The fuzzy rules used to make a correlation between incoming and outgoing signal, such that the output signal is express in terms of an input signal. These fuzzy rules are required to insert in a fuzzy logic controller block [17]. The fuzzy rules are set as if X then Y form. Where the condition C is composed of fuzzy inputs variable and fuzzy connectors (AND, OR, NOT), and the action Y is fuzzy output variable

Table 5 Monthly mean data of solar irradiance and temperature for cities in Central India

Month	Bhopal		Chhindwara		Raipur	
	S. Irrad.	Temp.	S. Irrad.	Temp.	S. Irrad.	Temp.
January	4.58	18.96	4.72	18.60	4.84	17.62
February	4.94	22.83	4.97	22.43	5.14	23.36
March	5.97	27.74	5.97	27.27	5.99	29.10
April	6.72	32.54	6.71	31.17	6.63	32.47
May	6.83	36.09	6.87	34.70	6.64	35.69
June	5.57	32.40	5.31	29.45	5.53	31.33
July	3.23	26.69	2.87	25.28	3.52	27.19
August	2.92	25.37	2.66	24.23	3.20	26.27
September	4.14	24.57	4.38	23.64	4.70	25.69
October	5.47	24.44	5.54	23.92	5.53	24.24
November	4.58	21.74	4.69	22.01	4.77	21.77
December	4.21	17.25	4.20	17.45	3.94	17.86

Table 6 Monthly mean data of solar irradiance and temperature for cities in Northeast India

Month	Agartala		Gangtok		Shilong	
	S. Irrad.	Temp.	S. Irrad.	Temp.	S. Irrad.	Temp.
January	4.21	15.53	4.18	5.88	4.01	12.86
February	4.81	20.76	4.32	8.02	4.63	16.02
March	5.53	25.53	4.00	10.78	5.22	19.69
April	5.29	26.91	3.77	13.10	4.50	21.98
May	4.24	27.27	3.86	15.22	3.73	23.57
June	4.11	27.98	3.69	18.04	3.62	25.27
July	4.42	28.33	3.41	18.51	3.57	25.79
August	4.64	28.24	3.69	18.31	4.36	25.51
September	4.82	28.03	3.76	17.29	3.82	24.50
October	4.37	25.40	4.18	12.57	4.31	21.00
November	4.57	22.34	3.84	9.70	4.27	17.49
December	3.72	18.69	3.32	6.22	3.59	14.52

[18]. These fuzzy rules are representing by a membership function. In the final step of defuzzification, the linguistic value of output signal, i.e., priority factor for site selection is converted into a crisp value. The fuzzy rule table with linguistic levels and AND operator is shown in Table 7. The membership function of solar irradiance (S) in blue color, temperature (T), and priority index (P) is shown by red color in Fig. 5.

Figure 6 shows the relationship between solar irradiance, temperature, and priority

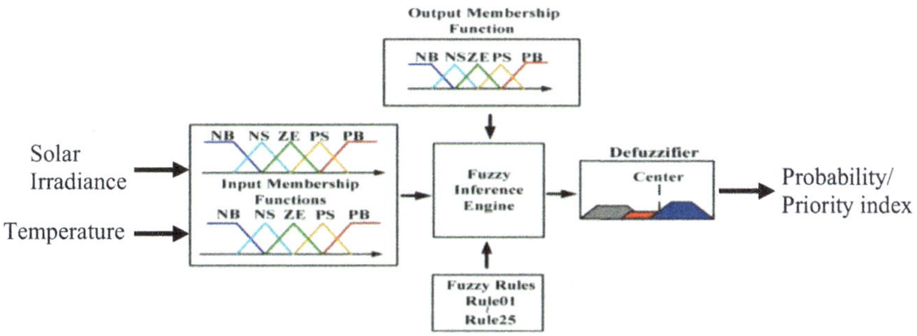

Fig. 4 Block diagram of fuzzy logic control

Table 7 Fuzzy control rules

S	T				
	VL	L	M	H	VH
VL	L	L	VL	VL	VL
L	L	L	L	VL	VL
M	M	M	M	L	L
H	VH	VH	H	H	H
VH	VH	VH	VH	H	H

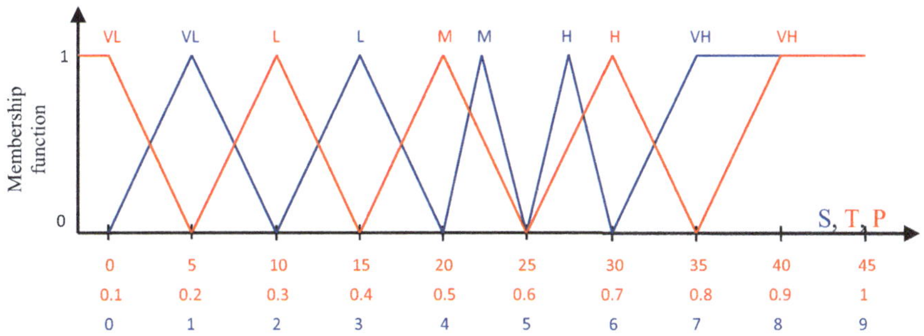

Fig. 5 Membership functions of input and output variables

factor for different values. To check the priority factor of each city, we apply solar irradiance and temperature input values as given in Tables 1, 2, 3, 4, 5, and 6 to fuzzy logic block in MATLAB. The priority factor of each three cites of six zones is analyzed.

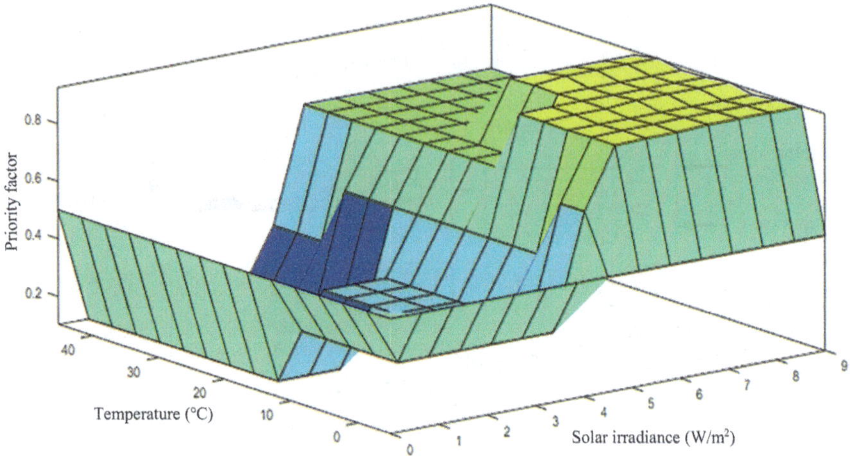

Fig. 6 3D plot of relation between solar irradiance and temperature with priority factor

6 Conclusion

The climate conditions impact PV performance, but this effect diverges from one zone to another, depending on geographical and topographical conditions. This study was conducted for solar irradiance and temperature effect on PV performance and optimization of site selection by using a fuzzy logic approach. The solar irradiance has a positive effect on PV performance; on the other hand, temperature has an inverse effect. Therefore, optimal site which has a high probability/priority factor requires higher solar irradiance and lower temperature condition. We appraised the solar irradiance and temperature data for different zones in India by considering these data as input to the fuzzy logic controller. It is seen that in north India Chandigarh has the highest priority factor, i.e., 0.66. In the South India zone, Bangalore has the highest priority factor of 0.76. In the eastern zone, Bhubaneswar has the highest priority factor, i.e., 0.68. In the western zone, Jodhpur is the highest priority site with a priority factor of 0.80. In central India, Chhindwara is the highest priority site with a priority factor of 0.72. In the north eastern zone, Agartala is the most optimal site with a priority factor of 0.60. So we can conclude that southern, western, and central India have favorable climate conditions for PV installation. Eastern below part of the northern region has also good weather condition, and northeast and upper part of the north region have less favorable climate conditions for PV installation.

References

1. Kazem, H.A., Chaichan, M.T.: Effect of environmental variables on photovoltaic performance-based on experimental studies. Int. J. Civ. Mech. Energy Sci. (IJCMES), **2**(4), 1–8 (2016)

2. Fuke, P., Yadav, A.K., Anil, I.: Performance analysis of different PV technologies under non-uniform outdoor conditions. In: 2020 International Conference on Power Electronics & IoT Applications in Renewable Energy and its Control (PARC), pp. 37–41 (2020)
3. Emetere, M.E., Akinyemi, M.L.: Weather effect on photovoltaic module adaptation in coastal areas. Int. J. Renew. Energy Res. **5**(3), 821–825 (2015)
4. Khan, G., Rathi, S.: Optimal site selection for solar PV power plant in an Indian state using geographical information system (GIS). Int. J. Emerg. Eng. Res. Technol. **2**(7), 260–266 (2014)
5. Karafil, A., Ozbay, H., Kesler, M.: Temperature and solar radiation effects on photovoltaic panel power. J. New Results Sci. **5**, 48–58 (2016)
6. Solangi, Y.A., Shah, S.A.A., Zameer, H., Ikram, M., Saracoglu, B.O.: Assessing the solar PV power project site selection in Pakistan: Based on AHP-fuzzy VIKOR approach. Environ. Sci. Pollut. Res. **26**(29), 30286–30302 (2019)
7. Badran, Q., Abdulhadi, E., Sarhan, H.: Solar sites assessment in Jordan using fuzzy logic. Appl. Solar Energy **44**(3), 211–217 (2008)
8. Moharil, R.M., Kulkarni, P.S.: Reliability analysis of solar photovoltaic system using hourly mean solar radiation data. Sol. Energy **84**(4), 691–702 (2010)
9. Sharma, N.K., Tiwari, P.K., Sood, Y.R.: Solar energy in India: Strategies, policies, perspectives and future potential. Renew. Sustain. Energy Rev. **16**(1), 933–941 (2012)
10. Al-Bashir, A., Al-Dweri, M., Al-Ghandoor, A., Hammad, B., Al-Kouz, W.: Analysis of effects of solar irradiance, cell temperature and wind speed on photovoltaic systems performance. Int. J. Energy Econ. Policy **10**(1), 353 (2020)
11. Pathak, P.K., Yadav, A.K., Alvi, P.A.: Advanced solar MPPT techniques under uniform and non-uniform irradiance: a comprehensive review. Solar Energy Eng. **142**(4), 040801–040826 (2020)
12. Fesharaki, V., Dehghani, J.M., Fesharaki, J.J., Tavasoli, H.: The effect of temperature on photovoltaic cell efficiency. In: Proceedings of the 1st International Conference on Emerging Trends in Energy Conservation–ETEC, Tehran, Iran, pp. 20–21 (2011)
13. Parishwad, G.V., Bhardwaj, R.K., Nema, V.K.: Estimation of hourly solar radiation for India. Renew. Energy **12**(3), 303–313 (1997)
14. Suganthi, L., Iniyan, S., Samuel, A.A.: Applications of fuzzy logic in renewable energy systems—a review. Renew. Sustain. Energy Rev. **48**, 585–607 (2015)
15. Serrano, N., Seraji, H.: Landing site selection using fuzzy rule-based reasoning. In: Proceedings of IEEE International Conference on Robotics and Automation, pp. 4899–4904 (2007)
16. Howard, A., Seraji, H.: A fuzzy rule-based safety index for landing site risk assessment. In: Proceedings of the 5th IEEE Biannual World Automation Congress, vol. 14, pp. 579–584 (2002)
17. Dhunny, A.Z., Doorga, J.R.S., Allam, Z., Lollchund, M.R., Boojhawon, R.: Identification of optimal wind, solar and hybrid wind-solar farming sites using fuzzy logic modelling. Energy **188**, 116056 (2019)
18. Errouha, M., Derouich, A., Motahhir, S., Zamzoum, O., El Ouanjli, N., El Ghzizal, A.: Optimization and control of water pumping PV systems using fuzzy logic controller. Energy Rep. **5**, 853–865 (2019)

Comparative Study of Two Teaching Methodologies Using Fuzzy Set Theory

T. P. Singh, Manoj Sahni, Ashnil Mandaliya, and Radhika Nanda

Abstract In this paper, a comparative study is done for two different teaching methodologies and a mathematical model is suggested. The teaching methodologies are reframed according to the current students demand and to improve the teaching techniques. For this study, two first year Mathematics subjects (I and II) are considered, in which Mathematics II was taught using classical technique, and the second one Mathematics I were taught using innovative teaching methodologies. For their comparison, a questionnaire was framed based on the satisfaction derived by the students studying the subject using two different techniques. It was filled by them answering various questions in the form of linguistic 2-tuple, which focused on the ability of the student in regards to conceptual learning, the application of the learned subject in their core field, and the enhancement of the knowledge based on the study of any particular subject. It was found that teaching done through innovative teaching method is more effective for the students rather than classical method which have been considered for comparison.

Keywords Linguistic 2-tuples · Fuzzy sets · Aggregation · Euclidean distance · Fuzzy divergence

1 Introduction

For the creation of innovative talent and for the future of each and every student, it is very important to develop a quality education system and for that it is desirable to create innovative as well as effective teaching methodologies. Time to time evaluation of quality of teaching is one of the tools, by which any University can refined

T. P. Singh · M. Sahni
School of Technology, Pandit Deendayal Petroleum University, Gandhinagar, Gujarat, India

A. Mandaliya (✉) · R. Nanda
School of Liberal Studies, Pandit Deendayal Petroleum University, Gandhinagar, Gujarat, India
e-mail: ashnilman@gmail.com

M. Sahni et al. (eds.), *Mathematical Modeling, Computational Intelligence Techniques and Renewable Energy*, Advances in Intelligent Systems and Computing 1287,
https://doi.org/10.1007/978-981-15-9953-8_44

their syllabus, teaching methods, and other related things from the students' perspective. Different students have different capability of understanding, thinking skill, and have different ideas for solution of a common problem. So, teacher should acquire those effective methods which are useful to all categories of students. To evaluate the teacher's ability or for the monitoring of the effectiveness of their teaching methodology, one of the easy method is to collect data from the students based on the basis of their understanding of the subject, their interest in the class which is clearly dependent on the teaching technique, how they are applying their class knowledge in the practical life and many more using questionnaires. But all this collected information or evaluated factors are not always very precise, in fact, it contains vague terms, and in that case, the data cannot be represented by classical sets.

To tackle such problems, Zadeh [1] introduced the concept of fuzzy sets in 1965. The concept is mainly based on the membership value of any particular factor within the set. Those values are not precise but are vague or fuzzy and vary from person to person thinking and decision-making skills. In case of development of effective education system, students' perspective is very important but the information collected from different students are also vague because of differing ability of understanding. In all decision-making problems in our real-life Zadeh's fuzzy set theory plays very important role. So, the need arose to extend the fuzzy set theory according to various situations. With the development of fuzzy sets, other generalizations are proposed such as intuitionistic fuzzy sets, interval-valued fuzzy sets, ortho-pair fuzzy sets, pythagorean fuzzy sets, and many more. A book by Zimmermann [2] shows the applications of fuzzy set theory in diverse fields.

In fact, one can observe that almost every day we have to face some situations, where we have to make some decision. In almost every decision-making problem, fuzzy theory is applied to accommodate the vagueness in the data. Fuzzy decision making is one of the popular research problems in recent years (see for instance, [3, 6, 7, 9–14], etc.). These decisions are made by different people working in different area in different context. For all decision-making problems, collective information is necessary and the solution of any decision-making problem, information are collected from the concerned persons, after that from those information, a concrete decision is made. This whole process is known as information aggregation. Due to different thinking skills and different thought process, one can find deviation of arguments in various decision-making situations. For the calculation of deviations, a distance operator or distance measure is used to find out the similarity between the collected information. Similarity measures are defined using various distance operators and its being used in different decision-making problems, such as in share market, pattern recognition, career determination, medical services, risk analysis, etc. In all these uncertain information processing, Shannon [15] entropy plays an important role, which was further generalized by Verma et al. [8]. One can find some new information measures and entropy in the papers [13, 14]. The concept of distances was first developed in fuzzy points of view by Dubois and Prade [4] in 1984 and for intuitionistic fuzzy sets, Szmidt et al. [5] derived an expression for distances. It was used in various applications of real-life situations. A lot of research papers are available in the literature using the concept of distances. Many research articles are published in education sectors also to improve the education system by finding loopholes of the system using fuzzy aggregation and distance operators [9–12].

In this paper, also, we have compared two teaching methods, i.e., classical and innovative teaching methods, based on the collection of response of the students using the concept of distance measure. The results are also verified by fuzzy symmetric divergence given by Verma et al. [8]. Using the concept of distances, it is seen that the innovative teaching method is more effective teaching method as compared to classical teaching method.

2 Preliminaries

2.1 *Fuzzy Sets [1]*

If X is a set whose elements are denoted by x, a fuzzy set $\bar{A}$ in X can be defined as a set of ordered pair $\bar{A} = \{(x, \mu(x))|x \in X\}$, where $\mu(x)$ is called the membership value of x in $\bar{A}$ which is a mapping from X to $[0, 1]$.

2.2 *Distance Measure [4]*

Distance measure is a kind of similarity measure, which measures the similarity between two objects. Considering two sets A and B as two arbitrary fuzzy sets having membership function $\mu_{\bar{A}}(x)$ and $\mu_{\bar{B}}(x)$ respectively, then the Euclidean and normalized Euclidean distance is defined as follows:

Euclidian Distance. The Euclidean distance is defined as

$$(\bar{A}, \bar{B}) = \sqrt{\sum_{i=1}^{n}\left(\mu_{\bar{A}}(x_i) - \mu_{\bar{B}}(x_i)\right)^2}.$$

Normalized Euclidian Distance. The normalized Euclidean distance is defined as

$$q(\bar{A}, \bar{B}) = \sqrt{\frac{1}{n}\sum_{i=1}^{n}\left(\mu_{\bar{A}}(x_i) - \mu_{\bar{B}}(x_i)\right)^2}$$

2.3 Aggregation Operator [2]

Let w be a n dimensional weighing vector, $w = \left[w_1, w_{2,} \ldots, w_n\right]$, such that

1. $w_i \in [0, 1]$
2. $\sum_i w_i = 1, \quad i = 1 \ldots n.$

A mapping $f_{\mathrm{wm}} : \mathbb{R}^n \to \mathbb{R}$ is a weighted mean operator of dimension n if

$$f_{\mathrm{wm}}(x_1, x_2 \ldots x_n) = \sum_i w_i x_i$$

2.4 Fuzzy Divergence [8]

The fuzzy divergence operator is defined as

$$D(A|B) = \frac{1}{n} \sum_{i=1}^{n} \left[\mu_A(x_i) \log \frac{\mu_A(x_i)}{\frac{\mu_A(x_i)+\mu_B(x_i)}{2}} + \nu_A(x_i) \log \frac{\nu_A(x_i)}{\frac{\nu_A(x_i)+\nu_B(x_i)}{2}} \right]$$

2.5 Symmetric Fuzzy Divergence [8]

The symmetric fuzzy divergence operator is defined as

$$D(A; B) = D(A|B) + D(B|A)$$

2.6 Generalized Symmetric Fuzzy Divergence [8]

The generalized symmetric fuzzy divergence operator is defined as

$$D_\lambda(A; B) = \frac{1}{n} \sum_{i=1}^{n} \left[\mu_A(x_i) \log \frac{\mu_A(x_i)}{\lambda \mu_A(x_i) + (1 - \lambda) \mu_B(x_i)} \right.$$

$$+\nu_A(x_i)\log\frac{\nu_A(x_i)}{\lambda\nu_A(x_i)+(1-\lambda)\nu_B(x_i)}\Bigg]$$

$$+\frac{1}{n}\sum_{i=1}^{n}\Bigg[\mu_B(x_i)\log\frac{\mu_B(x_i)}{\lambda\mu_B(x_i)+(1-\lambda)\mu_A(x_i)}$$

$$+\nu_B(x_i)\log\frac{\nu_B(x_i)}{\lambda\nu_B(x_i)+(1-\lambda)\nu_A(x_i)}\Bigg]$$

3 Work Done

In this paper, a questionnaire having six questions is formed and circulated to 69 students out of which 10 students were picked for evaluation whose CGPA are more than 8. The questions responses are divided into three categories (C1, C2, C3) and each category contains two questions (Q1, Q2), (Q3, Q4) and (Q5, Q6). Responses of the two questions (Q7, Q8), containing information regarding their attendance, and grades, are collected from their faculty members for tempering of the data. The questionnaire responses are collected in terms of linguistic 2-tuple; i.e., firstly, the responses scale is from 1 to 5, having the linguistic labels, very good, good, average, poor, and very poor, and a scale from -0.5 to 0.5 is further used to assess the appropriate choice. The responses collected from the students have been fuzzified and depicted in Tables 1 and 2 for subjects Mathematics I and Mathematics II, respectively.

3.1 The Set of Questionnaires that is Being Asked to the Students are Given Below

1. If asked about after a certain period of time, would you be able to recall and explain the topic?
2. How often did the method of teaching solve the problems you encountered?
3. How often did you feel the need to mug up the concepts, even after being taught in the class?
4. How often could you relate the concepts taught to you to different topics?
5. How often do you think you needed a more hands-on/practical approach to the topics taught in class?
6. How often do you think you figured out the real-life application of the topics, if they were not explained in the class?

The values of the first two questions (Q1, Q2) have been aggregated to form a single value for the first category C1. Similarly, category C2 has the aggregates of the two questions, i.e., Q3 and Q4, and category C3 has the aggregates of the two

Table 1 Converted responses from the students and their faculties of Ist year for mathematics I

	S1	S2	S3	S4	S5	S6	S7	S8	S9	S10
Q1	0.8	0.5	0.95	0.8	0.95	0.95	0.5	0.95	0.8	0.24
Q2	0.5	0.8	0.5	0.8	0.95	0.5	0.8	0.5	0.8	0.62
Q3	0.2	0.8	0.5	0.95	0.8	0.5	0.5	0.2	0.5	0.5
Q4	0.5	0.8	0.05	0.5	0.8	0.8	0.8	0.8	0.7	0.8
Q5	0.5	0.8	0.05	0.5	0.95	0.84	0.82	0.7	0.38	0.7
Q6	0.18	0.05	0.95	0.2	0.8	0.3	0.5	0.05	0.5	0.7
Q7	0.9	0.95	0.95	0.9	0.9	0.95	0.9	0.95	0.9	0.9
Q8	0.8	0.95	0.95	0.95	0.8	0.95	0.8	0.9	0.95	0.95

Table 2 Converted responses from the students and their faculties of Ist year for mathematics II

	S1	S2	S3	S4	S5	S6	S7	S8	S9	S10
Q1	0.8	0.2	0.8	0.5	0.95	1	0.2	0.95	0.88	0.5
Q2	0.5	0.5	0.5	0.5	0.5	0.5	0.8	0.5	0.5	0.7
Q3	0.2	0.8	0.5	0.8	0.95	0.5	0.8	0.2	0.5	0.3
Q4	0.5	0.8	0.5	0.5	0.8	0.8	0.8	0.95	0.76	0.42
Q5	0.5	0.5	0.5	0.5	0.5	0.84	0.8	0.7	0.8	0.8
Q6	0.2	0.5	0.95	0.2	0.95	0.5	0.2	0.2	0.5	0.5
Q7	0.9	0.95	0.95	0.9	0.9	0.95	0.9	0.95	0.9	0.9
Q8	0.8	0.95	0.95	0.95	0.8	0.95	0.8	0.9	0.95	0.95

questions, i.e., Q5 and Q6, and category C4 having means of last two questions, i.e., Q7 and Q8. The category 4 has been used for tempering the values of the categories C1, C2, and C3. The average value of category 4 is multiplied with each of the average value of categories 1, 2, and 3 and is depicted in Tables 3 and 4 for both Mathematics I and Mathematics II, respectively.

We presume that the weightage of each category can be different for Colleges and Universities across the globe. So, each category has been assigned weights corresponding to their outcomes, i.e., conceptual learning (O1), the ability to apply the selected mathematics subjects in their area of specialization (O2), and the enhancement of knowledge (O3), depending on the manner, the questions relate to the outcomes. The weight table for outcome against category has been shown in Table 5.

Table 3 Aggregation and tempering of responses using category 4 for Mathematics I

	S1	S2	S3	S4	S5	S6	S7	S8	S9	S10
C1	0.5525	0.5525	0.4250	0.6800	0.8075	0.6163	0.5525	0.6163	0.6800	0.3655
C2	0.2975	0.6800	0.2338	0.6163	0.6800	0.5525	0.5525	0.4250	0.5100	0.5525
C3	0.2890	0.3613	0.4250	0.2975	0.7438	0.4845	0.5610	0.3188	0.3740	0.5950

Table 4 Aggregation and tempering of responses using category 4 for Mathematics II

	S1	S2	S3	S4	S5	S6	S7	S8	S9	S10
C1	0.5525	0.2975	0.3613	0.4250	0.6163	0.6375	0.4250	0.6163	0.5865	0.5100
C2	0.2975	0.6800	0.2338	0.5525	0.7438	0.5525	0.6800	0.4888	0.5355	0.3060
C3	0.2975	0.4250	0.6163	0.2975	0.6163	0.5695	0.4250	0.3825	0.5525	0.5525

Table 5 Weight table for expected outcome against category

	C1	C2	C3
O1	0.6	0.3	0.1
O2	0.3	0.6	0.3
O3	0.1	0.1	0.6

The weights are being assigned in such a manner so that the total sum across row and column will remain as 1.

The following Tables 6 and 7 are created by the matrix multiplication of Tables 3 and 4 with Table 5, respectively, so as to map the students to the outcome. The aggregate of the three categories are also shown to see overall effect.

In Table 8, the ideal case is represented as 1, i.e., a perfect scenario, so that the distance from a single number is easily comparable for each element. The optimum values in Table 8 are the maximum values of each corresponding element in Tables 6 and 7, i.e., max [6, 7].

The distance is calculated using the Euclidian distance measure which is defined in preliminaries, for the two subjects Mathematics I and Mathematics II of first year each with the ideal and optimum set of values in Table 9. It is seen that the distance

Table 6 Multiplication of weight Table 5 with Table 3

	S1	S2	S3	S4	S5	S6	S7	S8	S9	S10
O1	0.4478	0.5079	0.4059	0.5589	0.7757	0.5704	0.5551	0.5079	0.5712	0.4531
O2	0.3732	0.6099	0.3103	0.6035	0.7247	0.5648	0.5534	0.4718	0.5474	0.5007
O3	0.3179	0.4760	0.3676	0.4314	0.7310	0.5181	0.5576	0.3804	0.4454	0.5593
Average	0.3797	0.5313	0.3613	0.5313	0.7438	0.5511	0.5553	0.4533	0.5213	0.5043

Table 7 Multiplication of weight Table 5 with Table 4

	S1	S2	S3	S4	S5	S6	S7	S8	S9	S10
O1	0.4505	0.3740	0.4250	0.3995	0.6290	0.6086	0.4505	0.5334	0.5712	0.5024
O2	0.3740	0.5398	0.3103	0.4888	0.6928	0.5797	0.5780	0.5164	0.5525	0.3919
O3	0.3230	0.4888	0.4760	0.3868	0.6545	0.5712	0.5015	0.4378	0.5508	0.4743
Average	0.3825	0.4675	0.4038	0.4250	0.6588	0.5865	0.5100	0.4958	0.5582	0.4562

Table 8 Ideal and optimum values for Tables 6 and 7

	S1	S2	S3	S4	S5	S6	S7	S8	S9	S10
Ideal	1	1	1	1	1	1	1	1	1	1
Optimum										
O1	0.4505	0.5079	0.4250	0.5589	0.7757	0.6086	0.5551	0.5334	0.5712	0.5024
O2	0.3740	0.6099	0.3103	0.6035	0.7247	0.5797	0.5780	0.5164	0.5525	0.5007
O3	0.3230	0.4888	0.4760	0.4314	0.7310	0.5712	0.5576	0.4378	0.5508	0.5593
Average	0.3825	0.5313	0.4038	0.5313	0.7438	0.5865	0.5553	0.4958	0.5582	0.5043

Table 9 Euclidean distance measure form ideal and optimum values

	Ideal		Optimum	
	Mathematics—I	Mathematics—II	Mathematics—I	Mathematics—II
O1	1.5009	1.6208	0.0701	0.2752
O2	1.5415	1.6106	0.0534	0.1759
O3	1.6879	1.6481	0.1708	0.1350
Average	1.5719	1.6198	0.0789	0.1642

calculated in both cases, i.e., from the ideal and optimum case is less for Mathematics I as compared to Mathematics II. Using the optimum value, a significant difference is observed when we use optimum method. The lesser distance calculated proves that the teaching methodology applied in teaching Mathematics I is better as compared to the teaching methodology applied in Mathematics II.

The same results is also seen when we use the generalized symmetric fuzzy divergence measure [8]. The generalized symmetric fuzzy divergence measure has been applied to the responses of students from Tables 6 and 7 for all the values of λ from 0 to 1 with a step of 0.1 (the values ranging from 0 to 1) neglecting the values of 0 and 1 for obvious reasons.

It is seen from Table 10 that the fuzzy divergence measure is more for Mathematics II as compared to Mathematics I which implies that the innovative teaching methods are more appropriate in the current scenario of teaching. It helps to develop the better remembering capacity of the students as well as help them to understand the topic better as compared to the classical method.

4 Conclusion

The following points have been concluded in this research paper:

1. Comparison of two teaching methodologies is done.
2. The responses were recorded from the students and their teachers, with respect to their attendance and their grades, which were then converted into fuzzy values.

Table 10 Generalized symmetric fuzzy divergence measure

	C1		C2		C3		Overall	
λ	Maths I	Maths II	Maths I	Maths II	Maths I	Maths II	Maths I	Maths II
0.1	0.007012	0.113658	0.004029	0.044461	0.041733	0.027238	0.018192	0.044582
0.2	0.005539	0.089611	0.003184	0.035100	0.032949	0.021509	0.014372	0.035200
0.3	0.004241	0.068543	0.002437	0.026863	0.025218	0.016464	0.011003	0.026941
0.4	0.003116	0.050368	0.001791	0.019738	0.018529	0.012096	0.008084	0.019795
0.5	0.002164	0.035025	0.001244	0.013714	0.012873	0.008403	0.005614	0.013753
0.6	0.001385	0.022473	0.000796	0.008786	0.008246	0.005382	0.003594	0.008809
0.7	0.000779	0.012687	0.000448	0.004949	0.004645	0.00303	0.002022	0.004961
0.8	0.000346	0.005666	0.000199	0.002204	0.002068	0.001348	0.000899	0.002209
0.9	0.000087	0.001425	0.000049	0.000552	0.000518	0.000338	0.000225	0.000553

3. The Euclidean distance measure is used to calculate the distances in Table 9 from both optimum and ideal distances, and it was further verified with generalized symmetric fuzzy divergence measure.
4. We can conclude that using innovative techniques, students can remember the concepts and retain their knowledge related to particular topic for a longer period of time.
5. The distance from an ideal selection (i.e. maximum limit 1) between the two methods do not have much difference but the distance from an optimum selection shows a different story altogether and have a much larger difference between the two methods which further implies that the innovative teaching method is more effective than the classical method.

References

1. Zadeh, L.A.: Fuzzy sets. Inf. Control **8**(3), 338–353 (1965)
2. Zimmermann, H.J.: Fuzzy Set Theory-and Its Applications, 4th edn. (2001)
3. Merigó, J.M., Casanovas, M., Liu, P.: Decision making with fuzzy induced heavy ordered weighted averaging operators. Int. J. Fuzzy Syst. **16**(3) (2014)
4. Dubois, D., Prade, H.: On distances between fuzzy points and their use for plausible reasoning. In: Proceedings of IEEE International Conference on Cybernetics & Society, Bombay, New Delhi, 30 Dec 1983–7 Jan 1984, pp. 300–303 (1984)
5. Szmidt, E., Kacprzyk, J.: Distances between intuitionistic fuzzy sets. Fuzzy Sets Syst. **114**, 505–518 (2000)
6. Mandaliya, A., Sahni, M., Verma, R.: Career Determination using Information Theoretical Measure and It's Comparison with Distances in IFS and PFS. Int. J. Math. Models Methods Appl. Sci. **13**, 28–34 (2019)
7. Sahni, M., Mandaliya, A., Sahni, R.: Evaluation of teachers' performance based on students' feedback using aggregator operator. WSEAS Trans. Math. **18**, 85–90 (2019)
8. Verma, R., Sharma, B.D.: On generalized intuitionistic fuzzy divergence (relative information) and their properties. J. Uncertain Syst. **6**(4), 308–320 (2012)

9. Oskouei, N.N., Saemian, F.: Analyzing and comparing the effects of two teaching methods, student-centered versus teacher-centered, on the learning of biostatistics at SBMU. J. Paramedical Sci. (JPS) **3**(4) (2012)
10. Moazami, F., Bahrampour, E., Azar, M.R., Jahedi, F., Moattari, M.: Comparing two methods of education (virtual versus traditional) on learning of Iranian dental students: a post-test only design study. BMC Med. Educ. **14**, 45 (2014)
11. Patil, S.A., Prasad, S.R.: Innovative methods of teaching & learning electronics engineering. J. Eng. Educ. Transform. (2015)
12. Napitupulu, D., Rahim, R., Abdullah, D., Setiawan, M.I., Abdillah, L.A., Ahmar, A.S., Simarmata, J., Hidayat, R., Nurdiyanto, H., Pranolo, A.: Analysis of student satisfaction toward quality of service facility. IOP Conf. Series J. Phys. Conf. Series **954**, 012019 (2018)
13. Bhandari, D., Pal, N.R.: Some new information measures for fuzzy sets. Inf. Sci. **67**, 209–228 (1993)
14. Wei, P., Ye, J.: Improved intuitionistic fuzzy cross-entropy and its application to pattern recognitions. In: International Conference on Intelligent Systems and Knowledge Engineering, pp. 114–116 (2010)
15. Lin, J.: Divergence measures based on Shannon entropy. IEEE Trans. Inf. Theory **37**, 145–151 (1991)

A Generalized Solution Approach to Matrix Games with 2-Tuple Linguistic Payoffs

Rajkumar Verma, Manjit Singh, and José M. Merigó

Abstract In real-world, game theory has been found great success in solving competitive decision-making problems more effectively. This paper develops a generalized approach to solve matrix games with payoffs represented by 2-tuple linguistic values. We define a pair of non-linear optimization models to solve this class of games, which include an attitudinal parameter λ. By considering different values of parameter λ, we can obtain different sets of optimal strategies corresponding to players as per our requirement. Finally, a numerical example is given to show the flexibility and applicability of the developed method.

Keywords Matrix games · Optimal strategies · Linguistic variable · 2-Tuple linguistic value

1 Introduction

Game theory provides an efficient tool to solve competitive decision problems by considering strategic interaction among rational decision-makers. After the pioneering work of Neuman and Morgenstern [1], a wide class of results have been obtained related to game theory with crisp payoffs and applied successfully in several areas, including economics, social sciences, business model, management sciences, and engineering [2–5]. In many real-life situations, it becomes challenging to get payoffs

R. Verma (✉) · J. M. Merigó
Department of Management Control and Information Systems,
University of Chile, Av. Diagonal Paraguay 257, 8330015 Santiago, Chile
e-mail: rverma@fen.uchile.cl; rkver83@gmail.com

M. Singh
Department of Mathematics, DCR University of Science and Technology,
Murthal, Sonepat 131039, India

J. M. Merigó
Faculty of Engineering and Information Technology, School of Information, Systems and Modeling, University of Technology Sydney, Sydney, Australia

M. Sahni et al. (eds.), *Mathematical Modeling, Computational Intelligence Techniques and Renewable Energy*, Advances in Intelligent Systems and Computing 1287,
https://doi.org/10.1007/978-981-15-9953-8_45

of a matrix game in terms of crisp values due to the presence of uncertainty or insufficient information about the problem data. To overcome this problem, fuzzy numbers [6] have been used to represent payoffs of matrix games. The fuzzy games were firstly studied by Aubin [7] and Butnariu [8]. Later, Campos [9] formulated some solution approaches for matrix games with fuzzy payoffs. Sakawa and Nishizaki [10] made a study on multi-objective zero-sum matrix games under fuzzy environment. Maeda [11] defined matrix equilibrium to fuzzy matrix games by using the fuzzy max order. Bector et al. [12] developed a fuzzy linear programming duality based approach to solving matrix games with fuzzy goals. Cevikel and Ahlatcioglu [13] developed two models for studying two-person zero-sum matrix games with fuzzy payoffs and fuzzy goals. Further, by using the duality theory of the fuzzy relation approach, Vijay et al. [14] presented a generalized model to study matrix games with fuzzy goals and fuzzy payoffs.

There are many situations in which the experts may feel more comfortable using linguistic variables to express their assessment information rather than numerical ones [15]. In 1975, Zadeh [16] proposed the concept of a linguistic variable (LV) and applied it to fuzzy reasoning. Over the last decades, a lot of work has been reported on linguistic variables and their applications in several areas [17–22]. In order to avoid the loss and distortion of information during the linguistic information process, Herrera et al. [23] proposed a 2-tuple linguistic representation model, which is a useful tool for handling decision-making problems with qualitative information. Because aggregation operators always play an essential role in the decision-making process, researchers have proposed many aggregation operators to aggregate 2-tuple linguistic information [24–29]. In 2014, Wei and Zhao [30] proposed some generalized probabilistic aggregation operators for aggregating 2-tuple linguistic information and discussed their application in decision making.

Recently, Singh et al. [31] were studied a two-player constant-sum matrix game with payoffs represented by 2-tuple linguistic values. They also developed a linguistic linear programming approach to solve such a class of games. It is worth mentioning that the solution method proposed by Singh et al. [31] is rigid in nature, which does not provide any flexibility to the user for considering different situations. It is also unable to recognize the attitudinal character of the players in the solution process. Therefore, the main objective of this work is to develop a new flexible approach to solve matrix games with 2-tuple linguistic payoffs. For doing so, the work uses the generalized probabilistic 2-tuple average (GP-2TA) operator to calculate the expected value of the game and formulates a pair of non-linear programming models with a flexible parameter to solve this class of games. Paper also gives a numerical example to illustrate the flexibility of the developed method.

The rest of the paper is organized as follows: Sect. 2 presents some preliminary results on LVs, 2TLVs, and the matrix game with 2-tuple linguistic payoffs. Section 3 The non-linear programming models are proposed for solution of the matrix games with payoffs represented by 2TLVs. In Sect. 4 a numerical example is considered to demonstrate the flexibility and effectiveness of the developed approach. Section 5 concludes the work.

2 Preliminaries

In this section, we briefly review some basic results related to LVs, 2TLVs, and the matrix game with 2-tuple linguistic payoffs , which will be used for further development.

2.1 *Linguistic Variable*

Zadeh [16] introduced the fuzzy linguistic approach to deal with qualitative information in real-world decision making situations. It represents qualitative aspects in term of linguistic values by means of LVs. Herrera and Martinez [23] defined the LVs as follows.

Definition 1 Let $\mathbb{P} = \{\mathfrak{p}_i \mid i = 0, 1, \ldots, t\}$ be a totally ordered discrete linguistic term set (LTS) with the odd cardinality, where and t is a positive integer. Any level $\mathfrak{p}_i$ represents a possible value for a linguistic variable. The LTS $\mathbb{P}$ should satisfy the following properties:

(i) Order relation: If $\mathfrak{p}_i \leq \mathfrak{p}_j \Leftrightarrow i \leq j$
(ii) Negation operator: $neg\,(\mathfrak{p}_i) = \mathfrak{p}_{t-i}$
(iii) Maximum operator: $\max\left(\mathfrak{p}_i, \mathfrak{p}_j\right) = \mathfrak{p}_i \Leftrightarrow i \geq j$
(iv) Minimum operator: $\min\left(\mathfrak{p}_i, \mathfrak{p}_j\right) = \mathfrak{p}_i \Leftrightarrow i \leq j$.

For example, a set of six linguistic terms can be defined as:

$$\mathbb{P} = \left\{ \begin{array}{l} \mathfrak{p}_0 = \text{extremely poor (EP)}, \mathfrak{p}_1 = \text{very poor (VP)}, \mathfrak{p}_2 = \text{poor (P)}, \mathfrak{p}_3 = \text{ medimu (M)}, \\ \mathfrak{p}_4 = \text{good (G)}, \mathfrak{p}_5 = \text{very good (VG)}, \mathfrak{p}_6 = \text{extremely good (EG)}, \end{array} \right\}$$

To preserve all the given information, Xu [19] extended the the discrete term set $\mathbb{P}$ to a continuous term set as $\mathbb{P}_{[0,t]} = \{\mathfrak{p}_i | \mathfrak{p}_0 \leq \mathfrak{p}_i \leq \mathfrak{p}_t, i \in [0, t]\}$.

2.2 *2-Tuple Linguistic Representation Model*

The 2-tuple linguistic representation model represents the linguistic assessment information by means of a 2-tuple $(\mathfrak{p}_i, \alpha_i)$, where $\mathfrak{p}_i$ is a LT from predefined LTS $\mathbb{P}$ and α_i is a numerical value (symbolic translation), and $\alpha_i \in [-0.5, 0.5)$.

Definition 2 [23] Let $\mathbb{P} = \{\mathfrak{p}_i \mid i = 0, 1, \ldots, t\}$ be a LTS and $\eta \in [0, t]$ be a real number which denotes the symbolic aggregation result. The 2-TLV that expresses the equivalent information with η can be obtained by the following function

$$\Delta : [0, t] \rightarrow \mathbb{P} \times [-0.5, 0.5),$$

where

$$\Delta(\eta) = (\mathfrak{p}_i, \alpha_i), \text{ such that } \begin{cases} i = \text{round}(\eta) \\ \alpha_i = \eta - i, \quad \alpha_i \in [-0.5, 0.5) \end{cases}$$

and round$(\cdot)$ is the usual round operation that assigns the closest integer number $i = \{0, 1, \ldots, t\}$ to η.

Definition 3 [23] Let $\mathbb{P} = \{\mathfrak{p}_i \mid i = 0, 1, \ldots, t\}$ be a LTS and $(\mathfrak{p}_i, \alpha_i)$ be a 2-TLV. Then, there is a function Δ^{-1}, defined by

$$\Delta^{-1} : \mathbb{P} \times [-0.5, 0.5) \rightarrow [0, t],$$
$$\Delta^{-1}(\mathfrak{p}_i, \alpha_i) = (i + \alpha_i) = \eta,$$

which converts a 2-TLV $(\mathfrak{p}_i, \alpha_i)$ into an equivalent numerical value $\eta \in [0, t]$.

Definition 4 [23] Let $(\mathfrak{p}_i, \alpha_i)$ be a 2-TLV and $\mathbb{P} = \{\mathfrak{p}_i \mid i = 0, 1, \ldots, t\}$. The negation operator for a 2-tuple is defined as follows :

$$neg\,(\mathfrak{p}_i, \alpha_i) = \Delta\left(t - \Delta^{-1}(\mathfrak{p}_i, \alpha_i)\right).$$

Remark 1 Based on Def. 3 and Def. 4, we can see that a linguistic term $\mathfrak{p}_d \in \mathbb{P}$ can be expressed by the 2-tuple $(\mathfrak{p}_d, 0)$.

Definition 5 [23] Let $(\mathfrak{p}_l, \alpha_m)$ and $(\mathfrak{p}_k, \alpha_n)$ be two 2-tuple linguistic values (2-TLVs). Then the lexicographic ordering between two 2-TLVs is defined as

(i) If $l < k$, then $(\mathfrak{p}_l, \alpha_m)$ is smaller than $(\mathfrak{p}_k, \alpha_n)$.
(ii) If $l = m$, then

   (a) if $\alpha_m = \alpha_n$, then $(\mathfrak{p}_l, \alpha_m) = (\mathfrak{p}_k, \alpha_n)$ and represent the same information,
   (b) if $\alpha_m < \alpha_n$, then $(\mathfrak{p}_l, \alpha_m) < (\mathfrak{p}_k, \alpha_n)$,
   (c) if $\alpha_m > \alpha_n$, then $(\mathfrak{p}_l, \alpha_m) > (\mathfrak{p}_k, \alpha_n)$.

In the following, we denote the 2-TLV by $\kappa_i = (\mathfrak{p}_i, \alpha_i)$ and the set of all 2-TLVs defined in LTS $\mathbb{P}$ by Ω. In 2014, Wei and Zhao [30] proposed the generalized probabilistic 2-tuple average (GP-2TA) operator as follows:

Definition 6 [30] Let $\kappa_i = (\mathfrak{p}_i, \alpha_i)$; $(i = 1, 2, \ldots, n)$ be a collection of n 2-TLVs and $GP2 - TA : \Omega^n \rightarrow \Omega$, if

$$GP2 - TA\,(\kappa_1, \kappa_2, \ldots, \kappa_n,) = \Delta\left(\left(\sum_{i=1}^{n} \gamma_i \left(\Delta^{-1}(\mathfrak{p}_i, \alpha_i)\right)^{\lambda}\right)^{1/\lambda}\right), \quad (1)$$

then GP2-TA is called the generalized probabilistic 2-tuple average operator, where $\gamma = (\gamma_1, \gamma_2, \ldots, \gamma_n)^{\mathrm{T}}$ is probabilistic weight vector of κ_i with $\gamma_i > 0$, $\sum_{i=1}^{n} \gamma_i = 1$ and λ is a parameter such that $\lambda \in (-\infty, +\infty)$.

Especially, when $\lambda = 1$, $\lambda \to 0$ and $\lambda = -1$, then the GP2-TWA operator given in Eq. (1) becomes the probabilistic 2-tuple average (P-2TA) operator, the probabilistic 2-tuple geometric (P-2TG) operator and the probabilistic 2-tuple harmonic average (P-2THA) operator, respectively.

2.3 Matrix Game with Payoffs Represented by 2TLVs

Assume that $S^m = (\varphi_1, \varphi_2, \ldots, \varphi_m)$ and $S^n = (\chi_1, \chi_2, \ldots, \chi_n)$ are sets of pure strategies for Player I and Player II, respectively. The vectors $x = (x_1, x_2, \ldots, x_m)^T$ and $y = (y_1, y_2, \ldots, y_n)^T$ are known as the mixed strategies of Player I and Player II,respectively, which satisfy $\sum_{i=1}^{m} x_i = 1, x_i \geq 0$ and $\sum_{j=1}^{n} y_j = 1, y_j \geq 0$. Here, $x_i\ (i = 1, 2, \ldots, m)$ denotes the probability of Player I choosing strategy $\varphi_i \in S^m$ and y_j is the probability of Player II selecting strategy $\chi_j \in S^n$. The sets of all mixed strategies for Player I and Player II are denoted by

$$X = \left\{ (x_1, x_2, \ldots, x_m) : x_i \geq 0, i = 1, 2, \ldots, m; \sum_{i=1}^{m} x_i = 1 \right\};$$

$$Y = \left\{ (y_1, y_2, \ldots, y_n) : y_j \geq 0, j = 1, 2, \ldots, n; \sum_{j=1}^{n} y_j = 1 \right\}.$$

Singh et al. [31] defined the two player constant-sum matrix game with 2-TL payoffs as follows.

A matrix game with payoffs represented by 2-TLVs can be represented by $2 - \mathfrak{TLG} = \left(S^m, X, S^m, Y, \mathbb{P}_{[0,t]}, \widetilde{\mathcal{K}}\right)$, where S^m and S^n are strategy sets as discussed above, $\mathbb{P}_{[0,t]} = \{\mathfrak{p}_i | \mathfrak{p}_0 \leq \mathfrak{p}_i \leq \mathfrak{p}_t, i \in [0, t]\}$ be a predefined continuous LTS with odd cardinality, $\widetilde{\mathcal{K}} = \left(\kappa_{ij}\right)_{m\times n}$ is the 2-TL payoff matrix of Player I against Player II, X and Y are mixed strategies spaces for Players I and II.

If Player I chooses a pure strategy $\varphi_i \in S^m$ and Player II chooses a pure strategy $\chi_j \in S^n$, then Player I gains a payoff $\kappa_{ij} = \left(\mathfrak{p}_{ij}, \alpha_{ij}\right)$ such that $\mathfrak{p}_{ij} \in \mathbb{P}$, $\alpha_{ij} \in [-0.5, 0.5)$. On the other hand, the Player II gains a payoff $neg\left(\kappa_{ij}\right) = \Delta\left(t - \Delta^{-1}\left(\mathfrak{p}_{ij}, \alpha_{ij}\right)\right)$. The payoff matrix corresponding to Player I is represented as

$$\widetilde{\mathcal{K}} = \left(\kappa_{ij}\right)_{m\times n} = \begin{array}{c} \\ \varphi_1 \\ \varphi_2 \\ \vdots \\ \varphi_m \end{array} \begin{array}{c} \begin{matrix} \chi_1 & \chi_2 & \cdots & \chi_n \end{matrix} \\ \begin{pmatrix} (\mathfrak{p}_{11}, \alpha_{11}), & (\mathfrak{p}_{12}, \alpha_{12}) & \cdots & (\mathfrak{p}_{1n}, \alpha_{1n}) \\ (\mathfrak{p}_{21}, \alpha_{21}) & (\mathfrak{p}_{22}, \alpha_{22}) & \cdots & (\mathfrak{p}_{2n}, \alpha_{2n}) \\ \vdots & \vdots & \vdots & \vdots \\ (\mathfrak{p}_{m1}, \alpha_{m1}) & (\mathfrak{p}_{m2}, \alpha_{m2}) & \cdots & (\mathfrak{p}_{mn}, \alpha_{mn}) \end{pmatrix} \end{array}.$$

If Player I plays the mixed strategy $x \in X$ and Player II plays the mixed strategy $y \in Y$, then the quantity $E(x, y) = x^{\mathrm{T}}\widetilde{\mathcal{K}}y = \Delta\left(\sum_{j=1}^{n}\sum_{i=1}^{m} x_i y_j \left(\Delta^{-1}\left(\mathfrak{p}_{ij}, \alpha_{ij}\right)\right)\right)$ is called the expected payoff of Player I by Player II.
The Player I (the maximizing player) and Player II (the minimizing player) will select their strategies according to the maximin and minimax principles, respectively. Then, we obtain

$$v^{-} = \max_{x\in X}\min_{y\in Y}\left\{x^{\mathrm{T}}\widetilde{\mathcal{K}}y\right\},$$
$$v^{+} = \min_{y\in Y}\max_{x\in X}\left\{x^{\mathrm{T}}\widetilde{\mathcal{K}}y\right\},$$

which are called the lower value and the upper value of the game with 2-tuple linguistic payoffs, respectively. The game $2-\mathfrak{TLG}$ has value if and only if $v^{+} = v^{-}$. In the next section, we develop a generalized approach based on mathematical programming to solve matrix games with 2-tuple linguistic payoffs.

3 A New Generalized Mathematical Programming Approach for Matrix Games with Payoffs Represented by 2-TLVs

Let $x = (x_1, x_2, \ldots, x_m) \in X$ and $y = (y_1, y_2, \ldots, y_m) \in Y$ be a pair of mixed strategies for Player I and Player II, respectively, then the expected payoff of Player I can be obtained based on the GP2-TA operator mentioned in Eq. (1) as

$$\mathfrak{E}(x, y) = x^{\mathrm{T}}\widetilde{\mathcal{K}}y = \Delta\left(\left(\sum_{j=1}^{n}\sum_{i=1}^{m} x_i y_j \left(\Delta^{-1}(\mathfrak{p}_{ij}, \alpha_{ij})\right)^{\lambda}\right)^{1/\lambda}\right), \ \lambda > 0. \tag{2}$$

Player II is interesting to select a mixed strategy $y \in Y$ so as to minimize $\mathfrak{E}(x, y)$, denoted by

$$\Theta = (\mathfrak{p}_{\Theta}, \alpha_{\Theta}) = \min_{y\in Y}\left(\Delta\left(\left(\sum_{j=1}^{n}\sum_{i=1}^{m} x_i y_j \left(\Delta^{-1}(\mathfrak{p}_{ij}, \alpha_{ij})\right)^{\lambda}\right)^{1/\lambda}\right)\right).$$

Since Θ is the function of x only. Hence, Player I should choose a mixed strategy $x^{*} \in X$ that maximizes the value of Θ, i.e.,

$$\Theta^{*} = (\mathfrak{p}_{\Theta^{*}}, \alpha_{\Theta^{*}}) = \max_{x\in X}\min_{y\in Y}\left(\Delta\left(\left(\sum_{j=1}^{n}\sum_{i=1}^{m} x_i y_j \left(\Delta^{-1}(\mathfrak{p}_{ij}, \alpha_{ij})\right)^{\lambda}\right)^{1/\lambda}\right)\right). \tag{3}$$

Similarly, Player I is interesting to select a mixed strategy $x \in X$ so as to maximize $\mathfrak{E}(x, y)$, represented by

$$\Psi = (\mathfrak{p}_\Psi, \alpha_\Psi) = \max_{x \in X} \left(\Delta \left(\left(\sum_{j=1}^{n} \sum_{i=1}^{m} x_i y_j \left(\Delta^{-1}(\mathfrak{p}_{ij}, \alpha_{ij}) \right)^{\lambda} \right)^{1/\lambda} \right) \right).$$

Here, Ψ is the function of y only. Then, Player II should choose a mixed strategy $y^* \in Y$ that minimizes the value of Ψ, i.e.,

$$\Psi^* = (\mathfrak{p}_{\Psi^*}, \alpha_{\Psi^*}) = \min_{y \in Y} \max_{x \in X} \left(\Delta \left(\left(\sum_{j=1}^{n} \sum_{i=1}^{m} x_i y_j \left(\Delta^{-1}(\mathfrak{p}_{ij}, \alpha_{ij}) \right)^{\lambda} \right)^{1/\lambda} \right) \right). \quad (4)$$

Note that here Θ^* and Ψ^* are represent the minimum gain and maximum loss for Player I and Player II, respectively. The mixed strategies x^* and y^* are called the maximin and the minimax strategies for Player I and Player II, respectively.
Now to obtain the maximin strategy x^* and the minimum gain Θ^* for Player I, we propose the following non-linear 2-tuple optimization model:
(**MOD 1**) $\max \{(\mathfrak{p}_\Theta, \alpha_\Theta)\}$

$$s.t. \begin{cases} \Delta \left(\left(\sum_{j=1}^{n} \sum_{i=1}^{m} x_i y_j \left(\Delta^{-1}(\mathfrak{p}_{ij}, \alpha_{ij}) \right)^{\lambda} \right)^{1/\lambda} \right) \geq (\mathfrak{p}_\Theta, \alpha_\Theta), \quad \text{for any } y \in Y \\ \mathfrak{p}_\Theta \in \mathbb{P}, \ \alpha_\Theta \in [-0.5, 0.5), \lambda > 0 \\ x_i \geq 0, \ \sum_{i=1}^{m} x_i = 1, \ i = 1, 2, \ldots, m, \end{cases} \quad (5)$$

where

$$(\mathfrak{p}_\Theta, \alpha_\Theta) = \min_{y \in Y} \left(\Delta \left(\left(\sum_{j=1}^{n} \sum_{i=1}^{m} x_i y_j \left(\Delta^{-1}(\mathfrak{p}_{ij}, \alpha_{ij}) \right)^{\lambda} \right)^{1/\lambda} \right) \right).$$

Since Δ is a strictly monotonically increasing function and $\lambda > 0$, then $\max \{(\mathfrak{p}_\Theta, \alpha_\Theta)\} \Leftrightarrow \max \left\{ \left(\Delta^{-1} (\mathfrak{p}_\Theta, \alpha_\Theta) \right)^{\lambda} \right\}$. Therefore, the (MOD 1) is transformed into the following optimization model:
(**MOD 2**) $\max \left\{ \left(\Delta^{-1} (\mathfrak{p}_\Theta, \alpha_\Theta) \right)^{\lambda} \right\}$

$$s.t. \begin{cases} \sum_{j=1}^{n} \sum_{i=1}^{m} x_i y_j \left(\Delta^{-1}(\mathfrak{p}_{ij}, \alpha_{ij}) \right)^{\lambda} \geq \left(\Delta^{-1} (\mathfrak{p}_\Theta, \alpha_\Theta) \right)^{\lambda}, \quad \text{for any } y \in Y \\ \mathfrak{p}_\Theta \in \mathbb{P}, \ \alpha_\Theta \in [-0.5, 0.5), \lambda > 0 \\ x_i \geq 0, \ \sum_{i=1}^{m} x_i = 1, \ i = 1, 2, \ldots, m, \end{cases} \quad (6)$$

Since Y is a finite and compact convex set, so, we can consider only the extreme points of the set. Taking $\Delta^{-1} (\mathfrak{p}_\Theta, \alpha_\Theta) = \mathfrak{J}_1$, then, (MOD-2) can be modified as follows:

(**MOD 3**) $\max\left\{(\mathfrak{J}_1)^{\lambda}\right\}$

$$s.t.\begin{cases}\sum_{i=1}^{m} x_i\left(\Delta^{-1}(\mathfrak{p}_{ij},\alpha_{ij})\right)^{\lambda}\geq(\mathfrak{J}_1)^{\lambda}, & j=1,2,\ldots,n\\ \mathfrak{J}_1\geq 0, x_i\geq 0,\ \sum_{i=1}^{m} x_i=1,\ i=1,2,\ldots,m,\ \text{and}\ \lambda>0. & \end{cases} \tag{7}$$

Similarly, the minimax strategy y^* and the minimum loss Ψ^* for Player II can be obtained by solving the following linear programming model given by:
(**MOD 4**) $\min\{(\mathfrak{p}_{\Psi},\alpha_{\Psi})\}$

$$s.t.\begin{cases}\Delta\left(\left(\sum_{j=1}^{n}\sum_{i=1}^{m} x_i y_j\left(\Delta^{-1}(\mathfrak{p}_{ij},\alpha_{ij})\right)^{\lambda}\right)^{1/\lambda}\right)\leq(\mathfrak{p}_{\Psi},\alpha_{\Psi}), & \text{for any } x\in X\\ \mathfrak{p}_{\Psi}\in\mathbb{P},\ \alpha_{\Psi}\in[-0.5,0.5),\lambda>0, & \\ y_j\geq 0,\ \sum_{j=1}^{n} y_j=1,\ i=1,2,\ldots,n. & \end{cases} \tag{8}$$

where

$$(\mathfrak{p}_{\Psi},\alpha_{\Psi})=\max_{x\in X}\left(\Delta\left(\left(\sum_{j=1}^{n}\sum_{i=1}^{m} x_i y_j\left(\Delta^{-1}(\mathfrak{p}_{ij},\alpha_{ij})\right)^{\lambda}\right)^{1/\lambda}\right)\right).$$

Since Δ is a strictly monotonically increasing function and $\lambda>0$, then $\min\{(\mathfrak{p}_{\Psi},\alpha_{\Psi})\}\Leftrightarrow\min\left\{\left(\Delta^{-1}(\mathfrak{p}_{\Psi},\alpha_{\Psi})\right)^{\lambda}\right\}$. Hence, (MOD 4) is reduced into the following optimization model:
(**MOD 5**) $\min\left\{\left(\Delta^{-1}(\mathfrak{p}_{\Psi},\alpha_{\Psi})\right)^{\lambda}\right\}$

$$s.t.\begin{cases}\sum_{j=1}^{n}\sum_{i=1}^{m} x_i y_j\left(\Delta^{-1}(\mathfrak{p}_{ij},\alpha_{ij})\right)^{\lambda}\leq\left(\Delta^{-1}(\mathfrak{p}_{\Psi},\alpha_{\Psi})\right)^{\lambda}, & \text{for any } x\in X\\ \mathfrak{p}_{\Psi}\in\mathbb{P},\ \alpha_{\Psi}\in[-0.5,0.5),\lambda>0, & \\ y_j\geq 0,\ \sum_{j=1}^{n} y_j=1,\ i=1,2,\ldots,n. & \end{cases} \tag{9}$$

Since X is a finite and compact convex set, so, it is sufficient to consider only the extreme points of the set. Assuming $\Delta^{-1}(\mathfrak{p}_{\Psi},\alpha_{\Psi})=\mathfrak{J}_2$, the (MOD-5) can be written as follows:
(**MOD 6**) $\min\left\{(\mathfrak{J}_2)^{\lambda}\right\}$

$$s.t.\begin{cases}\sum_{j=1}^{n} y_j\left(\Delta^{-1}(\mathfrak{p}_{ij},\alpha_{ij})\right)^{\lambda}\leq(\mathfrak{J}_2)^{\lambda}, & i=1,2,\ldots,m\\ \mathfrak{J}_2\geq 0, y_j\geq 0,\ \sum_{j=1}^{n} y_j=1,\ j=1,2,\ldots,n. & \end{cases} \tag{10}$$

The maximin and minimax strategies for Players I and II can be obtained by solving optimization models given in Eqs. (7) and (10) with the optimal values $\mathfrak{J}_1^*$ and $\mathfrak{J}_2^*$. Hence, the linguistic optimal values for Players I and II will be $\Delta^{-1}\left(\mathfrak{J}_1^*\right)$ and $\Delta^{-1}\left(\mathfrak{J}_2^*\right)$.

Remark 2 If $\lambda=1$, then (MOD 3) and (MOD 6) reduce to the optimization models for Players I and II as defined by Singh et al. [31].

Now, in order to demonstrate the flexibility and effectiveness of the developed approach, we consider a numerical example borrowed from Singh et al. [31].

4 Numerical Illustration

Example 1 Consider a matrix game with payoffs represented by 2-TLVs defined on LTS

$$\mathbb{P} = \left\{ \begin{array}{l} \mathfrak{p}_0 = \text{very poor (VP)}, \mathfrak{p}_1 = \text{poor (P)}, \mathfrak{p}_2 = \text{moderately poor (MP)}, \mathfrak{p}_3 = \text{ Fair (F)}, \\ \mathfrak{p}_4 = \text{moderately good (MG)}, \mathfrak{p}_5 = \text{good (G)}, \mathfrak{p}_6 = \text{very good (VG)} \end{array} \right\},$$

as follows:

$$\tilde{\mathcal{K}} = \begin{array}{c} \\ \varphi_1 \\ \varphi_2 \\ \varphi_3 \end{array} \begin{array}{ccccc} \chi_1 & \chi_2 & \chi_3 & \chi_4 & \chi_5 \\ \end{array} \begin{pmatrix} (\mathfrak{p}_4, 0) & (\mathfrak{p}_4, 0) & (\mathfrak{p}_3, 0) & (\mathfrak{p}_3, 0) & (\mathfrak{p}_4, 0.3) \\ (\mathfrak{p}_4, 0) & (\mathfrak{p}_3, -0.4) & (\mathfrak{p}_6, -0.1) & (\mathfrak{p}_4, 0) & (\mathfrak{p}_0, 0) \\ (\mathfrak{p}_3, 0.4) & (\mathfrak{p}_6, 0) & (\mathfrak{p}_4, 0) & (\mathfrak{p}_4, 0) & (\mathfrak{p}_4, 0) \end{pmatrix}.$$

Solution: Utilizing the (MOD 3) and (MOD 6) given in Eqs. (7) and (10), the following non-linear-programming optimizations models are obtained:

$$\max \left\{ (\mathfrak{J}_1)^{\lambda} \right\}$$

$$s.t. \begin{cases} \left(\Delta^{-1}(\mathfrak{p}_4, 0.0)\right)^{\lambda} x_1 + \left(\Delta^{-1}(\mathfrak{p}_4, 0.0)\right)^{\lambda} x_2 + \left(\Delta^{-1}(\mathfrak{p}_3, 0.4)\right)^{\lambda} x_3 \geq (\mathfrak{J}_1)^{\lambda}, \\ \left(\Delta^{-1}(\mathfrak{p}_4, 0.0)\right)^{\lambda} x_1 + \left(\Delta^{-1}(\mathfrak{p}_3, -0.4)\right)^{\lambda} x_2 + \left(\Delta^{-1}(\mathfrak{p}_6, 0.0)\right)^{\lambda} x_3 \geq (\mathfrak{J}_1)^{\lambda}, \\ \left(\Delta^{-1}(\mathfrak{p}_3, 0.0)\right)^{\lambda} x_1 + \left(\Delta^{-1}(\mathfrak{p}_6, -0.1)\right)^{\lambda} x_2 + \left(\Delta^{-1}(\mathfrak{p}_4, 0.0)\right)^{\lambda} x_3 \geq (\mathfrak{J}_1)^{\lambda}, \\ \left(\Delta^{-1}(\mathfrak{p}_3, 0.0)\right)^{\lambda} x_1 + \left(\Delta^{-1}(\mathfrak{p}_4, 0.0)\right)^{\lambda} x_2 + \left(\Delta^{-1}(\mathfrak{p}_4, 0.0)\right)^{\lambda} x_3 \geq (\mathfrak{J}_1)^{\lambda}, \\ \left(\Delta^{-1}(\mathfrak{p}_4, 0.3)\right)^{\lambda} x_1 + \left(\Delta^{-1}(\mathfrak{p}_0, 0.0)\right)^{\lambda} x_2 + \left(\Delta^{-1}(\mathfrak{p}_4, 0.0)\right)^{\lambda} x_3 \geq (\mathfrak{J}_1)^{\lambda}, \\ \mathfrak{J}_1 \geq 0, \lambda > 0, x_i \geq 0,\ i = 1, 2, 3 \text{ and } x_1 + x_2 + x_3 = 1. \end{cases}$$

or

$$\max \left\{ (\mathfrak{J}_1)^{\lambda} \right\}$$

$$s.t. \begin{cases} (4.0)^{\lambda} x_1 + (4.0)^{\lambda} x_2 + (3.4)^{\lambda} x_3 \geq (\mathfrak{J}_1)^{\lambda}, \\ (4.0)^{\lambda} x_1 + (2.6)^{\lambda} x_2 + (6.0)^{\lambda} x_3 \geq (\mathfrak{J}_1)^{\lambda}, \\ (3.0)^{\lambda} x_1 + (5.9)^{\lambda} x_2 + (4.0)^{\lambda} x_3 \geq (\mathfrak{J}_1)^{\lambda}, \\ (3.0)^{\lambda} x_1 + (4.0)^{\lambda} x_2 + (4.0)^{\lambda} x_3 \geq (\mathfrak{J}_1)^{\lambda}, \\ (4.3)^{\lambda} x_1 + (0.0)^{\lambda} x_2 + (4.0)^{\lambda} x_3 \geq (\mathfrak{J}_1)^{\lambda}, \\ \mathfrak{J}_1 \geq 0, \lambda > 0, x_i \geq 0,\ i = 1, 2, 3 \text{ and } x_1 + x_2 + x_3 = 1. \end{cases} \quad (11)$$

and

$$\min\left\{(\mathfrak{J}_2)^{\lambda}\right\}$$

$$s.t.\begin{cases}\left(\Delta^{-1}(\mathfrak{p}_4,0.0)\right)^{\lambda}y_1+\left(\Delta^{-1}(\mathfrak{p}_4,0.0)\right)^{\lambda}y_2+\left(\Delta^{-1}(\mathfrak{p}_3,0.0)\right)^{\lambda}y_3\\ \qquad +\left(\Delta^{-1}(\mathfrak{p}_3,0.0)\right)^{\lambda}y_4+\left(\Delta^{-1}(\mathfrak{p}_4,0.3)\right)^{\lambda}y_5\le(\mathfrak{J}_2)^{\lambda},\\ \left(\Delta^{-1}(\mathfrak{p}_4,0.0)\right)^{\lambda}y_1+\left(\Delta^{-1}(\mathfrak{p}_3,-0.4)\right)^{\lambda}y_2+\left(\Delta^{-1}(\mathfrak{p}_6,-0.1)\right)^{\lambda}y_3\\ \qquad +\left(\Delta^{-1}(\mathfrak{p}_4,0.0)\right)^{\lambda}y_4+\left(\Delta^{-1}(\mathfrak{p}_0,0.0)\right)^{\lambda}y_5\le(\mathfrak{J}_2)^{\lambda},\\ \left(\Delta^{-1}(\mathfrak{p}_3,0.4)\right)^{\lambda}y_1+\left(\Delta^{-1}(\mathfrak{p}_6,0.0)\right)^{\lambda}y_2+\left(\Delta^{-1}(\mathfrak{p}_4,0.0)\right)^{\lambda}y_3\\ \qquad +\left(\Delta^{-1}(\mathfrak{p}_4,0.0)\right)^{\lambda}y_4+\left(\Delta^{-1}(\mathfrak{p}_4,0.0)\right)^{\lambda}y_5\le(\mathfrak{J}_2)^{\lambda},\\ \mathfrak{J}_2\ge0,\lambda>0,y_j\ge0,\ j=1,2,3,4,5\ \text{and}\ y_1+y_2+y_3+y_4+y_5=1.\end{cases}$$

or

$$\min\left\{(\mathfrak{J}_2)^{\lambda}\right\}$$

$$s.t.\begin{cases}(4.0)^{\lambda}y_1+(4.0)^{\lambda}y_2+(3.0)^{\lambda}y_3+(3.0)^{\lambda}y_4+(4.3)^{\lambda}y_5\le(\mathfrak{J}_2)^{\lambda},\\ (4.0)^{\lambda}y_1+(2.6)^{\lambda}y_2+(5.9)^{\lambda}y_3+(4.0)^{\lambda}y_4+(0.0)^{\lambda}y_5\le(\mathfrak{J}_2)^{\lambda},\\ (3.4)^{\lambda}y_1+(6.0)^{\lambda}y_2+(4.0)^{\lambda}y_3+(4.0)^{\lambda}y_4+(4.0)^{\lambda}y_5\le(\mathfrak{J}_2)^{\lambda},\\ \mathfrak{J}_2\ge0,\lambda>0,y_j\ge0,\ j=1,2,3,4,5\ \text{and}\ y_1+y_2+y_3+y_4+y_5=1.\end{cases} \tag{12}$$

We can solve the non-linear optimizations models given in Eqs. (11) and (12) with the help of MATLAB software. The obtained optimal solutions for different values of λ are summarized in Table 1.

Futher, based on Table 1, the optimum value Θ^* for Player I; optimum value Ψ^* for Player II and the optimal value of the game $\mathfrak{E}(x^*, y^*)$ in terms of 2-TLVs are calculated and shown in Table 2, which depict that $\Theta^* = \Psi^* = \mathfrak{E}(x^*, y^*)\ \forall\,\lambda > 0$. It is also interesting to note that the our optimal solution for $\lambda = 1$ is coincide with the results obtained by Singh et al. [30]. Hence, we can conclude that our developed solution approach is more flexible and applicable in real-life competitive situations with 2-TL information.

5 Conclusion

In this work, we have studied matrix games with linguistic information and proposed a generalized approach for solving them. Paper has been defined a pair of nonlinear-programming models to solve matrix games with payoffs represented by 2-TLVs. Our developed models have a flexible parameter λ, which provides an ability to consider the attitudinal character of the players in solution process. It is worth mentioning that the developed method generalizes the solution approach of Singh et al. [31] and provides an ability to consider the attitudinal character of the players in solution process.

Table 1 ptimal solutions of the models given in Eqs. (11) and (12) taking different values of λ

λ	x^{*T}	$\mathfrak{I}_1^*$	y^{*T}	$\mathfrak{I}_2^*$
0.5	(0.3463,0.0592,0.5945)	3.6374	(0.5945, 0.0000, 0.0000, 0.3591, 0.0464)	3.6374
1	(0.3343, 0.1086, 0.5571)	3.6657	(0.5571, 0.0000, 0.0000, 0.3593, 0.0836)	3.6657
2	(0.3155, 0.1871, 0.4974)	3.7137	(0.4974 ,0.0000,0.0000,0.3646, 0.1380)	3.7137
5	(0.2802, 0.3357, 0.3841)	3.8122	(0.3841,0.0000, 0.0000,0.4022, 0.2137)	3.8122
7	(0.2631, 0.4014, 0.3355)	3.8548	(0.3355, 0.0000, 0.0000, 0.4365, 0.2280)	3.8549
10	(0.2393, 0.4796, 0.2811)	3.8989	(0.2812,0.0000 , 0.0000,0.4931, 0.2258)	3.8989
15	(0.1989, 0.5860, 0.2151)	3.9422	(0.2151,0.0000,0.0000,0.5886, 0.1963)	3.9422
25	(0.1232, 0.7515, 0.1253)	3.9790	(0.1254,0.0000, 0.0000,0.7515,0.1231)	3.9790
30	(0.0929, 0.8135, 0.0936)	3.9870	(0.0936, 0.0000, 0.0000, 0.8135, 0.0929)	3.9870

Table 2 The values of Θ^*, Ψ^* and $\mathfrak{E}(x^*, y^*)$ with different λ

λ	Θ^*	Ψ^*	$\mathfrak{E}(x^*, y^*)$
0.5	$(\mathfrak{p}_4, -0.3626)$	$(\mathfrak{p}_4, -0.3626)$	$(\mathfrak{p}_4, -0.3626)$
1	$(\mathfrak{p}_4, -0.3343)$	$(\mathfrak{p}_4, -0.3343)$	$(\mathfrak{p}_4, -0.3343)$
2	$(\mathfrak{p}_4, -0.2863)$	$(\mathfrak{p}_4, -0.2863)$	$(\mathfrak{p}_4, -0.2863)$
5	$(\mathfrak{p}_4, -0.1878)$	$(\mathfrak{p}_4, -0.1878)$	$(\mathfrak{p}_4, -0.1878)$
7	$(\mathfrak{p}_4, -0.1452)$	$(\mathfrak{p}_4, -0.1452)$	$(\mathfrak{p}_4, -0.1452)$
10	$(\mathfrak{p}_4, -0.1011)$	$(\mathfrak{p}_4, -0.1011)$	$(\mathfrak{p}_4, -0.1011)$
15	$(\mathfrak{p}_4, -0.0578)$	$(\mathfrak{p}_4, -0.0578)$	$(\mathfrak{p}_4, -0.0578)$
25	$(\mathfrak{p}_4, -0.0210)$	$(\mathfrak{p}_4, -0.0210)$	$(\mathfrak{p}_4, -0.0210)$
50	$(\mathfrak{p}_4, -0.0021)$	$(\mathfrak{p}_4, -0.0021)$	$(\mathfrak{p}_4, -0.0021)$

Acknowledgements The first author acknowledges the financial support from the Chilean Government (Conicyt) through the Fondecyt Postdoctoral Program (Project Number 3170556).

References

1. Von Neumann, J., Morgenstern, O.: Theory of Games and Economic Behavior. Princeton University Press (1953). https://doi.org/10.2307/j.cttlr2gkx
2. Leng, M., Parlar, M.: Game theoretic applications in supply chain management: a review. INFOR **43**(3), 187–220 (2005). https://doi.org/10.1080/03155986.2005.11732725

3. Kapliński, O., Tamošaitiene, J.: Game theory applications in construction engineering and management. Ukio Technologinis ir Ekonominis Vystymas **16**(02), 348–363 (2010)
4. Lambertini, L.: Game Theory in the Social Sciences, 1st edn. Routledge, London (2011). https://doi.org/10.4324/9780203818640
5. Rass, S., Schauer, S.: Game Theory for Security and Risk Management, 1st edn. Birkhäuser, Basel (2018)
6. Dubois, D., Prade, H.: Fuzzy Sets and Systems. Theory and Applications. Academic Press, New York (1980)
7. Aubin, J.P.: Cooperative fuzzy games. Math. Oper. Res. **6**(1), 1–13 (1981). https://doi.org/10.2307/3689261
8. Butnariu, D.: Fuzzy games: a description of the concept. Fuzzy Sets Syst. **1**(3), 181–192 (1978). https://doi.org/10.1016/0165-0114(78)90003-9
9. Campos, L.: Fuzzy linear programming models to solve fuzzy matrix games. Fuzzy Sets Syst. **32**(3), 275–289 (1989). https://doi.org/10.1016/0165-0114(89)90260-1
10. Sakawa, M., Nishizaki, I.: Max-min solutions for fuzzy multiobjective matrix games. Fuzzy Sets Syst. **67**(1), 53–69 (1994). https://doi.org/10.1016/0165-0114(94)90208-9
11. Maeda, T.: On characterization of equilibrium strategy of two-person zero-sum games with fuzzy payoffs. Fuzzy Sets Syst. **139**(2), 283–296 (2003). https://doi.org/10.1016/S0165-0114(02)00509-2
12. Bector, C.R., Chandra, S., Vijay, V.: Duality in linear programming with fuzzy parameters and matrix games with fuzzy pay-offs. Fuzzy Sets Syst. **146**(2), 253–269 (2004). https://doi.org/10.1016/S0165-0114(03)00260-4
13. Cevikel, A.C., Ahlatolu, M.: Solutions for fuzzy matrix games. Comput. Math. Appl. **60**(3), 399–410 (2010). https://doi.org/10.1016/j.camwa.2010.04.020
14. Vijay, V., Chandra, S., Bector, C.R.: Matrix games with fuzzy goals and fuzzy payoffs. Omega **33**(5), 425–429 (2005). https://doi.org/10.1016/j.omega.2004.07.007
15. Herrera, F., Alonso, S., Chiclana, F., Herrera-Viedma, E.: Computing with words in decision making: foundations, trends and prospects. Fuzzy Optim. Decis. Making **8**, 337–364 (2009). https://doi.org/10.1007/s10700-009-9065-2
16. Zadeh, L.A.: The concept of a linguistic variable and its application to approximate reasoning-I. Inf. Sci. **8**(3), 43–80 (1975). https://doi.org/10.1016/0020-0255(75)90036-5
17. Ben-Arieh, D., Chen, Z.: Linguistic-labels aggregation and consensus measure for autocratic decision making using group recommendations. IEEE Trans. Syst. Man Cybern Part A Syst. Hum. **36**(3), 558–568 (2006). https://doi.org/10.1109/TSMCA.2005.853488
18. Xu, Z.: Induced uncertain linguistic OWA operators applied to group decision making. Inf. Fusion **7**(2), 231–238 (2006). https://doi.org/10.1016/j.inffus.2004.06.005
19. Xu, Z.: A method for multiple attribute decision making with incomplete weight information in linguistic setting. Knowl. Based Syst. **20**(8), 719–725 (2007). https://doi.org/10.1016/j.knosys.2006.10.002
20. Li, D.F.: Multiattribute group decision making method using extended linguistic variables. Int. J. Uncertainty Fuzziness Knowl. Based Syst. **17**(6), 793–806 (2009). https://doi.org/10.1142/S0218488509006273
21. Verma, R., Sharma, B.D.: Trapezoid fuzzy linguistic prioritized weighted average operators and their application to multiple attribute group decision making. J. Uncertainty Anal. Appl. **2**, 1–19 (2014). https://doi.org/10.1186/2195-5468-2-10
22. Verma, R.: Multiple attribute group decision making based on generalized trapezoid fuzzy linguistic prioritized weighted average operator. Int. J. Mach. Learn. Cybern. **8**(6), 1993–2007 (2017). https://doi.org/10.1007/s13042-016-0579-y
23. Herrera, F., Martínez, L.: A 2-tuple fuzzy linguistic representation model for computing with words. IEEE Trans. Fuzzy Syst. **8**(6), 746–752 (2000). https://doi.org/10.1109/91.890332
24. Herrera, F., Martínez, L.: A model based on linguistic 2-tuples for dealing with multigranular hierarchical linguistic contexts in multi-expert decision-making. IEEE Trans. Syst. Man Cybern. Part B Cybern. **31**(2), 227–234 (2001). https://doi.org/10.1109/3477.915345

25. Dong, Y., Xu, Y., Yu, S.: Computing the numerical scale of the linguistic term set for the 2-tuple fuzzy linguistic representation model. IEEE Trans. Fuzzy Syst. **17**(6), 1366–1378 (2009). https://doi.org/10.1109/TFUZZ.2009.2032172
26. Wei, G.W.: Extension of TOPSIS method for 2-tuple linguistic multiple attribute group decision making with incomplete weight information. Knowl. Inf. Syst. **25**(3), 623–634 (2010). https://doi.org/10.1007/s10115-009-0258-3
27. Martínez, L., Herrera, F.: An overview on the 2-tuple linguistic model for computing with words in decision making: extensions, applications and challenges. Inf. Sci. **207**, 1–18 (2012). https://doi.org/10.1016/j.ins.2012.04.025
28. Wan, S.P.: Some hybrid geometric aggregation operators with 2-tuple linguistic information and their applications to multi-attribute group decision making. Int. J. Comput. Intell. Syst. **6**(4), 750–763 (2013). https://doi.org/10.1080/18756891.2013.804144
29. Liu, Y., Liu, J., Qin, Y., Xu, Y.: A novel method based on extended uncertain 2-tuple linguistic muirhead mean operators to MAGDM under uncertain 2-tuple linguistic environment. Int. J. Comput. Intell. Syst. **12**(2), 498–512 (2019). https://doi.org/10.2991/ijcis.d.190315.001
30. Wei, G., Zhao, X.: Methods for probabilistic decision making with linguistic information. Technol. Econ. Dev. Econ. **20**(2), 193–209 (2014). https://doi.org/10.3846/20294913.2014.869515
31. Singh, A., Gupta, A., Mehra, A.: Matrix games with 2-tuple linguistic information. Ann. Oper. Res. **287**(2), 895–910 (2020). https://doi.org/10.1007/s10479-018-2810-6

Chi-Square Similarity Measure for Interval Valued Neutrosophic Set

Ritu Sahni, Manoj Sahni, and Nayankumar Patel

Abstract The present paper deals with a similarity measure known as chi-square similarity measure which is applied to interval valued neutrosophic sets. Although many similarity measures are available in the neutrosophic set theory, we have proposed a new one and compared it with some existing similarity measures and found the results are in good agreement with the already defined measures. In this paper, we have applied the new chi-square similarity measure on the real-life problems based on the recognition of pattern and diagnosis of illness in medical field. Further, the compared results with different similarity measures are shown in the form of table. The importance of pattern recognition problem in various application areas such as in image processing and medical diagnosis is shown for single-valued neutrosophic sets (SVNSs) but not for interval valued neutrosophic sets (IVNSs). Results related to IVNSs ensure that chi-square similarity measure is indeed effective and in some cases gives better results than the existing similarity measure.

Keywords Measures of similarity · Cosine similarity · Chi-square measure · Neutrosophic sets · Interval valued neutrosophic sets

1 Introduction

In real-life scenario, the informations related to day-to-day situations is not always in the form of precise numerical data. Some imprecise factors or uncertainties are always associated with those informations. In order to deal with various types of uncertainties occurring in any model of real-life situations, Zadeh [1] the architect

R. Sahni (✉)
Institute of Advanced Research, Gandhinagar, Gujarat, India
e-mail: ritusrivastava1981@gmail.com

M. Sahni
Pandit Deendayal Petroleum University, Gandhinagar, Gujarat, India

N. Patel
Adani Institute of Infrastructure Engineering, Ahmedabad, Gujarat, India

M. Sahni et al. (eds.), *Mathematical Modeling, Computational Intelligence Techniques and Renewable Energy*, Advances in Intelligent Systems and Computing 1287,
https://doi.org/10.1007/978-981-15-9953-8_46

of the development of fuzzy theory gave this concept in the year 1965. Because of its tremendous applications, this theory is diversified in many ways. Initially, it is merely a generalization of classical set theory, in which the members are inside the set or outside the set, whereas this theory allows graded membership; i.e., all the members have some degree of membership in the defined universal set. If A is any set, then $\mu_A(x)$ denotes membership of element x in A and $\mu_A(x) \in [0, 1]$. With the initial development of fuzzy sets in 1960s, an extension of it were introduced by Atanassov [2], known as intuitionistic fuzzy set, which contains both membership and non-membership grades. This idea was further generalized to interval valued intuitionistic fuzzy sets [2, 3]. The idea behind this set is that sometimes situations arises where it is not possible to assign a crisp value to any membership function $\mu_A(x)$, so the concept of interval valued fuzzy set allows to take membership values from an interval rather than a single crisp value. It uses an interval value $\left[\mu_A^L(x), \mu_A^U(x)\right]$, with $0 \le \left[\mu_A^L(x), \mu_A^U(x)\right] \le 1$. In this way, many variants of these fuzzy concepts were further introduced by several researchers which generalize the fuzzy set theory and are helpful in real life scenario. All these concepts are capable enough to deal with incomplete informations. There are the situations, where indeterminate or inconsistent informations are arrived. For example, consider any patient suffering from toothache. One doctor advise that 90% chances is that patient undergoes with a tooth extraction through surgery, other says 90% chances that he/she does not require surgery. But the two doctors have different qualification, one is just cleared BDS and started practicing and other is MDS with many years of practice. Different doctors have different views based on their indeterminate knowledge. The impact of the final decision by a panel of doctors can be different as there is a possibility that they have the same values of acceptance. Fuzzy set theory is unable to solve this kind of problem. We can find similar problems in many areas, such as in diagnosis of disease, in the prediction related to weather forecasting, image processing, and many more. Smarandache [4] was the person who proposed neutrosophic set theory, which has capability to deal not only with incomplete and vague informations, but also with indeterminacy and inconsistent informations. Thus in this way this theory generalizes all the previous theories, viz classical crisp set theory [1], fuzzy set theory [1], interval valued fuzzy set theory [2], intuitionistic fuzzy set theory [2, 3], interval valued intuitionistic fuzzy set theory [2, 3], etc. In neutrosophic fuzzy set, there are three components, T, I, and F, representing truth value, indeterminacy value and false value, respectively, within the non-standard unit interval $]^{-}0, 1^{+}[$. From philosophical point of view, the theory is capable enough to provide the solution in each and every situation, but for scientific and engineering problems non-standard unit interval creates the problem, that means, for these problems interval must be specified. So for that purpose, single-valued neutrosophic set is defined by Wang et al. [5] in the year 2010. After that neutrosophic fuzzy set theory is applied in almost all the areas of mathematics including calculus [6], differential equations [7], graph theory [8], etc. We can find their applications in medical diagnosis [9], pattern recognition [10], multi-attribute decision making [11, 12], and many more. In these problems, similarity measures play a crucial role to identify the patterns between two sets of data or to compare the data collected in all disciplines.

Various fuzzy similarity measures such as cosine [13, 14], sqrt cosine [15], and many more (see for instance [11–20]) are defined in order to compare the fuzzy data. Out of which cosine similarity measure is the most popular one. This similarity measure is defined for fuzzy set, intuitionistic fuzzy set, interval valued intuitionistic fuzzy set, neutrosophic set, interval valued neutrosophic sets, etc. Some other trigonometric measures of similarity such as tangent [11], cotangent [16] are also defined for previously mentioned fuzzy sets and neutrosophic sets. But these similarity measures are not efficient enough. In fact Ren et al. [20] in 2019 shown that the existing similarity measures have shortcomings. They developed a new similarity measure for single-valued neutrosophic sets [4], which is based on Chi-square distance measure. They have provided some numerical examples to show that the proposed chi-square similarity is superior to other similarity measures available in the literature. They have used chi-square similarity measure in single-valued neutrosophic sets, where some real number $\mu_A(x) \in [0, 1]$ is used to represent truth, indeterminacy, and false membership, respectively, of fuzzy set A on some universal set X. But sometimes, it is not possible to define the real number $\mu_A(x)$ by only one single crisp value, so to overcome these problem neutrosophic interval sets came into existence, which uses an interval value $\left[\mu_A^L(x), \mu_A^U(x)\right] \subseteq [0, 1]$ with $0 \leq \mu_A^L(x) \leq \mu_A^U(x) \leq 1$ to represent the grade of membership, indeterminacy and non-membership of each of the element of the neutrosophic set A. This motivated us to generalize the chi-square measure of similarity for interval valued neutrosophic sets (IVNSs).

In this paper, we have defined chi-square measure of similarity for interval valued neutrosophic sets and the results obtained are compared with other existing measures of similarity and shown in the form of table. The real-life examples based on recognition of pattern and diagnosis of disease are used to show the importance of chi-square measure of similarity.

2 Basic Definitions

In this section, the basic definitions related to our work are given.

Definition 2.1 (*Neutrosophic Set (NS)*) [4]. Considering X to be any universal set and let $A = \{< x : T_A(x), I_A(x), F_A(x) >, x \in X\}$, where the mapping $T, I, F : X \to \left]^-0, 1^+\right[$ provide, respectively, truth, indeterminacy and the false values, respectively, of the element $x \in X$ to the set A satisfying the following condition

$$^-0 \leq T_A(x) + I_A(x) + F_A(x) \leq 3^+,$$

then A is called as neutrosophic set.

Here, it is defined that all the values are chosen from the subsets of interval $]^-0, 1^+[$ which are in the form of non-standard in nature. Considering the practical situations, this create problems in calculating value of belongingness (truth), the values of indeterminacy and the value of non-belongingness (false) of any object in the set U. To overcome this problem, Wang et al. [5] replaced interval $]^{-}0, 1^{+}[$ with interval [0, 1] and new definition is coined named as single-valued NS.

Definition 2.2 (*Single-valued Neutrosophic Set (SVNS)*) [5]. In SVNS, the interval $]^{-}0, 1^{+}[$ is replaced with [0, 1] and the functions T, I, F satisfy the condition: $0 \leq T_A(x) + I_A(x) + F_A(x) \leq 3$, for $x \in X$.

The concept of neutrosophic set is further generalized to interval valued neutrosophic set by Broumi and Smarandache [14].

Definition 2.3 (*Interval valued Neutrosophic Set*) [14]. Let X be any universal set and let A be the subset of X defined as:

$$A = \left\{ < x : [T_A^L(x), [T_A^U(x)], [I_A^L(x), [I_A^U(x)], [F_A^L(x), [F_A^U(x)] >, x \in X \right\},$$

where $T_A(x)$ denotes membership function, $I_A(x)$ is the indeterminacy function and $F_A(x)$ the false function and mapping of the function $T_A(x), I_A(x), F_A(x) : X \to [0, 1]$ satisfying the condition $0 \leq T_A(x) + I_A(x) + F_A(x) \leq 3$ where $T_A(x_j) = \left[T_A^L(x_j), T_A^U(x_j)\right]$, $I_A(x_j) = \left[I_A^L(x_j), I_A^U(x_j)\right]$, $F_A(x_j) = \left[F_A^L(x_j), F_A^U(x_j)\right] \subseteq [0, 1]$, then it is called an interval valued neutrosophic set.

To measure the difference between two sets of data, many measures of similarity are defined and one of the initial popular measures is cosine similarity measure.

Definition 2.4 (*Cosine Similarity*) [9]. Let $X = (x_1, x_2, \ldots, x_n)$ and $Y = (y_1, y_2, \ldots, y_n)$ are two n-dimensional vectors, then the cosine similarity for the vectors X and Y is given as:

$$\cos\theta = \frac{\sum_{j=1}^{n} x_j y_j}{\sqrt{\sum_{j=1}^{n} x_j^2}\sqrt{\sum_{j=1}^{n} y_j^2}}$$

In vector spaces, cosine similarity between two fuzzy sets A and B having membership values $\mu_A(x_i)$ and $\mu_B(x_i)$, respectively, is defined as follows:

$$C_F(A, B) = \frac{\sum_{j=1}^{n} \mu_A(x_j)\mu_B(x_j)}{\sqrt{\sum_{i=1}^{n} \mu_A(x_j)^2}\sqrt{\sum_{i=1}^{n} \mu_B(x_j)^2}}$$

In case of two IFSs, the cosine measure of similarity is defined as:

$$C_{\mathrm{IF}}(A, B) = \frac{\sum_{i=1}^{n}\left(\mu_A(x_j)\mu_B(x_j) + \nu_A(x_j)\nu_B(x_j)\right)}{\sqrt{\sum_{i=1}^{n}\left(\mu_A(x_j)^2 + \nu_A(x_j)^2\right)}\sqrt{\sum_{i=1}^{n}\left(\mu_B(x_j)^2 + \nu_B(x_j)^2\right)}}. \quad (1)$$

These similarity measures are also defined in the case of neutrosophic sets.

Definition 2.5 (*Cosine Similarity Measure for SVNSs*) [9]: Let X be the universal set and $A = \left\{ < x_j : T_A(x_j), I_A(x_j), F_A(x_j) >, x_j \in X \right\}$ and $B =$

$\{< x_j : T_B(x_j), I_B(x_j), F_B(x_j) >, x_j \in X\}$ are two SVNS in the universal set, where all the membership values lies in the closed interval 0 and 1 for any element in A and $T_B(x_j), I_B(x_j), F_B(x_j) \in [0, 1]$ for any $x_j \in X$ in B. The cosine similarity measure for two SVNSs is defined as

$$C_1(A, B) = \frac{1}{n}\sum_{j=1}^{n} \frac{T_A(x_j)T_B(x_j) + I_A(x_j)I_B(x_j) + F_A(x_j)F_B(x_j)}{\sqrt{(T_A(x_j))^2 + (I_A(x_j))^2 + (F_A(x_j))^2}.\sqrt{(T_B(x_j))^2 + (I_B(x_j))^2 + (F_B(x_j))^2}} \quad (2)$$

However, this cosine measure of similarity has some drawbacks [9]. Ye [9] in 2015 proposed two improved version of cosine measure of similarity for SVNSs in further definitions.

Definition 2.6 (*Improved Cosine Similarity measure for SVNSs*) [9]. The new improved cosine similarity measure is defined by Ye [9] as

$$C_2(A, B) = \frac{1}{n}\sum_{j=1}^{n} \cos\left[\pi \frac{\max(|T_A(x_j) - T_B(x_j)|, |I_A(x_j) - I_B(x_j)|, |F_A(x_j) - F_B(x_j)|)}{2}\right], \quad (3)$$

$$C_3(A, B) = \frac{1}{n}\sum_{j=1}^{n} \cos\left[\pi \frac{(|T_A(x_j) - T_B(x_j)| + |I_A(x_j) - I_B(x_j)| + |F_A(x_j) - F_B(x_j)|)}{6}\right] \quad (4)$$

Definition 2.7 (*Cosine Similarity Measure for IVNSs*) [9]. Let $A = \{< x_j : T_A(x_j), I_A(x_j), F_A(x_j) >, x_j \in X\}$ and $B = \{< x_j : T_B(x_j), I_B(x_j), F_B(x_j) >, x_j \in X\}$ are the two interval valued neutrosophic sets derived from the set $X = (x_1, x_2, \ldots, x_n)$, where $T_A(x_j) = [T_A^L(x_j), T_A^U(x_j)]$, $I_A(x_j) = [I_A^L(x_j), I_A^U(x_j)]$, $F_A(x_j) = [F_A^L(x_j), F_A^U(x_j)] \subseteq [0, 1]$ for any $x_j \in X$ in A and $T_B(x_j) = [T_B^L(x_j), T_B^U(x_j)]$, $I_B(x_j) = [I_B^L(x_j), I_B^U(x_j)]$, $F_B(x_j) = [F_B^L(x_j), F_B^U(x_j)] \subseteq [0, 1]$ for any $x_j \in X$ in B. Then

$$C_4(A, B) = \frac{1}{n}\sum_{i=1}^{n} \frac{\begin{pmatrix} T_A^L(x_j)T_B^L(x_j) + I_A^L(x_j)I_B^L(x_j) + F_A^L(x_j)F_B^L(x_j) \\ +T_A^U(x_j)T_B^U(x_j) + I_A^U(x_j)I_B^U(x_j) + F_A^U(x_j)F_B^U(x_j) \end{pmatrix}}{\sqrt{(T_A^L(x_j))^2 + (I_A^L(x_j))^2 + (F_A^L(x_j))^2 + (T_A^U(x_j))^2 + (I_A^U(x_j))^2 + (F_A^U(x_j))^2} \\ \sqrt{(T_B^L(x_j))^2 + (I_B^L(x_j))^2 + (F_B^L(x_j))^2 + (T_B^U(x_j))^2 + (I_B^U(x_j))^2 + (F_B^U(x_j))^2}} \quad (5)$$

Ye [9] in 2015 defined the following improved version of cosine measure of similarity for interval valued neutrosophic sets:

Definition 2.8 (*Improved Cosine Similarity Measure for IVNSs*) [9]: Assume that there are two interval valued neutrosophic sets $A = \{< x_j : T_A(x_j), I_A(x_j), F_A(x_j) >, x_j \in X\}$ and $B = \{< x_j : T_B(x_j), I_B(x_j), F_B(x_j) >, x_j \in X\}$, derived from the set $X = (x_1, x_2, \ldots, x_n)$, where $T_A(x_j) = [T_A^L(x_j), T_A^U(x_j)]$, $I_A(x_j) = [I_A^L(x_j), I_A^U(x_j)]$,

$F_A(x_j) = [F_A^L(x_j), F_A^U(x_j)] \subseteq [0, 1]$ for any $x_j \in X$ in A and $T_B(x_j) = [T_B^L(x_j), T_B^U(x_j)]$, $I_B(x_j) = [I_B^L(x_j), I_B^U(x_j)]$, $F_B(x_j) = [F_B^L(x_j), F_B^U(x_j)] \subseteq [0, 1]$ for any $x_j \in X$ in B. Then two improved cosine measure of similarity are defined in the following manner:

$$\begin{aligned} SC_1(A, B) = \frac{1}{n}\sum_{j=1}^{n} \cos\Big[\frac{\pi}{4}\Big(\big|T_A^L(x_j) - T_B^L(x_j)\big| \vee \big|I_A^L(x_j) - I_B^L(x_j)\big| \vee \big|F_A^L(x_j) - F_B^L(x_j)\big| \\ + \big|T_A^U(x_j) - T_B^U(x_j)\big| \vee \big|I_A^U(x_j) - I_B^U(x_j)\big| \vee \big|F_A^U(x_j) - F_B^U(x_j)\big|\Big)\Big] \quad (6) \end{aligned}$$

$$\begin{aligned} SC_2(A, B) = \frac{1}{n}\sum_{j=1}^{n} \cos\Big[\frac{\pi}{12}\Big(\big|T_A^L(x_j) - T_B^L(x_j)\big| + \big|I_A^L(x_j) - I_B^L(x_j)\big| + \big|F_A^L(x_j) - F_B^L(x_j)\big| \\ + \big|T_A^U(x_j) - T_B^U(x_j)\big| + \big|I_A^U(x_j) - I_B^U(x_j)\big| + \big|F_A^U(x_j) - F_B^U(x_j)\big|\Big)\Big] \quad (7) \end{aligned}$$

Broumi and Smarandache [14] presented the cosine similarity measure of interval valued neutrosophic sets as follows:

Definition 2.9 (*Cosine Similarity Measure for IVNSs*) [14]: Let A and B be two interval valued neutrosophic sets defined as $A = \{< x_j : T_A(x_j), I_A(x_j), F_A(x_j) >, x_j \in X\}$ and $B = \{< x_j : T_B(x_j), I_B(x_j), F_B(x_j) >, x_j \in X\}$ in the set $X = (x_1, x_2, \ldots, x_n)$, where $T_A(x_j) = [T_A^L(x_j), T_A^U(x_j)]$, $I_A(x_j) = [I_A^L(x_j), I_A^U(x_j)]$, $F_A(x_j) = [F_A^L(x_j), F_A^U(x_j)] \subseteq [0, 1]$ for any $x_j \in X$ in A and $T_B(x_j) = [T_B^L(x_j), T_B^U(x_j)]$, $I_B(x_j) = [I_B^L(x_j), I_B^U(x_j)]$, $F_B(x_j) = [F_B^L(x_j), F_B^U(x_j)] \subseteq [0, 1]$ for any $x_j \in X$ in B. Then

$$C_5(A, B) = \frac{1}{n}\sum_{i=1}^{n} \frac{\begin{pmatrix} (T_A^L(x_j) + T_A^U(x_j)) \cdot (T_B^L(x_j) + T_B^U(x_j)) \\ +(I_A^L(x_j) + I_A^U(x_j)) \cdot (I_B^L(x_j) + I_B^U(x_j)) \\ +(F_A^L(x_j) + F_A^U(x_j)) \cdot (F_B^L(x_j) + F_B^U(x_j)) \end{pmatrix}}{\sqrt{(T_A^L(x_j) + T_A^U(x_j))^2 + (I_A^L(x_j) + I_A^U(x_j))^2 + (F_A^L(x_j) + F_A^U(x_j))^2} \sqrt{(T_B^L(x_j) + T_B^U(x_j))^2 + (I_B^L(x_j) + I_B^U(x_j))^2 + (F_B^L(x_j) + F_B^U(x_j))^2}} \quad (8)$$

Definition 2.10 (*Similarity Measure for SVNSs*) [19]: Mandal and Basu [19] proposed two new similarity measures between sets A and B defined as

$$\begin{aligned} S_1(A, B) = \frac{1}{n}\sum_{x_j \in X}\Big[1 - \log_2\Big[1 + \frac{1}{4}[|T_A^L(x_j) + T_A^U(x_j) - T_B^L(x_j) - T_B^U(x_j)| \\ + 2|I_A^L(x_j) + I_A^L(x_j) - I_B^L(x_j) - I_B^U(x_j)| \\ +|F_A^L(x_j) + F_A^L(x_j) - F_B^L(x_j) - F_B^U(x_j)|]]\Big] \quad (9) \end{aligned}$$

and

$$S_2(A,B)=\frac{1}{n}\sum_{x_j\in X}\cos\left[\frac{\pi}{2}\cdot\frac{1}{4}\left[\left|T_A^L(x_j)+T_A^U(x_j)-T_B^L(x_j)-T_B^U(x_j)\right|\right.\right.$$
$$+2\left|I_A^L(x_j)+I_A^L(x_j)-I_B^L(x_j)-I_B^U(x_j)\right|$$
$$\left.\left.+\left|F_A^L(x_j)+F_A^L(x_j)-F_B^L(x_j)-F_B^U(x_j)\right|\right]\right] \quad (10)$$

Definition 2.11 (*Chi-square distance*) [20]. The chi-square distance measure of two n-dimensional real vectors $X=(x_1,x_2,\ldots,x_n)$ and $Y=(y_1,y_2,\ldots,y_n)$ are given as follows:

$$d(x,y)=\sum_{i=1}^{n}\frac{(x_j-y_j)^2}{x_j+y_j},$$

and the revised version of chi-square distance measure is given as:

$$d(x,y)=\sum_{i=1}^{n}\frac{(x_j-y_j)^2}{2+x_j+y_j}.$$

Ren et al. [20] defined following new neutrosophic measure of similarity based on revised chi-square distance definition.

Definition 2.12 (*Chi-square distance for SVNSs*) [20]: Let $X=\{x_1,x_2,\ldots,x_n\}$ be a universal set. The new neutrosophic fuzzy information measure for two given single value neutrosophic sets $A=\{<x:T_A(x_j),I_A(x_j),F_A(x_j)>,x_j\in X\}$ and $B=\{<x:T_B(x_j),I_B(x_j),F_B(x_j)>,x_j\in X\}$ is given as

$$S_{\mathrm{CH}_1}(A,B)=1-\frac{1}{2n}\sum_{i=1}^{n}\left[\frac{(T_A(x_j)-T_B(x_j))^2}{2+T_A(x_j)+T_B(x_j)}+\frac{(I_A(x_j)-I_B(x_j))^2}{2+I_A(x_j)+I_B(x_j)}\right.$$
$$\left.+\frac{(F_A(x_j)-F_B(x_j))^2}{2+F_A(x_j)+F_B(x_j)}+\left|m_A(x_j)-m_B(x_j)\right|\right] \quad (11)$$

where $m_A(x_i)=\frac{1+T_A(x_j)-F_A(x_j)}{2}$ and $m_B(x_i)=\frac{1+T_B(x_j)-F_B(x_j)}{2}$, $j=1,2,\ldots,n$.

Using the above definition for chi-square distance for SVNSs, we define the measure of similarity for interval valued neutrosophic sets as follows:

Definition 2.13 (*Chi-square Similarity Measure for IVNSs*). Assume that there are two interval valued neutrosophic sets $A=\{<x_j:T_A(x_j),I_A(x_j),F_A(x_j)>,x_j\in X\}$ and $B=\{<x_j:T_B(x_j),I_B(x_j),F_B(x_j)>,x_j\in X\}$ in the universe of discourse $X=(x_1,x_2,\ldots,x_n)$, where $T_A(x_j)=[T_A^L(x_j),T_A^U(x_j)]$, $I_A(x_j)=[I_A^L(x_j),I_A^U(x_j)]$, $F_A(x_j)=[F_A^L(x_j),F_A^U(x_j)]\subseteq[0,1]$ for any $x_j\in X$ in A and $T_B(x_j)=[T_B^L(x_j),T_B^U(x_j)]$, $I_B(x_j)=[I_B^L(x_j),I_B^U(x_j)]$,

$F_B(x_j) = \left[F_B^L(x_j), F_B^U(x_j)\right] \subseteq [0, 1]$ for any $x_j \in X$ in B. Then the measure of similarity for interval valued neutrosophic sets (IVNSs) is defined as follows:

$$S_{\mathrm{CH}_2}(A, B) = 1 - \frac{1}{2n}\sum_{i=1}^{n}\left[\frac{\left((T_A^L(x_j) + T_A^U(x_j)) - (T_B^L(x_j) + T_B^U(x_j))\right)^2}{2 + \left(T_A^L(x_j) + T_A^U(x_j)\right) + \left(T_B^L(x_j) + T_B^U(x_j)\right)}\right.$$
$$+ \frac{\left(\left(I_A^L(x_j) + I_A^U(x_j)\right) - \left(I_B^L(x_j) + I_B^U(x_j)\right)\right)^2}{2 + \left(I_A^L(x_j) + I_A^U(x_j)\right) + \left(I_B^L(x_j) + I_B^U(x_j)\right)}$$
$$+ \frac{\left((F_A^L(x_j) + F_A^U(x_j)) - (F_B^L(x_j) + F_B^U(x_j))\right)^2}{2 + \left(F_A^L(x_j) + F_A^U(x_j)\right) + \left(F_B^L(x_j) + F_B^U(x_j)\right)}$$
$$\left. + \left|m_A(x_j) - m_B(x_j)\right|\right] \quad (12)$$

where $m_i(x_j) = \frac{1+\left(T_j^L(x_j)+T_j^U(x_j)\right)-\left(F_j^L(x_j)+F_j^U(x_j)\right)}{2}$, $j = 1, 2, \ldots, n$ and $i = 1, 2, \ldots, m$.

3 Comparison of New Improved Chi-Square Measure of Similarity with Other Measures of Similarity

In order to study the ability of the newly defined chi-square measure of similarity for IVNSs, we have considered here the pattern recognition problems in interval valued neutrosophic environment. We have taken this problem in the form of numerical example given in the paper of Ye [9], where various previously defined cosine measures are used along with the newly defined improved chi-square measure of similarity for interval valued neutrosophic sets. Further, we have shown that our measure result matches with the already known results and it helps to deal with diagnosis of disease problems.

Example 1 Consider two interval valued neutrosophic sets A and B in X. The values of the measure of similarity between these two IVNSs are calculated using chi-square measure of similarity and the results obtained are compared with the previously several measures of similarity available in the literature [9, 14, 19, 20]. The comparison of pattern recognition is done using Eqs. (5), (7), (8), (9), (10), and (12), and results are shown in Table 1.

From Table 1, it is seen that chi-square measure gives better results in some cases as compared to similarity measures or at least gives equivalent results as per the existing similarity measures.

Table 1 Pattern recognition problem

	Case 1			Case 2			Case 3			Case 4			Case 5		
A	0.3	0.2	0	0.3	0.2	0.4	0.3	0.2	0.4	0	1	0	0.9	0.9	1
	0.5	0.4	0.1	0.5	0.4	0.5	0.4	0.3	0.5	1	0	0	1	1	0
B	0.3	0.2	0	0.4	0.2	0.3	0.6	0.4	0.8	1	0	1	1	1	0
	0.5	0.4	0.1	0.5	0.4	0.5	0.8	0.6	1	0	1	1	0	1	1
$C_4(A, B)$ (eq. 5)	1			0.989473684			1			0			0.651338947		
$S_1(A, B)$ (eq. 9)	1			0.929610672			0.277533976			0.415037499			0.649502753		
$S_2(A, B)$ (eq. 10)	1			0.996917334			0.522498565			0.707106781			0.908143174		
$C_5(A, B)$ (eq. 8)	1			0.994475138			1			0.577350269			0.954032718		
$SC_2(A, B)$ (eq. 7)	1			0.998629535			0.852640164			0			0.669130606		
Chi-square $S_{CH_2}(A, B)$ (eq. 12)	1			0.947297297			0.768359404			0			0.69149948		

3.1 Medical Diagnosis Using Improved Chi-Square Similarity Measure and Its Comparison with Other Cosine Similarity Measures

In the medical diagnosis problems, the data available to the medical practitioner may contain a lot of vague or fuzzy informations. The diagnosis elements may have different membership of truth, indeterminacy and falsity values; so that they can be easily defined in terms of interval valued neutrosophic set. Thus, all the similarity measures defined above can be used as a tool for medical diagnosis.

The method for medical diagnosis is as follows:

Let $A = \{A_1, A_2, \ldots, A_n\}$ be the set of diagnosis and $S = \{S_1, S_2, \ldots, S_m\}$ be the set of symptoms. Suppose from a patient (P), we collect informations with all the symptoms from a set S. The informations collected from the patient (P) related to diagnosis (A) having the symptoms (S) are represented in the form of IVNSs. Considering the principle of pattern recognition based on degree of similarity with a maximum value, that is to find the proper diagnosis A_i, for the patient P, in medical diagnosis problem, we use the formula

$$i = \arg \max_{1 \leq i \leq n} \{C_N(A_i, P)\}$$

To demonstrate the application of improved chi-square similarity measure and the comparison with other similarity measure, we have taken following example:

Example 2 Considering the medical diagnosis problem taken by Broumi and Smarandache [14], in which it is given that if the patient has temperature value between 0.5 and 0.7, then it represents severity/certainty of the truth value. If the value is from 0.2 to 0.4, then somehow it is indeterminable. If it is between 0.1 and 0.2, then it is sure that temperature has no relation with the main disease. The values related to one patient and one symptom is given in the following Table 2. Similarly, the relation of a patient with the headache values in terms of IVNSs can be defined.

The values of the patient in terms of IVNSs are defined as $P = \{< x_1, [0.5, 0.7], [0.2, 0.4], [0.1, 0.2] >, < x_2, [0.2, 0.3], [0.3, 0.5], [0.3, 0.6] >\}$. The diagnosis $A_i (i = 1, 2, 3)$ is represented by interval valued neutrosophic numbers related to all symptoms as follows:

Table 2 Pattern recognition in medical diagnosis

	Case 1				Case 2		
P (patient)	0.5	0.2	0.1	P (patient)	0.2	0.3	0.3
T (temperature)	0.7	0.4	0.2	H (headache)	0.3	0.5	0.6

$$A_1 = \{< x_1, [0.5, 0.6], [0.2, 0.3], [0.4, 0.5] >, \\ < x_2, [0.2, 0.6], [0.3, 0.4], [0.6, 0.7] >\}$$

$$A_2 = \{< x_1, [0.4, 0.5], [0.3, 0.4], [0.5, 0.6] >, \\ x_2, [0.3, 0.5], [0.4, 0.6], [0.2, 0.4]\}$$

$$A_3 = \{< x_1, [0.6, 0.8], [0.4, 0.5], [0.3, 0.4] >, \\ < x_2, [0.3, 0.7], [0.2, 0.3], [0.4, 0.7] >\}$$

Table 3 shows the ranking for the diseases A_1, A_2, A_3 from which the person can suffer. It is predicted that the person is suffering from Typhoid disease A_3 as it has highest similarity with the symptoms depicted in Table 2. The various similarity measures are used, and each of them shows the same result.

4 Conclusion

In this research paper, a new improved chi-square measure of similarity is defined for interval valued neutrosophic sets. Its application is shown in recognition of pattern and diagnosis of disease problems. The results obtained are compared with those of several existing measures of similarity. In future work, chi-square similarity measure can be applied to several other fields of science, engineering, and management.

Table 3 Calculation of medical diagnosis

	Case 1			Case 2		
A_1 (viral fever)	0.5	0.2	0.4	0.2	0.3	0.6
	0.6	0.3	0.5	0.6	0.4	0.7
A_2 (malaria)	0.4	0.3	0.5	0.3	0.4	0.2
	0.5	0.4	0.6	0.5	0.6	0.4
A_3 (typhoid)	0.6	0.4	0.3	0.3	0.2	0.4
	0.8	0.5	0.4	0.7	0.3	0.7
Various similarity measures	$C_5(A, B)$ (Eq. 8)	$C_4(A, B)$ (Eq. 5)	$SC_2(A, B)$ (Eq. 7)	Chi-square $S_{CH_2}(A, B)$ (Eq. 12)	$S_1(A, B)$ (Eq. 9)	$S_2(A, B)$ (Eq. 10)
$C_N(A_1, P)$	0.898878291	0.887829942	0.949100575	0.735124552	0.594056095	0.841943573
$C_N(A_2, P)$	0.856053911	0.853402075	0.947583192	0.725564469	0.53133588	0.794720264
$C_N(A_3, P)$	0.965455219	0.95109961	0.975328029	0.906484113	0.671287385	0.911957484
Ranking	$A_3 > A_1 > A_2$	$A_3 > A_1 > A_2$	$A_3 > A_1 > A_2$	$A_3 > A_1 > A_2$	$A_3 > A_1 > A_2$	$A_3 > A_1 > A_2$
Difference	0.109401308	0.097697536	0.027744838	0.180919644	0.139951505	0.11723722

References

1. Zadeh, L.A.: Fuzzy sets. Inf. Control **8**(3), 338–353 (1965)
2. Atanassov, K.T.: Intuitionistic fuzzy sets. Fuzzy Sets Syst. **20**(1), 87–96 (1986)
3. Atanassov, K., Gargov, G.: Interval valued intuitionistic fuzzy sets. Fuzzy Sets Syst. **31**(3), 343–349 (1989)
4. Smarandache, F.: Neutrosophic set, a generalization of intuitionistic fuzzy sets. Int. J. Pure Appl. Math. **24**, 287–297 (2005)
5. Wang, H., Smarandache, F., Zhang, Y.Q., Sunderraman, R.: Single-valued neutrosophic sets. Multispace Multistructure **4**, 410–413 (2010)
6. Son, N.T.K., Dong, N.P., Son, L.H.: Towards granular calculus of single-valued neutrosophic functions under granular computing. Multimed Tools Appl. **79**, 16845–16881 (2020)
7. Sumathi, I.R., Antony, C., Sweety, C.: New approach on differential equation via trapezoidal neutrosophic number. Complex Intell. Syst. **5**, 417–424 (2019)
8. Şahin, R.: An approach to neutrosophic graph theory with applications. Soft Comput. **23**, 569–581 (2019)
9. Ye, J.: Improved cosine similarity measures of simplified neutrosophic sets for medical diagnosis. Artif. Intell. Med. **63**, 171–179 (2015)
10. Guo, Y., Cheng, H.D.: New neutrosophic approach to image segmentation. **42**, 587–595 (2009)
11. Mondal, K., Pramanik, S.: Neutrosophic tangent similarity measure and its application to multi attribute decision making. Neutrosophic Sets Syst. **9**, 80–87 (2015)
12. Ye, J., Fu, J.: Multi-period medical diagnosis method using a single valued neutrosophic similarity measure based on tangent function. Comput. Methods Programs Biomed. **123**, 142–149 (2016)
13. Ye, J.: Cosine similarity measures for intuitionistic fuzzy sets and their applications. Math. Comput. Model. **53**, 91–97 (2011)
14. Broumi, S., Smarandache, F.: Cosine similarity measure of interval valued neutrosophic sets. Neutrosophic Sets Syst. **5**, 15–20 (2014)
15. Sohangir, S., Wang, D.: Improved sqrt-cosine similarity measurement. **4**, 25 (2017)
16. Flora, S.C., Arockiarani, I.: Cotangent similarity measures of vague multi sets. Int. J. Pure Appl. Math. **120**(7), 155–163 (2018)
17. Alcin, A.S., Kilic, H.S., Caglayan, N.: An integrated model with interval valued neutrosophic sets for the selection of lean and sustainable suppliers, intelligent and fuzzy techniques in big data analytics and decision making. In: INFUS 2019. Advances in Intelligent Systems and Computing, vol 1029. Springer, Cham (2019)
18. Ye, J.: Vector similarity measures of simplified neutrosophic sets and their application in multicriteria decision making. Int. J. Fuzzy Syst. **16**(2), 204–211 (2014)
19. Mandal, K., Basu, K.: Improved similarity measure in neutrosophic environment and its application in finding minimum spanning tree. J. Intell. Fuzzy Syst. **31**, 1721–1730 (2016)
20. Ren, H.P., Xiao, S.X., Zhou, H.: A chi-square distance based similarity measure of single valued neutrosophic set and applications. Int. J. Comput. Commun. Control **14**(1), 78–89 (2019)

Author Index

M. Sahni et al. (eds.), *Mathematical Modeling, Computational Intelligence Techniques and Renewable Energy*, Advances in Intelligent Systems and Computing 1287,
https://doi.org/10.1007/978-981-15-9953-8